World Sustainability Series

Series editor

Walter Leal Filho, Hamburg, Germany

More information about this series at http://www.springer.com/series/13384

Walter Leal Filho · Ricardo Noyola-Cherpitel
Pedro Medellín-Milán · Valeria Ruiz Vargas
Editors

Sustainable Development Research and Practice in Mexico and Selected Latin American Countries

Springer

Editors
Walter Leal Filho
School of Science and the Environment
Manchester Metropolitan University
Manchester
UK

Pedro Medellín-Milán
Coordinación de la Agenda Ambiental
Universidad Autónoma de San Luis Potosí
San Luis Potosí
Mexico

Ricardo Noyola-Cherpitel
Programas Multidisciplinarios de Posgrado en Ciencias Ambientales
Universidad Autónoma de San Luis Potosí
San Luis Potosí
Mexico

Valeria Ruiz Vargas
School of Science and the Environment
Manchester Metropolitan University
Manchester
UK

ISSN 2199-7373 ISSN 2199-7381 (electronic)
World Sustainability Series
ISBN 978-3-319-70559-0 ISBN 978-3-319-70560-6 (eBook)
https://doi.org/10.1007/978-3-319-70560-6

Library of Congress Control Number: 2017957674

Printed on acid-free paper

This Springer imprint is published by Springer Nature
The registered company is Springer International Publishing AG
The registered company address is: Gewerbestrasse 11, 6330 Cham, Switzerland

Preface

Many universities in Mexico and in Latin America perform high-quality research on matters related to sustainable development. Yet, there are relatively few publications where a multidisciplinary overview of research efforts and projects across the region is documented, and where researchers from across the spectrum of the natural and social sciences have had the opportunity to document and promote research methods, the results of empirical research or exchange ideas about ongoing and future research initiatives focusing on sustainable development.

It is against this background that the book "Sustainable Development Research and Practice in Mexico and Selected Latin American Countries" has been produced. It is the outcome of a specialist event organised by the Universidad Autónoma de San Luis Potosí, Mexico; Manchester Metropolitan University, UK; and HAW Hamburg, Germany, under the auspices of the Inter-University Sustainable Development Research Programme (IUSDRP) and a number of institutions of higher education active in this field, from across the region.

The book has three aims:

(a) to provide researchers at universities in Mexico and Latin America performing research on issues pertaining sustainable development with an opportunity to document their work (i.e. curriculum innovation, empirical work, activities, case studies, practical projects);
(b) to support the exchange of information, ideas and experiences acquired in the execution of research projects, successful initiatives and good practice;
(c) to introduce methodological approaches and projects and provide a platform for further cooperation.

This book entails contributions from researchers in the field of sustainable development in the widest sense, from business and economics, to arts, administration, environment, languages and media studies, with a focus on "Sustainable Development Research".

Thanks to its nature, this publication is expected to contribute to the further development of this fast-growing field.

We thank the authors for sharing their knowledge and know-how, and the many reviewers who have assisted with the peer review of the papers. We hope this book will further support the development of sustainability research in Latin America.

Manchester, UK — Walter Leal Filho
San Luis Potosí, Mexico — Ricardo Noyola-Cherpitel
San Luis Potosí, Mexico — Pedro Medellín-Milán
Manchester, UK — Valeria Ruiz Vargas
Winter 2018

Contents

A Strategic Differentiation Proposal for a Cruise Tourism Offer, Built on the Cultural Identity of the Host Community of Puerto Progreso, Yucatan

Argelia Salazar Estrada

Abstract The purpose of this paper is to present a strategic approach which builds on the cultural identity of Puerto Progreso, Yucatan, in order to increase the destination competitiveness and differentness, increase the market share of the Caribbean cruise tourism, and increase the local profit. An applied research was carried out using a qualitative approach, first to determine the competitiveness state of Puerto Progreso and to define which cultural elements of its local culture could be leveraged, second, to propose a differentiation strategy that could increase the destination competitiveness. It was concluded that the cultural identity of a host community is an essential element of an effective strategy to differentiate the destination while integrating the local community in the attention of the cruise tourism. The proposed strategy could be replicated by following certain guidelines.

Keywords Competitiveness · Cruise port management · Host community
Cultural identity · Strategic differentiation

1 Introduction

A cruise ship is defined as a vessel that offers transport from port of call to port of call, which presents a unique closed social and service setting in which experience is both created and judged (Brejla and Gilbert 2014). The ship is presented as a resort destination itself and it is sold as the primary destination (Wood 2000).

The cruise market is continuously innovating with bigger and improved ships, with a better recreational offer on board, with better naval architecture and engineering, differentiating continuously their offer. It is a dynamic market led by three commercial groups: Carnival Corporation & Plc, Royal Caribbean Ltd., and Star cruises. Together, they control 88% of the worlds cruise berths (UNWTO 2008).

A. S. Estrada (✉)
Anahuac Mexico University, Av. Universidad Anahuac #46,
Lomas Anahuac, Huixquilucan, Edo. de Mexico CP 52786, Mexico
e-mail: argeliasalazaredt@gmail.com

W. Leal Filho et al. (eds.), *Sustainable Development Research and Practice in Mexico and Selected Latin American Countries*, World Sustainability Series,
https://doi.org/10.1007/978-3-319-70560-6_1

This competitive position allows them to obtain advantageous benefits when negotiating with destinations and countries, especially with the ones that are not attractive enough for the cruise passengers.

The Caribbean region is the most important cruise area in the Americas, since it is included in the 33.7% of itineraries (CLIA 2016), which results very relevant for Mexico, because it has three touristic ports in the area: Cozumel, the most important port in the Caribbean Sea (FCCA 2013), Majahual and Puerto Progreso. On the other hand, the United States of America is the major emitting market of the world (FCCA 2013), and it is the main market for Mexico too.

Puerto Progreso is within the so-called Mayan World; it is located in the northern part of the Yucatan Peninsula. Its location is strategic for maritime commerce and cruise tourism; it allows ships to gain access to the Caribbean Sea, the Gulf of Mexico and the Atlantic Ocean. Also, it is near to the North and South continental maritime routes crossing. Progreso is the nearest port to Florida, Galveston, Houston, New Orleans, Biloxi, Mobile and Tampa (API Progreso and SCT s.f.).

It is worth mentioning the relevance of this area from an environmental point of view too. Puerto Progreso and its surroundings are part of the Priority Terrestrial and Maritime Regions for Biodiversity Conservation as well as Ramsar Wetlands and Bird Conservation Areas (AICA in Spanish), and some of the nearby municipalities, e.g. Celestun, are included too in UNESCO Biosphere Reserves.

Puerto Progreso has received cruise ships since 1998. From a regional perspective, in 2008 it received the second largest amount of cruise passengers, and between 2010 and 2011 it was home port for Pullmantur Cruise Line. In other words, it was attractive to cruise ships and competitive relative to other ports in the Caribbean region. Nevertheless, from 2008 to 2014 Puerto Progreso experienced a 33.75% decline in arrivals, compounded by a dramatic reduction of the number of cruise lines visiting the port to just one: Carnival Cruise Line.

This situation could be explained by important international events that affected the global tourism industry around those years, specifically: the oil price increase in 2008 to historical levels; the 2009 financial crisis in the USA; the 2011 economic crisis in Europe, and locally, the wave of crime incidents in Mexico that aroused in 2008.

Additionally, Sectur and SCT (2008), and Cestur (2011) diagnosed that the touristic offer of the Mexican ports is limited (meaning few options) and lacks a significant differentness that could attract cruise passengers. Therefore, the Mexican ports can be easily replaced in the cruise itineraries.

By the same token, at the beginning of this research, it was found that between 2003 and 2013 the shore excursions emphasized predominantly tours to archeological sites like Chichen Itza, at the expense of Puerto Progreso. A reduction or elimination of the cultural and the identity elements, and the ecotourism and gastronomical attractions in the shore excursions offer was found too, regardless Progreso port proximity to natural reserves and the availability of ecotourism cooperatives working in the area (García and Xool 2012; García et al. 2011).

With all the above mentioned, we can perceive that the Progreso port has a good location, near one of the most important ports of the world, and within the cruise ship routes departing from and reentering to the ports of the Gulf of Mexico and Florida, but its touristic offer is not differentiated, even when the area is rich in natural and cultural resources. This is why we thought the cultural identity could be used to differentiate the destination and to improve the utilization of touristic assets available in the region, so the cruise destination could be more attractive to cruise lines and there could be an improvement of the cruise tourism profitability. With this in mind, and since the differentiation strategy is a competitive strategy, it was also important to examine the competitive situation of the cruise destination.

In this context, the research question is:

How can the cultural identity of Puerto Progreso be leveraged to differentiate the destination, improve its touristic competitiveness and produce economic benefits for the local community?

2 Literature Review

In general, the research in cruise tourism management focuses on its economic contribution (Brida and Zapata 2008; Business Research and Economic Advisors 2015; Dawson et al. 2016; Lester and Weeden 2004; Thurau et al. 2013; Seidl et al. 2005; Macpherson 2008) and the passenger vacation experience on board (Forgas-Coll et al. 2014; Mackay and Kerstetter 2005). Most of this research aims at understanding the tourist's motivations and drivers of satisfaction before and during the trip, like quality of service, perceived value, passenger's personality, as well as the trust that passengers have on a cruise brand. The research reviewed does not consider how the cultural elements and the identity of host communities contribute to the passenger's overall satisfaction and his/her future recommendation of the visited destination.

The strategic differentiation and derived advantage of cruise vacations are addressed by taking into account the cruise line brand identity (Lemmetyinen and Go 2010), elements like the shipscape (ship landscape), the servicescape, and how the combination of these variables influence the experience of the passenger as he/she relates to them (Kwortnik 2008). Likewise, the influence of the brand awareness as a motivator and how it is related to the satisfaction is also studied (Lemmetyinen et al. 2016).

A research regarding the influence of the excursion package on destination satisfaction and influence over the passengers' post-purchase behavior, like intention to return and WoM—Word of Mouth (Parola et al. 2014), concluded that cruise passengers sample destinations to visit them in future vacations, but Larsen and Wolff (2016) showed evidence that this argument was incorrect.

Cruise tourism to conservation areas has been studied considering the potential economic benefit of the cruise activity to fund environmental preservation programs, sustainable development of tourism destinations (Chan-Cob and Santín 2003) and the integration of ecotourism to the cruise touristic offer. Other study concluded that there are economic benefits for the cruise lines and the visited destination, that is, higher revenue for the host community and a rich/memorable passenger experience in an exotic and well-preserved natural environment (Thurau et al. 2013).

Finally, the host community has been studied to determine its perceptions about how and how much the communities have achieved a better standard of life derived from cruise tourism, and the community's attitude towards cruise passengers (Brida et al. 2012, 2014; Chiappa et al. 2016), concluding that inhabitants of the destination perceived that the activity produced economic benefits which helped improve the community overall conditions, and therefore it is welcomed as long as it remains beneficial, but no further interest from the host community nor from the visitor perspective has been explored.

This research studies the competitiveness of a cruise port, located in an environmental relevant area, and proposes a strategic differentiation based on the cultural identity of the host community, which also enables their integration to the management of cruise tourism.

3 Cruise Tourism and Competitiveness

In Mexico, the cruise tourism is managed through the Public Policy for Cruise Tourism (Sectur and SCT 2004). This policy mandates that the elaboration of touristic products should promote the competitive development and improvement of the destinations, strengthening the integration of natural and cultural attractions, maximizing benefits, and promoting the development of host communities and the return of visitors.

Competitiveness, in this context, is defined as the ability of the destination to attract tourists and cruise ships, and participate effectively in value chains within the cruise industry inside a competitive environment, increasing participation in various market segments and attracting productive and profitable investments (Sectur and SCT 2008). The comparative and competitive advantages of a cruise destination are determined by the following seven factors:

1. The situation concerning the location of the outbound market, the cruise routes, and the provision of tourist attractions.
2. Availability of urban infrastructure and support services.
3. Availability of port infrastructure and services.
4. Port and touristic costs.
5. Size and dynamism of the market.
6. Management and organization of the activity.
7. Institutional and regulatory framework and policies aimed to develop the cruise industry (Sectur and SCT 2008).

The differentiation variable is not mentioned, in spite of being one of the main elements of the competitive strategy used by the cruise lines—in addition to other variables like competitive costs and the cruise ship which is considered as a destination in itself—and being part of the generic competitive strategies proposed by Porter (2012).

In this regard, it is observed that the identity of a touristic destination can be used as a sustainable competitive advantage, given that destination resources are difficult to imitate, since they are usually found either in their unique natural environment (climate, wildlife or landscape) or in their cultural heritage: places that reflect their host community roots in terms of historical wealth, religion or cultural expressions such as arts, architecture and design (Aaker 2001, Johnson and Scholes 1999 in Govers and Go 2004). Nature, culture, and landscape are the support of identity traits that strengthen and singularize the touristic product in an increasingly competitive environment. These are the elements with which tourism and territory managers promote the development of a destination. It is, for this reason, that host communities must be considered as an active and influential element, and not only as an object of curiosity (Requejo 2007).

Destinations can build a competitive advantage through core competencies, like exclusive capabilities in a host community, which may be deemed unique to attract visitors (Govers and Go 2004). Incidentally, the collective identity has been used too as a base line for promoting tourism, since it is composed of a network of meanings that refer to the territory and its inhabitants, where the tourist destination is located, through which the destination can be recognized, understood and enjoyed by the travelers (Uña and Villarino 2011).

4 The Cultural Identity of the Host Community

The cultural identity of a host community can be defined as a collective identity, which is based on the cultural values of a group; these can be customs, traditions, knowledge, innovations, religion, language, shared history, social relationships, beliefs, rites, ceremonies and moral values, in addition to obligations and rights. Therefore, the cultural identities that are part of the collective identity arise from the culture, and both conceptions change and evolve (Alejos 2006; Giménez 2007; Memmi 1999; Molano 2006; United Nations Education Scientific and Cultural Organization, Man and the Biosphere Programme s/f; Uña and Villarino 2011).

In this frame of reference, the identity elements of the host community that were identified during the literature review are as follows: (a) animality—the belief that all natural and supernatural beings are part of the same culture that underlies "natural" differences, the human being is part of the forest, and the animals have culture and language too, while natural, cultural and supernatural places are part of the same space (Alejos 2006; Estrada 2009); (b) the places of identity (Franco 1999; Quiroz 2009)—places that the community recognizes as representative of their identity; (c) the traditions (Hervik 2002; Toledo et al. 2007); (d) the religion (Alejos 2006; Estrada 2009); (e) the local products (Toledo et al. 2007; Uña and Villarino 2011); (f) the Mayan language (Hervik 2002; Flores 2003) and (g) the cosmogony (Franco 1999; Alejos 2006; Estrada 2009; Hervik 2002)—a beliefs system of the collective history of the natives of a region, based on ancestral thinking, it defines the ways people give meaning to reality and the notion of time and space.

5 Methodology

The purpose of this paper is to present a strategic approach which builds on the cultural identity of Puerto Progreso, Yucatan, in order to increase the destination competitiveness and differentness, increase the market share of the Caribbean cruise tourism, and increase the local profit. An applied research was carried out using a qualitative approach, first to determine the competitiveness state of Puerto Progreso and to define which elements of its local culture could be leveraged to design an effective differentiation strategy, second, to propose the differentiation strategy.

In order to examine the competitive situation of the cruise destination, fourteen interviews were conducted among key stakeholders who were selected based on the guidelines proposed by Sectur and SCT (2008), which indicates that the management of touristic ports in Mexico is done through a Local Committee for Cruise Management, which is integrated by a representative of the Ministry of Transport and Communications, a representative of the port administration, representatives from State and Local Government, representatives from touristic and local venues and their associations, and the cruise lines representatives.

In Puerto Progreso, the Local Committee for Cruise Management is integrated by the port administration (Administración Portuaria Integral de Progreso S.A. de C.V.), representatives from the Cruise Tourism Coordination of the Tourism Promotion Ministry of Yucatan, and representatives from the Town Council.

The interviewed key stakeholders included a representative of the General Ports Coordination and Merchant Shipping of the National Ministry of Transport and Communications, the President of the Tourism Business Council of Yucatan, a representative of the Cruise Tourism Coordination of the Tourism Promotion Ministry of Yucatan, the Tourism and Market Director in the Town Council, Puerto Progreso Commercialization Manager, the Cruise Terminal Operation Manager, one the General Manager of Turitransmerida (receptive and cruise tour operator), and the Port Manager of Aviomar Adventours Cruise Excursions company.

A questionnaire was designed and applied in order to revise the responsibilities of the Local Committee, and the destination level of competitiveness, using the competitive factors proposed by Sectur and SCT (2008) previously mentioned. By this means also, the touristic offer within the shore excursions available in Progreso was profiled, so were the identity elements currently available in the touristic offers within the area of influence of the cruise ships, the cruise demand, and the relationships within the cruise industry among key stakeholders in Progreso.

The approach used to determine the identity elements of the host community that could be used as elements of differentness where to base the strategy was as follows. First, the location of the touristic resources and tourist attractions available within the cruise tourism influence area was done based on the literature review, the cruise lines web pages, the destination web pages, the port reports, blogs, etc.

Likewise, it was deemed necessary to define the area of influence of Puerto Progreso, it was concluded that the area encompassed 165 km around the port and was delimited by the furthest travel distance from the cruise terminal to an attractive site to which cruise passengers are willing to travel, in this case, Chichen Itza, based on the information on shore excursions offered from 2003 to 2013, and confirmed later by the interviewed tour operators.

The tourism resources and tourist attractions were located on a map using the Digital Map tool (InegiV6, see Fig. 1). In order to confirm the availability of the resources and tourist attractions, and to characterize them, a modified and more holistic version of the inventory of tourism resources and tourist attractions proposed by Martínez (1999) was used to register the touristic equipment, services and facilities availability in place within the location of every tourist attraction and tourism resource.

Additionally, this instrument was used to record the presence of the identity and live culture elements and manifestations like handicrafts, rituals, typical food and beverages, dance and music, textiles, alternative medicine, and languages, in this case, Mayan language. Twenty-one municipalities were visited and included in the analysis.

The inventory results were used to evaluate the quality, differentiation and the presence of the identity and cultural elements of each tourist resource and tourist attraction. This allowed rating them using a scale of one to five (5 = very good, 4 = good, 3 = average, 2 = bad, and 1 = lacking).

Also, the questionnaire used in the interviews contained a section where the interviewee was asked to rate the importance of the touristic venues within the region, and the potential of each venue to be integrated successfully to the cruise touristic offer.

The outcomes of the inventory and survey were utilized to design a strategy to reposition Puerto Progreso by building on the key findings of the research to make recommendations to improve the competitiveness of the cruise port, and to construct a proposal of differentiated touristic products.

Fig. 1 Tourism resources and tourist attractions location. *Source* Prepared by the author

The touristic products were conceived to build and sustain a meaningful differentiation of the destination via the methodology proposed by Chias (2005), and leveraging the elements of destination identity and cultural manifestations (handicrafts, music, etc.), while ensuring that the host community had an active role in the Committee and as shore excursion service providers too (eg. ecotourism cooperatives), and to improve the competitive position of the cruise port with a differentiated offer of touristic products.

Regarding the limitations of this work, it should be mentioned that the study of the demand for Puerto Progreso's cruise tourism was only defined using government documents and key stakeholder's interviews, the motivations of these visitors to come to Progreso were not revised in this research, the relationship between this port and other ports in the region and the interactions among them wasn't revised neither.

Likewise, the field study was not as extensive as desired, since it was not possible to interview representatives from the National Ministry of Tourism, from Carnival Cruise line, its general agent, among other key actors.

6 Findings

6.1 Competitive Situation

Puerto Progreso is located at 6.5 km from the coast. The port administration company, Administración Portuaria Integral de Progreso S.A. de C.V. (API Progreso) is responsible for managing all aspects related to the port facilities; these include a cruise terminal that can receive three cruise ships at the same time (the dock is 330 m long and has nine meters of draft); a passenger terminal that has a commercial area with two bars, a pool and a crew club; WIFI service for free; and free transport service to land.

The tourist attractions of the studied cruise destination are listed in Table 1, yet it was found that not all of them are attractive to the cruise visitor or can be visited by them, e.g. the National Park Arrecife Alacranes.

Competitive issues regarding the location of the port and the cruise ship and passengers' terminal facilities design

- The location of the port is far from the coast, this prevents the port from offering enough fresh water and electricity to cruise ships, and from having a quicker transport system to land.
- The facilities design difficults the access of tour operators to the passenger terminal, especially for those who do not have contracts or arrangements with the cruise line.
- The passenger terminal has a reduced space and not enough staff, this is the main obstacle to become a home port; in addition, it lacks signpost and maps to help the visitor find its way around the port.

Table 1 Tourist attractions in Puerto Progreso

Classification	Attractions
Natural sites	• National Park Arrecife Alacranes • State Reserve Cienegas y Manglares de la Costa Norte de Yucatan • Ecologic Reserve El Corchito • Canoeing track • Beach
Museums and historical cultural manifestations	• Downtown area • Progreso lighthouse • International East Promenade (Puerto Progreso) • Chocolate Pier/Fishermen Pier • Casa del Pastel
Folklore	• Carnaval del Mar (2nd week of February) • San Telmo celebration (August 1st) • Hanal Pixan (October 31st through November 2nd)
Programmed events	• Rally Maya Mexico • Solidaire du Chocolat (April, every two years departing from Progreso port) • Marine Day (May 1st) • "Toh" Yucatan Bird Festival (November 23rd to 25th) • Kite surf events and other aquatic sport events

Source Prepared by the author based on the results of interviews with key stakeholders

Competitive issues regarding the cruise tourism management

- The Local Committee for Cruise Management does not have the local community/tour operators input. That is, the committee is integrated only with port managers and State and Local Government representatives. It presents no integration between local entrepreneurs, tour operators, ecotourism cooperatives, State government, Local government, port administration, cruise line representatives and cruise line association representatives, preventing it from issuing recommendations to improve the cruise tourism management as well as from having an integral and a long time frame destination management strategy.
- The focus of this committee is the promotion of the destination internationally and in the process it does not take into account the local tour operators, leaving them on their own to negotiate with cruise lines and creating an intense competition among them.
- API Progreso, the company that manages the port, presides over the committee, and uses the same strategy to manage the services of the port, which is based on low prices and a diversified portfolio of services, as for the touristic management it is done in the same way. The cruise lines expect and demand low prices on a constant basis, but a differentiated portfolio of touristic products.

Awareness of local attractions appeal

- There is a lack of awareness among tour operators, port managers, taxi drivers, tour guides and in general, host community about the level of appeal of local cultural and natural resources of Puerto Progreso and its surroundings, and how recognizing and leveraging this appeal can contribute to improve the host community competitiveness and eventually the standard of life of its inhabitants.
- The touristic offer is perceived as differentiated, but this does not derive from the Puerto Progreso tourist attractions, but from those located in the Yucatan Peninsula, in general.

On the Economic side of the subject

- Puerto Progreso receives only two cruise ships on a weekly basis since 2013, a dynamic that prevents the generation of economies of scale that might allow the improvement of the tours offer. A larger demand of tours would allow to provide proper maintenance to the available touristic transport and equipment (like bicycles, surf tables, kayaks, among others) and to integrate some of the eco-tourism cooperatives working in the area.
- Cruise tourism demand is perceived as a low-spending one, Brea (2015) reports that the average spending of cruise tourists was $40.65 per person regardless of the type of the tour on land taken. The average expenditure decreased from 2007 to 2014 by 23.43% and spending on handicrafts decreased by 47.84%.
- A lacking utilization of assets provided by the ecotourism cooperatives prevents the touristic offer to be differentiated and the natural resources, which are abundant in this port, to be fully exploited.
- The time spent by cruises on this port (6 h approximately) is not enough to visit different and more representative touristic venues. It was observed that the time spent at the port is twice as much on carnival season. Several key stakeholders, e.g. the port administration, tour operators like Aviomar, have also observed that cruise passengers associate Puerto Progreso's Carnaval del Mar with the New Orleans carnival and tourists seem to have a great experience. Larger cruise ships arrive between February and March to take advantage of the mystical phenomenon known as the "Descent of Kukulkan" in Chichen Itza's castle, which can be seen a few days before and after the equinox before dawn, and this cruise ships remain too for a longer time at Puerto Progreso.

On service and infrastructure

- A lack of continuous training of the staff that works in tourist venues was found during the interviews. Interviewees mentioned that the owners of the touristic venues were not interested in improving their employees' capabilities because the additional training would encourage them to look for a better job. By the same token, the study also revealed a lack of interest of key stakeholders like restaurant entrepreneurs and masseuses on port, and the taxi drivers in general, in learning to communicate in English—which, by the way, is recognized as a critical skill in the tourism industry all around the globe—, this was pointed out by the Cruise Tourism Coordinator in the Tourism Promotion Ministry of Yucatan.

- Lastly, the touristic infrastructure lacks adequate maintenance as well as the beach. The Cruise Tourism Coordinator in the Tourism Promotion Ministry of Yucatan also stated that once the construction of the facilities was completed and handed into the local government for its administration, the maintenance seems to be the last priority of the authorities, since they perceived that the touristic activity does not generate relevant economic benefit to the port.

7 Identity and Cultural Elements to Be Enhanced

The Yucatan Peninsula is part of the territories inhabited by the Mayan people; their traditions, their holidays, their language, and the use and protection of natural resources reflect the Mayan culture. Historically, this is the place where the crater of the Chicxulub meteor is located. The belief is that this meteor caused the disappearance of 75% of the planet species 66 million years ago.

Puerto Progreso and its surroundings were too very important in the 19th century because of the henequen exploitation. Also, the Progreso municipality is part of several land and sea priority regions where special efforts are made to preserve specific natural species of those regions.

Table 2 presents Puerto Progreso tourism resources and tourist attractions with distinctive identity elements. It is worth mentioning that in the available attractions for cruise passengers, tourist does not find local handicrafts, and the handicrafts being offered belong to the center of the country, e.g. Mariachi hats, Teotihuacan handicrafts, tequila, among others. However, they do find the offer of sea species, i.e. coral (Oculina diffusa and Mycetophyllia sp), starfishes and seashells (belonging to Xancus angulatus, Strombus costatus, Strombus gigas, Cassis madacascariensis, Turbinella angulata y Melongena bispinosa species; the first three are permanently forbidden to be fished and commercialized since 1994 in Mexico). Table 3 shows the representative elements of cultural identity that were found around the Puerto Progreso municipality.

8 Discussion

The proposed identity and cultural elements within the tourism resources and tourist attractions that should be enhanced, in order to convey a differentiated offer to cruise tourists, are shown in Table 4. The enhancement could be accomplished by involving ecotourism cooperatives to secure the proper execution of the activities that produce a memorable experience for the tourists. The analysis of the available resources and tourist attractions led to define eleven touristic products. The analysis

Table 2 Tourism resources and tourist attractions within the cruise tourism influence area with more distinctive identity elements

Cultural identity element	Attraction (Municipality)	Touristic resource (Municipality)
Animality	• Puerto Progreso, Yucatán	• Dzilam de Bravo • Noc-Ac Estate (Dzityá) • Canoeing track (Puerto Progreso) • Ecologic Reserve El Corchito (Puerto Progreso)
Places of identity	• Izamal • Gran Museo del Mundo Maya (Merida) • Merida • Puerto Progreso	• Cultural center (Puerto Progreso) • Noc-Ac Estate (Dzitya) • International East Promenade (Puerto Progreso) • Progreso market • Carnaval del Mar (Puerto Progreso) • Motul de Carrillo Puerto
Traditions	• Izamal • Gran Museo del Mundo Maya (Merida) • Merida • Puerto Progreso	• Pumpkin with cane syrup (Puerto Progreso) • Sweet potato with coconut (Puerto Progreso) • Compote of cocoyol (Puerto Progreso) • Candy of ciricote (Puerto Progreso) • Noc-Ac Estate (Dzityá) • Sea bass in Maccum (Puerto Progreso) • Grouper in Tikin Xic (Puerto Progreso) • Pozole with coconut (Puerto Progreso) • Cultural center (Puerto Progreso) • International East Promenade (Puerto Progreso) • Progreso market • Carnaval del Mar (Puerto Progreso) • Motul de Carrillo Puerto • Museo de Arte Popular de Yucatán (Mérida) • Yucatecan Trova (Puerto Progreso)
Religiosity	• Izamal • Gran Museo del Mundo Maya (Merida) • Mérida • Puerto Progreso	• Pozole with coconut (Puerto Progreso) • International East Promenade (Puerto Progreso) • Progreso market • Carnaval del Mar (Puerto Progreso) • Motul de Carrillo Puerto • Museo de Arte Popular de Yucatán (Mérida)
Local products	• Izamal	• Pumpkin with cane syrup (Puerto Progreso)

(continued)

Table 2 (continued)

Cultural identity element	Attraction (Municipality)	Touristic resource (Municipality)
	• Gran Museo del Mundo Maya (Mérida) • Mérida • Puerto Progreso	• Sweet potato with coconut (Puerto Progreso) • Compote of cocoyol (Puerto Progreso) • Candy of ciricote (Puerto Progreso) • Grouper in Tikin Xic (Puerto Progreso) • Sea bass in Maccum (Puerto Progreso) • International East Promenade (Puerto Progreso) • Progreso market • Carnaval del Mar (Puerto Progreso) • Museo de Arte Popular de Yucatán (Mérida) • Pozole with coconut (Puerto Progreso) • Yucatecan Trova (Puerto Progreso)
Cosmogony	• Gran Museo del Mundo Maya (Mérida) • Puerto Progreso	• International East Promenade (Puerto Progreso) • Pozole with coconut (Puerto Progreso)

Source Prepared by the author based on the results obtained from the inventory of tourism resources and tourist attractions and interviews

also allowed defining the narratives that could enhance the cultural and identity elements. It is worth clarifying that the analysis was based on Chias (2005) methodology, putting special attention to the availability of identity elements within the resources.

It is important that members of the host community be an active player in the planning, management, and delivery of touristic services. This implies that they should be involved in the development of plans, which today are led by the committee in charge of managing the Cruises operations.

On the other hand, current executives in charge of this committee need to realize that Puerto Progreso offer should be built on its distinctive identity elements; these include the biodiversity of the region where the tourist attractions are located and tourist services are provided, local handicrafts, local rituals, and festivals, among others.

It is also important that a regulatory framework to regulate the deals between cruise lines and local tours operators is created, established and enforced, aimed to secure mutual benefits for the cruise and the local providers.

It is essential to create and promote better practices to manage cruise tourism in order to compete on a fair basis at the cruise destination or to negotiate conjointly with cruise lines and cruise associations. This could be accomplished by integrating

Table 3 Representative elements of cultural identity in Puerto Progreso

Fields	Representative elements
Denominations and designations	• Entrance to the Mayan World • Port of Peace
Activities and y events that shaped Puerto Progreso	• Aquatic sports (Sail, Windsurf) • Solidaire du Chocolat (sailboat regatta) • Henequen fiber exploitation and exportation
Representative places	• Downtown area • Puerto Progreso lighthouse • Pier • International East Promenade and beach • La Ría de Progreso tourist inn • Ecologic Reserve El Corchito
Celebrations	• Carnaval del Mar
Gastronomy and enology	• Montejo beer • Ceviche (octopus, fish, and shellfish) • Grouper in Tikin Xic • Octopus cooked in its ink • Xtabentún
Handicrafts made of	• Coconut • Shell
Representative species	• Flamingos

Source Prepared by the author based on the results of interviews with key stakeholders

a joint front with other port authorities in the Caribbean region. This would lead to a stronger negotiating position to stop cruise managers from taking advantage of its bargaining power, derived from the number of tourists they carry, from their associative capabilities, and the lack of this capacity within and among cruise destinations.

The marketing strategy should consider the differences among the different cruise lines; for example the destination could identify and target new markets and companies that usually stay in port for at least one night, so while visiting Puerto Progreso, the cruise passengers could take advantage of special land excursions, as well as special events or activities that are tailor-made to meet their expectations; for example, shore excursions could include a visit on the first day to Chichen Itza and on the second day to Izamal, a very important touristic venue as well, that displays an important amount of identity elements and cultural manifestations.

Building tourist products that are rooted on local identity traits may render higher benefits to the local community since these would represent a differentiated offer and experience that hopefully would command a higher price. For example, Puerto Progreso can develop a unique ecotourism strategy, leveraging local ecotourism cooperatives that offer environmental preservation experiences.

Table 4 Touristic product grid

Identification		Carnaval del mar	Mero en Tikin Xic	Puerto Progreso	Reserva Ecológia El Corchito	Parseo Montejo	cenote Xla Kaj	Gran museo del Mundo Maya	Mérida	Ceiba	Museo del Pueblo Maya	Parque Nacional Dzibilchaltún	Cenote Papakal	Cenote Chakjinikché	Cenote Bolonchoojol	Cenote Chelentún	Ecomuseo del chocolate	Zona arqueológica	Zona arqueológica	Hacienda Noc-Ac	Hacienda Temozon	Refugio Faunistico	Izamal	Sotuta de Peón	Chichén Itza
Location (Municipality)		Progreso	Progreso	Progreso	Progreso	Mérida	Mérida	Mérida	Mérida	Mérida	Mérida	Mérida	Cuzamá	Cuzamá	Cuzamá	Cuzamá	Santa Elena	Santa Elena	Santa Elena	Dzityá	Umán	Celestun	Izamal	Sotuta de Peón	Tinúm
Resource/attraction	Touristic resource	1	1		1					1		1										1		1	
	Tourist attractions			1		1	1	1	1		1		1	1	1	1	1	1	1	1	1		1		1
Diferentiated	Natural Resource (3-0)			3	3		3			3		3	3	3	3	3	3			3	3	3			
	Cultural Resource (3-0)	3	3	2		3		3	3	3	3	3				0	3	3	3	3	3		3	3	3
	Live cultural elements (3-0)	3	2	2		2	2	2	3	2	2	2					3	2	2	2	2	1	3	2	1
Live cultural elements available in place	Handicrafts (0-1)	1		1		1		1	1		1	1					1	1	1	1	1	1	1	1	1
	Rituals (0-1)	1							1	1							1						1		
	Typical food and beverages (0-1)	1	1	1		1			1			1					1	1	1	1	1	1	1	1	
	Dance/music (0-1)	1		1					1		1												1		
	textiles (0-1)	1				1		1	1			1					1				1		1		
	Alternative medecine (0-1)										1										1				
	Languaje (0-1)	1							1			1					1								
Cultural identity elements to be enhanced	animality			1	1												1			1		1			
	Identity places	1		1	1	1	1	1	1	1		1	1	1	1	1	1	1	1	1	1	1	1	1	1
	traditions	1	1	1				1	1	1	1						1			1	1		1	1	
	religiosity	1		1				1	1			1											1		
	Local products	1	1	1				1	1		1						1				1	1	1	1	1
	Cosmogony			1			1	1		1	1	1	1	1	1	1	1	1	1					1	1
Factible activities to be done in place	Do	1	1	1	1	1	1		1		1	1	1	1	1	1	1	1	1	1	1	1	1	1	1
	Know and learn	1	1	1	1	1	1	1	1	1	1	1	1	1	1	1	1	1	1	1	1	1	1	1	1
	See and rest	1		1	1	1	1	1	1	1	1	1	1	1	1	1	1	1	1	1	1	1	1	1	1
	Buy	1	1	1		1		1	1		1	1					1	1	1		1	1	1	1	1

Simbology

Maximum rate
Medium rate
Minimum rate

Within the rainforest the Mayan people eat chocolate
History and architecture in
Kayacs and flamingos in Laguna
Zoo and Cenote in a Estate
Springwaters within the Mangroove area
About Candy preparation
A grouper in Progreso
Blue blooded octopus
Rainforest, Turks and Cenotes
Flamingos and photographic zoofary
Pirats and Henequen in Sisal port

Source Prepared by the author based on research results

9 Conclusions

The main lessons of this paper are as follows:

The cultural identity of a host community is an essential element to secure an effective differentiated touristic offer. This strategy should take into account the following factors: the identity traits must be analyzed and must not to be stereotyped, homogenized or banalised; its use needs to be integrated into a promotional plan and must involve public and private actors who participate in the activities, so that the quality and integrity of those features is maintained.

The above factors would allow the development of goods and services that guarantee the protection of the environment, support local projects, create mechanisms to integrate new touristic products with a fair distribution of the economic benefits.

In this scenario, the most important implication for public policy is the need for leaders to have a long-term vision, which builds on a differentiated positioning, cooperation, and public-private sectors investments to successfully develop the cruise activity in Puerto Progreso, in particular, and in the Mexican cruise ports, in general.

Finally, it is essential to create and promote better practices to manage cruise tourism in order to compete on a fair basis at the cruise destination or to negotiate conjointly with cruise lines and cruise associations. It is proposed that a national regulatory framework is created and implemented in order to regulate the deals between cruise lines and local tours operators.

Future research on cruise destinations in Mexico may focus on diagnosing the existence of mechanisms to enhance economic and social benefit for local communities, and if necessary, the creation of mechanisms that allow to better manage the segment, to create and integrate differentiated touristic services, and to maintain destinations operating optimally.

References

Administración Portuaria Integral de Progreso S. A. de C.V. (API Progreso) & Secretaría de Comunicaciones y Transportes (SCT). (s/f). Programa Maestro de Desarrollo de Puerto Progreso y Puertos Pesqueros de Yucatán, 2009–2014. www.puertosyucatan.com/qs/pmdp.htm

Alejos JO (2006) Identidad Maya y globalización. Estudios de Cultura Maya 28:57–71

Brejla P, Gilbert D (2014) An exploratory use of web content analysis to understand Cruise Tourism Services. Int J Tourism Res, 16(September 2012), 157–168. doi:https://doi.org/10.1002/jtr

Brida J, Zapata S (2008). The impacts of the cruise industry on tourism destinations. In: U. of M. B. Research Unit on Sustainable Development (Ed.), Sustainable tourism as a factor of local development Organized by GRISS—Research Unit on Sustainable Development, University of Milano Bicocca. Monza, Italy, pp 7–9. http://ssrn.com/abstract=1298403

Brida JG, Chiappa G, Meleddu M, Pulina M (2012) The perceptions of an island community towards cruise tourism: a factor analysis. Tourism 60(1):29–42

Brida JG, Chiappa G, Del Meleddu M, Pulina M (2014) A comparison of residents' perceptions in two cruise ports in the mediterranean sea. Int J Tourism Res 16:180–190. doi:10.1002/jtr.1915

Business Research and Economic Advisors (BREA) (2015) Economic contribution of Cruise Tourism to the destination economics volume II. http://www.f-cca.com/downloads/2015-cruise-analysis-volume-2.pdf. 2015, 15 de octubre

Centro de Estudios en Turismo (Cestur) (2011) Estudio Sobre Productos y Servicios Diferenciados en los Principales Destinos de Cruceros en México. Cestur, México

Chan-Cob JA, Santín B (2003) Potencialidades del turismo para financiar la conservación y el desarrollo sustentable del Caribe mexicano, el caso de los cruceros turísticos en la isla de Cozumel. In: V Congreso Nacional de Investigación turística. Sectur, pp 1–16

Chiappa G, Lorenzo-Romero C, Gallarza M (2016) Host community perceptions of cruise tourism in a homeport: a cluster analysis. J Destination Mark Manage. www.elsevier.com/locate/jdmm

Chias J (2005) El negocio de la felicidad, desarrollo y marketing turístico de países, regiones, ciudades y lugares. Pearson Educación, S.A, Madrid

Cruise Lines International Association (CLIA) (2016) 2016 Cruise industry outlook. http://www.cruising.org/docs/default-source/research/2016_clia_sotci.pdf

Dawson J, Stewart EJ, Johnston ME, Lemieux CJ (2016) Identifying and evaluating adaptation strategies for cruise tourism in Arctic Canadá. J Sustain Tourism 24(10):1425–1441. doi:10.1080/09669582.2015.1125358

Estrada AC (2009) Naturaleza, cultura e identidad: reflexiones desde la tradici—n oral maya contempor‡nea. Revistas UNAM. 4. www.revistas.unam.mx. 2014, 20 de marzo

Flores JM (2003) Psicología, subjetividad y cultura en el Mundo maya actual, una perspectiva Crítica. Int J Psychol 37(2):327–340. ISSN: 0034-9690. http://www.psicorip.org/Resumos/PerP/RIP/RIP036a0/RIP03724.pdf. 2015, 20 de agosto

Florida-Caribbean Cruise Association (FCCA) (2013) Cruise industry overview—2013. EE.UU.: Autor. http://www.f-cca.com/downloads/2013-cruise-industry-overview.pdf. 2013, 23 de abril

Forgas-Coll S, Palau-Saumell R, Sánchez-García J, Caplliure-Giner EM (2014) The moderating effects of the cruise line brand The role of trust in cruise passenger behavioral intentions. Int J Cult Tourism Hospitality Res Iss Tourism Themes 52(2):1346–1367. doi:http://dx.doi.org/10.1108/MD-09-2012-0674

Franco, SR (1999). Etnicidad y jerarquía: la formación histórica de la sociedad poliétnica en Yucatán. Dimensión Antropológica. 17:37–65. http://www.dimensionantropologica.inah.gob.mx/?p=1194. 2015, 3 de marzo

García A, Xool M (2012) Turismo alternativo y desarrollo en la costa de Yucatán. En Turismo, globalización y sociedades locales en la península de Yucatán. http://www.pasosonline.org/Publicados/pasosoedita/PSEdita7.pdf

García A, Xool M, Euán JI, Munguía A, Cervera MD (2011) La costa de Yucatán en la perspectiva del desarrollo turístico. México: Secretaría de Medio Ambiente y Recursos Naturales, Corredor Biológico Mesoamericano México y Comisión Nacional para el Conocimiento y Uso de la Biodiversidad. http://www.biodiversidad.gob.mx/corredor/cbmm/DOC/costa_yucatan_desarrollo_turistico.pdf

Giménez G (2007) Estudios sobre la cultura y las identidades sociales. México: CONACULTA-ITESO. http://filosinsentido.files.wordpress.com/2013/07/gilberto-gimc3a9nez-estudios-sobre-la-cultura-y-las-identidades-sociales.pdf. 2014, 12 de marzo

Govers R, Go FM (2004) Cultural identities constructed, imagined and experienced: a 3-gap tourism destination image model. Tourism 52(2):154–182

Hervik P (2002) Mayan People Within and Beyond Boundaries: Social Categories and Lived Identity in Yucatán. Harwood Academic Publishers, Amsterdam. http://kuscholarworks.ku.edu

Kwortnik RJ (2008) Shipscape influence on the leisure cruise experience. Int J Cult Tourism Hospitality Res 2(4):289–311. doi:http://dx.doi.org/10.1108/17506180810908961

Larsen S, Wolff K (2016) Exploring assumptions about cruise tourists' visits to ports. Tourism Manage Perspect 17:44–49. doi:10.1016/j.tmp.2015.12.001

Lemmetyinen A, Go F (2010) Building a brand identity in a network of Cruise Baltic's destinations: a multi-authoring approach. Brand Manage 17(7):519–531. doi:10.1057/bm.2010.5

Lemmetyinen A, Dimitrovski D, Nieminen L, Pohjola T (2016) Brand awareness as a moderator in motivation-satisfaction relation Cruise destination. Tourism Rev Int J Retail Distrib Manage Iss Tourism Rev 71(4):245–258. doi:10.1108/TR-07-2016-0027

Lester J-A, Weeden C (2004) Stakeholders, the natural environment and the future of Caribbean Cruise Tourism. Int J Tourism Res 6:39–50. doi:10.1002/jtr.471

Mackay C, Kerstetter D (2005) Casting off an exploration of Cruise ship space, group tour behavior, and social interaction. J Travel Res 43:368–379. doi:10.1177/0047287505274650

Macpherson C (2008) Golden goose or Trojan horse? Cruise ship tourism in Pacific development. Asia Pac Viewpoint 49(2):185–197. doi:10.1111/j.1467-8373.2008.00369.x

Martínez T (1999) Manual para la identificación, formulación y evaluación de empresas de turismo rural en México. Centro de Estudios en Turismo, México

Memmi A (1999) Las fluctuaciones de la identidad cultural. Política y cultura, 11, 127–141. http://www.redalyc.org/pdf/267/26701107.pdf

Molano OL (2006). La identidad cultural, uno de los detonantes del desarrollo territorial, Proyecto Desarrollo Territorial con Identidad. Territorios con identidad cultural MIMSP. https://ecaths1.s3.amazonaws.com/historiaregional/1854370848.identidad-cultural-uno-de-los-detonantes-del-desarrollo-territorialPARAREGIONAL.pdf

Parola F, Satta G, Penco L, Persico L (2014) Research in transportation business and management destination satisfaction and cruiser behaviour: the moderating effect of excursion package. RTBM 13:53–64. doi:10.1016/j.rtbm.2014.11.001

Porter ME (2012) Ventaja competitiva: creación y sostenimiento de un desempeño superior. Grupo Editorial Patria, México

Quiroz H (2009) Turismo, arquitectura e identidad urbana: el caso de tres ciudades en la Península de Yucatán, México. Topofilia. Revista de Arquitectónica, Urbanismo y Ciencias Sociales. 1 (3). http://topofilia.net/coloquio09quiroz.html

Requejo J (2007) La diferenciación y el reforzamiento de la identidad basados en la naturaleza, la cultura y el paisaje como factores de competitividad turística. Estudios turísticos 172–173 (1):115–120

Secretaría de Comunicaciones y Transportes (SCT), & Secretaría de Turismo (Sectur) (2004). Política Pública de Cruceros en México. México: Autor. http://sedetur.qroo.gob.mx/normatividad/federal/politica%20publica%20de%20cruceros%20en%20mex.pdf. 2013, 3 de junio

Secretaría de Turismo (Sectur), & Secretaría de Comunicaciones y Transportes (SCT) (2008) Agendas de competitividad para impulsar o mejorar la actividad de cruceros en Puertos de México. Sectur and SCT, México

Seidl A, Giuliano F, Pratt L, Castro R, Majano AM (2005) Cruise Tourism and Community Economics in Central America and the Caribbean: The Case of Costa Rica. Working paper by The Latin American Center for Competitiveness and Sustainable Development (CLACDS) of INCAE. http://www.incae.edu/EN/clacds/publicaciones/pdf/cen664.pdf

Thurau B, Seekamp E, Carver AD, Lee JG (2013) Should cruise ports market ecotourism? A comparative analysis of passenger spending expectations within the Panama Canal Watershed. Int J Tourism Res 17:45–53. doi:https://doi.org/10.1002/jtr.1965

Toledo VM, Barrera N, García E, Alarcón P (2007) Manejo y uso de la biodiversidad entre los mayas yucatecos. Biodiversitas 70:10–15. Recuperado de www.biodiversidad.gob.mx/Biodiversitas/Articulos/biodiv70art3.pdf

Uña M, Villarino M (2011) Configuraciones de identidad en territorios del turismo, condiciones generales en Galicia. Cuadernos de turismo 27:259–272

United Nations Education Scientific and Cultural Organization, Man and the Biosphere Programme, & Ministerio de Medio Ambiente y Medio rural y Marino, G. de E. (s/f). Red Mundial de Reservas de Biósfera: Lugares de Desarrollo Sostenible 2010. UNESCO. http://unesdoc.unesco.org/images/0020/002070/207051s.pdf

UNWTO (2008) Turismo de cruceros: situación actual y tendencias. Madrid: Autor. http://www.e-unwto.org. 2013, 11 de abril

Wood RE (2000) Caribbean cruise tourism: globalization at sea. Ann Tourism Res 27(2):345–370. doi:10.1016/S0160-7383(99)00073-0-272

UrBis: A Mobile Crowdsourcing Platform for Sustainable Social and Urban Research in México

S. Ruiz-Correa, E. E. Hernandez-Huerfano, L. Alvarez-Rivera, V. E. Islas-López, V. A. Ramirez-Sanchez, M. González-Abundes, Ma. de L. Hernández-Castañeda, E. Carrillo-Sanchez, R. Hasimoto-Beltrán and I. Plata-Ortega

Abstract Recently, the United Nations Human Settlements Program (UN-HABITAT) stated that cities in Mexico are steadily expanding, sometimes outstripping the rise in population by two-fold due to urban sprawl, and that 85% of the Mexican population will live in towns and cities by 2050. México also has booming young populations, who are not only the inheritors of significant social

---

S. Ruiz-Correa (✉) · E. E. Hernandez-Huerfano · L. Alvarez-Rivera · M. González-Abundes · Ma. de L. Hernández-Castañeda · E. Carrillo-Sanchez · I. Plata-Ortega
Youth Innovation Laboratory (You-i Lab), Instituto Potosino de Investigación Científica y Tecnológica, Camino a la Presa de San José 2055, San Luis Potosí, Slp, Mexico
e-mail: salvador.ruiz@ipicyt.edu.mx

E. E. Hernandez-Huerfano
e-mail: emilio.hernandez@ipicyt.edu.mx

L. Alvarez-Rivera
e-mail: leonardo.alvarez@ipicyt.edu.mx

M. González-Abundes
e-mail: mariana.abundes@gmail.com

Ma. de L. Hernández-Castañeda
e-mail: lourdes.hernandez@ipicyt.edu.mx

I. Plata-Ortega
e-mail: ivonne.plata@ipicyt.edu.mx

V. E. Islas-López
Instituto Tecnológico Integral, Emiliano Zapata 231, León Gto, Mexico
e-mail: viktor_yzlas@hotmail.com

V. A. Ramirez-Sanchez
Escuela de Sistemas, Unidad Torreón, de la Universidad Autónoma de Coahuila (UA de C), Blvd. Harold R. Pape, Monclova, Coah, Mexico
e-mail: victor00320@hotmail.com

R. Hasimoto-Beltrán
Centro de Investigación en Matemáticas (CIMAT), Jalisco S/N Col. Valenciana, CP: 36023 Guanajuato, Gto, Mexico
e-mail: hasimoto@cimat.mx

W. Leal Filho et al. (eds.), *Sustainable Development Research and Practice in Mexico and Selected Latin American Countries*, World Sustainability Series,
https://doi.org/10.1007/978-3-319-70560-6_2

and urban problems, but also the most promising source of solutions. UrBis is a technological platform that was developed to help people in México, particularly youth, document, characterize and reflect about the social and urban problems they face. The platform supports an integrative approach that combines mobile crowdsourcing, social technologies and community practices to effectively document, characterize, reflect about socio-urban issues, and to develop possible solutions through the combined effort of citizens. The purpose of this paper is to share our action-research experience with UrBis in engaging youth populations in the field, outlining our findings after analyzing collected data in an effort to determine how our technological-based approach supports the study of urban environments in five cities located across the country.

Keywords Information technologies · Mobile crowdsourcing
Sustainable community practices · Action-research · Socio-urban studies

1 Introduction

In its most recent report on the state of the world cities, the United Nations predicts that by the year 2030, more people in every region of the world will live in urban more than in rural areas (UN-HABITAT 2012). The report states that cities in México are steadily expanding, sometimes outstripping the rise in population by two-fold due to urban sprawl, and that 85% of the Mexican population will live in towns and cities by 2050. The urbanization process in México is a phenomenon not only characterized by the growth of urban populations, but also by complex changes in aspects such as the demographic composition; the scale, density, and functionality of urban settlements; the lack of employment; different modes of governance, societal values, as well as the composition of ethnic, social and cultural groups (UN-HABITAT 2012). For these reasons, Mexican youth, which accounts for 26% of the total population (INEGI 2010), is not only the inheritor of significant socio-urban challenges but also the most promising source of sustainable solutions.

The rising question is how can we incentivize young people in Mexico to become more aware of their urban environment? How can we help them gain a deeper understanding and insights about the social and urban problems they encounter? How can we empower them create effective citizen-based solutions that foster the collective participation of society?

UrBis is a technological platform which was developed to help people, particularly youth, document, characterize and reflect about the social and urban problems they face. The platform supports an integrative approach (called SenseCityVity) that combines mobile crowdsourcing, social technologies and sustainable community practices to effectively document, characterize, reflect about socio-urban issues, and to develop potential solutions through the collective participation of citizens (Castells and Himanen 2014; Ruiz-Correa et al. 2014, 2017).

UrBis enables deployment of *mobile crowdsourcing* experiments in the form of *Urban Data Challenges* (UDCs) co-designed by our action-research team and community of young people, focused on:

- collecting and mapping georeferenced images, audio, video, text and surveys; as well as analyzing, appropriating, and
- creatively using the collected data for community reflection and development of potential solutions for sustainable development.

The purpose of this paper is to share our action-research experience with UrBis in engaging youth populations in the field, summarizing our findings after analyzing the data in an attempt to determine how our technological-based approach supports the study of urban environments in five important Mexican cities: Cortazar, León, Guanajuato, Mérida and Torreón.

2 An Integrative Approach

UrBis is a digital platform that integrates community practices, mobile crowdsourcing for data collection, and mixed methods for analysis (both quantitative and qualitative) to advance the understanding of how we can help citizens in Mexico better document, characterize, and reflect on urban concerns. At the same time, we aim to generate previously unavailable data resources and create suitable conditions for transforming the collected data into insights about social and urban issues, with the aim that citizens can develop community-oriented solutions to problems that matter to them. Beyond the creation of maps, our approach encourages the use of the collected data for educational endeavors with emphasis in youth groups. Our work helps demonstrate that integrating existing methods and best practices in ubiquitous computing and human computer interaction with the understanding of a specific social context can result in a participatory framework that exploits phone-based sensing to let everyday citizens survey and articulate urban challenges. Mobile technology enable participants share their experiences and reflections in a variety of ways, thus facilitating the interaction with government institutions that support their efforts. Our approach has six main characteristics, which enabled us conduct urban studies in five cities possessing distinctive urban features:

- It can be easily implemented in the field due the technological flexibility provided by UrBis platform.
- Facilitates the collection of multimedia data that complements traditional data sources such as social media streams.
- Fosters the participation and education of populations from different social segments.
- It is applicable in a wide variety of urban environments.

- It integrates sustainable community practices.
- Facilitates data dissemination and interaction leading to synergies between citizens and government institutions.

3 Mobile Crowdsourcing

Crowdsourcing refers to a sourcing model in which contributions from Internet users are gathered to accomplish a specific task. *Mobile crowdsourcing* involves activities conducted through mobile devices such as smart phones, which allows for real-time data gathering with greater reach and accessibility. This technique enables participatory sensing of urban spaces by combining the collective participation of citizens and sustainable community practices. In the context of our research experience such practices include social service activities, and volunteer work organized with the help of teachers, members of community groups, and NGOs, among others. Mobile crowdsourcing techniques facilitate sustainable socio-urban research for two reasons. First, as we show here, the same technological platform can be used to engage in different action-research projects. Reusing previously developed technology significantly lowers the implementation cost of a new project. Second, target populations are directly engaged in the development and implementation of the research experience. As a consequence, the outcome directly relates to solutions of problems that matter to them, increasing the likelihood of successful results. The integration volunteer's action and local community practices lead to a sustainable research process due to a steady interest of volunteers in finding effective solutions.

4 Urban Laboratories: Cortazar, Guanajuato, León, Mérida and Torreón

Guanajuato is a touristic city in central Mexico, and the capital of a state of the same name. The city occupies a valley, forming a complex network of narrow roads, pedestrian alleys, and stairways running uphill. Guanajuato is a historical city and a UNESCO world heritage site, with a vibrant tourism industry that is centered on the city's historical downtown area and several large art festivals. The city of *León* is a business and industrial hub in the state of Guanajuato that drives a large part of the economical activity of the state. León has a strong leather industry, offering products both to the national and international markets. León also receives a large number of visitors. *Mérida*, the capital of Yucatán state, is the largest historical city in the Yucatán peninsula. Mérida is located in the southeast of the country, 22 miles away from the Gulf of México. The city, funded in 1542 by the Spanish crown, is today one of the most vibrant and traditional cities in south México.

Table 1 UrBis platform was used to conduct 6 UDCs in 5 cities across México

City (population)	UDC	Social actors involved
Cortazar (34,500)	SCV-Cortazar	Instituto de la Juventud Guanajuatense, and Universidad Politécnica de Guanajuato (UPG)
Guanajuato (170,000)	SCV-Guanajuato	Centro de Estudios Científicos y Tecnológicos (CECYTE) Guanajuato, and local volunteers
León (1,500,000)	SCV-León	ITSI León, Universidad Iberoamericana León, and local volunteers
León (1,500,000)	SCV-Lobo	Instituto de la Juventud Guanajuatense, Casa del Adolescente de León, and Instituto Municipal de la Juventud de León
Mérida (777,616)	SCV-Mérida	Universidad Autónoma de Yucatán, Campus de Arquitectura, Hábitat, Arte y Diseño (CAHAD), and Unidad Académica Bachillerato con Interacción Comunitaria (UABIC)
Torreón (634,629)	SCV-Torreón	Escuela de Sistemas Unidad Torreón de la UA de Coahuila

Each challenge was lead by our action-research team in collaboration with participant local institutions and volunteers

Torreón is located near the southwest border of the northern state of Coahuila, within the so-called Laguna region. Torreón is one of Mexico's most important economic and industrial centers neighboring Matamoros, Gómez Palacio, and Lerdo municipalities. Finally, *Cortazar* is a small city located in Guanajuato state. Cortazar is an industrial hub located within the main industrial corridor of the state. It also has agro-industrial infrastructure, which employs 15% of the economically active population of the region.

UrBis platform enabled our action-research team conduct UDCs in these cities (Table 1). Each challenge was lead by our team in collaboration with local actors that supported our fieldwork. Two UDCs were organized in León city with the aid of independent groups of volunteers. Actors participating in the challenges include students, teachers and school authorities of universities and technical highs schools, officials from government institutions, and local volunteers. The goals of each UDC varied according to the needs and interests of the local communities (Table 2).

5 UrBis Architecture

The Urbis platform enables mobile crowdsourcing tasks through two main software modules. The first module consists of an Android mobile application that allows UDC users to capture georeferenced multimedia content (pictures, videos, audios, text and surveys). Users' collected data are automatically collated, synchronized and uploaded into the platform (Fig. 1). The other module consists of a back-end

Table 2 The goals of each UDC vary according to the needs of the local communities

UDC (year; duration in months)	Goals summary (no. volunteers; years of age)
SCV-Cortazar (2016; 3)	Use participatory sensing to document and characterize the urban environment of Cortazar City; (20; 18–22)
SCV-Guanajuato (2015; 14)	Use participatory sensing to document and characterize the urban environment of Guanajuato City. Make use of the collected data for artistic creation. Develop recommendations that could be useful for develop public policies to improve the city's urban environment; (200; 16–65)
SCV-León (2015; 6)	Use participatory sensing to examine how youth perceive their urban environment. Engage youth in a research experience as a means to enrich their studies and professional practice. Incorporate data and information collected in the field within classroom activities; (50; 18–22)
SCV-Lobo (2016; 4)	Support institutional efforts to diagnose the state young people in situation of social exclusion and risk of social violence in León city; (20; 18–40)
SCV-Mérida (2015; 3)	Develop a participatory sensing experiment lead by youth to document the positive and negative aspects of the urban environment in Mérida city (80; 14–25)
SCV-Torreón (2016; 2)	Identify and document problems in pavement and sewers in of Torreón city to provide timely information on the state of the city's transit routes (40; 18–22)

and front-end servers. The back-end server provides a set of PHP web services that use JSON technologies for data sharing among software modules. The front-end is a user-friendly UrBis website (HTML, CSS and JavaScript languages) that facilitates registration and administration of UDCs and users.

A designated user administrates each UDC. The administrator has privileges to access the visualization tools and gives permissions for other registered users to participate in the challenge as either data collectors or observers (Fig. 2). UrBis data integrity is kept through a secure non-relational database that allows great flexibility for storing and managing UDCs data. Our data collection process uses best practices to satisfy the requirements of anonymized data management, including those related to personal data safety (Laurilla et al. 2013).

6 UrBis Deployment and Data Collection

Each UDC was conducted trough four main activities that enabled the integration of volunteers work supported by mobile and social technologies: recruit a population of volunteers, co-design a mobile crowdsourcing experiment (that is, an UDC), implement the UDC, encourage participating volunteers to reflect about the positive

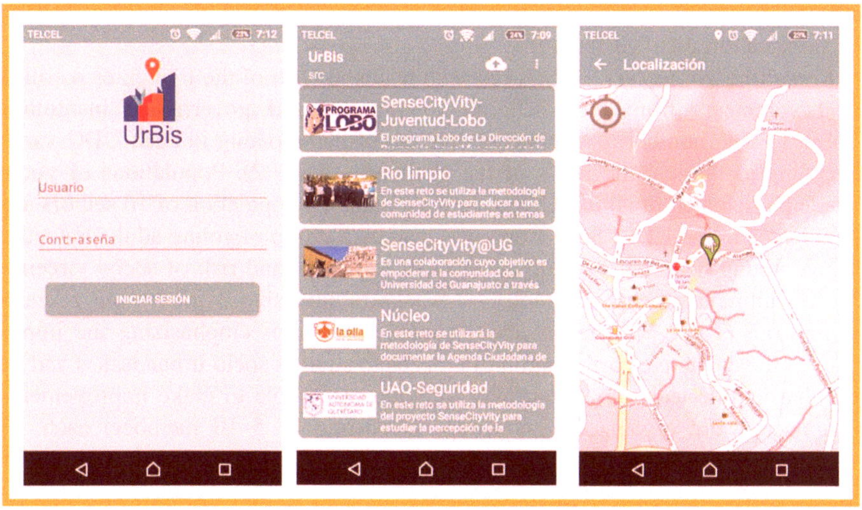

Fig. 1 UrBis mobile application supports data collection for multiple UDCs. The application allows users capture georeferenced multimedia content (pictures, videos, audios, text and surveys). Data collected by the user are automatically collated, synchronized, and uploaded into a back-end server. The figure shows screen shots of the application: (left) UrBis access screen, (center) list of available UDCs of a user, and (right) geographic location of the user during data collection

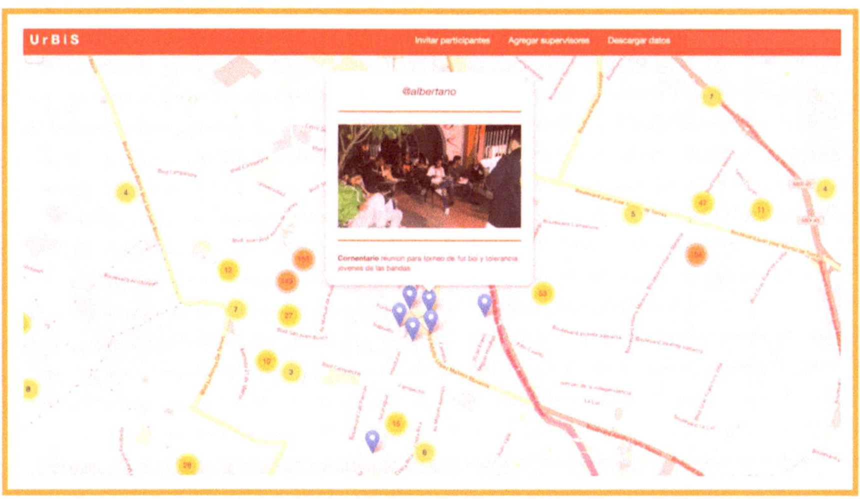

Fig. 2 UrBis front-end is a user-friendly website that allows users to administrate UDCs and visualize georeferenced data collected during fieldwork. The figure shows a León city area of León where data from one of the UDCs were collected. Colored circles indicate evidence (i.e. data) clusters. A friendly graphical interface allows users inspect the evidence within clusters (pictures, videos, images, text, and surveys)

and negative aspects of the data collected, and develop potential solutions based on the information distilled from it.

Recruiting volunteers. Action-research teams in each of the five cities recruited local volunteers from high schools, universities and government institutions (Table 1). The number of volunteers and age of participating in each UDC varied according to the specific goals of the challenge (Table 2). Populations of young people participated in all our UDCs. In Guanajuato, a population of 16 seniors also participated in the UDC. SCV-Lobo was lead by a group of young adults trained to engage young people in situations of social exclusion and risk of social violence. The recruitment process for took 3–4 weeks of intensive work by our research action-team (Table 3). The process included workshops emphasizing the importance of the collective participation of citizens to address socio-urban issues and the role that mobile technology can play to empower people to make improvements. Volunteers were organized into groups consisting of 5–10 members each. To incentivize student participation, registered volunteers were allowed to accumulate the time spent in UDC activities toward their social service requirement. A total of 480 registered volunteers participated in the six UDCs.

Co-designing the experiment. For each UDC, our action-research team and the participating volunteers designed the mobile crowdsourcing experiment.

Table 3 Recruitment workshops were conducted for each UDC

UDC	Recruitment period in weeks	Recruitment workshops	Participants	Codesign workshops	Social media
SCV-Cortazar	2	1	Students, school authorities, government officials	1	Facebook
SCV-Guanajuato	4	4	Students, teachers, school authorities, parents, and seniors	4	Facebook and Twitter
SCV-León	4	4	Students, teachers, and school authorities	4	Facebook and WhatsApp
SCV-Lobo	2	1	Students, government officials	2	No
SCV-Mérida	4	2	Students and teachers	2	No
SCV-Torreón	3	2	Students and teachers	2	WhatsApp

Workshops were lead by our action-research team

The co-design was conducted during workshops that led to discussions about mobile and social media technologies for the common good, ethics, data privacy, personal safety, urbanism, and basic techniques in photography. During the last workshop, participating volunteer groups developed a chronogram of the activities that would be conducted during the UDC. Groups were instructed to center their attention on urban challenges affecting the city landscape focusing on both positive and negative aspects. Therefore, each group defined one, two or three urban problems they wanted to document and the specific approach to conduct their investigation. Some of the urban concerns highlighted during the co-design workshops ranged from garbage in the streets and alleys to crime and vandalism, worn public infrastructure, graffiti, and the prevalence of alcoholism and drugs. In the specific case of SCV-Lobo, participants conducted a survey designed to give a basic characterization of youth facing social exclusion issues in impoverished neighborhoods of León city. In SCV-Lobo, the UrBis platform was used to apply the survey to groups of marginalized youth that agreed to contribute voluntarily. Besides the surveys, pictures, videos and audio interviews were collected in the field.

Implementing the UDC. Urban Data Challenges had duration between 2 and 12 months (Table 2). Challenges mainly consisted of safe data collection activities. Fore example, in Guanajuato city, fieldwork was mainly conducted on weekends. On these weekend gatherings, members of our action-research team led a review session to reinforce the key ideas discussed during the workshops and to check each group's work plans for data collection. This check was necessary to ensure that student groups didn't visit unsafe parts of the city. After the review session was finished, each group moved independently to document the various parts of the city using mobile phones. Smartphones were given to each team; however, volunteers also used their own phones for data collection after they downloaded the UrBis application. During the UDC, students groups explored and documented various routes within the city, which can be seen on an animated map at http://bit.ly/1ZS6yij. Another example is SCV-Torreón, in which volunteers worked in areas of the city that were safe to document. The city was previously subdivided into twenty square regions that were explored during weekdays for documenting potholes and clogged sewers in streets and sidewalks. Many of the documented potholes are known to previously caused street accidents and damaged cars. In Mérida city the data challenge took place in eight sectors of the city that surround the high school campus that hosted the UDC (Fig. 3). Only volunteers that have specialized training to interact with marginalized youth conducted SCV-Lobo. The interaction took place following a protocol approved by government experts with more than 25 years of experience helping young people in situations of social exclusion and risk of social violence. The routes explored by volunteers can be seen an animated map at http://bit.ly/2lEJ2tK. Four of the UDCs integrated the use of social media to maintain communication with participant groups while collecting additional images through a dedicated channel (Fig. 4).

Fig. 3 This figure shows a collage of pictures taken during the SCV-Mérida UDC

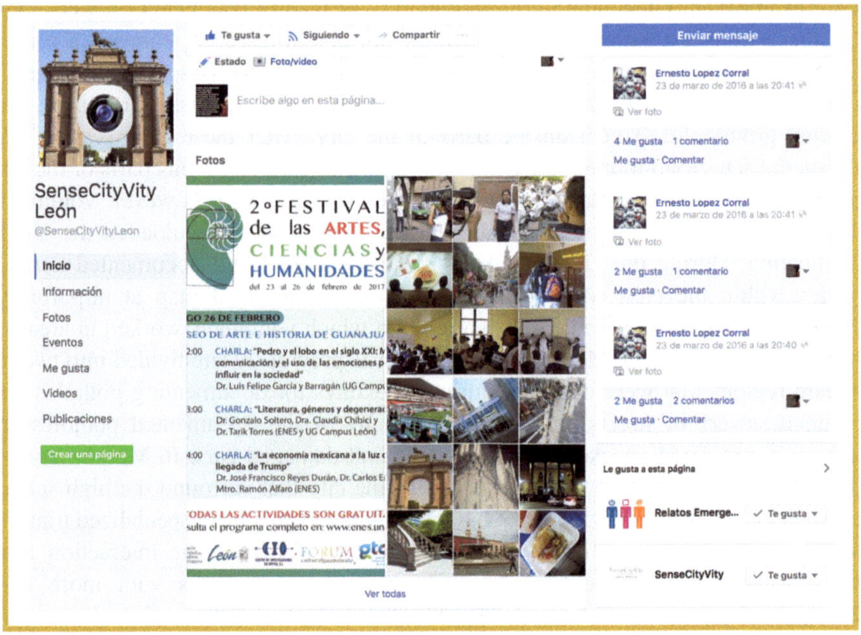

Fig. 4 UDCs integrated the use of social media to maintain communication with participant groups. SCV-León Facebook page is shown in the figure

7 Data Analysis

Data collected during each UDC was analyzed with quantitative and qualitative methods.

A statistical characterization of the collected data was conducted through the calculation of descriptive statistics, the computation of several types of choropleth maps (v.gr. heat maps, cluster maps, and torque maps) (Slocum et al. 2014) and hypothesis testing of georeferenced data (Haining 2003). Recorded interviews were analyzed by means of the qualitative methods described in (Creswell 2014).

Our action-research team shared the information distilled from collected data with participating volunteers as a means to incentivize reflection and discussions about positive aspects of the urban landscapes explored and possible solution to the socio-urban problems detected. UDCs leaders elaborated a report describing their work and potential solutions to these problems.

Although the general methodology used by all UDCs is the same, their goals are different and therefore, the nature of the information distilled from data for each UDC is quite different. For example, SCV-León UDC emphasized the importance of documenting urban perception and awareness of participant volunteers thorough qualitative techniques. On the other hand, UDCs in Torreón and Mérida focused in documenting the urban environment through quantitative measures. SCV-León, SCV-Cortazar and SCV-Guanajuato used both quantitative and qualitative methods to distill information from collected data.

8 Constraints and Limitations

Our research on socio-urban environments was constrained to specific areas of mid-sized and small cities in México that that were safe for data collection activities to take place during the UDCs. Cities in México suffer from a wave of violence that limits the use of our approach to study urban environments affected by violence/criminal activities. Except for SCV-Guanajuato, the number of participating volunteers in each UDC was limited to less than 100 people. This limitation was adopted as a means to facilitate the organization of volunteers within the institutions that supported the implementation of the UDCs.

9 Results

The collected data of each UDC is rich in content. It is enabling a number of qualitative and quantitative studies, also briefly discussed here.

9.1 Collected Data

Pictures were used as the principal means for documenting issues. A total of 12,426 geolocalized images were collected during the six data challenges (Table 4). The least number of pictures were collected in Mérida (347), and the largest number in Guanajuato (7000). A total of 215 videos of urban scenes were collected in SCV-Cortazar, SCV-Guanajuato, and SCV-Torreón (Table 4). Fifty-eight geolocalized audio-recorded interviews were collected in the filed (SCV-Cortazar, SCV-Guanajuato, and SCV-Lobo). Geolocalized surveys were conducted in SCV-Cortazar (129) and SCV-Lobo (44).

9.2 Mapping Evidence

From a quantitative viewpoint, one of the first tasks was to create heat maps for each UDC to identify areas where participants collected data. We created these maps by applying density estimation techniques to GPS locations embedded in the geolocalized data. Four heat maps are shown in Fig. 5 corresponding SCV-Guanajuato, SCV-León, SCV-Cortazar, and SCV-Lobo UDCs. Colored regions in red indicate higher data density compared to those colored in blue. The heat map corresponding to SCV-Guanajuato (top-left) shows regions in red with higher incidence of urban issues. Heat maps and sample pictures from the other UDCs are not shown due to lack of space. UDCs data sets are heterogeneous and rich in content, which depicts features of each city related to the context of the goals of each UDC. For instance, SCV-Lobo UDC contains data related to the urban context of impoverished areas in León city, which includes images/videos depicting graffiti, murals (religious images) and urban sites. These data contrast with those of SCV-Torreón UDC, which is very homogeneous, and mainly portray deteriorated street infrastructure.

Table 4 Data collected during the UDCs conducted in 5 cities

UDC	Pictures	Videos	Audios	Surveys
SCV Cortazar	894	86	9	129
SCV-Guanajuato	7000	85	20	–
SCV-León	1357	–	–	–
SCV-Lobo	1816	40	29	44
SCV-Mérida	347	–	–	–
SCV-Torreón	1012	4	–	–

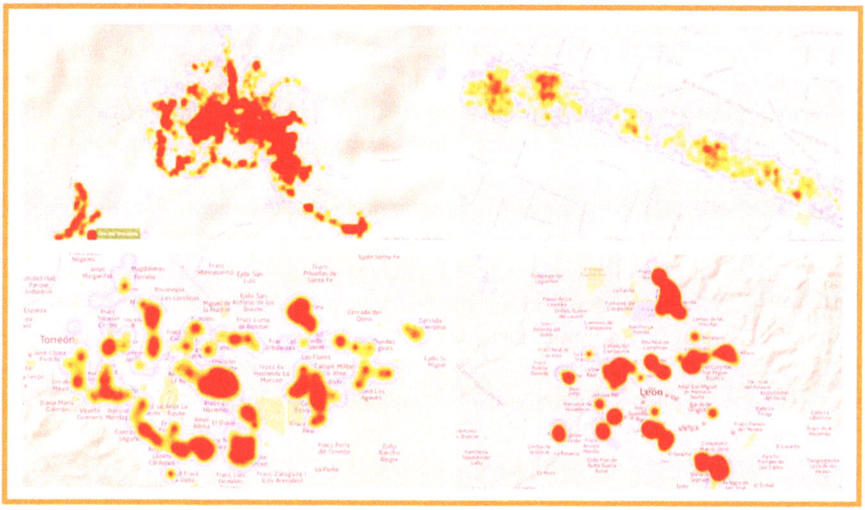

Fig. 5 Evidence heat maps computed for: (top-left) SCV-Guanajuato, (top-right) SCV-León, (bottom-left) SCV-Torreón, and (bottom-right) SCV-Lobo UDCs. Colored regions in red indicate higher data density compared to those colored in blue. The heat map corresponding to SCV-Guanajuato shows data corresponding to urban issues such as garbage, deteriorated infrastructure, inaccessibility issues, etc. Therefore, regions colored in red indicate of higher incidence of urban problems. Most of the SCV-León data were collected in the main street leading to the historic center

9.3 Videos of Urban Scenes

UDC participants in Guanajuato, León and Cortazar, recorded 211 videos to capture urban issues (such as a large wall with painted graffiti on a bridge or tunnel) or to show relevant events (such as a bumpy ride in a city bus or a pedestrian crossing a dangerous street). Collected videos last between close 6–10 s on average. About one third of these videos include comments from the person recording it, which makes these videos valuable to understand how volunteers feel about their city. This kind of fieldwork, which requires direct interaction with people on the street, offered a unique opportunity for participants to experience of the dynamics of the urban landscape. It also let them reconnect to the tangible reality of their city through the lens of their mobile device.

9.4 Emerging Themes

Themes captured by participants were manually coded from audio interviews, videos of urban scenes and surveys captured in SCV-Guanajuato, SCV-Lobo, and SCV-Cortazar UDCs.

SCV-Guanajuato UDC. Three categories were used to code emerging themes: (a) city image—showing littered garbage, graffiti, street dogs, and so on; (b) infrastructure—showing infrastructural problems, ranging from insufficient street lights and garbage containers to inadequate street access; and (c) quality of life—showing everything from vandalism, alcoholism, and drugs to insufficient public transportation.

In Fig. 6, we plot the number of participants who described a specific problem according to these three categories. This plot suggests which problems are perceived as more prevalent. Interviewed citizens agree that garbage, crime, and insecurity are the most pressing urban problems. Other problems that were also mentioned include: overpopulation, pots in streets, the lack of adequate sidewalks, the large number of people that work in the city but lives outside thus creating serious traffic problems. Perceptions also change according to the age. The senior population strongly agreed that many of the city's problems are caused by the lack of civic education and the lack of job for youth. Families are having a difficult time to create the appropriate environment to raise children in a suitable environment where human values, culture, and civism are the most important topics to teach to the new generations. The lack of work for young people is also a serious problem in the city. This problem has lead youth to engage into criminal activities and addictions. One of the interviewed participants mentioned that "[...] the lack of education is the source of crime, if parents do not teach their children to be respectful, to be honest, how do we expect not to have problems with violence when these children grow up?"

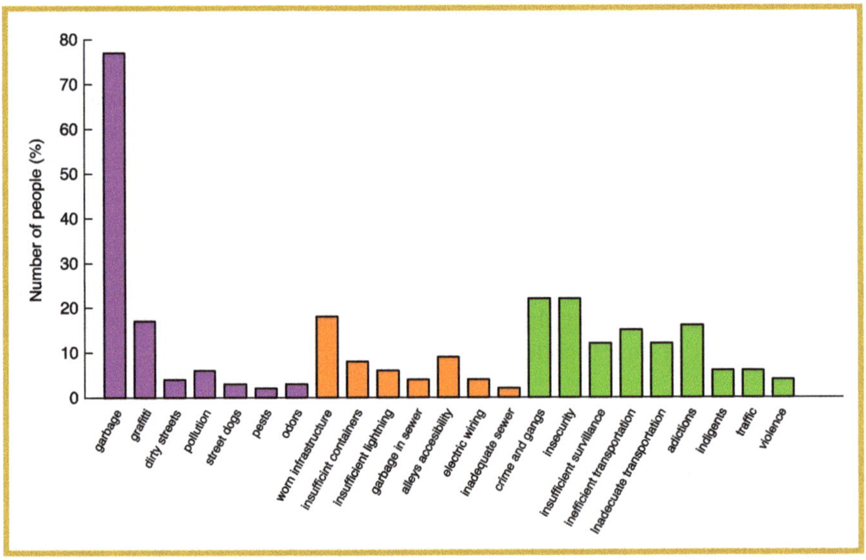

Fig. 6 Urban problems detected in Guanajuato City, which are further divided in three categories: the city image (purple), infrastructure issues (orange), and quality of city life (green)

These reflections agree with the fact that Guanajuato state has one of the largest population of youth that do not work or study: 540,000 youngsters (15–17) according to the data reported by the Instituto Nacional de Geografía y Estadística (INEGI 2010). On the other hand, youth are more concerned about problems affecting their immediate interaction with their urban environment (garbage, insufficient and inadequate transportation, insecurity, etc.) The views expressed by interviewed youth and seniors provide key insights on the origin of the observed urban problems, as well as the key issues that should be addressed. Education, strengthening of the family structure and creative ways of creating jobs for youth are some of the topics that should be considered to improve the socio-urban environments.

SCV-Lobo UDC. Three main themes emerged from the interviews: (a) the quality of life in the city, (b) worn on non-existing urban infrastructure, (c) social exclusion (lack of jobs), and (d) safety issues. People are very concerned about the lack of jobs for youth and the lack of care from their parents. These are two of the main causes for young people enter a gang or create new one. In general, a gang controls an urban space such a street or a block, which becomes a safe space for their members. Oftentimes, territory control is one of the main causes of violence among gangs. Social exclusion issues such as the lack of urban services and infrastructure exacerbate the conditions leading to a large number of gangs in León City. Although the exact number of street gangs in the city is not known, government officials from the Instituto Municipal de la Juventud León estimate that this number could reach over fifteen hundred with an average number of 20 members per band (personal communication by Martha Hernández). The pilot survey conducted during the UDC was aimed at characterizing street gangs in critical neighborhoods of León City such as Las Joyas, San Juan Bosco, and 10 de Mayo (Fig. 5). Table 5 summarizes the results of the survey applied to 44 gangs. In general, the surveyed gangs are composed mostly of males with more than double the proportion of women. The bands have an average of 25 members with a maximum average age of 20 years and a minimum average of 15. Children under 16 created some recently formed gangs.

Table 5 Results of a survey applied to 44 León city street gangs

	Mean	Median	Standard deviation	Min	Max
Number of males per band	21.18	20	9.67	10	50
Average number	8.05	7	6.48	0	30
Number of females per band	30.63	25	18.24	11	120
Number of band members	20.37	8	50.89	0.25	30
Age of the band	16.65	15	8.48	1	38
Average age of the youngest band members	20.27	20	10.66	2	40
Average age of the oldest band members	21.18	20	9.67	10	50

SCV-Cortazar. This challenge focused on documenting the perception a group of young people with regard to educational opportunities within the Cortazar City. The challenge considered the application of a brief survey and documentation of the urban environment where the surveys were applied (http://bit.ly/2m8xbF8). Young people 15–18 years of age participated in the survey, which consists of three key questions: (a) What is your main goal in life? (b) What are the main obstacles that force young people give up their studies? (c) What do young people need to go back to school? With regard to the first question 61.1% of the interviewed subjects aim at finishing university studies and become professionals, 19.8% are more interested in finding a good job, and the remaining 19.1% are focused on subjects like sports, music, and marriage. With regard to the second question, 50.3% believe that the main reason why students give up their studies is the lack of economic resources, 23.7% believe that young people lose the interest in continuing their studies due to a variety or reasons (such as social exclusion), and the remaining 30% believes that the main reason is related to family issues, the lack of abilities, antisocial behaviors, or premature pregnancy. With regard to the third question, most interviewed subject agree that there is urgent need to achieve an economic turn that results from actions conducted at institutional level, which leads to more resources, support and incentives for youth to continue their academic endeavors.

10 Urban Data Challenge Outputs

Beyond the collection and analysis of a novel and rich set of socio-urban data, UDCs resulted in an educational experience for participating volunteers and members of our action-research team. These experiences lead to a reflection and the articulation of potential solutions to the social and urban issues observed. With the aid of local actors, our action-research team wrote a formal report describing the details of each UDC. For SCV-Guanajuato, SCV-Cortazar, and SCV-Lobo UDCs, the reports were presented to local authorities and government officials as a means to promote synergies between citizen and government to find practical solutions to issues that matter to citizen through development of suitable public policies.

One of the findings of our work is that that there are social and urban problems that can only be addressed by government-lead solutions. However, we also found that there are many social and urban issues that can be directly addressed by citizens themselves. One example is the issue of garbage in Guanajuato City, which can be effectively addressed if citizens make a better use of the city's garbage facilities. The collective action of citizens can catalyze the government actions to solve pressing issues.

For SCV-Mérida, SCV-León and SCV-Torreón UDCs, the reports were presented to local school authorities to get further support for expanding the studies to larger areas of the city. For instance, SCV-Merida report is an 83-page document that employs urban ecology techniques to characterize eight sectors of a large Mérida City neighborhood. The document includes a diagnostic study based on social, environmental, and infrastructural variables measured through various techniques, including mobile crowdsourcing. This document serves as a guide to conduct urban studies with the collective participation of citizens over larger areas of Mérida city. It can also be used as a model to conduct studies in other Mexican cities.

Finally we emphasize that the use of social media played a fundamental role in the creation of a communication channel with communities that continuously follow the information generated during the UDCs. Two notable examples are the SCV-León and SCV-Guanajuato Facebook pages, which are followed by a large audience since their creation in early 2015 (http://bit.ly/2lKqgBl and http://bit.ly/2lECpr3).

11 Conclusion

In this paper we have described an action-research experience in engaging youth populations in the field with the help of a mobile crowdsourcing methodology aimed at supporting socio-urban studies in the Mexican cities of Cortazar, León, Guanajuato, Mérida and Torreón. Our methodology combines mobile computing tools, social technologies and community practices to effectively document, characterize, reflect about socio-urban issues, and to develop possible solutions through the combined effort of citizens. Our mobile crowdsourcing platform (UrBis) enables deployment of field experiments in the form of Urban Data Challenges co-designed by our action-research team and community of young people. Our experiments focused on: (a) collecting and mapping georeferenced images, audio, video, text and surveys; and (b) creatively using the collected data for community reflection and development of potential solutions for sustainable development.

The results of the UDCs experiences described here suggest that the use UrBis platform can empower citizens and foster synergies with government organizations to document, characterize, reflect and propose solutions to socio urban issues. We have shown that UrBis facilitates the collection of rich multimedia data, fosters the participation and education of populations from different social segments, is applicable in a wide variety of urban environments, integrates sustainable community practices (such as the social service school requirements), and facilitates dissemination of data and information for a more effective interaction between citizens and government institutions. Preliminary results of ongoing UDCs conducted in Querétaro, San Luis Potosí, Guanajuato, and two cities in Colombia (data not shown due to lack of space) provide further support to our claims. The results of these new UDCs will be reported in a future paper.

Collection of multimedia content in Mexican cities had proven necessary in our action-research work. Platforms such as Foursquare, Google Street Map and Google Street View provide images of cities in México. However, there are many urban sites that are digitally invisible in these platforms. In a preliminary study, we demonstrated that in some areas of Guanajuato, as much as half of the data from

GSV have limitations (Ruiz-Correa et al. 2014). For this reason, the tools provided by UrBis facilitate mobile crowdsourcing studies where documentation of the urban sites and the perception of citizens about their socio-urban environment are of prime importance.

There are various related works, including civic reporting systems, youth and community engagement programs, open mapping initiatives, and social media. Due to lack of space, these topics are not discussed here. The interested reader can consult the following references (Santani et al. 2015; Ruiz-Correa et al. 2017).

Acknowledgements UrBis was funded in part by el Consejo Nacional de Ciencia y Tecnología project 247802. We acknowledge the enthusiastic participation of CECYTE Guanajuato community, the Instituto de la Juventud Guanajuatense, the Instituto Municipal de la Juventud de León, and the Lobo Program staff members and volunteers.

References

Castells M, Himanen P (2014) Re-conceptualizing development in the global information age. Oxford University Press. ISBN: 9780198716082. doi:10.1093/acprof:oso/9780198716082.001.0001

Creswell JW (2014) Research design qualitative, quantitative and mixed methods approaches. Sage Publications Inc., Thousand Oaks. ISBN 9781452226101

Haining R (2003) Spatial data analysis theory and practice. Cambridge University Press, Cambridge. ISBN 0521774373

Instituto Nacional de Estadística y Geografía (INEGI) (2010) Encuesta Nacional de Juventud, http://www.inegi.org.mx. Accessed 20 Jan 2017

Laurilla JK et al (2013) From big smartphone data to worldwide research: the mobile data challenge. Pervasive Mobile Comput 9(6):752–771

Ruiz-Correa S, Santani D, Gatica-Perez D (2014) The young and the city: crowdsourcing urban awareness in a developing country. In: Proceedings first international conference on IoT in urban space (Urb-IoT) Rome, Italy, pp 74–79

Ruiz-Correa S, Santani D, Ramirez-Salazar B, Ruiz-Correa I, Olmos-Carrillo C, Sandoval-Mexicano BC, Arcos-García AH, Hasimoto-Beltran R, Gatica-Perez D (2017) SenseCityVity: mobile crowdsourcing, urban awareness and collective action in Mexico. IEEE Pervasive Comput 16(2):44–53. doi:10.1109/MPRV.2017.32

Santani D, Ruiz-Correa S, Gatica-Perez D (2015) Looking at cities in Mexico with crowds. In: DEV '15 Proceedings of the 2015 annual symposium on computing for development, pp 127–135. ISBN: 978-1-4503-3490-7

Slocum TA, McMaster RB, Kessler FC, Howard HH (2014) Thematic cartography and geovisualization, 3rd edn. Prentice Hall Series in Geographic Information Science, USA

United Nations Human Settlements Program (UN-HABITAT) (2012). State of the world's cities 2010–2011: bridging the Urban Divide

Author Biographies

Salvador Ruiz-Correa is director of the You-i Lab.

Emilio Ernesto Hernandez-Huerfano You-i Lab—WebApps.

Leonardo Alvarez-Rivera You-i Lab—MobileApps.

Victor Emmanuel Islas López is Professor at ITSI León.

Victor Andrés Ramirez Sánchez is Professor at the UA de C.

Mariana González-Abundes is member of the Ce Mobili.

Ma. de Lourdes Hernández Castañeda is a postdoctoral fellow at IPICYT.

Elizabeth Carrillo-Sánchez is researcher at You-i Lab.

Rogelio Hasimoto-Beltrán is Professor at CIMAT.

Ivonne Plata Ortega is a postdoctoral fellow at IPICYT.

The Living Interculturality of Chiapas to Recreate the Modern University Towards Sustainability Horizons

Felipe Reyes-Escutia

Abstract Naturalizing modernity has enslaved our thinking, our imagination, our actions and our proposals, subjecting them to their univocal rationality. Regions like Chiapas are essential to imagine and build sustainable realities and horizons for Humanity due to the biocultural diversity of its peoples and its powerful expressions of life, dignity, hope and intercultural dialogue. This paper does not present a research report but the epistemic construction achieved by the team of the Environmental Education and Sustainability Laboratory of the University of Sciences and Arts of Chiapas during the journey with native and mestizo peoples oriented towards community sustainability between 1996 and 2017. Throughout this time, different modern theoretical and methodological positions were assumed and confronted with local realities, discovering their insufficiency to understand the space-time-symbolic complexity of human communities and their global-local biocultural processes, and to propose strategies of sustainability. We have observed that the disciplinary and rationalist epistemic framework does not correspond to the complex, intercultural, diverse, spiritual, loving and deeply alive universes of the Chiapas peoples. In their cosmogonies, gnoseologies, territories and social models we recognize fundamental references for the construction of a sustainable Humanity: respectful identities with life, solidarity, intercultural dialogue, understanding and loving understanding with the Earth and with the peoples, life. With this learning, we propose strategic lines towards the recovery of the University as a fundamental institution for the civilizational construction, in the key of sustainability.

Keywords University · Modernity crisis · Interculturality · Sustainability Education

F. Reyes-Escutia (✉)
Instituto de Ciencias Biológicas, Universidad de Ciencias y Artes de Chiapas,
Libramiento norte poniente 1150, C.P. 29039 Tuxtla Gutiérrez, Chiapas, Mexico
e-mail: felipe.reyes@unicach.mx

W. Leal Filho et al. (eds.), *Sustainable Development Research and Practice in Mexico and Selected Latin American Countries*, World Sustainability Series,
https://doi.org/10.1007/978-3-319-70560-6_3

1 Introduction

Modern Western civilization is unsustainable. It needs new forms of knowledge and understanding of the world, of living in it and of coexistence between cultures. The University, for its vocation to knowledge, its critical thinking and its responsibility in the training of professionals, thus acquires a fundamental meaning and an indispensable role in the imagination and configuration of horizons for Humanity. The University is currently faced with the challenge of undertaking, as it did in its origins, a profound civilizing process.

In Western civilization, Reason and Man remain centers of an apprehensible, dominable and knowable universe. In this regard, Steger (1974) states: "Excessive departmentalization has contributed to the compartmentalisation of scientific practice based on unidisciplinarity, and can also produce, from the sociological point of view, the perverse effect of transforming knowledge into goods sold and bought according to the needs of the market".

The university has to transcend modern scientific paradigms as an institutional reference. It must build spaces for the encounter of diverse forms of knowledge, understanding and construction of Worlds and Humanity. In order to aspire to imagine and construct new and different realities, others are fundamental to the University, but not this one. The University should deeply review its identity and project as a social and academic institution. The university education has to recover cultural diversity, respect for the land, a clear planetary awareness and an ethical framework founded on life (Reyes-Escutia 2008). Thus, the dialogue and living diversity emerges to embrace, imagine and create identities and utopias. The modern University finds fertile ground, a source of life-giving water in the painful and hopeful interculturality that throbs in the towns of Chiapas.

Faced with the hegemony of univocal thinking, the peoples of Chiapas offer epistemic diversity and dialogic universes, from the adversity and pain of modern marginality. Faced with the objectification of the world, the mestizo and native peoples identify themselves, organize, dialogue and build knowledge from the sacredness of life to which they belong. They build sustainability every day, in adversity. To understand, to enter and to dialogue with these universes is essential for the University.

2 The University in Chiapas: Decolonizing Our Thinking to Understand Intercultural Realities

For over 20 years we have been able to observe how the approach based on economic development has generated economic dependence and deteriorated the autonomy of peasant peoples; We have seen how official social programs have provoked cultural dependence, abandonment of peasant identities, denial of one's own biocultural heritage, and have encouraged the migration of young people to the city or “to the North”; We ourselves have been part of academic projects that fragment the problems, affirm the ignorance and guilt of the peasants and offer technological solutions. And, in all cases, environmental degradation and the poverty of people living with forests, rivers or seas were not halted.

In this paper we present the concepts that we have built and integrated from the learning that these peasants have given us, generous, to whom we form the University, to recreate it, to recreate us.

Identity and worlds. Understanding the universes that flow with vitality in the thought, body and spirit of the peoples of Chiapas is not an adventure that can be undertaken from the exclusionary poverty of modern reason or from the aseptic exercise of discipline. It is necessary to liberate and broaden our intelligence to understand the complex unity we are, the poetic consciousness of belonging and to assume that we are not disconnected fragments of intellect, sensation, emotion and spirit.

To impose a single and exclusive notion, a single paradigm of the world and society denies and/or submits and/or marginalizes the authenticity, validity and civilizing value in the planetarity of each complex and vital territory-nation-history-culture system. In the configuration of each biocultural universe national and community identities are vivified and validated.

Identity is the point of departure and destination; it helps us walk in the current planetary crisis. We must not reduce it to individuality or cultural totalitarianism or to civilizing colonial imposition. On the contrary, we must understand and assume it as the beginning of dialogue, to be and to form community. As Villoro (1998) states, identity is an intersubjective representation that is shared by the majority of all the members of a people—or community—which constitutes a collective Self.

In the face of the modernist denial of recognizing the different as normal, the consciousness of strengthening the cultural-historical identities of each community emerges as a fundamental condition for imagining, exercising, communicating, loving, enriching and defending what one is; to dialogue lovingly and to build poeticized communities with/in the others. To Be Us, in short (Reyes-Escutia 2017, in press).

On the other hand, in the hegemonic process of colonialist cultures that understand equality as homogenization, the diversity of biocultural universes in the world is being weakened by the deterioration of local cultural and historical identities, by the weakening of their capacity and vigor for dialogue and for the global construction of Humanity. Weak identities can be subdued easily, and colonized.

In this process, what identities does the University—as an academic and social institution with civilizing responsibility—seek in education, science and culture?

To recognize the intangible landscapes that bring us closer to an answer, we refer to a concrete experience:

> In 1999, in the design of the sixth curricular project for the training of biologists in Chiapas, the integration of an environmental education seminar was proposed. The proposal was rejected, on the grounds that it did not represent a scientific field, despite the fact that a significant number of biologists perform educational work in communities and in protected natural areas of the State. The following ensuing case illustrates the consequences of this decision:

A group of biologists was hired by the Ministry of Environment and Natural Resources to conduct environmental education workshops in rural populations established in several Biosphere Reserves. Without the necessary training due to the refusal of the seminar described above, the biologists prepared a plastic recycling workshop. They started the workshop in the morning by asking the children to bring empty plastic bottles. Noticing the confusion of the children, the biologists clarified: an empty soda bottle. Then, with this clarification, the children ran home from their school without walls. When they arrived they said to their mothers: "the teacher says to give me money to buy a soda." Without being totally convinced the mothers gave the necessary money to buy the soda.

We must stop here to understand what happened beyond the immediately sensory-observable:

1. Biologists proposed plastic recycling because it was the only activity they could recognize with meaning and environmental value.
2. The community had no problem with plastic waste.
3. The workshop was not designed to correspond with the realities and problems of the community nor with its environmental expression.
4. Mothers gave money to their children to buy soft drinks because they were forced to comply with school activities in order to receive the economic support from the federal government.
5. Money is scarce in the community.
6. To quench their thirst—and also their hunger!—people in the community ingests a drink of pre-Hispanic origin, based on corn and cocoa, called pozol, obtained from their knowledge and practices on their territory.
7. Pozol is a cultural, technological and nutritional heritage of the community.
8. By inducing, from ignorance of community life, the substitution of pozol. consumption in favor of the purchase of bottled soft drinks affects the cultural, technological and nutritional heritage of the community and, at the same time, promotes the cultural, economic, and technological and communitarian dependence model associated with modernity. In addition, the decision favors the development of disfavorable dental and gastrointestinal conditions in a community lacking medical and dental services.

The modern coloniality exposed in the expressions of identity presented by the biologists, thus, affected the cultural processes of a community whose identities are being submitted to the univocal, vertical and hegemonic structures and practices imposed from the school and the modernizing national institutions.

On the other hand, communities with strong cultural identities, deep roots in their history and in their territory, and with visions and paths with their own utopias and light, are genuinely authentic and whole. They express consciously what they are and want. For example, we can retrieve the dialogues developed at the First Encounter of the Zapatistas and the Con-SCIENCES for Humanity, held in San Cristobal de las Casas, Chiapas, Mexico; from December 26, 2016 to January 4, 2017. These are the words of *Sub-Commander insurgent Galeano* on the first day of the Meeting (SubGaleano-EZLN 2017):

> And for you to know us as zapatistas, we will begin by telling you what we do not want and what we do want.
>
> We do not conceive of knowledge as a symbol of social status or measure of intelligence.
>
> …
>
> We do not want to go to university, we want the university to rise in our communities, to teach and learn together with our people.
>
> …
>
> We want schools to be set up for the training of scientists, not workshops disguised as schools, which teach only trades at the service of capitalism…
>
> We want scientific studies, not just technical ones.
>
> …
>
> We do not want to enter scientific and technological competitions, those that so enthuse the universities …; we want to learn and do science and technology to win the only worthwhile competition: life against death.
>
> We do not want to go to big cities and get lost there. And it is not for lack of knowledge, or for the lack of intelligence, or a lack of money.
>
> It's because we do not want to stop being what we are. And we are originative peoples, originative people say. And what makes us what we are, it is our land, our people, our history, our culture and, as Zapatistas, our struggle.
>
> In short, we want to understand the world, to know it. Because only if we know it, can we make a new one, a bigger one, a better one.

We can then understand that within the efforts of recovery and strengthening of the historical-cultural identities of human communities the following possibilities emerge:

1. Give substance to the intercultural encounter.
2. Awaken the collective desire to build better worlds, our own worlds.
3. Imagine and build loving, poetic, supportive, and sustainable Humanity.
4. Imagine and build civilization in dialogue diversity.
5. Recognize ourselves as different beings but in dialoguing communities.
6. Understand our belonging to life.
7. Recover and strengthen our cultural heritage in diversity.
8. Support responsible forms of life and coexistence with Mother Earth.
9. Build viable, just, and dignified societies, and
10. Recover our complex Humanity that integrates spirit, intellect and body.

The construction of identities from and in dialogue—with the land, with life, in time, between people, towns and nations—forms community, it gives rise to and sustains concrete, unique, differentiated, and complex biocultural processes; it builds nation, heritage, imaginary and Utopia. Communities sprouted and cultivated in the creative dialogue between territory, history and culture are the basis of human diversity and the sustenance of the vast and fundamental cultural heritage, essential now to imagine a Humanity that we begin to recognize and that we have to be to achieve our Good-living.

In this way, modern institutions, and especially the University, must understand that each human community is unique and constructed with or from its territory, history and culture. The systems of unitary and totalizing knowledge, rationalities and epistemologies—universalizing—do not correspond to this condition of identity and diversity. In the face of the planetary crisis, urgent is the emergence of gnoseologies, systems and knowledge practices that are based on this principle of dialogue, understanding and constructing of society and Humanity to build and weave emerging identities, realities and horizons. To recover hope has to be the reason and purpose of an unpostponable University to be built.

A university project that assumes identity in the diversity Humanity can only be configured and understood from articulated referents of complexity, transdisciplinarity, interculturality, dialogue, solidarity, spirituality, critical subjectivity, historicity, uncertainty, belonging, and community. This aspiration is impossible for the cosmogonic, philosophical, and epistemic approaches of an objectivist, dualistic, reductionist or fragmentary character. Recreating the modern institution that is the University is an indispensable and urgent task. The forging of Solidarity Humanity, must be its beginning, a path and destiny that gives us birth and gives utopia bringing us closer to the possibility of an intercultural civilization.

Belonging to Life to be Humanity. Every cultural project shapes, vivifies, problematizes and recreates identities that are manifested in the human expressions guiding the social processes of each people. However, in the philosophical referent of being and being in modernity, the notion of belonging to sacred universes or entities in their relation to the Earth is not recognized, identified or assumed. Earth is an object, a thing. Our superior civilization cannot belong to it and this is not congruent with its dualistic and colonial nature. So, in the modern project, one wonders, to what does Humanity belong? To what does thought belong? To what does the university belong?

It is possible to draw a landscape that helps recognize the need for such questions. Under the modern food model and according to the Food and Agriculture Organization of the United Nations (FAO 2015), there are now 795 million hungry people in the world; according to the report “Global Food: waste not, want not” (IME 2013), 30–50% of the food produced in the world is now wasted: an ethical contradiction, to be sure. Moreover, three quarters of the 795 million people who suffer from hunger live in rural communities, that is, where food is produced. In the same sense, Ortiz and Cummins (2011) point out that in 1% of the global richest population one finds concentrated the same economic wealth as 56% of the poorest population. Then, from this absurdity, we must inquire: What civilization is this that only generates inequality and hunger? What is the notion of Humanity, world and territory that assumes and builds this reality? What has been the role of knowledge, culture, science and technology—paradigms of the modern project—in the configuration of this crisis? And, then, how does the modern university participate in this critical process?

Education, especially in universities, guides the processes of continuity and social change. In this practice they assume and apply notions of well-being and progress aimed at privileging individuality over communality, capitalist materiality at the expense of a spirituality that favors belonging to life, competition and confrontation over dialogue and cooperation, violence over peace and reason over love. The result of this current project of civilization is the crisis of Humanity, expressed in severe planetary environmental deterioration, in human poverty and inequity.

Neoliberal modernity does not assume as a general principle or practice its belonging to Life. Life and its expressions are merchandise, resource and object in its economic-social rationality; these have become natural capital, objects of study, and natural resources in the hegemonic academic rationality. The modern university assumes these notions to define and build Humanity, world, society, culture, and territory. Man is a subject, as are Life and object. No possibility of belonging exists here.

The throbbing experience of the communities and nations that give life to their principles of being and being from a deep sense of belonging to life cultivate a spirituality that expresses itself in a loving and poetic life with the Earth and gives body to a vigorous project of civilization. The principle of belonging gives sense and horizon to the dialogue between identity and utopia that leads to sustainability —beyond its technical definition—in Latin American academic thinking, a notion close to the Good Living of originative nations and peoples that interweave with the Earth. There are margins and horizons of hope in many ways.

The possibility of a university that leads us to Good-living dwells in the notion of belonging that originative peoples assume within the systems that integrate society, world and knowledge. In this sense, Padilla (2014) helps us to recognize a new life in diversity by describing the poetic and simple life of the originative women and peasants of Chiapas, who recognize Earth as the mother of life, of food, of their identities and vital links:

> The grandmothers of Coquiteel and Corostic say: "Before, rituals were performed … people prayed for rain, but they also asked the animals that lived in the milpa, like ants and moles, not to eat the milpa; that's why they brought an offering in their caves, … they told the animals that they could share the food with them.

This simple cultural practice is a deep and complex living expression of a cosmogony that is based on belonging to the Earth and Life, which translates into Good Living: a vast and diverse agricultural system that dialogues with ecological processes (the Maya milpa). It is to harvest food free of agrochemicals, because to the Maya the animals with which they coexist in their crops are not a plague. This means that their soils remain fertile and are not contaminated; its production is organic because no chemical fertilization is needed. In turn, this mode of thought-living represents lower health risks for people as it provides a varied and healthy diet meaning less economic dependence because no fertilizers or pesticides are required. Agricultural practices are oriented to food sovereignty. In this way, they strengthen their community identities and their links with the territories and their biocultural landscapes. Poetic, civilizing, vigorous and dialogic exercise as well as belonging, identity, imagination, and utopia are achieved.

But not only in the originative peoples other living worlds exist. There are also mestizo peoples who build worlds, belonging to the Earth, as related in the following work experience:

A group of university students came to a community in the Sierra de Chiapas to conduct an environmental education workshop because they knew that deforestation is a serious problem in this region. They guided the workshop to explain the importance and ecological function of trees. When the students concluded their explanation about photosynthesis, a peasant stood up and said, "I understood, then the trees and we are ourselves!" The students, trapped in their fragmentary and objective rationality, did not understand why the peasant affirmed such a thing. Puzzled, they asked him to explain.

-Yes, you say that the tree releases the oxygen we breathe in the air and we give it that thing you said (carbon dioxide). Well, there it is! When I go to rest under the shade of the tree and I breathe its oxygen is in me and my coal is in it. At that moment, the tree and I are ourselves (testimony of the author).

Thus, Belonging, as a philosophical principle of Humanity, brings to the philosophical notion of Being the transcendent possibility of building Humanity, society, civilization and woven worlds into a Self that recognizes us and understands us in/with the world. This is a fundamental condition to imagine paths for Humanity that is based on dignity, solidarity and the biocultural viability for any civilization project.

The university must understand the cosmogonic universes of non-modern peoples, those which understand that Humanity belongs to the hypercomplexed cosmic process by which our planet exists and to the biospheric processes that make life flourish on Earth. We are an expression of the universe and of life. This understanding is naturalized in the cosmogonies and cultural processes of the originative peoples of Chiapas and Central America, as can be seen in the collection of books collected and known as *Chilam balam* on Mayan civilization.

In these cosmogonies, Humanity gives life to civilization processes with notions of world, society, territory and life that determine identities, imaginaries and utopias in concrete territories. Each nation builds a social project: food systems, forms of social organization, models of knowledge and technology, systems of beliefs and values, models of coexistence between peoples, are set up with land and life. Understanding this diversity of civilizing expressions in different territories represents for the University the challenge and opportunity to be the institution that Humanity needs to transcend this planetary crisis.

The university, in its origins, arises to guide a civilizing process during the Middle Ages as a result of a long process of social and cultural reorganization (Tünnermann 2003). The University, today, must recover its historical responsibility born thousand years ago: to build Humanity and civilization.

And thus we come to knowledge: referent centering on being and doing university. We need to understand what knowledge belongs to. We can find an answer in reminding those 795 million hungry people in the world who are inscribed in a world food system defined more by market and capital criteria than by Humanity. We can also recover the concept of knowledge blindness proposed by Morin and Kern (1993), who states:

> True rationality, open by nature, dialogues with a reality that resists it. It operates an incessant going and coming between the logical instance and the empirical instance; It is the fruit of the argumented debate of ideas and not the ownership of a system of ideas. A rationalism that ignores beings, subjectivity, affectivity, life, is irrational. Rationality must recognize the side of affection, of love, of repentance. True rationality knows the limits of logic, of determinism, of mechanism… True rationality is recognized by the ability to recognize its shortcomings.
>
> … Now, we must know that in every society, including the archaic, there is rationality both in the making of tools, strategy for hunting, knowledge of plants, animals, terrain as there is in myth, magic, religion. In our Western societies, there is also the presence of myths, magic, religion, including the myth of a providential reason and also including a religion of progress. We begin to be truly rational when we recognize the rationalization included in our rationality and when we recognize our own myths among which lie the myth of our all-powerful reason and that of guaranteed progress.

There is a notion of knowledge to be built, one that belongs to—and makes emerge—an intercultural Humanity, a notion that interweaves the rationalities, humanities, worlds, identities and utopias of the peoples of the world to generate understandings that recognize us humans from Belonging to life and to the cosmos; a complex notion that recovers poetically the complex multi and interdimensionality that gives body and life to our human being; a notion that does not fragment our intellectual-emotional-spiritual-material-communal integrality. This is a notion of knowledge to understand, and to understand us, to sustain a loving being, and a belonging to Life.

3 Utopia: Towards Sustainability Since the Transformation of the University

Possible worlds, society drawn in the collective imagination, feelings-thoughts that emerge in different latitudes, nourish the hope of a utopian landscape for our civilization. Actions, speeches and poetic words meet and dialogue to transform and transform themselves. Thus, we find the Latin American and poetic word of Eduardo Galeano (1998), which allows us to depict this "other" world: dignified, spiritual, communitarian and loving:

> "What if we rave for a while?
>
> Let's stop looking so hard at the infamy
>
> to guess another possible world:…"
>
> The right to delirium (1998)

Building Humanity requires a fine, articulating, careful, subtle, delicate thread that integrates into the same tissue the awareness and praxis of our identity and belonging. In this weaving, performed with the dreams and hopes from the community and the territory to forge a utopia from critical and complex thinking, we cannot fail to see the complexity of a modernized Humanity that has been impoverished and subdued in the naturalization of the objectification of the world, which has embroidered positivist scientific knowledge, still hegemonic, in the canvases of reason. However, in order to build a dignified life, a free, solidary and happy Belonging Humanity, it is not permissible to remain enslaved by this poor notion of reason, knowledge and life; it is not possible to hide the beauty that emerges from the color, the musicality and the serenity that is only accessible from the spirituality, the emotionality and communality that, interconnected with the intelligence, make us human. There the utopia: to be Humanity loving belonging to life.

The nature of Utopia is to be light that illuminates, hope that sustains, dreams that lead. Utopia mobilizes, encourages the Being to belong to a community, a family, a nation; to life. The spirituality that gives life to the belonging of a people is intangible, unavoidable for objective scientific knowledge; however, it is essential to understand the complex unity that we are belonging to and, with that understanding, transform (us), and recreate ourselves.

And as a sign that these other worlds exist, are real, live intensely and can feed our imagination and our thoughts, is Chiapas. This South, this tropic that we are, that still dreams and lives with the pain of daily exclusion and incomprehension, offers the color of the historical fabrics of the liveliness and complex interculturality of peoples—cultures—who assume and know that they belong, live and deeply dialogue with the worlds and with the beings that inhabit them, with whom they share time and space. The vigorous worldviews and utopias of the original and mestizo peoples of Chiapas express a deep and structured enchantment with life. They allow us to recognize other universes to create and recreate new civilization; open doors to the intellectual, spiritual, cultural and social integration of our humanity to spin poetic future with roots in life, and polychrome landscapes in diversity and hope.

To achieve the University we need, it is essential to undertake a creative, collective, dialogic exercise to build this place that currently does not exist: that place called utopia, where one dreams, thinks, aspires, desires, and evokes. Utopia for the university is interculturality, to exist and progress in a loving dialogue between cultures, histories, societies and territories as an inalienable principle, in each expression of what we now call substantive functions. Drawing these landscapes of utopia and tracing paths for human walking is today a task to give body to the University, even in the sea of difficulties, contradictions, blindness, dogmas, inertia, poverty and resistance that coexist in this institution. The task of constructing Utopia must be the nature and daily process of the university's living body, in order to guide and sustain hope in the contradictory, convulsive, conflictive and exhausting realities.

The modern university's living body, modernized, without roots, without territory and aimless cannot continue walking. It has to transform and underpin the belonging that enlivens and gives hope, in the collective dreams of each people of the world to be with an identity in the diversity and, then, to build a reality and utopia for the global human planetary process, that is current in every territory. Current in the culture, prevailing in the encounter between the nations, this identity is clarifier for the hope imbued in the process of constructing university and daily Utopia.

Hobsbawn (1998) invites us: "Let us dream of the future. There are many reasons for that. Historians, like other human beings, are entitled to have their idea of a future desirable for Humanity, to fight for Humanity and to be encouraged if they discover that History seems to go where they want, as occurs sometimes. In any case, it is not a good sign of the way of the world if men lose confidence in the future…"

And then, yes, another world is possible. To nourish hope and give certainty to the way, we must know that these "different" worlds already exist, alive, fighting, being and belonging poetically. Modern civilization ignores them, does not recognize that in the dialogue with them lies their only possibility of a dignified future.

4 Learning by Way of Conclusions

We consider that the use of knowledge does not take life as a center and as reason. In order to build sustainable Humanity and civilization, the necessary biocultural diversity exists, and dialogue between cultures, gnoseologies and utopias is possible and expressed every day. Undertaking this construction is imperative and the transformation of the University is the key.

Chiapas makes it possible to recognize for the University a body of pathways for understanding-intervention to build "other/our" paths, to understand the true being of our peoples, from our realities, worlds and humanities that are painfully naturalized as modern. Reyes-Escutia and Quintero-Reyes (2017, in press) recognize the following strategic lines in their work *The New School in Chiapas*. These are:

1. **The possibility of building a civilizational proposal that will recover us in life from the University, in the South and the Tropics that is Chiapas, must denaturalize the modern as a reality, a rationality, and a gnoseological structure and utopia**. This requires dialogue between the unitary and partial modernity with the diversity and belonging to life that vivify other communities. It also requires a reformulation of the systems of knowledge, beliefs, knowledge and values of modern institutions, especially educational and scientific institutions, which have the fundamental responsibility of forming a gnoseology and a process of decolonization from the dialogical and intercultural. Necessary between the peoples of the world and their systems of knowledge and life, it is essential to build Humanity with local force and a planetary viability. The formation of subjects that build knowledge of their and other`s global-local value is a first, important, step.
2. **Local cultural identities must be recognized, strengthened and dialogued.** For the intercultural construction of knowledge and sustainable societies it is fundamental to strengthen local identities and cultures. In the cultural diversity of the world lies the possibility of transcending the civilization crisis. It is imperative for the University to sustain and seek Good-living in interculturality.

3. **Diversity, processes and biocultural heritage must be appropriated profoundly.** This task requires the transformation of institutions, their processes, structures and academic and normative frameworks in an effort to inscribe them profoundly in the dialogue of solidarity, in the construction of societies, in the recovery, respect and promotion of diversity and in the processes by which societies can Be with their territory.
4. **The recovery of our multidimensional Humanity as a principle and purpose of the University must be understood and internalized.** To understand our complex realities, it is necessary to imagine-build healthy and dialogic communities with the Earth, among peoples and between reason and spirit. We must build the reality we need, where poverty, marginalization and hunger do not take place, do not make sense, are not normalized, accepted and institutionalized, even if only in assistance-giving institutions.
5. **A hologramatic and recursive reading of history is necessary and essential.** Recovering Morin's Hologramatic Principle (Morin 2003), a substantial consciousness of a broad temporality that recovers and re-graphs the past as *living* history to understand current realities and to imagine and build futures must be achieved; only then can history and a present be seen as the sustenance of social change and continuity, which depending on the matter of the subject can be present dialogically diversely.
6. **The need to educate in and for interculturality, and the recovery of notions of process, belonging, conflict, dialogue and commonality to achieve the urgency of being Humanity in a profound and poetic oneself**. These elements as well as the transposition of, in our construction of utopia, material and economic developments towards the happiness in the Good-living are necessary to provide and exist with the adequate referents for the understanding of complex realities, and for the intercultural social construction. Starting from communitarian identities and utopias and, always, in the glocal dialogical solidary, it is important to make knowledge, word and life.

In the painful but hopeful dialogue between an expired modernity and a Humanity to be built, the University-from the living and dialogical universes in Chiapas-can, thus, be imagined, constructed, be and belong. Alive nations and territories that live deeply in diversity and interculturality learning to be Humanity —like those expressed in Chiapas—represent the only possibility to imagine, redefine and recreate that University, to make it another and true expression of a loving constitutive life. Being a university that talks and understands belonging to the palpitate of life, is to be hope in the belonging to Utopia, to be intercultural in order to belong to and interact with poetic Humanity. That, certainly, is the University we need and toward which we work currently in Chiapas, Mexico.

In this way of dialogue and collective learning, we can affirm from Chiapas that: To assume, among communities, nations and civilizations, in our social, spiritual, economic, political and knowledge systems as Intercultural Humanity pertaining to Life constitutes the germinal architecture for an Epistemology of sustainability.

References

Galeano E (1998) *Patas arriba. La escuela del mundo al revés.* Siglo XXI, México

Hobsbawm E (1998) Sobre la historia. Colección crítica. Editorial Grijalbo-Mondadori, Barcelona

IME (2013) *Global food waste not, want not.* The Institution of Mechanical Engineers, London

Morin E, Kern AB (1993) *Tierra Patria.* Kairós, Barcelona

Morin E (2003) *Introducción al pensamiento complejo.* Gedisa, Barcelona

FAO (2015) *El estado de la inseguridad alimentaria en el mundo.* Food and Agriculture Organization of the United Nations, Roma

Ortiz I, Cummins M (2011) *Global inequality: beyond the bottom billion.* A rapid review of income distribution in 141 countries. UBIEF, New York

Padilla A (2014) Nosotras, Chiapas, El amor a la madre tierra en sus rituales. In: La Jornada del Campo núm. *78*. Retrieved from: http://www.jornada.unam.mx/2014/03/15/cam-amor.html. (15 March 2014)

Reyes-Escutia F (2008) Universidad pública y sustentabilidad. Chiapas: Aportes para la resignificación social y académica de la universidad. En: Reyes-Escutia F, Bravo MT (eds) Educación Ambiental para la sustentabilidad en México. Aproximaciones conceptuales, metodológicas y prácticas.UNICACH-UNAM-ANEA, México

Reyes-Escutia F (2017, in press) *Construir un Nosotros con la Tierra. Aportes latinoamericanos para la descolonización del pensamiento ambiental.* UNICACH-Editorial Ítaca, México

Reyes-Escutia F, Quintero-Reyes Y (2017, in press) El Espíritu de la sustentabilidad y el deseo de trascender la razón, desde el corazón de los profesores en la Nueva Escuela Chiapaneca. En: *La Nueva Escuela Chiapaneca. Universos, Humanidades, Diàlogos y Utopías.* Secretaría de Educación del Estado de Chiapas, México

Stéger HA (1974) *Las Universidades en el desarrollo social de la América Latina.* Fondo de Cultura Económica, México

SubGaleano-EZLN (2017) Subcomandante Galeano's letter to scientists. Retrieved from: http://enlacezapatista.ezln.org.mx/2016/12/26/algunas-primeras-preguntas-a-las-ciencias-y-sus-conciencias/ . (6 March 2017)

Tünnermann BC (2003) La Universidad Latinoamericana ante los retos del siglo XXI. Unión de Universidades de América Latina, México

Villoro L (1998) *Estado plural, pluralidad de culturas.* UNAM/Paidós, México

Sustainable Development for Farmers Transforming Agroindustrial Wastes into Profitable Green Products

Noé Aguilar-Rivera and Teresita de Jesús Debernardi-Vázquez

Abstract The world population is facing significant changes in the supply of food, feed, fuel and fiber from agroindustrial byproducts. However, poverty, climatic and environmental effects, loss of productivity of major agroindustrial activities, low-use of byproducts, land-use change, biodiversity loss and high water consumption are among some of the factors that contribute to reducing the profitability of traditional agribusinesses. The development of sustainable low-cost technologies, which are simple for rural farmers to apply for byproduct valorization is, therefore, a key option to transform traditional agribusiness in sustainable value chain. This paper proposes, to develop in farms, three productive activities, to achieve the socioeconomic and environmental sustainability with byproducts, as raw material, from the sugarcane, coffee, banana, corn and citrus fruit agroindustries in Mexico, integrating innovative techniques reported in the literature, traditional knowledge, experience of farmers and analytical methods: (a) biodrying for removal of water in byproducts, causing a decrease in volume and an increase in calorific value, (b) production of compost, vermicompost and bocashi and (c) cultivation and harvest of edible mushrooms. These activities have had a positive impact on the sustainability and profitability of farmers in Mexico by reducing the amount of mineral fertilizers used in fields, obtaining rural fuel and generating income; additionally, food is generated for humans and livestock from the production of edible mushrooms.

Keywords Agroindustrial byproducts · Sustainable technology · Farms

N. Aguilar-Rivera (✉) · T. de Jesús Debernardi-Vázquez
Facultad de Ciencias Biológicas Y Agropecuarias, Universidad Veracruzana, Km. 1 Carretera Peñuela Amatlan de Los Reyes S/N, C.P. 94945 Córdoba, Veracruz, Mexico
e-mail: naguilar@uv.mx

W. Leal Filho et al. (eds.), *Sustainable Development Research and Practice in Mexico and Selected Latin American Countries*, World Sustainability Series,
https://doi.org/10.1007/978-3-319-70560-6_4

1 Introduction

Agroindustrial byproducts and wastes present complex widespread problems with considerable socioeconomic and environmental consequences in developed countries. However, there is great interest in the use of agroindustrial byproducts because they contain large amounts of sugars and lignocellulosic biomass that make them potentially excellent raw materials for sustainable biotechnological and conventional technological transformation processes, resulting in the production of new added-value products with specific properties (Cañete-Rodríguez et al. 2016; Kusch et al. 2015; Galanakis 2012).

Byproducts may constitute as much as 70% of crops after harvest or agroindustrial processing and considerable attention has been placed on converting them into sustainable products. Treatment of solid waste generated from agricultural and agroindustrial production activity is another serious problem in developing countries such as Mexico, where the main agroindustrial crops (Table 1) are produced under the rainfed system and low irrigation, are highly vulnerable to climatic change and other environmental and anthropic effects and with high generation of by-products due to the lack of agro-industrial technology (Figs. 1 and 2)

Therefore, technologies of recycling, reuse and sustainable use of by-products, could reduce contamination and spaces required for disposal. Therefore, it is necessary to evaluate major challenges and technologies to determine the most realistic

Table 1 Main agro industrial crops of Mexico (SIAP 2016)

Crop	Producing municipalities (#)	Harvested acreage (ha)	Production (t)	Yield (t/ha)
Corn	2343	7,099,723.8	24,694,046.25	3.48
Pastures	864	2,560,399.04	50,923,935.93	19.89
Sorghum	637	1,658,673.66	5,195,388.74	3.13
Bean	1810	1,555,131.7	969,146.28	0.62
Wheat	549	819,928.09	3,710,706.27	4.53
Sugar cane	267	758,607.94	55,396,061.34	73.02
Coffee	489	664,885.1	1,026,251.98	1.54
Forage maize	492	539,116.78	13,660,717.7	25.34
Alfalfa	821	384,375.63	32,575,005.78	84.75
Orange	457	318,379.6	4,515,520.33	14.18
Mango	392	182,680.07	1,775,506.77	9.72
Avocado	543	166,944.96	1,644,225.86	9.85
Lemon	545	160,083.58	2,326,068.34	14.53
Chile	741	148,688.77	2,782,340.75	18.71
Cotton	45	133,232.3	593,439.49	4.45
Nut	247	79,080.23	122,714.05	1.55
Banana	218	77,548.95	2,262,028.25	29.17

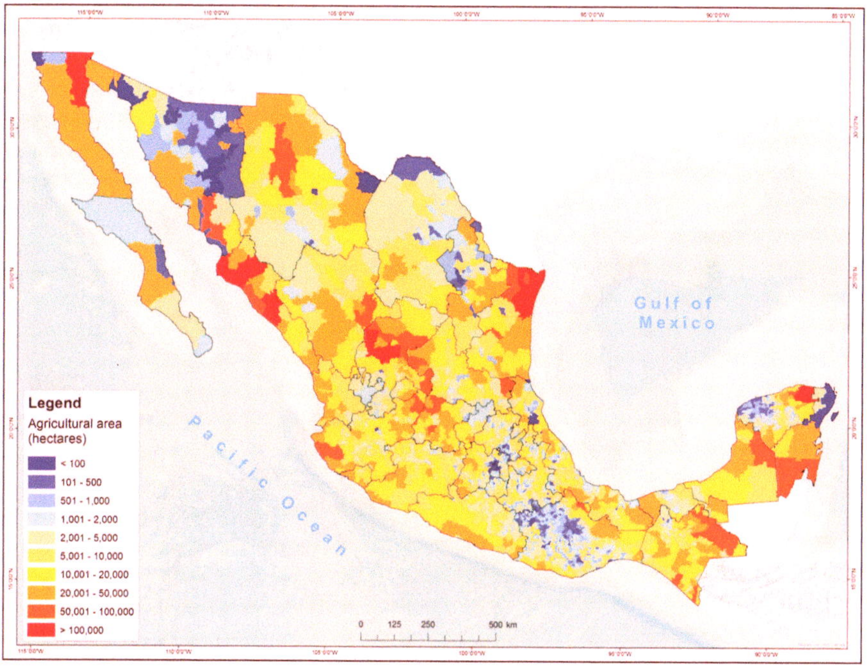

Fig. 1 Most important crops in Mexico (SIAP 2016)

options for the use of byproducts and construction of a sustainable future based on the environmental rationality. Certain agroindustrial byproducts may be used directly as animal feed or manure while byproducts from the sugar, rice, corn and soy agroindustries, among others, can in general be transformed into the 4Fs (Food–Fiber–Fuel–Feed) or raw material for sustainable process design, green engineering and biorefineries considering competitiveness according to Hanes and Bakshi (2015), as well as both, the physical amount of byproduct produced and the environmental impacts. Therefore, new technologies, traditional knowledge, experience, analytical methods from physical and chemical sciences and frameworks, to quantify the environmentally sustainable and socioeconomically feasible potential of agroindustrial residues to generate novel and greener products, bring opportunities for the design of processing value chains, as the main target of sustainability in farms. Mirabella et al. (2014); Koutinas et al. (2014); Pfau et al. (2014) and Vandermeersch et al. (2014) carried out an extensive review of the possible uses of byproducts and wastes under the bioeconomy concept in order to transform them into resources for production of new products and energy, applying industrial ecology and eco-innovative approaches.

Therefore, the aim of this research was to integrally evaluate the use of byproducts to help achieve sustainability strategies in farm and agroindustrial activities by determining their potential as lignocellulosic substrates, with the use of

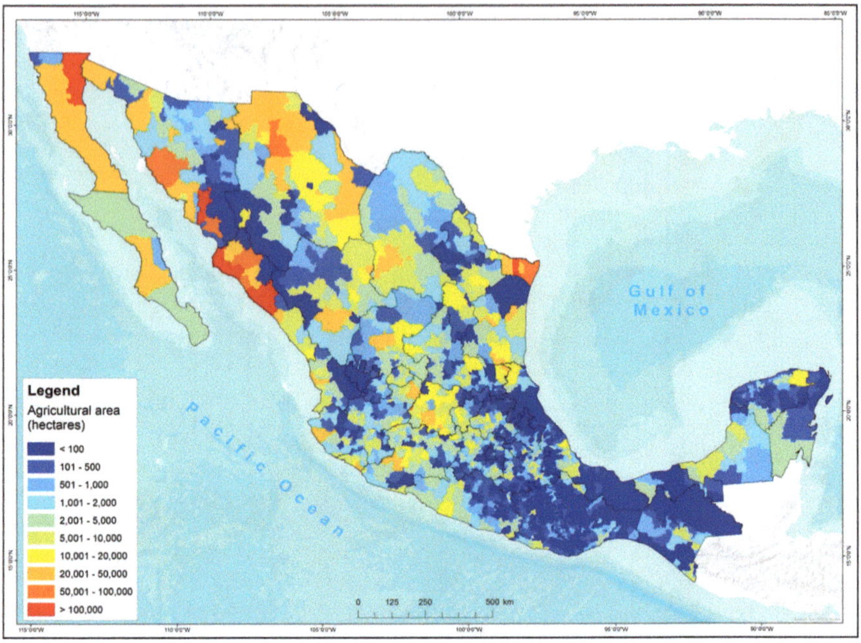

Fig. 2 Irrigated crops in Mexico (SIAP 2016)

analytical methods, low-cost processing technologies, traditional knowledge and experience of farmers, to obtain greener products: solid biofuel, compost and edible mushrooms (*Pleurotus ostreatus*).

2 Biofuel from Biodrying of Byproducts

There are several treatments for final waste disposal, but some of them require processes that entail high implementation costs. More recently, incineration has attracted global attention as an effective way of reducing the amount and toxicity of organic wastes, in addition to recovering energy from them despite their high moisture content (Ma et al. 2016).

Biodrying is a self-heating process in which the drying is carried out by the biological heat that is produced during the in situ decomposition of the organic matter, which offers an alternative for waste handling in terms of feasibility and costs (Tom et al. 2016). The objective of biodrying is to use the heat generated by the metabolic functions of the microorganisms present in the wastes, maintaining their calorific value which allows storing them as solid fuel (Sugni et al. 2005; Dominczyk and Ledakowicz 2014), where the main advantage is the reduction of mass and atmospheric emissions of CH_4, CO_2, SO_2 and NO_x (Suksankraisorn et al. 2010).

Biodrying is performed aerobically, in which the convective evaporation process is used to reduce the water content in the substrate, with minimal aerobic degradation (Velis et al. 2009).

In the particular case of residues from the citrus agroindustry, the large volume of crop residues and byproducts generated during juice extraction, particularly in the case of Valencia orange (*Citrus sinensis*), and rising waste disposal costs have led to increased interest in using these materials (Siles et al. 2016) (Figs. 3 and 4)

The main byproducts used are fresh citrus pulp, citrus silage, dry pulp, molasses, citrus peel liquor, citrus activated sludge and, to a lesser extent, waste or surplus harvest fruit (Okino Delgado and Fleuri 2016; Mamma and Christakopoulos 2014; Fava et al. 2013). Figure 5 presents the fresh composition of different citrus fruits obtained by analytical techniques.

In the figure above it can be seen that, except for the Mexican lemon, the highest proportion in weight for citrus fruits corresponds to the husk and bagasse, followed by the juice that is the product of interest. This type of byproduct contains a high moisture content, which hinders its final disposal as it tends to generate leachates and bad odors. Figures 6 and 7 show the moisture percentage in the husk and bagasse and the pulp of the evaluated citrus fruits.

In biodrying, the variables that affect the process include the initial moisture of the substrate and the airflow. The latter can be manipulated to control the

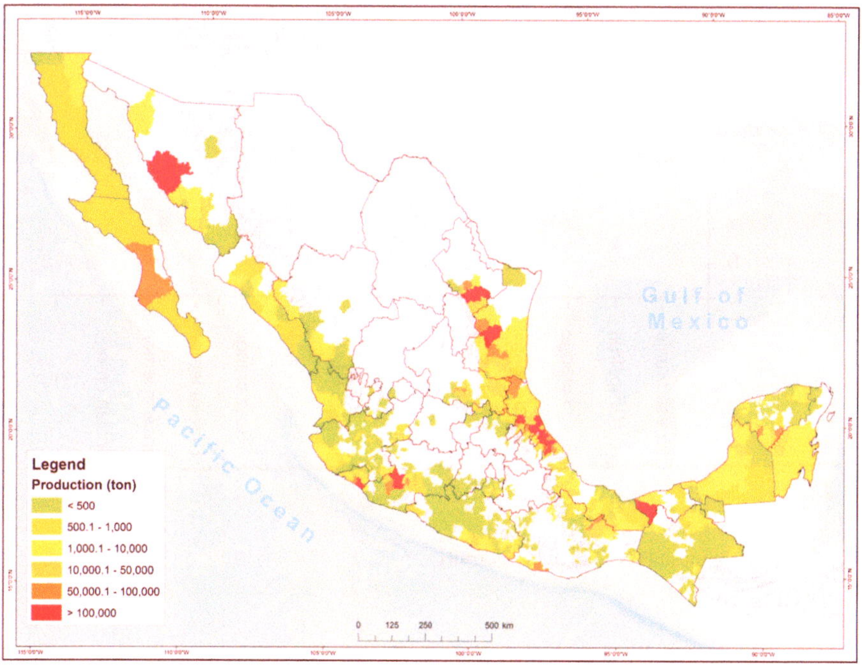

Fig. 3 Citrus in Mexico (SIAP 2016)

Fig. 4 Citrus and wastes

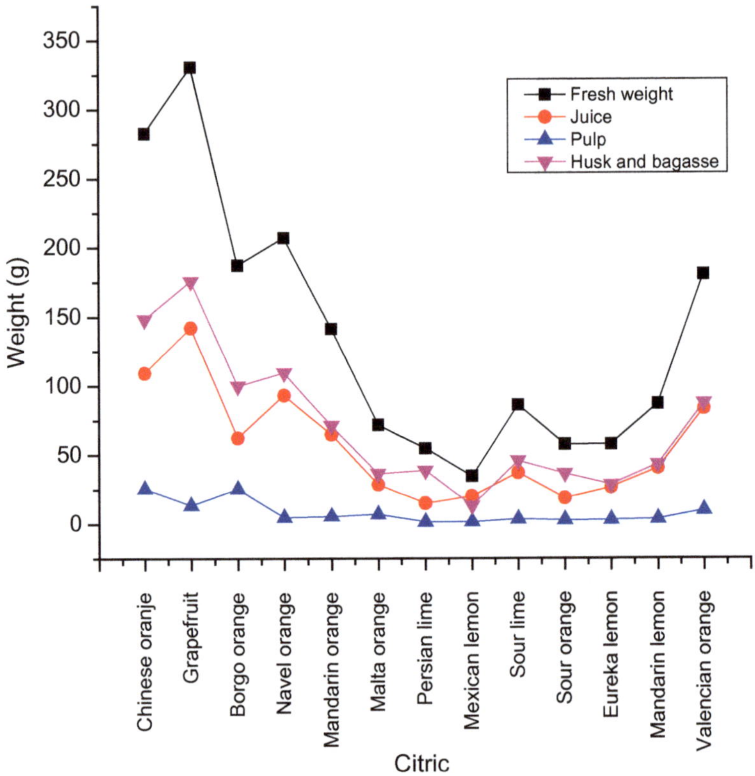

Fig. 5 Fresh composition of some citrus fruits

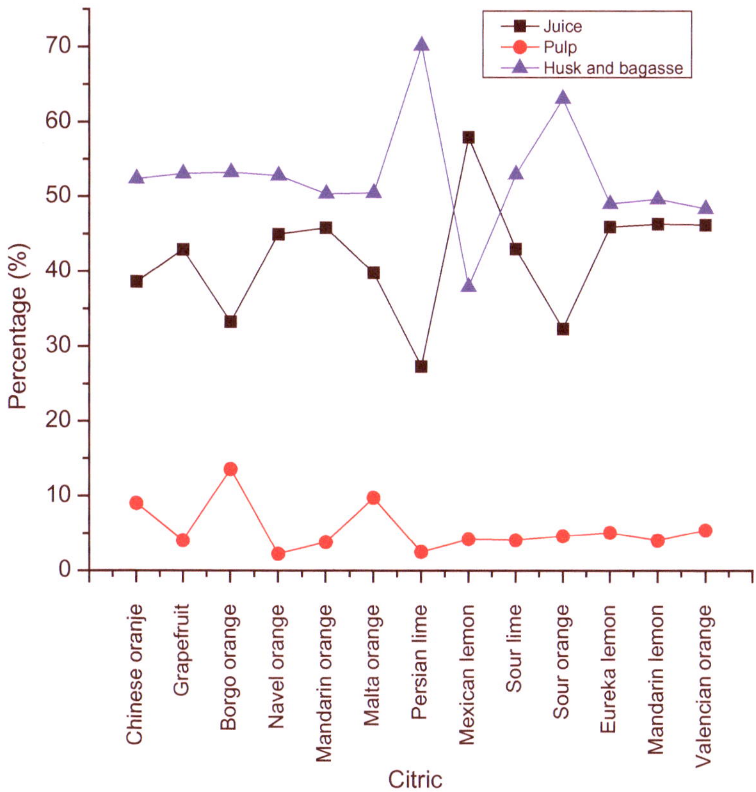

Fig. 6 Composition of citrus fruit byproducts (%)

temperature in the substrate matrix, as it affects the air dew point and biodegradation kinetics (Velis et al. 2009). In other words, the effect of air flow and turning frequency in sludge biodrying, by increasing the air flow, is to obtain higher water removal efficiency, while Cai et al. (2013) found that forced aeration controls the temperature of the pile and improves evaporation; thus, the loss of water is the key factor in sludge biodrying.

Moisture is a critical parameter in biodrying, since it influences the complex biochemical reactions associated with the growth of microorganisms and the biodegradation of organic matter that occurs in the process (Winkler et al. 2013). Thus, the excess prevents the transfer of oxygen and the microbial activity is reduced, which impedes the development of the process; in the contrary case, if the moisture is very low, the microbial activity becomes too slow and thus affects the drying process. According to Yang et al. (2014), the optimum moisture level for the biodrying process to be performed efficiently is in the range of 50–70%.

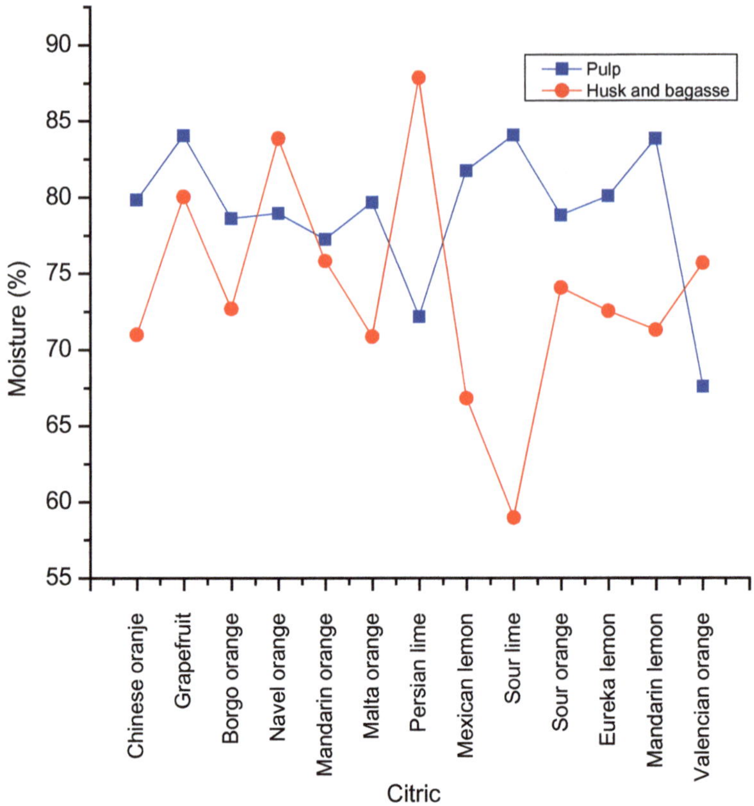

Fig. 7 Moisture in the husk and bagasse and pulp of different citrus fruits (%)

He et al. (2013) found that temperature is one of the key variables in biodrying, as a temperature increase in the matrix increases air temperature, which improves its ability to remove water from the substrate.

In the sustainable process at the farm level, a static biodrying pile with a rectangular pyramidal structure was used; in it, 9 points were established for the permanent monitoring of the inside temperature. As a mechanism to maintain the aeration process in the substrate matrix, a forced aeration system was used, providing a volume of 25 L/min. The aeration process was performed every 2 h with a volume of 125 L/min from 8:00 a.m. to 4:00 p.m. during the time of the experiment (Fig. 8a, b).

For temperature monitoring, an RTD data acquisition system connected to a Snap Pac System Opto 22, which operated 24 h a day, was used. To prevent the substrate from getting wet as a consequence of weather changes, the experiment was carried out in a greenhouse-type drying tunnel. The materials used for preparing the pile are listed in Table 2.

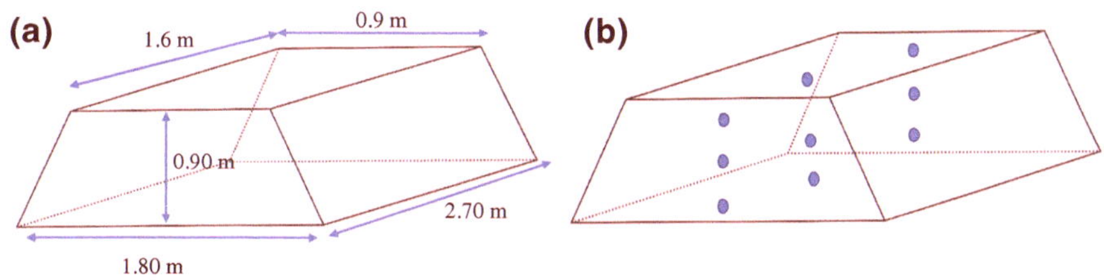

Fig. 8 Geometry of the biodrying pile and sampling points

Table 2 Materials for biodrying

Waste	Amount (kg)	Percentage (%)
Orange peel	200	56.0
Mulch (pruning waste)	88.9	24.9
Grass	68.3	19.1
Total	357.2	100

Grass and mulch were used as structuring materials in order to favor the formation of a porous matrix inside the pile that allowed air flow.

In semistatic biodrying piles, four phases can be observed in the evolution of the temperature: initiation phase, thermophilic phase, second heating phase and decay phase (Cai et al. 2013). This is due to the fact that when turning the materials that make up the pile, its interior is homogenized and the gradients in the concentration of oxygen, available water and nutrients for the microbial activity in it, which are affected by leaving the pile static, are changed; although there are changes in the behavior of the temperature, it remains more homogeneous throughout the process. The behavior of the temperature in the center of the biodrying pile is shown in Figs. 9, 10 and 11; the data used for making the graph correspond to the temperatures recorded at 10:00 am. A uniform distribution is observed in the central layer of the pile, with temperatures ranging from 23 °C, which was the ambient temperature on the day the pile was formed, to 67 °C, which was the maximum value reached in this layer.

According to Colomer-Mendoza et al. (2013) and Cai et al. (2013), there is a reduction in the volume of the pile as a result of the loss of water in the substrate matrix. This phenomenon is due to the evaporation of water by a convective mechanism, i.e., the air entering the matrix in a forced manner carries a certain amount of water that favors the hydration or dehydration of the matrix; moreover, the higher the temperature of the incoming air, the greater its capacity to remove water from the substrate However, if this phenomenon occurs the drying of the materials occurs in a convective way and not due to the effect of the heating caused by the metabolism of the thermophilic microorganisms which are a mixture of yeasts, bacteria and fungi.

While in the middle section of the pile the higher temperature lasted longer, on the sides it decreased as the process progressed. This phenomenon can be attributed

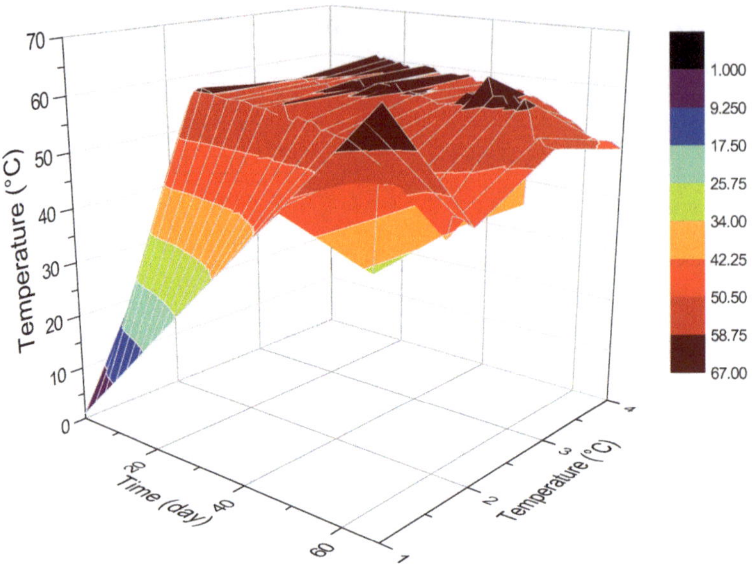

Fig. 9 Temperature in the central section of the biodrying pile

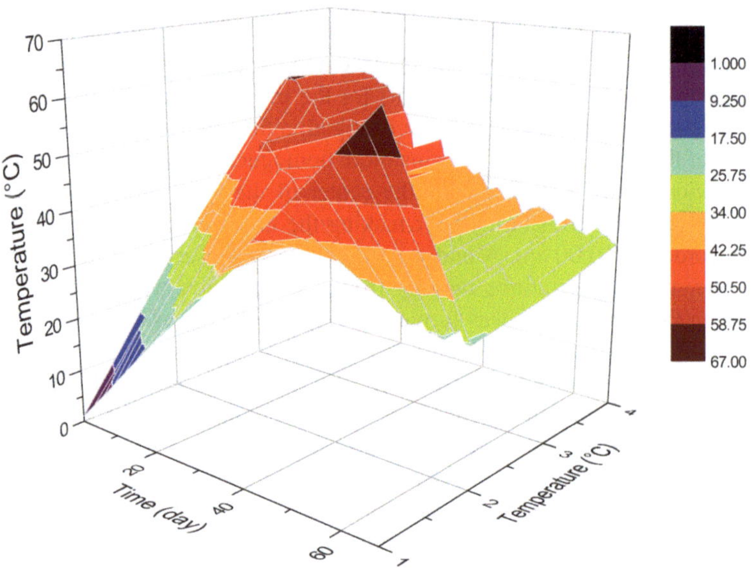

Fig. 10 Temperature in the left section of the biodrying pile

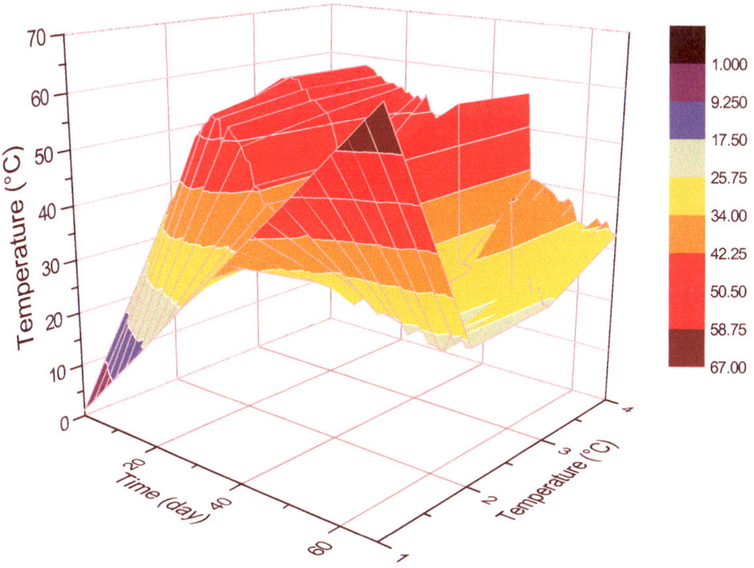

Fig. 11 Temperature in the right section of the biodrying pile

to the fact that, because the pile was not turned, the higher moisture content was concentrated in the center; the material that was located on the surface caused an insulating effect that helped to maintain moisture and allow the process to last up to 67 days, while the normal process with periodic turnings lasts for 20–30 days (Fabián et al. 2012).

It was observed that the maximum temperature was between 60 and 67 °C, decreasing until reaching the ambient temperature, once the microorganisms present no longer have available water to continue the substrate degradation process (Fig. 12).

According to the evolution of the substrate moisture and the temperature behavior in the pile, it can be observed that the moisture is directly linked to the activity of the microorganisms, since as the temperature decreases it tends to stabilize. It is desirable that the biodrying process be carried out in short periods of time in order that the substrate is degraded as little as possible and maintains its calorific value to be incinerated or stored as biofuel for energy cogeneration. Although this stabilized waste can be deposited directly in waste confinement areas for final disposal without the problem of leachate generation (Fabián et al. 2012), another advantage of the process is that there is a reduction in its volume and mass of up to 74%, which facilitates its transport and storage and reduces final treatment costs. Table 3 presents the results of the laboratory analysis according to Mexican standard NOM-021-SEMARNAT (2000) for chemical composition.

The final product presented a humidity of 12.31%, pH 4.92, organic matter 94.24%, which is important for its use as a rural fuel because it guarantees that the

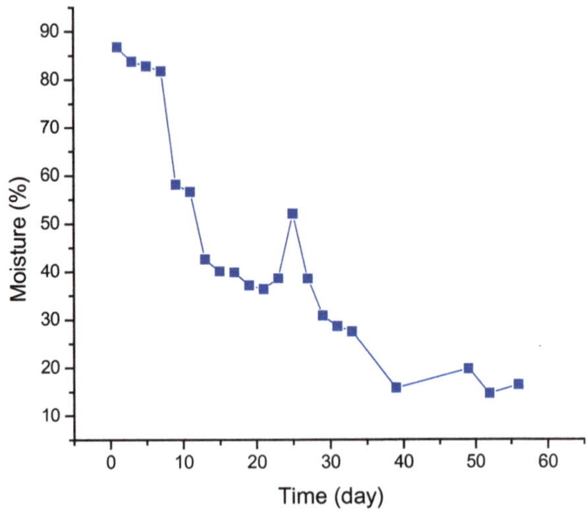

Fig. 12 Decrease in substrate moisture (%) (moisture balance method)

Table 3 Chemical composition of biodried biofuel from citrus

Variable	Unit	Value
Humidity	(%)	12.31
pH		4.92
Soil electrical conductivity	dsm^{-1}	1.45
Ash	(%)	5.76
Organic matter	(%)	94.24
Total carbon C	(%)	54.64
Total nitrogen N	(%)	1.13
C/N		48.38
Ca	(%)	1.544
Mg	(%)	0.186
Na	(%)	0.059
K_2O	(%)	0.934
P_2O_5	(%)	0.273
Fe	(%)	0.0448
Cu	(%)	0.0007
Zn	(%)	0.0034
Mn	(%)	0.0193

ash content at the end of the incineration will be low, with respect to the nitrogen content of 1.13% and carbon of 54.664% with a C/N ratio of 48.38 which is desirable for biodried fuels from biomass which are destined for incineration, favoring a low NOx formation Dominczyk and Ledakowicz (2014).

3 Edible Mushroom Production from Byproducts

Processing of agroindustrial plantations generates wastes in the field as leaves, stalks, straws or bagasses which necessitate the search, as substrates, for new alternatives for effective production as the cultivation of edible mushroom. In fact, cultivation of *Pleurotus* mushrooms is considered to be a simple, low cost and environmentally friendly technology for the utilization of rural and agroindustrial residues in developing countries (Mohamed et al. 2016; Kazemi Jeznabadi et al. 2016; Yang et al. 2016; Corrêa et al. 2016).

Several substrates were evaluated for edible mushroom (*Pleurotus ostreatus*) cultivation, namely byproducts derived from sugarcane, coffee, banana and corn plantations or processing plants (some of the main agroindustries in Mexico) (Fig. 13).

Fig. 13 Lignocelullosic by products from agroindustries

Pleurotus ostreatus, also known as oyster mushroom or white rot fungi, is one of the most common mushroom production systems worldwide. Mushroom cultivation provides direct bioconversion of solid wastes into edible biomass (Philippoussis 2009). The *Pleurotus* species requires a short growth time, compared to other mushrooms. Its fruiting body is not often attacked by pests and diseases and it can be grown in a simple and cheap way, with high yield, wider substrate utilization, sporelessness, wide temperature and chemical tolerance, as well as environmental bioremediation (Bellettini et al. 2016). For this process, substrate preparation is the most critical step (Vieira and de Andrade 2016).

The byproducts, intended for use as substrates after being collected, were completely dried under the sun, chopped, washed, soaked in cold water for at least 24 h to obtain a water content of 65–75%, sterilized with conventional heat treatment by immersion in hot water at 85 °C and deposited in a plastic container before being inoculated by hand with spawn. The average temperature was 24–30 °C and relative humidity was 50–65%. In order to determine suitable substrates and suitable ratios for the cultivation of oyster mushroom, the colonization and fructification times, yield, biological efficiency and productivity according to the literature were studied (Gaitán-Hernández and Salmones 2008; Mata et al. 2013; Héctor et al. 2013; Sözbir et al. 2015; Sardar et al. 2016).

Biological efficiency (BE) is determined as the ratio between the fresh edible mushroom or *Pleurotus* fruiting body weight produced and the dry weight of the substrate, while the rate of production (P) is calculated by the biological efficiency by the total number of days of evaluation, being considered from the first day of incubation of the spawn in the substrate until the last day of harvest. The yield (R) is determined as the weight of the produced edible mushroom from the wet weight of the substrate used, expressed as a percentage (Table 4).

$$\text{BE}\ (\%) = \frac{\text{Fresh edible mushroom (g)}}{\text{dry weight of the substrate}} \times 100$$

$$\text{Productivity} = \frac{\text{BE}\ (\%)}{\text{Days of evaluation}} \times 100$$

$$\text{Yield}\ (\%) = \frac{\text{Fresh edible mushroom (g)}}{\text{Wet weight of the substrate}} \times 100$$

The first harvest occurred after 21 days of incubation for corn leaves, and maize tops and the last after 49 days for banana leaves. Mushroom weight, cap diameter and stipe length development in oyster mushroom were significantly affected by substrate types. Productivity varied from 0.41 for leaves and coffee stem plant to 3.64% for maize leaves. Similarly, the highest yield value, 20%, was obtained with banana leaves and the lowest, 10%, with bagasse from mills. The water used in the washing and pasteurizing stages can be used in crop irrigation.

Several substrates had a comparatively higher mycelial growth rate, a shorter total colonization period, a larger cap diameter (4.5–15 cm) and a shorter stipe length (3.5–7 cm), mostly in leaf substrates. The rest had a smaller mushroom cap

Table 4 Productivity and physical characteristics of edible mushroom

Substrate	Days of treatment (incubation to harvest)	Productivity (%)	Yield (%)	Average cap diameter (cm)	Average stipe length (cm)
Bagasse from mills	22	1.36	5.5	6	3.8
Bagasse trapiche/bagasse mills (50/50)	22	1.82	5.88	5.8	3.7
Bagasse pith	23	1.74	8.89	5.3	4.9
Banana leaves	49	2.55	20	15	5.7
Banana stem	40	0.7	5.92	9	5.3
Banana trash	25	2.6	12.04	9.5	7
Coffee husk	22	1.14	6.98	8	4.3
Corn cob	28	2.2	15.42	10.9	4.5
Corn stalk	40	3.13	13.89	9	4
Corn leaves	21	2.78	8.75	6.5	4
Leaves and coffee stem plant	47	0.41	10.14	4.5	4.5
Maize flower	27	1.85	8.77	5	4.2
Maize tops	21	2.78	7	6	3.7
Maize leaves	22	3.64	10	4.5	3.7
Sugar industry mixture	23	2.52	8.53	6	3.5
Sugarcane flowers	22	2.5	9.91	7	3.8
Sugarcane tops	22	2.5	10.09	6.3	5
Sugarcane trash	22	2.27	7.69	6.3	3.7

diameter and a relatively long stipe length, which are undesirable marketing characteristics but suitable for self-consumption or processing as protein flour, or dried mushrooms, etc.

Besides, by making various substrate mixtures, BE varied significantly from 19% for leaves and coffee stem plant to 125% for corn stalks and banana leaves (Fig. 14).

At least 75% of the evaluated substrates (BE greater than 50%) are viable for the production of the mushroom *Pleurotus* on farms, especially in view of its low contamination in trials and its abundance, availability and diversity throughout the agricultural year as byproducts. Therefore, the analyzed byproducts are considered competitive substrates for the sustainable production of edible mushroom (Fig. 15).

Mushrooms obtained from substrates can be dried using direct solar radiation to avoid rapid deterioration before marketing. This processing technology also allows producers to increase the value added to mushrooms (as a high protein product),

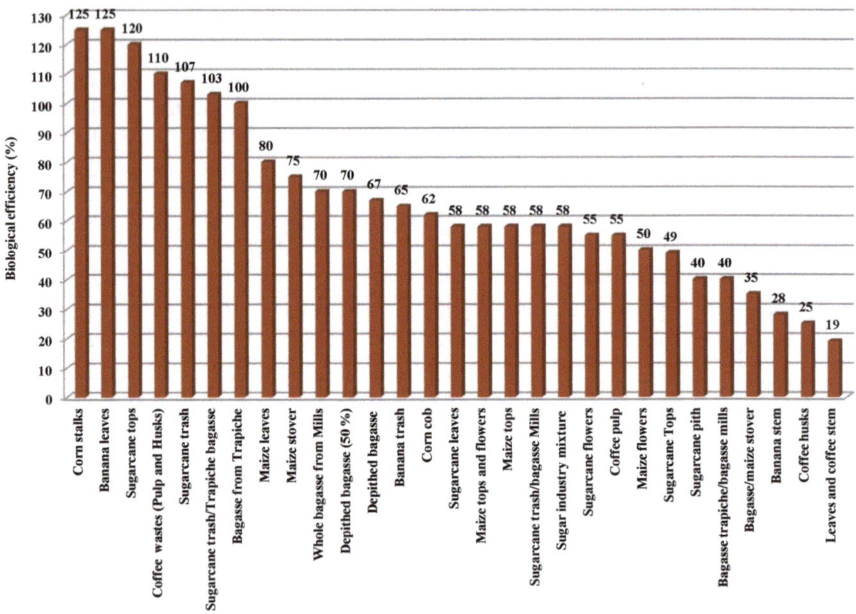

Fig. 14 Biological efficiency of treatments

Fig. 15 Stages of edible mushroom production

standardize quality, highlight certain food properties, and develop marketing strategies at local or national level. The spent substrates can be used as survival livestock feed or mixed with a daily ration as a source of fiber or composting. The containers can be recirculated in the next production cycle.

4 Production of Compost, Vermicompost and Bokashi

The production of manures, as an organic matter source from plantation or agroindustry byproducts, can be an alternative source of income and an option to recover soils with depleted nutrients and physical properties due to intensive use for growing plantation crops (Saldaña et al. 2014).

Composting, vermicomposting and bokashi all involve the biochemical process of aerobic decomposition of organic solid waste. They represent a low cost and environmentally-friendly waste management option involving several solid and liquid byproducts, manure from various farms and livestock, fibers, microorganisms and yeasts or earthworms that transform on-farm organic waste materials into a stable humic material, as a soil conditioner generating a useful product for farmers (Misra et al. 2016). In developing countries, conventional composting can be divided into three differentiated stages based on the temperature of the material in the process, mostly filter mud from sugar industry or sewage treatment sludge: initial mesophyll stage with a duration of 2–3 days, where the temperature follows an ascending behavior that starts with the ambient temperature and rises up to 40 °C; thermophilic stage, which has a variable duration, where the temperature increases up to 75 °C and the final mesophyll or maturation stage in which the temperature falls from 40 °C to ambient temperature. At this stage the biological stability of the material is reached (Morales et al. 2016; Van Fan et al. 2016 and Prado et al. 2013).

Sahu and Brahmaprakash (2016) discussed trends in the formulation of new organic fertilizers for agriculture. In addition, Ramos Agüero and Alfonso (2014) summarize some features related to the employment of organic manures, placing special emphasis on the development and production of fermented bokashi-type manures and their use in agriculture; moreover, composting methods, scales, and constraints have been recently reviewed by Misra et al. (2016); Qdais and Al-Widyan (2016); Pandey et al. (2016) and Lim et al. (2016).

The production of organic fertilizers from byproducts of the sugarcane, coffee, citrus and edible mushroom agroindustries can provide an alternative source of income and sustainable fertilizers for crops. Organic manures in the form of compost, vermicompost (with earthworm) and bokashi were produced according to Saranraj and Stella (2014) and Saldaña et al. (2014), and evaluated for their nutritional value according to Mexican standard NOM-021-SEMARNAT (2000) for compost analysis.

Organic manures obtained from byproducts of the local agroindustries were analyzed for their nutritional content in ten trials. The most important difference among products (Table) is the low N in filter mud (T1) and waste from the cultivation of oyster mushrooms and filter mud, bagasse and coffee (T5 and T9 respectively), which resulted in these having the lowest C/N ratios. The most attractive products for soil application are T2, T3, T6 and T7, which are high in calcium, contain a considerable amount of potassium (especially T6 and T7) and have a balanced pH. The low pH of T2, T9 and T10 is reflected in their low cation and phosphorous contents. In general, the addition of vermicompost by itself or in combination with other organic products (except the waste from the oyster mushrooms) increased the nutritional value of filter mud (Table 5 and Fig. 16).

T1 filter mud; T2 filter mud compost; T3 filter mud vermicompost; T4 filter mud vermicompost and sugarcane trash; T5 filter mud vermicompost and waste from cultivation of oyster mushrooms; T6 filter mud vermicompost and livestock manure; T7 bokashi filter mud, bagasse, $Ca(OH)_2$, yeast and molasses; T8, compost orange waste, filter mud and bagasse; T9 filter mud, bagasse and husk and coffee spent ground; T10 coffee pulp vermicompost.

Although there are differences due to the type of soil, management practices, chemical composition of byproducts and production techniques, the organic fertilizers obtained from agroindustrial byproducts can generate additional income in the crop areas; moreover, the manure encourages soil structure, fertility and nutrient availability to plants, and reduces the pH, promoting good crop development and productivity in farms in the short and medium term depending on the type of crop (Diacono and Montemurro 2015).

According to the productive options evaluated, Alwi et al. (2014) concluded that the 'engineers of the future' will have the responsibility of addressing the entire spectrum of sustainability constraints and push factors, including the economic, environmental, social and multi-generational dimensions generating *bioproducts* within the 4Fs and multidisciplinary approaches (Fig. 17)

5 Conclusions

Byproducts are produced in sugarcane, banana, corn, coffee and citrus plantations, among others, and in several agroindustries in large amounts and without any specific treatment and use. Therefore, to develop alternatives for the conversion of these wastes into value-added products through sustainable technologies in farms is a priority objective in developing countries such as Mexico, where the sustainability and the environmental, social and economic rationality is a slow process. In this work three high-value products were obtained with traditional knowledge of farmers, sustainable technologies and analytical methods: solid biofuel from citrus residues; organic fertilizers with various chemical properties for agriculture, combining various agroindustrial and livestock products; and edible mushroom from individual byproducts and mixtures thereof. These bioproducts can be obtained directly by farmers and contribute to their energy, food and economic sustainability and help them recover soils eroded by monocultures. Moreover, specialized

Table 5 Average nutrient content of organic manures from agroindustrial byproducts

Variable	Unit	T1	T2	T3	T4	T5	T6	T7	T8	T9	T10
pH		7.52	5.49	7.32	7.79	7.27	7.63	8.28	9.22	5.54	6.82
Soil electrical conductivity	dsm^{-1}	0.958	2.374	1.079	1.105	1.076	1.628	1.806	4.450	1.324	1.3
Ash	(%)	47.670	45.850	56.96	55.03	50.01	61.98	62.12	33.910	62.10	32.43
Organic matter	(%)	52.330	54.150	43.04	44.97	49.99	38.02	37.88	66.090	37.90	67.57
Total carbon C	(%)	30.354	31.409	24.965	26.085	28.996	22.053	21.976	38.335	21.984	39.194
Total nitrogen N	(%)	0.678	1.680	1.99	1.45	0.51	1.68	1.53	1.49	0.72	1.52
C/N		44.77	18.69	12.55	17.98	56.85	13.12	14.36	25.73	30.53	25.79
Ca	(%)	5.030	2.756	5.266	5.388	4.849	4.599	18.279	4.319	1.070	3.975
Mg	(%)	0.773	0.290	0.635	0.662	0.610	0.603	0.726	0.803	0.360	0.522
Na	(%)	0.0314	0.0300	0.029	0.040	0.029	0.043	0.042	0.083	0.056	0.068
K_2O	(%)	0.211	0.150	0.184	0.179	0.156	0.342	0.322	2.244	0.270	0.651
P_2O_5	(%)	3.947	2.796	3.384	3.956	3.711	3.431	2.839	2.721	0.387	3.332
Fe	(%)	0.5465	0.8961	1.088	0.898	0.7310	0.6990	0.5980	0.0170	0.5074	1.22
Cu	(%)	0.0078	0.0055	0.0156	0.0204	0.0051	0.0026	0.0047	0.0033	0.0025	0.0064
Zn	(%)	0.0216	0.0238	0.0227	0.0224	0.0324	0.0121	0.0204	0.0115	0.0087	0.0227
Mn	(%)	0.1659	0.1415	0.2227	0.2502	0.1950	0.1830	0.2205	0.0501	0.1070	0.1404

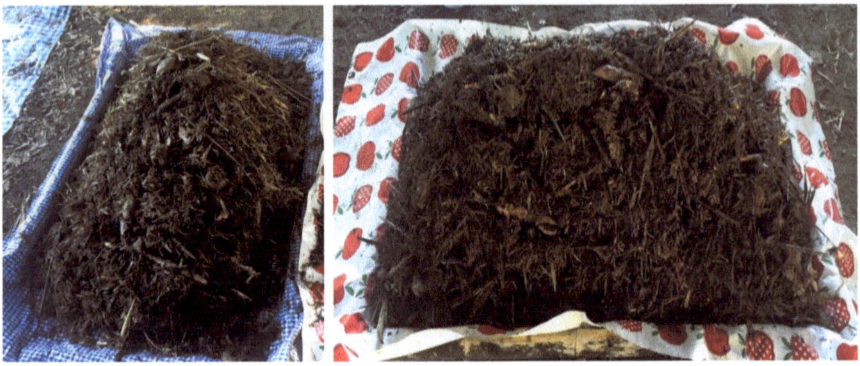

Fig. 16 Organic manure from Mexican byproducts

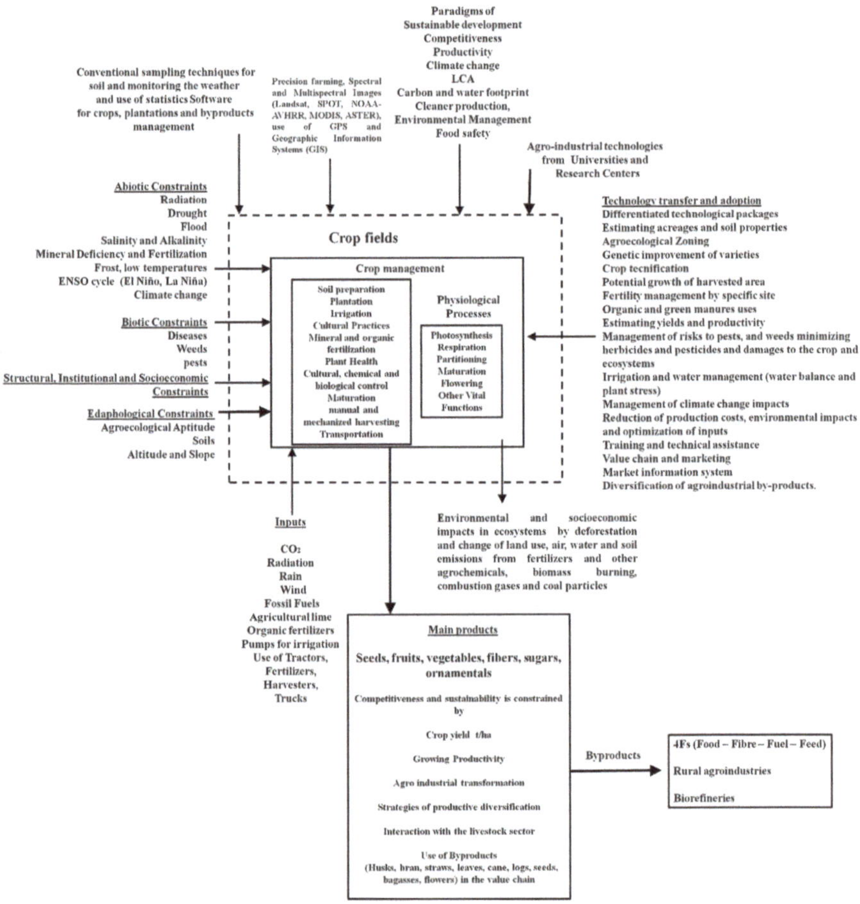

Fig. 17 Integration agroindustrial of byproducts

machinery is not required for processing and this process will potentially help to reduce the emission of greenhouse gases into the atmosphere. However, future works should be addressed to the implementation of sustainable development in agroindustries, which is a continuous process of cultural awareness and paradigm change by stakeholders (farmers, agricultural extensionists, environmentalists, policy makers, industry owners), to achieve success.

References

Alwi SRW, Manan ZA, Klemeš JJ, Huisingh D (2014) Sustainability engineering for the future. J Clean Prod 71:1–10

Bellettini M.B, Fiorda F.A, Maieves H.A, Teixeira G.L, Ávila S, Hornung P.S, Ribani R.H (2016) Factors affecting mushroom Pleurotus spp. Saudi J Biol Sci. In press http://dx.doi.org/10.1016/j.sjbs.2016.12.005

Cai L, Chen TB, Gao D, Zheng GD, Liu HT, Pan TH (2013) Influence of forced air volume on water evaporation during sewage sludge bio-drying. Water Res 47(13):4767–4773

Cañete-Rodríguez AM, Santos-Dueñas IM, Jiménez-Hornero JE, Ehrenreich A, Liebl W, García-García I (2016) Gluconic acid: properties, production methods and applications—an excellent opportunity for agro-industrial by-products and waste bio-valorization. Process Biochem 51(12):1891–1903

Colomer-Mendoza FJ, Herrera-Prats L, Robles-Martinez F, Gallardo-Izquierdo A, Piña-Guzman AB (2013) Effect of airflow on biodrying of gardening wastes in reactors. J Environ Sci 25 (5):865–872

Corrêa RCG, Brugnari T, Bracht A, Peralta RM, Ferreira IC (2016) Biotechnological, nutritional and therapeutic uses of Pleurotus spp. (Oyster mushroom) related with its chemical composition: a review on the past decade findings. Trends Food Sci Technol 50:103–117

Diacono M, Montemurro F (2015) Effectiveness of organic wastes as fertilizers and amendments in salt-affected soils. Agriculture 5(2):221–230

Dominczyk L Krzystek, Ledakowicz S (2014) Biodrying of organic municipal wastes and residues from the pulp and paper industry. Drying Technol 32(11):1297–1303

Fabián RM, Elizabeth M, Teodoro ES, Belem P, Carmen CM, Francisco J, Enrique DP (2012) Biodrying under greenhouse conditions as pretreatment for horticultural waste. J Environ Prot 3:298–303

Fava F, Zanaroli G, Vannini L, Guerzoni E, Bordoni A, Viaggi D, Talens C (2013) New advances in the integrated management of food processing by-products in Europe: sustainable exploitation of fruit and cereal processing by-products with the production of new food products (NAMASTE EU). New Biotechnol 30(6):647–655

Gaitán-Hernández R, Salmones D (2008) Obtaining and characterizing Pleurotus ostreatus strains for commercial cultivation under warm environmental conditions. Sci Hortic 118(2):106–110

Galanakis CM (2012) Recovery of high added-value components from food wastes: conventional, emerging technologies and commercialized applications. Trends Food Sci Technol 26(2):68–87

Hanes RJ, Bakshi BR (2015) Sustainable process design by the process to planet framework. AIChE J 61(10):3320–3331

He P, Zhao L, Zheng W, Wu D, Shao L (2013) Energy balance of a biodrying process for organic wastes of high moisture content: a review. Drying Technol 31(2):132–145

Héctor SV, Evaristo EV, Luis CA (2013) Sustainable production of oyster mushroom (*Pleurotus ostreatus*) in Chiapas, Mexico. World Acad Sci, Eng Technol, Int J Biol, Biomol, Agric, Food Biotechnological Eng 7(7):534–537

Jeznabadi EK, Jafarpour M, Eghbalsaied S, Pessarakli M (2016) Effects of various substrates and supplements on king oyster (*Pleurotus eryngii*). Compost Sci Utilization, 1–10. https://www.researchgate.net/profile/Mehrdad_Jafarpour/publication/309964957_Effects_of_Various_Substrates_and_Supplements_on_King_Oyster_Pleurotus_eryngii/links/58354b1708ae102f0739f8f2.pdf [Accessed 31 May 2017]

Koutinas AA, Vlysidis A, Pleissner D, Kopsahelis N, Garcia IL, Kookos IK, Lin CSK (2014) Valorization of industrial waste and by-product streams via fermentation for the production of chemicals and biopolymers. Chem Soc Rev 43(8):2587–2627

Kusch S, Udenigwe C.C, Cavinato C, Gottardo M, Micolucci F (2015) Value-added utilization of agro-industrial residues. Adv Food Biotechnol, 415–425, Wiley, p 752

Lim SL, Lee LH, Wu TY (2016) Sustainability of using composting and vermicomposting technologies for organic solid waste biotransformation: recent overview, greenhouse gases emissions and economic analysis. J Clean Prod 111:262–278

Ma J, Zhang L, Li A (2016) Energy-efficient co-biodrying of dewatered sludge and food waste: synergistic enhancement and variables investigation. Waste Manag 56:411–422

Mamma D, Christakopoulos P (2014) Biotransformation of citrus by-products into value added products. Waste Biomass Valorization 5(4):529–549

Mata G, Hernández RG, Salmones D (2013) Biotechnology for edible mushroom culture: a tool for sustainable development in Mexico. WIT Trans State-of-the-art Sci Eng 64:484–506

Mirabella N, Castellani V, Sala S (2014) Current options for the valorization of food manufacturing waste: a review. J Clean Prod 65:28–41

Misra RV, Roy RN, Hiraoka H (2016) On-farm composting methods. UN-FAO, Rome, Italy, p 51

Mohamed MF, Refaei EF, Abdalla MM, Abdelgalil SH (2016) Fruiting bodies yield of oyster mushroom (*Pleurotus columbinus*) as affected by different portions of compost in the substrate. Int J Recycl Org Waste Agric 5(4):281–288

Morales AB, Bustamante MA, Marhuenda-Egea FC, Moral R, Ros M, Pascual JA (2016) Agri-food sludge management using different co-composting strategies: study of the added value of the composts obtained. J Clean Prod 121:186–197

NOM-021-SEMARNAT (2000) México. http://biblioteca.semarnat.gob.mx/janium/Documentos/Ciga/libros2009/DO2280n.pdf

Okino Delgado CH, Fleuri LF (2016) Orange and mango by-products: agro-industrial waste as source of bioactive compounds and botanical versus commercial description—a review. Food Rev Int 32(1):1–14

Pandey PK, Cao W, Biswas S, Vaddella V (2016) A new closed loop heating system for composting of green and food wastes. J Clean Prod 133:1252–1259

Pfau SF, Hagens JE, Dankbaar B, Smits AJ (2014) Visions of sustainability in bioeconomy research. Sustainability 6(3):1222–1249

Philippoussis A.N (2009) Production of mushrooms using agro-industrial residues as substrates. In: Biotechnology for agro-industrial residues utilisation, pp 163–196. Springer, Netherlands

Prado RDM, Caione G, Campos CNS (2013) Filter cake and vinasse as fertilizers contributing to conservation agriculture. Appl Environ Soil Sci 2013:1–8

Qdais HA, Al-Widyan M (2016) Evaluating composting and co-composting kinetics of various agro-industrial wastes. Int J Recycl Org Waste Agric 5(3):273–280

Ramos Agüero D, Alfonso ET (2014) Generalities of the organic manures: Bokashi's importance like nutritional alternative for soil and plants. Cultivos Tropicales 35(4):52–59

Sahu P.K, Brahmaprakash G.P (2016) Formulations of Biofertilizers—approaches and advances. In: Microbial inoculants in sustainable agricultural productivity, pp 179–198. Springer, India

Saldaña MI, Gómez-Álvarez R, Rivera-Cruz MDC, Álvarez-Solís JD, Pat-Fernández JM, Ortiz-García CF (2014) The influence of organic fertilizers on the chemical properties of soil and the production of Alpinia purpurata. Ciencia e Investigación Agraria 41(2):215–224

Saranraj P, Stella D (2014) Composting of sugar mill wastes: a review. World Appl Sci J 31:2029–2044

Sardar H, Anjum MA, Nawaz A, Ejaz S, Ali MA, Khan NA, Raheel M (2016) impact of various agro-industrial wastes on yield and quality of pleurotus sajor-caju. Pak J Phytopathology 28 (1):87–92

SIAP (2016) Anuario Estadístico de la Producción Agrícola. http://infosiap.siap.gob.mx/aagricola_siap_gb/icultivo/index.jsp

Siles JA, Vargas F, Gutiérrez MC, Chica AF, Martín MA (2016) Integral valorisation of waste orange peel using combustion, biomethanisation and co-composting technologies. Biores Technol 211:173–182

Sözbir GD, Bektas I, Zulkadir A (2015) Lignocellulosic wastes used for the cultivation of pleurotus ostreatus mushrooms: effects on productivity. BioResources 10(3):4686–4693

Sugni M, Calcaterra E, Adani F (2005) Biostabilization–biodrying of municipal solid waste by inverting air-flow. Biores Technol 96(12):1331–1337

Suksankraisorn K, Patumsawad S, Fungtammasan B (2010) Co-firing of Thai lignite and municipal solid waste (MSW) in a fluidised bed: effect of MSW moisture content. Appl Therm Eng 30(17):2693–2697

Vandermeersch T, Alvarenga RAF, Ragaert P, Dewulf J (2014) Environmental sustainability assessment of food waste valorization options. Resour Conserv Recycl 87:57–64

Van Fan Y, Lee CT, Klemeš JJ, Bong CPC, Ho WS (2016) Economic assessment system towards sustainable composting quality in the developing countries. Clean Technol Environ Policy 18 (8):2479–2491

Velis CA, Longhurst PJ, Drew GH, Smith R, Pollard SJ (2009) Biodrying for mechanical–biological treatment of wastes: a review of process science and engineering. Biores Technol 100(11):2747–2761

Tom AP, Pawels R, Haridas A (2016) Biodrying process: a sustainable technology for treatment of municipal solid waste with high moisture content. Waste Manag 49:64–72

Vieira FR, de Andrade MCN (2016) Optimization of substrate preparation for oyster mushroom (Pleurotus ostreatus) cultivation by studying different raw materials and substrate preparation conditions (composting: phases I and II). World J Microbiol Biotechnol 32(11):190. doi:10.1007/s11274-016-2152-y

Yang B, Zhang L, Jahng D (2014) Importance of initial moisture content and bulking agent for biodrying sewage sludge. Drying Technol 32(2):135–144

Yang D, Liang J, Wang Y, Sun F, Tao H, Xu Q, Wan X (2016) Tea waste: an effective and economic substrate for oyster mushroom cultivation. J Sci Food Agric 96(2):680–684

Winkler MK, Bennenbroek MH, Horstink FH, Van Loosdrecht MCM, Van de Pol GJ (2013) The biodrying concept: an innovative technology creating energy from sewage sludge. Biores Technol 147:124–129

Strategies for Guiding Community Organizations in Sustainable Development: The Case of Monteria's Urban Areas

Jorge Rafael Villadiego Lorduy, Dennis Paul Huffman Schwocho, Yhonattan Mendez Nobles and Stalyn Yasid Guerrero Gómez

Abstract The purpose of this paper is to analyze the strategies used by community organizations dedicated to the sustainable development of urban areas in Monteria City. Three specific objectives were proposed in order to guide the designing of organization manuals capable of improving the functions and results obtained by these social institutions. The first was aimed at the diagnosis of current community organizations themselves and the perceptions that social and institutional actors had of these. The second was to determine the basic elements for designing organizational manuals of this type and the third was the elaboration of the manual itself. The diagnostic phase suggested that, on the one hand, the educational backgrounds of community leaders was very poor. Only 56% had primary studies. On the other hand, community organizations, in general, lacked focus and knowledge of local environmental management procedures. Perceptions among public and private areas of these institutions were not very favorable as well. For the most part, Community Action Boards (CABs) were considered to be only a part of the political machinery in the region. In addition, a major weakness of these was the lack of commitment and assumed responsibility by their members, towards sustainable development. Furthermore, among the main aspects of the manual, six basic elements were identified: two complementary ones (a Plan-Do-Check-Act Cycle, and systemic, holistic, sustainable development approaches), and four cen-

J. R. Villadiego Lorduy (✉)
Universidad Pontificia Bolivariana at Monteria, Cordoba 230003, Colombia
e-mail: Jorge.villadiegol@upb.edu.co

D. P. Huffman Schwocho
Instituto Tecnológico de Costa Rica, Cartago 56225, Costa Rica
e-mail: huffmandennis534@gmail.com

Y. Mendez Nobles
Universidad Pontificia Bolivariana at Montería, Córdoba 230002, Colombia
e-mail: yhonattan.mendez@upb.edu.co

S. Y. Guerrero Gómez
Fundación Salud Mental para Todos at Montería, Córdoba 110931, Colombia
e-mail: Syguerrerog@unal.edu.co

W. Leal Filho et al. (eds.), *Sustainable Development Research and Practice in Mexico and Selected Latin American Countries*, World Sustainability Series,
https://doi.org/10.1007/978-3-319-70560-6_5

tral lines of thought (territorial profiles, existing community groups, the training of the members of these social organizations and their interaction in solving environmental problems). After analyzing those elements, we proceeded to elaborate an organizational manual which contemplated the different relationships required between each of these elements in order to develop successful strategies for community actions in sustainable development. It was concluded that non-formal environmental education projects should be the basis for successful interactions between community organizations, government agencies and non- governmental social institutions for effective sustainable development of urban areas.

Keywords Community Action Boards (CABs) · Community organizations Profiles · Perceptions · Organizational manuals

1 Introduction

A community organization can be defined as a way to structure community actions for dealing with problems that concern them. Ross (1978, p. 65) defined community organization as how a community identifies its needs and objectives by hierarchy, seeking resources for treatment, and acting upon them. For the Permanent Formation Institute (INSFOP, for its acronym in Spanish) and the Special Program for National Nutritional Food Security (PESANN /FAO 2008, p. 10) a "community organization is when a group of people unite to study problems affecting their community and seek ways of solution". They also pointed out that an organized group seeking the solution to its problems must consider two underlying aspects: the pursuit of their objectives and the promotion of their interests.

As Mercer and Ruiz (2004, p. 293) stated: "A fundamental basis for integration, cohesion and joint action is the existence of common objectives, aspirations, problems and needs". There are two types of community organizations: Social and productive. Social organizations are organized to solve problems in a community, and productive organizations are created to obtain economic resources for manufacturing.

On the other hand, the relevance of communities to organize themselves lays on the fact that through the organization, values and abilities of their members, solutions can be given to problems and needs occurring in their territories. Some advantages for organizing a community are: (1) Use and effort of collective resources, (2) leadership development, (3) generation of opportunities for project management, etc. (INSFOP and PESANN /FAO 2008, p. 11). As important as it is to organize socially, community participation is also a key element in articulating group actions for the efficient work of a community structure and therefore it must always be considered during the process of the organization and afterwards.

Legally speaking, communal action in Colombia, or its first attempts, go back to the Law 19 of 1958. But a great leap occurred with the political constitution of 1991, because it allowed for citizen participation in public decisions by providing communities with mechanisms for participation at the individual and community levels. Currently, the Law 743 of 2002 and the Decree 2350 of 2003 regulate matters regarding the constitution of communal groups. For Duque and Gaitan (Duke and Gaitán 2001, p. 45) the social, economic, political and cultural context of the country has reflected a very strong tendency to recognize the community as an element of local, regional and national development in the last 15 years. Monteria is not unaware of this matter. In the urban areas, there are both social and productive organizations. Nevertheless, the interest of this study was focused on understanding how social organizations named Communal Action Boards (CABs) were organized into Community Association Boards (ASOCOMUNALES for its acronym in Spanish).

The City has 159 Community Action Boards (CABs) and two Community Association Boards, but they lack serious planning for sustainable development in their territories and even guidance that could allow them to organize their selves considering the characteristics of their communities where they were settled. For this reason, it was considered necessary to formulate a community organizational manual supported by four elements and phases: Environmental profiles, organizational structures, training and interaction in problem solving.

Adjacent to the manual, there were two complementary elements, one related to the approaches to be considered when organizing communities, and another one associated with continuous improvement. The work was carried out starting with the establishment of profiles of current social organizations (CABs) and the perceptions that social and institutional actors had concerning community organizations. In the analysis of profiles and perceptions, basic foundations were identified and a manual for the communal organizations pertaining to the urban areas of the City was elaborated.

2 Methodology

In order to carry out this research, two types of methods were defined: The first one of an explanatory type for the purpose of analyzing how processes of community organizations in the City had been developed, considering social, political and economic events that determined the configuration of community structures in the municipal seat, and the profiles of current communal organizations, as well as the perceptions that social and institutional actors, immersed in processes of communal organizations, had of these. The purpose of this method was to get a general idea of the successes and failures of community organizations for sustainable development in Colombia in order to determine a basis for designing an organizational manual that could improve the efficiency and effectiveness of these organizations.

The second was a descriptive statistical method using percentage estimates and absolute values based on the data obtained from the application of a questionnaire (previously validated by experts). The instrument was designed to obtain information concerning the profiles of community organizations by defining the organization's approach, type of structure, time needed to carry out community activities, roles of the organization within the community and other similar variables. Linked to this, was the application of the questionnaire with a semi-structured interview carried out with actors from both public and private sectors. The purpose of this method was to understand the correlation between the perceptions of those actors involved in organizations and institutions dedicated to promoting and improving community actions of conservation and preservation of the environment in Colombia and their ability to organize these actions.

From the public sector, two officials were interviewed from the Mayor's office and the Cordoba Government each. From the social point of view, two community leaders were consulted (they became references due to their work within communities) along with two representatives of non-governmental organizations (NGOs) that have been involved in this type of community activities from the past. Just like the questionnaire, the interview format was also validated by experts.

A sample size of 18 Community Action Boards was set up for a population of 159 organizations. For the profile of the organization a member of the board, preferring the president of the institution, was interviewed. The sample design implemented was of a simple random type. The software for the descriptive statistical analysis was an R version 3.0.2. Statistical data obtained from the application of these instruments which were consolidated into a matrix, implemented by Excel 2007.

To formulate this community organization manual, several studies of social organizations were considered: Strengthening Manual for Social Organizations (Ministerio de Desarrollo Social 2015), Community Organization (INSFOP—PESANN /FAO 2008), Agenda 21 and Processes of Social Participation (Marchonni 2002), community development (Chacon 2010) and others, aimed at responding to a social-environmental context and the promotion of sustainable development within the study.

3 Location of the Study Area

Monteria is located at 8°46′10 "north latitude, 75°53′00" west longitude with respect to the Greenwich meridian, at a height of 20 m above sea level. It borders on the north with the Canalete, Los Cordobas, Puerto Escondido, San Pelayo and Cerete municipalities, on the south with the Tierralta and Valencia municipalities, on the east with Planeta Rica and San Carlos, and on the west with the Department of Antioquia (see Fig. 1).

Fig. 1 Monteria geographical location. *Source* Google Earth (2015)

4 Limitations of this Study

Although the methodology of this study permitted us to better understand the workings and results obtained in the present day community organizations in Colombia for sustainable development due to the perceptions and professional profiles of those actors in position to exercise authority in decision making concerning actions and distribution of resources dedicated to conserving and preserving the environment, the major limitation of this research was its qualitative nature that depended upon the social image of the processes and products of sustainable development. We were unable to determine causal relationships between non-formal environmental education projects and the interactions between the main three actors of sustainable development in Colombia: community organizations, governmental agencies and non-governmental groups (NGOs). This would have to be left for future research projects in this field of scientific study.

5 Results and Analysis

5.1 *Profiles and Perceptions*

Monteria is a middle sized city with a projected population, in June 30, 2015, of 441,301 inhabitants (National Department of Statistics DANE 2015). The urban area is divided into 5 (five) zones: Center, South, North, East and West and has 9 communes spread throughout (Villadiego et al. 2014). From the community organizations' viewpoint, the city has two large communal associations that the Communal Action Boards (CABs) bring together. The first one, ASOCOMUNAL, has the largest number of associates (189) since it gathers both urban and rural CABs. The second one known as ASOCOMUSEIS has within its aggregates 14 entities, approximately, which belong only to the urban area In total, at the city seat, there are 159 Community Action Boards legally constituted and registered in the Departmental Secretary of the Interior and Citizen Participation (see Table 1).

Geographically, most of the Communal Action Boards (CABs) in the urban area of Cordoba's capital are concentrated in very popular sectors. It is estimated that 95% of these organizations are distributed in neighborhoods such as Canta Claro,

Table 1 Census of community organizations

Census of community organizations	
Community associations	Total CABs involved
ASOCOMUNAL	189
ASOCOMUSEIS	14
Community Action Board (CABs)	Total CABs involved
CABs urban area	159
CABs rural area	103

Source Departmental secretary of interior and citizen participation, November (2015)

Villa Jimenez, Paz del Norte, Mocari, Camilo Torres, Santander, Santa Fe, La Granja, Edmundo López, Los Robles del Norte, El Paraíso, Galilea, El Dorado, Pastrana Borrer y La Vid, among other neighborhoods with similar social-economic characteristics.

The education backgrounds of the leaders of this type of corporations was marked by inequality. Only 38% (7 individuals) had a university education, in contrast with 56% (10 people) with only a basic education that went from primary to secondary school, only 6% had technical studies (see Fig. 2). The poor academic formation of the leaders who oversaw the CABs could influence their ability to understand their surroundings, especially in the non-development of competences. On the one hand, critical thinking such as those proposed by Facione (2007) and, on the other hand, an argumentative and propositional type in search for solutions to local environmental problems were considered to be lacking. It was highlighted that although more than 50% did not have higher education, there were some people who, over time, had gained experience and knowledge in the fine print on matters of communal organizations.

CABs were usually unaware of the importance of applying an environmental approach to their organizations. So much was the failure on that matter that 72% of

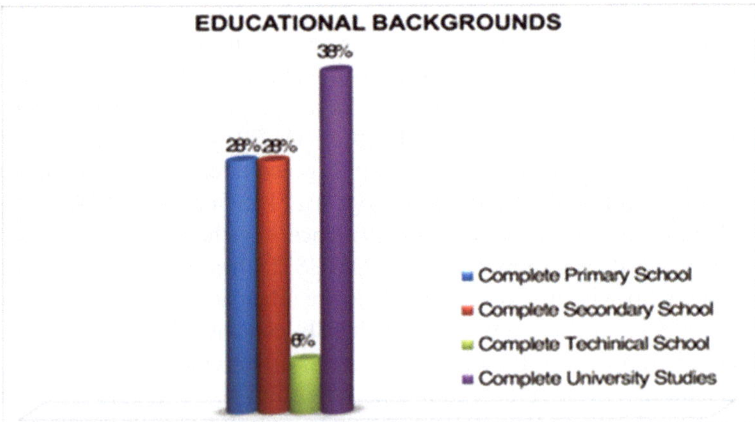

Fig. 2 Educational backgrounds. *Source* The authors

the presidents surveyed assured that they did not carry out a diagnostic or environmental profile of their communities. Thus, the question was: How could these social organizations propose solutions to their needs, environmental problems and conflicts in their sector if they did not have knowledge of the current state in which they were? Doubts arose on the quality of the "diagnoses" that the remaining 22% had made. In such communities, several social-environmental problems were visible, which, considering a well-developed environmental profile, many solutions could have already been proposed by them. There were also few environmental projects that the CABs developed or had developed since their legal constitution, barely 39% had advanced on any, and very sporadically on some environmental projects, among them, the cleaning of canals, tree planting, and educational campaigns could be highlighted.

A community that ignores the mechanisms of citizen participation, (tools with which it asserts its rights) is a defenseless organization, easy to subdue and manipulate. 67% of respondents were not aware nor had made use of these basic legal instruments in community management and, above all, in community environmental management where it was necessary to ensure a healthy environment. 94% of those interviewed said that the government had not provided support or training in community development. CABs did not feel strongly concerning this issue. Essentially, the organizational structure within was vertical (78% had this structure), giving the greater responsibility to the president of the organization and lessening the importance that other members could have had (see Table 2).

Table 2 Organizational profile matrix

Organization profile		
Questions	Answers	
What is your organization approach?	Environmental (0%)	
	Social (94%)	
	Political (6%)	
	Indigenous (0%)	
	Afro (0%)	
Did you develop a prior diagnosis of the socioeconomic and environmental conditions of the communities belonging to the organization?	Yes (28%)	No (72%)
How do you define the structure of your organization?	Vertical (78%)	Horizontal (22%)
Do the organization members have knowledge of mechanisms of citizen participation?	Yes (33%)	No (67%)
Does the organization receive training from the government on community development issues?	Yes (6%)	No (94%)
Does the organization develop or has developed environmental management or education projects in the communities they belong to?	Yes (39)	No (61%)

Source The authors

When asked: What was the procedure they developed to organize at the community level? The response was unified, the first step was to hold meetings to socialize the idea of constituting the CABs, then the nomination of candidates for the positions and committees, then the drafting of statutes and finally, their subsequent registration before the Chamber of Commerce and the Secretary of the Interior and Citizen Participation. It may also be pointed out that CABs originated more from unprompted reasons than from actual planning processes that lead to efficient and resilient social structures. At the same time, they were not given the true importance of building an environmental fabric in their territories. The main reasons for organizing at the community level were focused on gaining benefits for the communities and having representation in the local government and state institutions.

The role of CABs in the communities and, for the opinions given by their legal representatives, was concerning organization, management of projects, and giving guidance to populations facing a problem. Likewise, as a main strategy or mechanism for solutions, they emphasized meetings. However, there were other mechanisms or strategies that could have been considered and far more efficient ones than simple meetings, such as developing methodologies that could have included: analyzing, evaluating and reducing the problem, finding the root of the problem, evaluating methods and actions for solutions, developing plans of action, documenting and monitoring. Individuals or groups that did not develop skills in analysis and problem solving ended up experiencing, as Guerra (sf) pointed out, the phenomenon of tunnel vision, that is, focusing on the problem rather than on the solution.

Among weaknesses presented by the CABs, as organizations, was the lack of commitment and responsibility of their affiliates. The leaders who were interviewed said that few members worked for the good of their communities. Board presidents thought that shallow politics were the main setback concerning these institutions, in which CABS should be allied to a political party in order to manage actions of service in their neighborhoods. Without this aspect, support would not have been enough for the development of projects. Furthermore, the private sector didn't efficiently provide full support either. Therefore, much of the work and effort ended up being inconclusive.

Some members of community association highlighted the strength and persistence of some leaders to continue developing community activities despite low wages, or sufficient help, on the one hand, from the benefited communities, and, on the other, from the politicians in turn. The spirit of improvement that the members of the CABs had been viewed by their visible heads as an opportunity to strengthen these social organizations in a structural way, as well as the interest in serving the communities, to improve the conditions of life and their surroundings.

Regarding the perceptions that social and institutional actors belonging to public and private sectors had on 9 communal organizations questions, some interesting elements were found. Respondents were asked whether in their organization, strategies for planning, implementing policy, functions, activities and measures of community environmental management were applied, or if they encouraged joint work with local associations to solve local environmental problems. Based on their answers, opinions did not change much in terms of the fact that little had been done in the public and private sectors to implement or promote policies or actions of environmental management. Only the City of Monteria had derived "sustainable projects" in high-impact populations under the 2032 plan ("Sustainable City").

Both public (officials from the City and the Government) and private (community leaders and NGOs) actors considered that social organizations (CABs—ASOCOMUNALES) could be a means for the solution of local environmental problems, since these entities could generate actions of environmental management for solving problems in their communities. However, when asked about their perception of community organizations that promoted local environmental management, opinions were divided and centered on two considerations: (1) Not favorable because, at least in the urban area of Monteria, there were very few CABs that promoted environmental management and that they were not generating solutions to environmental problems, besides being poorly visible; (2) favorable because they believed that the promotion of local environmental management was already positive in itself. Nonetheless, most of the interviewees pointed out that social organizations that promoted environmental management could have had a greater positive and functional impact than those that did not work from this framework, because they became the main element that generated possible solutions to problems in their territories.

In the same vein, when asked if they had knowledge of successful experiences that had been carried out by community organizations promoting local environmental management, very few were pointed out except for the Communal Action Boards of the Pradera, Santander and Buenavista neighborhoods which had campaigned for tree planting, environmental education and the application of citizen participation mechanisms to solve environmental issues. It was striking that when asked about which elements to consider for the formulation of a community organizational manual, the interviewees agreed that training was fundamental for community organizations, indicating topics ranging from citizen participation mechanisms, organizational structures, to the protection of the environment among others.

There was a full consensus when asked whether they believed that Communal Boards emphasized key issues such as community development, empowerment and ownership. The answer was a resounding yes and they associated this stagnation with politicking which permeated all community organizations. Also, they agreed on the idea that CABs were mainly part of the political machinery of the City. Some of the factors that pointed towards this phenomenon were:

1. Low economic and educational levels of members of the CABs and the communities to which they belonged.
2. Corrupt community leaders seeking self-interests.

In short, the perception that people had from the public and private sectors, whether these were developed from the framework of environmental management or not, was unfavorable. They were not perceived as serious social structures able to manage local sustainable development, but as associations servile to the political interests of local leaders, totally losing the purpose for which they were created, that is, to seek solutions to the problems of their communities. Likewise, there was almost no environmental approaches to the actions of these organizations. They lacked knowledge of the environment and on how to carry out or request actions from institutions aimed at protecting the environment where they lived.

5.2 *Community Organization Manual*

A manual is a guidebook containing the steps to carry out an action properly. In the case of the proposed manual (see Fig. 3), the aim was to establish a way in which a community-based organization in the urban area of Monteria should have been efficiently structured. For this purpose, four key elements and articulators associated with an equal number of phases, and two complementary elements of the manual, were defined. The first complementary element refers to the PDCA cycle (Plan, Do, Check and Act) framed in the continuous improvement of both the manual itself and of the community organization. For Rincon (1998, p. 51) PDCA was a management concept that streamlined the relationship between man and processes, and sought to control them by setting standards.

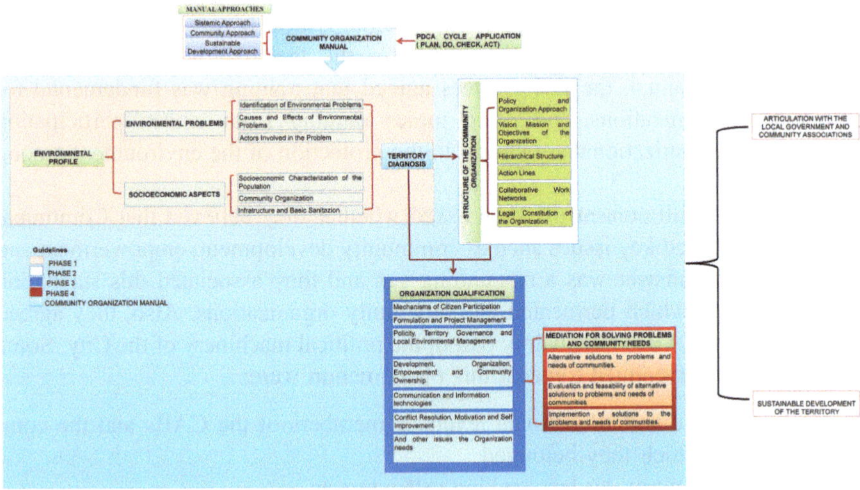

Fig. 3 Proposed community organization manual. *Source* The authors

The PDCA concept leads to continuous improvement of the processes of any organization (Moreira 2006). On the other hand, one of the main problems of social organizations in the urban area was, that they did not generate processes of continuous improvement neither inside nor outside of the organization, becoming structures that were neither functional nor resilient to the social-environmental changes that occur today, without omitting the absence of guidelines based on the realities and particularities of the communities that could have contributed to their efficient organization.

The second complementary element mentioned the approaches that must be considered when structuring a social organization based on the application of the manual in question. For Villadiego et al. (2013, p. 94), the community approach seeks to involve people in the analysis of the environmental problems of their territories. Likewise, from the systemic approach and according to Covas (2004, p. 3) all environmental problems necessarily have a systemic constitution, considering it as an organized whole, composed of parts that interact with each other. The sustainable development approach allowed us to understand the interrelation among the components of the environment: Social, economic and natural, as well as the establishment of the potential of the region.

The fundamental elements to consider in the document were, in first instance, the establishment of an environmental profile of the territory where communities wanted to organize themselves. The above was related to the development of the diagnostic phase to have a real-time radiograph of the current state of the territory: Name, neighborhood, commune, etc. The environmental profile analyzed variables related to the social-economic aspects of the populations interested in community structuring and included the identification and analysis of existing environmental problems. For Prado (2000) a profile provides an organization (government, municipality, institutions and/or organized civil society) with the necessary guidelines for the treatment of the environment, as well as the objectives and policies to develop joint actions between each one of the actors in that organization.

Sequentially, from the profile and its diagnostic phase, the structure of the community organization was developed. This included the definition of the policy and approaches on which the organization would work, its vision, mission and objectives that were key components in every organizational structure, especially in the communal type. Enz et al. (2012) recommended that when formulating a mission, to consider answers to questions such as: Who are we? What do we do? How are we going to do it? And what actors should we relate with?

Equally important was the definition of the hierarchical structure of the organization (i.e. communication, definition of responsibilities and authority within the organization). There were two types of hierarchical structures: Horizontal and vertical. The Communal Action Boards (CABs) in Monteria had a structure of the vertical type. There is a visible head and, on this rested the greater responsibility and the decision making. This type of hierarchy removed leadership and responsibilities from other figures that could have integrated communal boards. The aim would have been a redistribution of responsibilities among all members of the institution. Likewise, the definition of lines of action for the communal organization allowed to clearly establish areas of work to be executed, specifying them and allowing the creation of collaborative networks that facilitated, through specialized structures, the development of these.

Finally, once each of the above mentioned points were defined, the next step was to continue with the legal constitution of the organization before the Chamber of Commerce and its respective registration in the Departmental Secretary of the Interior and Citizen Participation. Generally, and as previously mentioned, community organizations in the City arose more often from passionate issues than from planning processes with a strong component of local and sustainable environmental management. In short, CABs were born on paper and died on paper after minutes of being registered. They began with the last step proposed here to achieve an efficient organizational structure, the elaboration of statutes (very probably copied from other CABs). Then they were registered before the Chamber of Commerce.

Training processes in an organizational structure allowed CABs to be more competitive and updated on the realities surrounding them. Therefore, the third element of the manual formulated was training. It was necessary and vital that communal organizations should periodically train their members in basic and vital issues such as: (1) Citizen participation mechanisms, (2) project formulation and management, (3) policy, territorial governance and local environmental management, (4) development, organization, empowerment and community ownership, (5) information and communication technologies, and (6) conflict resolution. The training phase was structured considering both the environmental profile and the organizational structure, so that once finished, it could reach the fourth and last element phase of the manual to be considered, solving the problems and needs of communities. This included three aspects: (1) Solutions to the problems identified in the environmental profile, (2) evaluation and feasibility of the proposed alternatives, and (3) implementation of solutions to the social problems of communities. Finally, and as Marchonni (2007) pointed out, the community process was one of organization, development or both. It would only advance if there were political willingness (including administrators and population) and technical-scientific support (human and economic resources). The future was becoming more complex, where more knowledge and applied science was needed to provide better living conditions for the communities. Although this manual emerged from a theoretical-conceptual construct supported by contextual diagnosis, the evaluation pointed out the needs of the urban area of the City to be efficiently organized at the community level. However, implementation or validation of the manual by means of pilot test was subject to what Marchonni had proposed above: Political will and technical and scientific capacity.

6 Conclusions

CABs are fundamental social structures in achieving sustainable development for a city or town and, from this study, the following conclusions were obtained:

1. The individuals who made up Community Action Boards were people characterized by low income and poor levels of education, few opportunities for employment and personal development. Mostly they were from a low social-economic stratum.
2. It was a priority to work in an articulated way between Community Action Boards in search of solutions to local problems of the City, both at the urban and rural levels, framed in the concept that environmental problems and environment per se had no political-administrative boundaries. Likewise, the exchange of experiences was an enriching factor for improving the way of operating and, in turn, feedback for themselves as key structures of the territory for achieving environmental sustainability.
3. One of the main weaknesses presented by the Community Action Boards of Monteria's urban areas (without omitting rural organizations) was that these were formed more as passionate and spontaneous actions to organize groups responding to political aims, rather than as structures designed on a diagnosis, characterizing the community itself and the individuals therein, and above all, to the environmental problems and needs to which solutions must be given. The purpose of CABs was to contribute to sustainable development and efficient environmental management of the territory, but the leaders of the organizations were not clear on what entailed achieving sustainable development and territorial environmental management.
4. The organizational structure of the Communal Action Boards was vertical, that is, greater responsibility was centered on a visible head, thus lessening the importance of other individuals who were part of the organization and the role they could have developed inside and outside the same organization. The training processes in this type of organizations were very poor, especially in relation to the formulation of projects and the capacity to manage them.
5. Elements such as territory profiles, structuring of organizations based on environmental profiles, the training of community organization members (especially in fundamental issues such as sustainable development, citizen participation mechanisms, formulation and management of projects, communication and information technologies, construction of environmental networks, organizational structure) and the solution of problems and needs were highly relevant factors when designing an efficient community organizational manual.

These conclusions gave us reason to believe that educational environmental training had to be the basis for promoting and guiding the actions of community organizations for sustainable development in which effective interactions between community organizations, government agencies and local non-governmental groups depended upon organizational strategies designed and carried out with manuals to guide the actions of all three social-economic sectors of specific communities.

Acknowledgements Special thanks to Doctor Efrain Diaz Aguilar, community leader who collaborated with the development of this research by contacting presidents of the Community Action Boards of Monteria's urban areas.

References

Chacon M (2010) Desarrollo comunitario [Community development]. Rev Educación y Experiencias Educativas, 29. Andalucía, Spain http://www.csi-csif.es/andalucia/modules/modense/revista/pdf/Numero_29/M_DOLORES_CHACON_1. Last Accessed 9 Dec 2015

Covas O (2004) La educación ambiental a partir de tres enfoques: Comunitario, sistémico e interdisciplinario [Environmental education based on three approaches: Community, systemic and interdisciplinary]. Rev Iberoamericana Educación 34:1–7. Centro de Altos Estudios Universitarios de la Organización de Estados Iberoamericanos, Spain http://www.rieoei.org/rie_contenedor.php?numero=eduamb5&titulo=Educaci%F3n%20Ambiental%20a%20partir%20de%20tres%20enfoques:%20comunitario,%20sist%E9mico%20e%20interdisciplinario. Last Accessed 9 Dec 2015

Departmental Secretary of Interior and Citizen Participation (2015) Dirección de Desarrollo Social (DDS), Colombian Government http://english.mininterior.gov.co/ubicación/bogotá-de?page=351. Last Accessed 19 Nov 2015

Duke P, Gaitán C (2001) Una mirada prospectiva a las metodologías de intervención comunitaria, responsabilidad de todos. [A Prospective look at methodologies of community intervention, responsibility of all]. Consejo Nacional de Trabajo Social, Bogotá, Colombia

Enz A, Franco V, Spagnuole V (2012) Manual de comunicación para organizaciones sociales: Hacia una gestión estratégica y participativa [Communication manual for social organizations: Towards strategic and participatory management]. Community Civil Association, Buenos Aires, Argentina

Facione P (2007) Pensamiento crítico: ¿Qué es y por qué es importante? [Critical thinking: What is it? and why is it important?]. Loyola University, Chicago, USA https://www.eduteka.org/PensamientoCriticoFacione.php. Last Accessed 9 Dec 2015

Google Earth (2015) Google maps/google earth. Internet http://www.google.com/es/earth/download/gep/agree.html. Last Accessed 9 Dec 2015

Guerra G (sf). La solución de problemas [Troubleshooting]. Universidad Carlos III, Madrid https://portal.uc3m.es/portal/page/portal/cultura_y_deporte/eses/historico/1415/orientacion3/bienestar_emocional/gestion_emociones/Soluci%F3n%20de%20problemas.PDF . Last Accessed 9 Dec 2015

Instituto de Formación Permanente INSFOP y el Programa Especial para la Seguridad Alimentaria Nutricional Nacional PESANN/FAO (2008) Organización comunitaria. [Institute for Ongoing Training (INSFOP) and the Special Program for National Nutritional Food Security PESANN / FAO (2008) Community Organization]. Nicaragua http://www.fao.org/3/a-as496s.pdf. Last Accessed 19 Nov 2015

Marchonni M (2002) Las agendas 21 y la evolución de los procesos de participación social. Sostenibilidad ¿para qué y para quién? [The agendas 21 and the evolution of the processes of social participation. Sustainability for what and for whom?]. Revista Sostenible, 4, Spain http://www.raco.cat/index.php/Sostenible/article/view/261899. Last Accessed 9 Dec 2015

Marchonni M (2007) Organización y desarrollo de la comunidad: La intervención comunitaria en las nuevas condiciones sociales. [Organization and community development: Community intervention in the new social conditions]. España http://extension.uned.es/archivos_publicos/webex_actividades/4698/acomunitariaponencia13b.pdf. Last Accessed 9 Dec 2015

Mercer H, Ruiz V (2004) Community participation in health management: an evaluation of the UNI Program experience. Interface—Comunic; Health. Educ 8(15):289–302. Sao Paulo Brasil http://www.scielosp.org/pdf/icse/v8n15/a08v8n15.pdf . Last Accessed 19 Nov 2015

Ministerio de Desarrollo Social (2015) Dirección de Desarrollo Social (DDS). Colombian Government http://www.drp.gov.co/programas/desarrollo-social/Pagnias/desarrollo-social.aspex. Last Accessed 19 Nov 2015

Moreira M (2006) Administración de procesos en instituciones de formación. [Management by processes in information institutions]. Revista Acimed 14(5). Revista de Información de la

Salud, Cuba http://bvs.sld.cu/revistas/aci/vol14_5_06/aci11506.htm. Last Accessed 9 Dec 2015

National Department of Statistics DANE (2015) Municipal and departmental population projections 2006–2020. Colombia http://www.dane.gov.co/files/investigaciones/poblacion/proyepobla06_20/ProjectionMunicipalities2005_2020.xls. Last Accessed 19 Nov 2015

Prado F (2000) Perfil ambiental de Santa Cruz de la Sierra, Bolivia [Environmental profile of Santa Cruz de la Sierra, Bolivia]. Editor Gobierno Municipal de Santa Cruz de la Sierra, SINPA. Bolivia http://repub.eur.nl/pub/32228/SINPA%2013%20Prado%20F%20%282000%29%20Profile%20Environmental%20de%20Santa%20Cruz%20de%20la%20Sierra.pdf. Last Accessed 9 Dec 2015

Rincon R (1998) Los indicadores de gestión organizacional: Una guía para su definición [The indicators of organizational management: a guide for its definition]. Revista Universidad EAFIT 34 (111):43–59. Universidad EAFIT, Colombia http://publicaciones.eafit.edu.co/index.php/revista-universidad-eafit/article/view/1104/996. Last Accessed 9 Dec 2015

Ross M (1978) Organización comunitaria colección asistencia social [Community organization social assistance collection]. Spanish Federation of Schools of the Church of Social Assistants, Madrid, p 65

Villadiego J et al (2013) Aspectos teóricos a considerar para la generación de un modelo no formal y participativo de educación ambiental orientado a la protección y conservación de los Humedales Bañó y Los Negros [Theoretical aspects to consider for the generation of a non-formal and participatory model of environmental education oriented towards the protection and conservation of the Bañó and Los Negros wetlands]. Revista Tecnologia en Marcha 26 (4):92–99. Costa Rica http://revistas.tec.ac.cr/index.php/tec_marcha/article/view/1587. Last Accessed 12 Dec 2015

Villadiego J et al (2014) Dinamica espacial e impactos económicos de los usos del suelo: Estudio de caso aplicado al Canton Norte de Monteria, Colombia [Spatial dynamics and economic impacts of land use: A case study applied to the north canton of Monteria, Colombia]. Editorial Académica Española. Germany, p 137 (ISSN 978–3-8443-4484-4)

Pedagogic Proposal Focused on Sustainable Development: Fracking, a Matter of Active Debate at Present in the Argentine Patagonia

Alida M. Abad, Cecilia E. S. Alvaro and Norma Sbarbati Nudelman

Abstract This research focuses on the development of a pedagogic unit on hydrocarbons for the last year of high school students that links the contents of science with social, technological and environmental aspects using a global approach (STES). This proposal emphasizes an holistic view of school contents, selecting a controversial topic such as oil extraction by hydraulic fracture known as "fracking" that arouses a deep debate due to its social implications. This exploratory research uses a qualitative methodology focused on a case study. Empirical evidence is gathered using classroom observations, student surveys and teachers′ interviews. Due to the size of the sample limited to a small group of students, this study does not admit generalizations, it only shows a tendency. This paper is a contribution to future research on STES approach in High School.

Keywords Pedagogic strategy · Fracking · Sustainable development
Environment

A. M. Abad (✉)
Facultad de Ciencias de la Educación, Universidad Nacional del Comahue,
Irigoyen 2000, 8324 Cipolletti, Río Negro, Argentina
e-mail: marinali2005@gmail.com

C. E. S. Alvaro (✉)
Depto de Química, Facultad de Ingeniería, Universidad Nacional del Comahue,
Buenos Aires 1400, Neuquén, Argentina
e-mail: silvana_alvaro@hotmail.com

C. E. S. Alvaro
Instituto de Investigaciones y Desarrollo en Ingeniería de Procesos, Biotecnología y
Energías Alternativas, CONICET–Universidad Nacional del Comahue,
Buenos Aires 1400, 8300 Neuquén, Argentina

N. S. Nudelman
Facultad de Ciencias Exactas y Naturales (UBA), Depto Química Orgánica,
Ciudad Universitaria Buenos Aires, Pab 2, Piso 3 Avda Cantilo s/n,
1428 Buenos Aires, Argentina
e-mail: sbarbati04@gmail.com

W. Leal Filho et al. (eds.), *Sustainable Development Research and Practice in Mexico and Selected Latin American Countries*, World Sustainability Series,
https://doi.org/10.1007/978-3-319-70560-6_6

1 Introduction

Growing concerns about a threatened environment, conflicts, inequities, poverty, ideological extremes, and consumerism are all indicative of a pressing need to reflect on the global status quo and to find constructive and long-term, sustainable strategies for planet and people. The need to give the younger generation "a better deal" for helping to shape a sustainable world is being embraced by several international associations, NGOs, and multiple sectors of the whole society (Lueddeke et al. 2017). Environment and society are synergistic in any environmental modeling; the reaction of society should be included, and the uncertainties are more of socio-economic nature than biophysics (Lubchenco 1998).

There is a widespread recognition that this century citizen should have a basic understanding of the fundamental concepts and procedures of science and technology because of their ever increasing relevance in his/her common life (Lok 2010). Nevertheless, teachers usually complain on the lack of interest in science of their students. To reverse students' lack of interest in science, several innovative pedagogies are being implemented. Among them the Science, Technology, Environment and Society (STES) and the Inquiry-Based Science Education (IBSE) approaches, are among the most successful methodologies for the teaching of STEM (Science, technology, engineering, mathematics) disciplines (Nudelman 2014).

The STES approach that links science to technology, society and environment has been applied by several research projects since the 1990s and continues to be of interest in science teaching today (Aikenhead 1992; Ramsey 1993; Jiménez-Aleixandre and Gallástegui 2011; Zoller 2013). In the IBSE approach the students are proposed a scientific problem or phenomena and working in a 4-members team they should look for experimental evidences of the STEM fundaments that could explain the observation (Harlen 2010; Nudelman 2016). In both pedagogical approaches the authors agree on the importance of structuring science contents with current problems that may be significant for the students. As a very important example, it is worthwhile mentioning that the need for including the society's impact is an essential requirement in any research on the economy of Global Climate Change when considering the efforts for reducing emissions (Sandmo 2015; Chisari and Miller 2016a, b).

Argentine Norpatagonia has one of the main reservoirs of unconventional hydrocarbons that must be extracted by hydraulic fracture. Fracking is a technique that allows the extraction of gas and oil from the subsoil by deep drilling and injection of high pressure water with the addition of numerous chemicals. This technique is severely questioned due to environmental implications, although it is one of the main sources of income for the region. It is also subject of controversial debates in different areas of society so students are familiar with the problem due to the repercussion in the mass media and social networks. Therefore, addressing this situation from a didactic proposal is interesting and highly convenient to develop critical judgment and environmental awareness.

For achieving the mitigations expected by 2030 (COP21 2015) education is considered a key issue since the young students are being mentioned as the real "agents of change" (Battro et al. 2017).

The aim of this paper is to discuss STES approach in Chemistry classes, selecting a hot topic such as "fracking" to foster skills such as critical analysis and argumentation and make the students aware of their own responsibility as active citizens.

2 Content and Work Development

The public image of Chemistry is fairly lowing down in the popularity stakes, especially because of the contention of the potential harm, to human health and the environment, caused by the application of chemistry. Environmental issues are the subject of countless books, articles, speeches and even sermons, and appear to constitute by far the biggest social challenge chemistry has ever had to face (Russell 2000). This bad image could be justified in some cases due to some mistakes, undesirable side-effects and accidents that have occurred. To face the challenge and overcome the unwelcome opposition, a more environmentally friendly new chemistry has being developed in the academy and in some industrial sectors since almost two decades ago (Anastas and Kirchhoff 2002). This kind of Chemistry is well accepted by the students and lead to a better understanding between environmentalists and chemists, for the benefits of the whole society (Nudelman 2004).

When Chemistry is taught by the traditional way, students get bored by the accumulation of laws, formulae, nomenclature, and so on, that (they presume) has nothing to do with their real life.

Particularly, in The Organic Chemistry Curriculum of almost any High School program, Hydrocarbons is one of the main units, usually teachers place emphasis on disciplinary contents, (structures, nomenclature, sources, etc.) leaving aside other aspects that allow a more global approach. The present proposal focuses not only in scientific aspects like properties and structures of hydrocarbons but also intends to take into account the differences between fracking and other methods for oil extraction, with regard to the environmental impacts.

The country's main geological formation called Vaca Muerta is located in Northern Patagonia, Argentine. It is a shale oil reserve of global importance and its extraction requires fracking. The technique for extracting petroleum and natural gas from non-conventional reservoirs exploits the hydrocarbons accumulated in the pores and fissures of certain sedimentary rocks. These rocks are of fine or very fine grain, generally of lime type, which permeability prevents the migration of hydrocarbons to large reservoirs. It is necessary to drill wells that occupy wide areas; millions of gallons of water must be injected with chemical and toxic substances in order to extract them. The discovery of the shale oil reservoir raises big controversy in different fractions of society. The exploitation of this resource, should bring, a great economic growth to the region but it also has important environmental implications that should be jointly considered (Brock and Taylor 2004). This technique is also the subject of active debate between those who are aware of the need to apply this methodology and their opponents, very often not well informed about the technological aspects. This research seeks to provide evidence about how the design and placement of a different teaching proposal with an IBSE and STES approach, can improve science teaching. The work proposal prioritizes the use of tools, such as searching and contrasting technical information, search for evidence, critical analysis, argumentation, etc. to discuss viability and consequences of using a highly questioned hydrocarbon extraction technology. Evaluation of the proposal is made by surveys, interviews and class observation. The field work of this research is carried out within the framework of a thesis for Chemistry teachers' training.

3 Teaching Sustainable Development in High School: Argumentation in Science Classes Using a STES Approach

The High School students are aware of these facts that surely will affect their lives in the near future. Very often several newspapers, mass media and social networks refer to the subject exposing very different points of view, usually depending on their own interests. It is expected that an informed citizen should have its own critical opinion on the application of technologies that could affect human life and/or the environment (Binmore 2011). The High School teachers, especially in the science classes, should help students to critically analyze the large body of information, understand the fundamentals of science and technology involved in the suggested oil extraction methodologies to build their own rational point of view. First of all, it is necessary to know about the different types of hydrocarbons, their structures and properties and how they are incidental in the diverse extraction methods in order to have enough background to discuss the topic seriously.

The STES approach that links Science, Technology and Society with environment seems suitable for this purpose, since it gives the chance to apply science contents to everyday problems and make decisions as citizens committed to their social environment. Duran (2013) emphasizes that it is fundamental to understand the centrality of sustainability ethics as a set of values that promote the understanding of the complex interactions between society and environment, taking into account the interrelationship of ecological, cultural, technological, economic and social aspects. We agree with the statement by Ramsey (1993) "From the perspective of social responsibility, scientific training should give rise to students who can and do participate in the resolution of social issues related to science. This means that students are willing and able to act effectively as citizens using the values and skills that come from both science and democracy". In STES science teaching, science content is certainly taught but from a social perspective that allows them to have an up to date view of science possibilities and limitations. To ensure sustainability of natural resources must be a priority in science teaching. From the perspective of sustainability, any relevant, generated or acquired knowledge that is put into action in the STES context should be guided by the idea of social responsibility (Zoller 2012). That is why choosing a relevant issue to discuss in class helps students develop different skills like argumentation in order to analyze data, draw conclusions and communicate effectively. When the students use IBSE methodology they look for experimental evidence, contrast different explanations of a phenomenon, theory, or model by evaluating evidence and drawing conclusions, they are using argumentation (Nudelman 2016). Definition for this practice is quite clear. We conceive argumentation as a mean to evaluate knowledge statements, hypotheses conclusions or theories based on available evidence. What we understand today as an effective science learning goes beyond understanding and using scientific concepts and models. It also includes participating in scientific practices. These practices correspond to the way scientific community works, creating new knowledge (Jiménez-Aleixandre and Gallástegui 2011).

Both, from the curricular documents that come from the Argentine Ministry of Education, called Priority Learning Issues, (NAP in Spanish) and from didactic research, argumentation is valued as a relevant competence in Science teaching. Pedrinaci (2012) says it contributes to develop responsible citizenship, able of participating in social decisions using critical thinking. The use of fracking, suspected of producing high pollution in watercourses, is a good point to start a debate. It is necessary to consider the non-renewable character of oil, to compare pollution generated by the different methods of oil extraction, sustainability in resource management, and be aware of economic-social impact. Jiménez-Aleixandre and Gallástegui (2011) associate argumentation to many purposes of science teaching, that help students develop the following skills:

- improve learning processes.
- responsible citizenship, able to participate in social decisions by exercising critical thinking.
- participate in scientific practices that expand scientific competences and allow an understanding of the complex nature of science.

Undoubtedly, the promotion of IBSE and STES-focused education for sustainability in science education, at all levels, raises the issue of education versus indoctrination. In this context, science teachers' job is not to tell the students what to think, but rather to develop their own thinking (Qablan et al. 2011).

4 Research Methodology

This is an exploratory study based on a qualitative methodology.

The researcher takes the role of participant observer in the present case study, to obtain information about the students' perception about the motivation, significance and importance of STES approach in chemistry classes. The emphasis of this didactic proposal is placed on communicative skills such as argumentation and critical judgment development. Cognitive processes developed are collected through analysis of student outputs. This didactic strategy will allow students:

- Relate disciplinary contents about hydrocarbons with technological, social and environmental aspects.
- Develop informative, communicative and argumentative skills.
- Critical thinking for the detection and analysis of manifest trends in different sources of information.
- Creativity to design their productions in different communication formats that make the community aware of relevant environmental issues.

This strategy is tested with thirty two senior students of a private high school.

After learning usual science contents like properties, structure of hydrocarbons and fractional petroleum distillation, students are asked to search scientific literature and identify recent articles that address the problem of fracking to exploit shale oil resource. After discussing in class several articles and different visions of the problem the students are organized into groups with different number of students each, in order to be part of a role playing process where they have to develop arguments in favor and against fracking and communicate their conclusions to peers. For a convenient support they have to seek information from different sources like non-governmental organizations, environmental forums in Internet, oil companies, government offices, and analyze the information with critical thinking.

Group 1 (10 students) have the task to prove that fracking is a convenient solution for the region and bring up reliable information from different sources. They have few minutes to present their idea using Power point or Prezi in sub-groups of two students.

Group 2 (10 students) have to support the opposite idea in the same way as group one.

Group 3 (5 students) have to record all relevant information in the debate for a report.

Group 4 (2 students) present each speaker, moderate and keep discussion under control, same chance to speak with equal exposure time. Read the final conclusions.

Group 5 (5 students) have to play the role as reporters, take pictures and record some main ideas and then write an article for the school on-line newspaper.

Students from another course of the same school, with their science teacher, are invited to play the role of audience and choose which they think, is the most convincing posture.

Once everyone had the chance to speak, and defend their judgment, the audience composed of peers express their verdict based on what they heard and considered more credible. They write down their position and hand the paper to the moderators who take their time to order the answers and then read the conclusions aloud, ending the session.

After the unit is finished all students answer surveys that have to do with learning and motivation. An interview is made to the science teachers who participated as audience.

5 Results and Conclusions

92% of the students answered that they liked the way the unit was taught.
3% didn't like it.
5% didn't answer.
Of those who answered yes,
20% said they liked because they learned new things about oil extraction.
25% because the new skills are very important for their future.
30% care for social responsibility.
15% considered an important issue because they live in a state whose main wealth is oil and need to make use of the resource in a sustainable way.
5% liked role playing and argumentation as a technique that allows learning and making decisions.
5% did not answer.

Students were positive about their learning process and were aware of advantages of argumentation. They feel worried about their future and also responsible for making decisions that will preserve environment and its natural resources. They also know that oil means money and the state depends largely on this resource.

Class observation reveals that students took role playing with great concern and studied more than in a traditional class to be able to communicate conclusions. The reports and the newspaper article were considered by the teacher as very satisfactory.

Interview made to the High school science teacher who participated in the audience revealed that she was surprised by the quality of the arguments made up by students and by the realism shown during the role playing.

We can finally say that developing science classes that ensure the use of scientific competencies such as argumentation in a curriculum with a STES approach, allows us teach science in a more profitable way and promote sustainability.

Sustainability science is a rapidly evolving area that challenges professionals and academics from all scientific disciplines to apply their expertise to promote sustainability (Clark and Dickson 2003).

Lozano and Watson (2013) prevent that participation of chemists in the collaborative field of Sustainability science is critical for developing and implementing sustainable stratagems for dealing with complex global dilemma.

We believe that is true not only for Chemists but also for Science teachers who are educating the new generations to make decisions that compromise the future of society (Nudelman 2017). It is urgent to update the science curricula to emphasize sustainability and involve students as the real "agents of change" for achieving the United Nations Sustainable Development Goals, UN SDGs, proposed for 2030 (Nudelman 2016).

This proposal aims to be a contribution to deepen the disciplinary contents and rescue competences such as writing and speaking in science. It also highlights the importance of disseminating regional environmental issues in the community using different communication formats.

Analysis of surveys show a substantial increase in student motivation by this type of proposal, focused on real issues that involve technological as well as social and environmental knowledge. The scientific procedures applied in the classroom, such as multiple variables analysis, well-founded conclusions and argumentation were mentioned by the teachers as very important. They also considered that the performance of the students was much better than in the traditional science classes.

Like all case studies, this research only shows what happened with this particular group of students and marks a trend that may be insured by future and larger studies.

Acknowledgements The authors gratefully acknowledge financial support from Universidad Nacional del Comahue (Grant No I221-UNCo) and to the student Ludmila Pereyra for her field work.

References

Aikenhead G (1992) The integration of STS into science education. Theory Prac 31(1):27–35

Anastas PT, Kirchhoff MM (2002) Origin, current status and future challenges of green chemistry. Acc Chem Res 35:864–870

Battro AM, Léna P, Sánchez Sorondo M, von Braun J (2017) Children and the sustainable development: ecological education in a globalized world. Pontificia Academy of Sciences, Springer, New York

Binmore R (2011) Rational decisions. Princeton University Press, Princeton

Brock W, Taylor MS (2004) Economic growth and the environment: a review of theory and empirics. NBER Working paper 10854. National Bureau of Economic Research, Cambridge, United States

Chisari O, Miller S (2016a) Climate change and migration: a CGE analysis for two large urban regions of Latin America. DB-WP-659. Inter-American Development Bank, Washington DC, pp 1–49, Feb 2016

Chisari O, Miller S (2016b) CGE Modelling alternative structural specifications for the evaluation of carbon taxes. Simulations for economics of Latin America and the Caribbean. In: Bryant T (ed) The WSPC reference on natural resources and environmental policy in the era of global change, vol. 3. World Scientific

Clark W, Dickson N (2003) Sustainability science: the emerging research program. Proc Natl Acad Sci 100(14):8059–8061

COP21 (2015) http://www.bbc.com/mundo/noticias/2015/51211_cumbre_clima_paris_cop21_ac. doi:10.1007/978-3-319-47130-3 Accessed 28 May 2016

Duran D (2013) Proyectos ambientales y sustentabilidad. Lugar Editorial, Buenos Aires, Argentine

Harlen W (2010) Principles and big ideas of science education. Ashfort Colour Press Ltd, Gosport Hants, Great Britain

Jiménez-Aleixandre MP, Gallástegui JM (2011) Argumentación y uso de pruebas construcción, evaluación y comunicación de explicaciones en física y química. In: Caamaño A (ed) (Coord) Didáctica de la Física y la Química. GRAO, Barcelona, Spain, pp 121–139
Lok C (2010) Science for the masses. Nature 465:416–418
Lozano R, Watson M (2013) Chemistry education for sustainability: Assessing the chemistry curricula at Cardiff University. Educación química 24(2), 184–192 Universidad Nacional Autónoma de México, Ciudad de Mexico
Lubchenco J (1998) Entering the century of the environment: a new social contract for science. Science 279:491–497
Lueddeke GR, Kaufman GE, Lindenmayer JM, Stroud CM (2017) Preparing society to create the world we need through "One Health" education SEEJPH. Posted April 2017. doi:10.4119/UNIBI/SEEJPH-2017-142
Nudelman NS (2004) Química Sustentable. Universidad Nacional del Litoral, Santa Fe, Argentine
Nudelman NS (2014) Innovative pedagogy for the teaching and learning of science-inquiry based science education (IBSE). J Sci Educ 15:36–38
Nudelman NS (2016) "Green" processes of "clean" production for a sustainable civilization. The environmental problem in the society, health and economics. Mundial S.A, Buenos Aires, Argentine, pp 89–108
Nudelman NS (2017) The urgency of science education reform: the urgent need of a science education reform in argentine. Revista Iberoamericana de Ciencia, Tecnologia y Sociedad, CTS 12:161–178
Pedrinaci E. (2012) Aprender ciencias es, en buena medida, aprender a leer, escribir y hablar ciencia. In: 11 ideas clave: el desarrollo de la competencia científica. GRAO, Barcelona, Spain, pp 147–168
Qablan A, Southerland SA, Saka Y (2011) My job isn't to tell them what to think: the fear of indoctrination and how it shapes education for sustainable development. Electron J Sci Edu 15 (2). http://ejse.southwestern.edu/issue/view/757. Accessed 28 June 2016
Ramsey J (1993) The science education reform movement: implications for social responsability. Sci Edu 77(2):235–258
Russell CA (2000) Chemistry, Society and environment: a new history of the bristish chemical industry. Royal Society of Chemistry, Cambridge
Sandmo A (2015) The public economics of climate change. Norwegian school of economics. http:/ssrn.com/abstract.2693142. Accessed 16 May 2017
Zoller U (2012) Science education for global sustainability: what is necessary for teaching, learning and assessment strategies? J Chem Educ 89:297–300
Zoller U (2013) Emergent topics on chemistry education, chemistry education and sustainability. Educación Química 24(2), 207–214. Universidad Nacional Autónoma de México, Ciudad de México

Author Biographies

Mg. Alida Marina Abad has a degree as Chemistry Teacher at National University of Buenos Aires (1982). Professor and researcher at the National University of Comahue, (UNCo). Faculty of Sciences of the Education. Natural Science Technology and its didactics. Author of numerous articles and book chapters on the didactics of science at different educational levels. Currently working on technologies applied to education and sustainable chemistry.

Prof. Cecilia E. Silvana Alvaro (Ph.D., Chemistry) Professor and researcher at the National University of Comahue at Faculty of Engineering and Faculty of Environmental Sciences (UNCo). Has a degree on Biochemistry at National University of Rosario, Master in Chemical Sciences at

National University of Comahue (2000) and Ph.D. degree in Chemistry at National University of South (2009). She has directed ten graduate thesis, three research projects and currently directs one at the National University of Comahue. Author of more than forty articles on Fundamental Organic Physicochemistry, Quantum Chemical Calculations and Enviromental Pretroleum Contamination, one book and a chapter on Organic Chemistry Education. Member of Institute for Research and Development in Process Engineering, Biotechnology and Alternative Energies, CONICET—National University of Comahue.

Prof. Norma Sbarbati Nudelman (Ph.D., Chemistry) Professor and researcher at the National University of Buenos Aires (UBA) and CONICET. Has a degree in Chemistry at National University of Buenos Aires (1957) and a Ph.D. degree in Chemistry at National University of Buenos Aires (1962). She had directed a numerous research projects in Organic Chemistry and twenty-four doctoral thesis. Actually is full Professor at Faculty of Exact and Natural Sciences. Chair of IANAS SEP (Inter American Network Academies of Science). Argentine Focal point of IANAS Women for Science. Honorary Member of IUPAC. Member of editorial board of “Enviromental Technology” and “Physical Organic Chemistry”.

A Low-Cost and Low-Tech Modular Solar Still as a Reliable and Sustainable Water Source

Gregor Zieke

Abstract The principle of solar water purification or solar distillation is one of the few feasible possibilities of drinking water treatment for many regions of the world, which suffer from groundwater contamination by fluorides and/or heavy metals. The implementation of an innovative low-cost, low-tech design of a modular solar distillation system is presented. It covers all aspects of sustainability, in the social, natural and economical dimensions: apart from facilitating dissemination by economic small scale production, which uses broadly available materials, the modularity of the described system permits its easy adaptation to the required drinking water output. It is completely user-serviceable and does not require additional energy sources. A distillate yield of up to 3.9 L per day and square meter is achieved. All analyzed contaminants of the raw water were effectively removed in the distillation process. Furthermore, a partial remineralization of the distillate is achieved by the use of a concrete trough in the module, thus making it suitable for human consumption.

Keywords Solar still · Low cost drinking water treatment · Low technology Fluorosis

1 Introduction

Adequately purified drinking water is a scarce resource—not only in Mexico, but also in many other countries of the world. Due to various infrastructural and technical factors, the tap water in Mexico in general cannot be considered as safe for human ingestion without special treatment. Furthermore, recent studies confirm the existence of arsenic and fluoride above the levels established in the National Mexican Norm for Potable Water (Castellanos Coutiño 2000) in at least 36

G. Zieke (✉)
Centro Cultural Alemán de San Luis Potosí, A.C., Av. Scop 900, Colonia Jardín, C.P, 78270 San Luis Potosí, SLP, Mexico
e-mail: gzb@protonmail.com

W. Leal Filho et al. (eds.), *Sustainable Development Research and Practice in Mexico and Selected Latin American Countries*, World Sustainability Series,
https://doi.org/10.1007/978-3-319-70560-6_7

Mexican aquifers, most of them located in the central and northern parts of the country (Reyes-Gómez et al. 2013). It is estimated that in Mexico about 6 million people are affected by a combination of both contaminants in the drinking water supply (Alarcón-Herrera et al. 2013), which therefore requires specialized treatment, such as Reverse Osmosis. Although the availability of drinking water treated by the latter is high in urban areas (bottled water), in rural areas the people rely mainly on untreated well-water for ingestion, thus suffering from health problems such as diarrhea in mild cases or *fluorosis* in severe ones. Up to date, there exists no affordable and independent solution to reliably purify the aforementioned water resources in small-scale installations, for example in households. One feasible approach is the distillation of the affected water, which is capable of completely removing nearly all known contaminants (Derickson et al. 1992). Commercially available electrical distillers impose a relatively high initial investment cost (starting at about 300 USD per unit at the time of redaction of this article) and have the disadvantage of continuously arising electricity costs. Using solar energy for this purpose is an ideal option for sunny regions in the world—especially for Mexico which has a high solar potential throughout nearly all of its territory. As for the construction of the solar water distiller, there exist various concepts; the simplest being the *solar still*. It requires only solar radiation for the process, therefore being an energy autonomous technology, which integrates perfectly the three aspects of sustainability: It can be constructed from simple, economic and durable materials, thus requiring only an initial inversion, which redeems in a short time span. For the same reason, it can easily be constructed and disseminated. Due to its simple working principle, operation and maintenance can be done by the user and no special spare parts are required. At last, it helps to conserve the natural water resources by tolerating a wide range of raw water without the need for pre-treatments, which promotes the use of surface- and gray-waters. This paper features the development and the proof of concept of a semi-automated modular solar distillation system, which integrates the aforementioned aspects into small solar still units, which are produced by a concrete casting process. The modularity facilitates the scaling of the system in order to cater for the users' drinking water needs in rural, as well as in urban areas.

2 State of the Art

Up to date, the only seemingly working and reproducible solar still is the commercially available *Rainmaker*™, which costs 489 USD and incorporates high-tech materials (SolAqua LLC 2017). The solar stills described by Foster and Eby-Martin (2001) and Ojinaga Santana and Foster (2001), which were implemented in communities in Northern Mexico, also include materials which are not easily accessible on the Mexican market. The latter characterizes the dilemma of small-scale solar stills in general: while functional solar stills can only be established in communities via financing schemes, which create technologic and economic dependencies from the manufacturer, in case of self-construction, the proposed materials are either not durable or contaminate the distillate, as described by Kinzer and Crane (2006). The present investigation is based on the necessity for a solar distillation system, which combines the aspects of economic construction and simple maintenance, semi-automated operation, as well as satisfactory yield and acceptable water quality (remineralization) with the application of durable and common materials which do not affect the user's health.

3 Methodology

The design of the solar still module integrates various already proven enhancements of the basic working principle, which are described in the following sections. A mold for concrete casting is created in order to produce six prototype modules, which are arranged in a system setup afterwards. The prototypes vary in the amount of incorporated insulation material (recycled polystyrene food trays) and the way of raw water preheating (with or without solar collector); the influence of these factors on the distillate yield is analyzed. In addition, the energy efficiencies are determined, in order to facilitate the comparison with existing technologies. Water quality measurements are conducted in order to corroborate the working principle and to determine the level of distillate remineralization. A short list of used materials and their cost completes the analysis. Whereas the physical implementation of the solar still system is described in the paper, no long-term studies or practical user experiences have been made with the technology yet.

4 Working Principle

As physical working principle, the *single basin, single-slope solar still* is used, which is viable for locations with a latitude equal or greater than 20°, according to Kalidasa Murugavel et al. (2008). It permits the use of a wide variety of raw water and has the simplest design of all solar distillers—without the need for additional energy (pumps) or other devices (filters). Also, in comparison to a *single basin, double-slope solar still*, this design achieves higher energy efficiencies—up to 10% above the latter (Al-Hayeka and Badran 2004). This is an important point, as solar stills in general suffer from low energy efficiencies between 25 and 45% (McCracken and Gordes 1985; Noble 2014) and therefore a low distillate output of approximately two to five liters per day per square meter of irradiated surface. The achievable maximum energy efficiency for the technology of single-basin, single-slope solar stills is estimated to be 60%, based on data by Quaschning (2005). In order to reach this efficiency, the basic design was improved with the interventions described in the following paragraphs. Figure 1 shows a cross section of a solar still module.

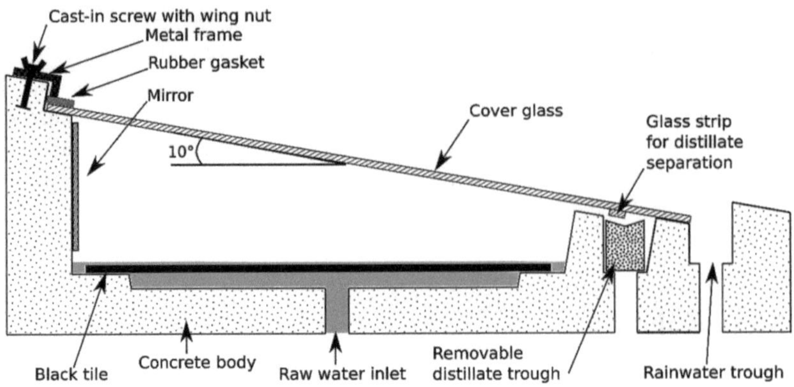

Fig. 1 Cross section of the modular solar still

5 Raw Water Dosage

Badran (2007) recommends maintaining the water level in a solar still as low as possible in order to enhance its yield. With a low raw water mass in the evaporation basin, the operating temperature of the evaporation process is elevated to the range between 50 and 80 °C and the distillate output rises significantly. The lowest raw water depth documented in literature is 2 cm (Badran and Al-Tahaineh 2005; Kalidasa Murugavel et al. 2008), which also led to the best performance results in comparison to stills with a higher water depth. In the modular solar still, the water level is decreased to approximately 5 mm as operating height. This is accomplished by using a precise floating valve in an external leveling tank which supplies the raw water to the modules via the physical principle of communicating vessels (see *System Design*). The latter also avoids desiccation of the evaporation basin which causes the incrustation of salts and minerals.

6 Evaporation Basin

The use of a removable standard ceramic tile as evaporation surface has the following advantages over the black painted metallic evaporation basins commonly used in solar stills:

- No problems of thermal expansion or warping in the absence of raw water, as observed by Zieke (2011).
- High resistance to corrosive substances.
- Easily cleanable due to enameled surface.
- Black pigments burnt into tile surface; no spalling of black paint or chemical degradation of the latter, as observed in a painted aluminum basin by Zieke (2011).

- Similar specific heat capacity as aluminum [tile: approx. 1 kJ/(kg*K), aluminum: 0.91 kJ/(kg*K)], thus similar thermal behavior.

Whereas common solar stills only have one raw water layer which gets heated and evaporated, the modular solar still has two—one layer above the black tile which is heated up rapidly by the incoming solar radiation, and another layer below the latter which heats up more slowly and serves as a thermal reservoir for nocturnal distillation.

7 Passive Energy Augmentation Techniques

Another way of raising the efficiency of the distillation process is the installation of a mirror at the rear wall of the distillation chamber, in order to reflect non-inciding solar radiation back to the evaporation basin. Abdallah et al. (2008), as well as Al-Hayeka and Badran (2004) describe a yield gain for this measure in the magnitude of 20–30%. For this purpose, the modular solar still has a small sheet of mirror glass installed (see Fig. 1). Further extension of this principle is implemented in the experimental system setup in form of an adjustable reflector made from galvanized sheet metal, which also serves as a protective cover for the modules in case of severe thunderstorms (see Fig. 7). By seasonally adjusting its inclination, a distillate gain from at least 23% is expected (Tanaka 2009). At the same time, the secondary reflector compensates for the smaller inclination of the condensation cover—the general recommendation is to adjust its inclination to the latitude of the place of project implementation for annual yield optimization (Singh and Tiwari 2004), i.e. approximately 22° for the city of San Luis Potosí. On the other hand, McCracken and Gordes (1985) emphasize that the distance between evaporation basin and condensation cover should be minimal in order to maximize the distillate yield. Therefore, the inclination of the cover glass was adjusted to 10°, in order to minimize the air volume inside the still. The smaller air volume lowers the losses of water vapor re-condensation, which occurs when the water vapor cools down in the air volume of the still before reaching the condensation cover.

8 Cover Glass and Distillate Deviation

Ghoneyem and Ileri (1997) determined that the distillate production rate of a solar still is inversely proportional to the thickness of the applied cover glass, which has to facilitate the dissipation of heat from the condensation process. Therefore, the applied glass sheet has a thickness of 3 mm as a compromise between distillate yield and mechanical stability. In order to deviate the condensed water into the collection trough, as well as to avoid contamination of the latter from raw water

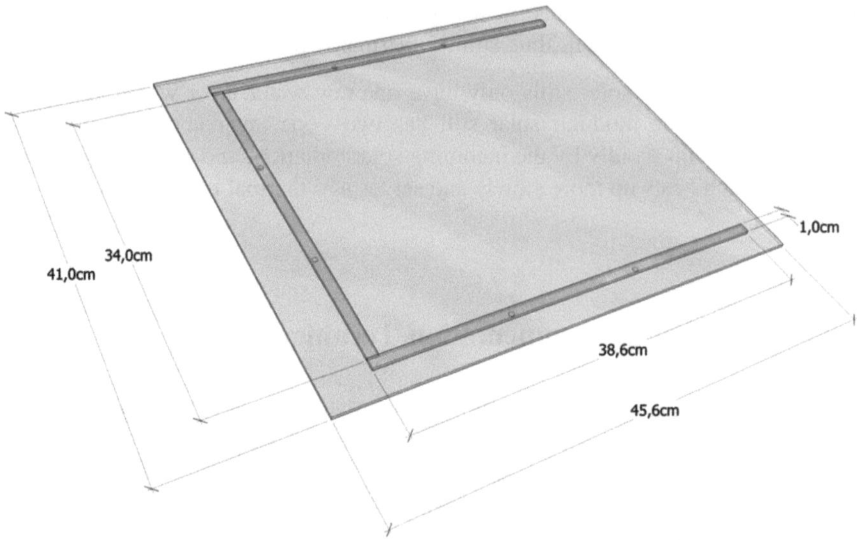

Fig. 2 Cover glass with glass strips on the inside for distillate deviation

rising at the interior walls (due to capillary effects of the concrete), a 1 cm—wide glass strip is adhered to the inside of the cover glass (see Fig. 2).

Although the strip is attached using standard silicone caulk, it is assured that the latter does not squeeze out in the subsequent clamping and drying process, in order to avoid its contact with the distilled water later on. The reason for this is that some commercially available silicones contain petroleum distillates which impose a hazard to human health when ingested. Commonly used materials for distillate deviation are either food grade silicone or rubber gaskets, but these materials are either expensive and/or difficult to obtain in rural areas and developing countries. In so far, the present approach is innovative, as it avoids the contact of the distilled water with potentially hazardous materials. For the same reason, the rubber gasket sealing the cover glass is placed on its outer surface (visible in Fig. 1). Apart from protecting the still from contamination and avoiding vapor leakage from the inside, the rubber gasket also inhibits the lateral rainwater runoff into the distillation chamber. Another advantage of the proposed design is the secure separation of rainwater from distillate by extending the cover glass towards a dedicated rainwater collection trough.

9 Construction Materials

Whereas the body of most of the solar stills used in small-scale installations is made from either sheet metal or wood, the present design uses a common composition of concrete with the relation of 1:1.4:1.6 (cement: gravel: sand). This has the advantage of exclusive use of economic and widely available materials which are estimated to last at least 20 years in continuous use and do not expose weathering problems, such as warping or rusting. At the same time, the use of special surface treatments is not necessary, which lowers the construction costs and avoids the incorporation of additional chemicals. Experiences using concrete for solar still construction exist up to date mainly for large-scale installations of deep basin stills. Therefore, the use of concrete for a shallow-basin, small-scale installation can be seen as a novelty. A layer of thermal insulation can easily be integrated into the concrete body during the casting process. For the described prototype, recycled polystyrene food trays are integrated into the bottom and side walls of the module (see Fig. 5). This provides a very economical way of insulation and makes use of an abundant waste material at the same time. The distillate collection trough can either be made of concrete or any other composition of petrous materials, in order to fulfill certain requirements of distillate remineralization.

As an economic material for the raw water inlet, as well as for the distillate and rainwater outlets, CPVC piping is used. By using screwable connectors, a piping network to interconnect the inlets and outlets of various solar stills can easily be established (see Fig. 3).

10 Distillate Remineralization

Since previous water quality measurements of a solar still determined an average pH value of 4.6 ± 0.05 (Zieke 2011), the distillate collection trough of the prototypes will be made from the same concrete mixture as the body, as a first

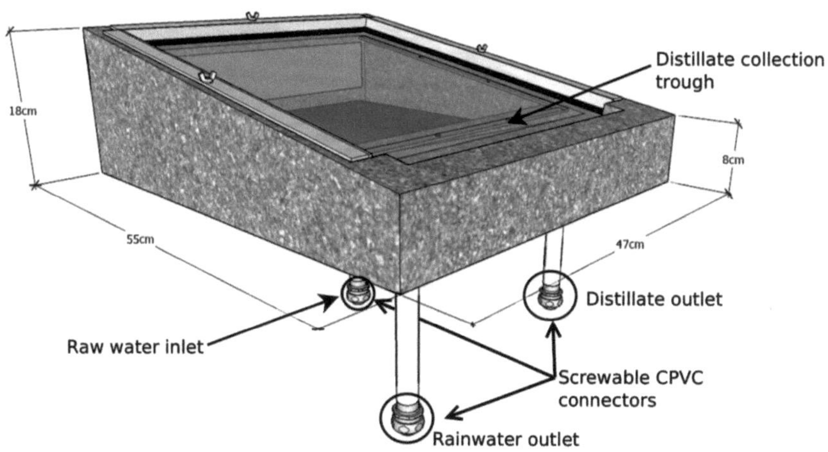

Fig. 3 Dimensions and connections of the solar still module

approach towards pH-adjustment and remineralization with calcium and magnesium carbonate. Due to the slow dynamics of solar distillation, the future integration of either a calcite or dolomite bed as described by Luptáková and Derco (2015) and Biyoune et al. (2017) into the distillate trough seems feasible.

11 Fabrication Procedure

In order to facilitate the production and reproduction of the modular solar still, the manufacturing process is based on the use of two completely dismountable molds: one mold for the body of the modular solar still and another mold for the distillate trough. By casting the trough in a separate mold, it can easily be reproduced, for example as a spare part. As construction material, standard steel panels and profiles were used with the purpose of providing an economic and at the same time durable material which permits a high precision and facilitates frequent castings without having to replace parts of the mold.

Figure 4 shows the components of the mold for the body of the modular solar still. The segmentation of the mold permits easy transportation to different places of project implementation. All parts of the mold are joined by screws and wing nuts in order to facilitate the assembly and disassembly without the need for special tools. Before the casting of the concrete takes place, the mold is treated with food-grade vegetable oil as a concrete release agent. After the casting process, the mirror is attached to the rear wall of the evaporation chamber. At this point, the rest of the components (distillate trough, black tile, cover glass, rubber gasket and metal frame) are added. Figure 5 illustrates the incorporation of polystyrene food trays in the casting process and shows one completed prototype module, which weighs approximately 50 kg.

12 System Design

It is estimated that one solar still module is capable of producing approximately 0.3–0.7 L of distilled water per day under Mexican climate conditions, considering an active area of approximately 0.11 m^2 [based on 3–6 L/m^2d, after Zieke (2011)]—the latter resulting from the dimensions of the tile (33 $\times$ 33 cm). This quantity is not sufficient for human consumption and has to be augmented by connecting various modules in parallel as illustrated in Fig. 6. In this manner, the distillate yield can easily be adjusted to the drinking water needs of the place of implementation.

The raw water is fed to the leveling tank either by gravity from another manually filled reservoir or directly from a pressurized water line (not drawn in the diagram). A floating valve maintains the water level at a constant height inside the tank. By means of communicating vessels, the raw water level in the tank and in the solar

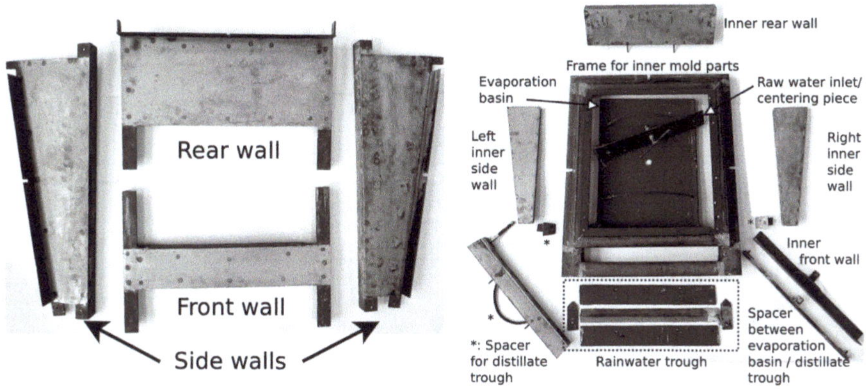

Fig. 4 Components of the mold for the modular solar still—outer frame (left) and inner structures (right)

Fig. 5 Polystyrene insulation in base and walls while casting and assembled solar still module

still modules is guaranteed to be equal. As long as the latter are perfectly leveled, one leveling tank can be used to supply a large number of them with raw water (installation of the modules in *strings*—similar to PV modules). The drain at the lowest point of the installation facilitates the routine cleaning of the modules and the raw water lines. It is mandatory to provide an emergency overflow in the raw water supply line(s) to avoid flooding of the distillate trough in case of malfunction of the floating valve in the leveling tank and to facilitate maintenance. The modules' distillate and rainwater outlets can practically be combined in two separate lines which feed the corresponding catchment tanks. Badran and Al-Tahaineh (2005) state that by using a simple solar collector to preheat the raw water, the distillate yield can be augmented by approximately 36%. Therefore, it is recommended to integrate a small-sized solar collector into the installation, where the raw water quality is adequate in terms of turbidity and dissolved solids (to avoid clogging).

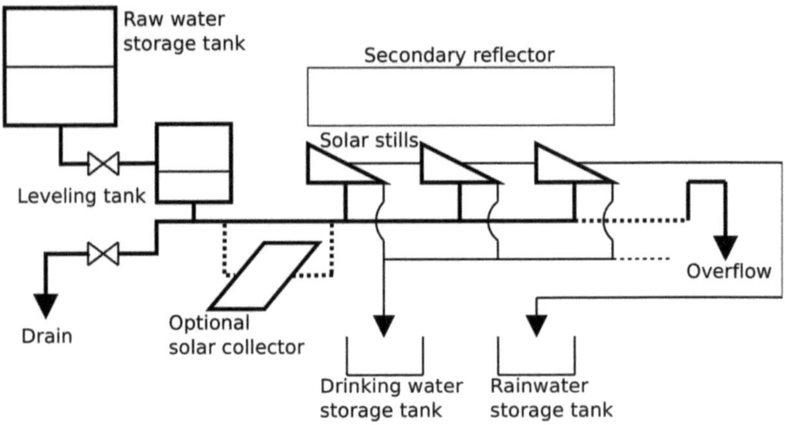

Fig. 6 Connection diagram for various modular solar stills

Although being an optional component, the integration of a secondary reflector into the system is advantageous in terms of yield maximization (redirection of sunlight) and protecting the units at the same time (as mentioned in *Passive Energy Augmentation Techniques*).

13 Operation and Maintenance

It is estimated that the proposed system can run on a semi-automated base, thus requiring manual intervention only in case of maintenance. Factors for the degree of possible automation of the system are the size of the installation, the kind of water supply (manual feeding or connection to a public water line) and the type of storage for the distilled water (manually drained collection vessel or fixed storage tank connected to the water outlets). During operation, the only tasks which have to be fulfilled manually are the adjustment of the inclination of the secondary reflector (if applied) and the cleaning of the evaporation basins of the installed modules—approximately once a month. The latter can also be done by *flushing* the entire system. Hanson et al. (2004) report satisfactory results for this technique which avoids the removal of the cover glass for cleaning the evaporation basin. The outer surface of the modules' cover glasses only has to be cleaned occasionally, depending on the climatological conditions (dust, sand). Apart from the aforementioned interventions, it might be necessary to manually lower the secondary reflector to horizontal position in case of extreme climatological conditions, in order to avoid damage of the modules' cover glasses and to lower the wind resistance of the entire assembly.

14 Experimental Setup

In order to optimize the proposed modular solar still towards distillate yield, an experimental system setup as shown in Fig. 7 is made.

Six prototype modules are mounted on a metal rack; the raw water dosage is done by a leveling tank which has a blackened metal surface, in order to facilitate raw water preheating throughout the day. Situated directly at its right side is the raw water overflow. The following three constructional variations are examined:

1. Body made entirely of concrete.
2. Body with polystyrene insulation below the evaporation basin.
3. Body with polystyrene insulation below the evaporation basin, as well as in the rear and lateral walls.

In the first configuration, the three modules from the left side work solely with the secondary reflector as passive energy augmentation technique; in the second configuration, the raw water is preheated by a solar collector, as illustrated in Fig. 8. The raw water is fed to the collector at its lowest point and is heated on its way to the outlets by flowing through blackened 13 mm copper pipes, which are fixed to a blackened metal sheet as absorber (approx. 0.7 m^2 of active area). An important detail is the integrated heat reservoir in form of cobblestones below the absorber, which facilitates nocturnal distillation by accumulating heat during the day.

Due to the dry climate during the measurement period, the rainwater output of the modules was not quantified.

All of the modules are supplied with tap water from the network of the Autonomous University of San Luis Potosí. During the measurement period of two weeks, the physical parameters listed in Table 1 were determined in intervals of 10 min.

In order to compare the performance of the modules with other systems, the energy efficiency is determined by the following equation [adapted from McCracken and Gordes (1985)]:

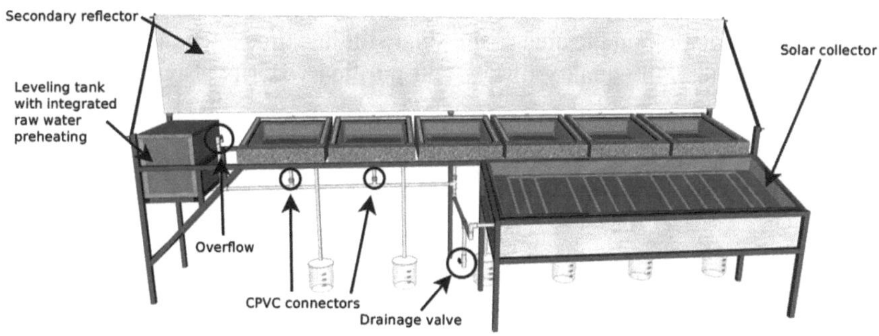

Fig. 7 Configuration of the experimental setup

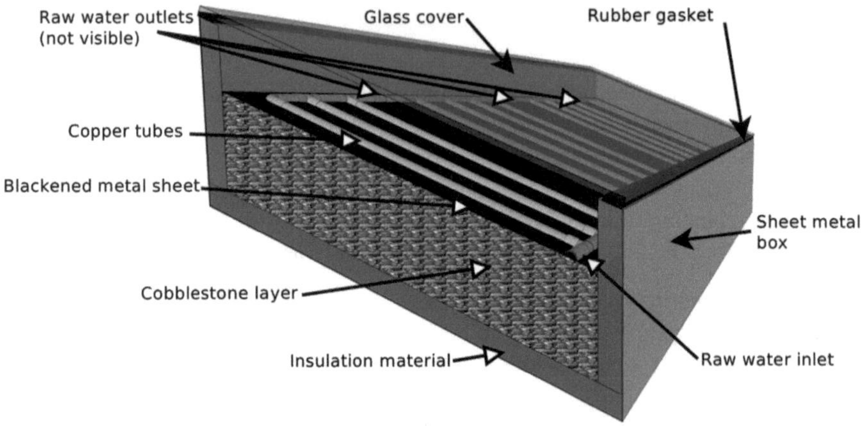

Fig. 8 Design of a simple solar collector for raw water preheating

Table 1 Determined physical parameters

Physical parameter	Unit
Raw water temperature in leveling tank	[°C]
Raw water temperature at module inlet	[°C]
Raw water temperature on tile surface	[°C]
Temperature on inside of cover glass	[°C]
Surface temperature of solar collector	[°C]
Temperature in cobblestone bed	[°C]
Ambient temperature	[°C]
Solar radiation	[W/m^2]
Distilled water output	[mL]

$$\eta_{DS} = \frac{\Delta H_{evap} * m_{D,i}}{Q_{sol,i}}$$

Where η_{DS} is the energy efficiency of the solar still, ΔH_{evap} the enthalpy of vaporization at a specific temperature [taken from Keenan and Hegemann (1978)], $m_{D,i}$ the mass of produced distillate in a time interval and $Q_{sol,i}$ being the amount of incident solar radiation on the area of the solar still in a time interval. Apart from the physical parameters, the quality of raw and purified water is also determined.

15 Results and Discussion

From the installment of the experimental setup until the conclusion of the measurement series (a total of two months), no signs of material deterioration or water/vapor leakage have been observed in the prototypes. All of the employed modules worked without errors; their main physical characteristics are resumed in Table 2.

Table 2 Physical parameters of solar still modules (PS = polystyrene, SC = solar collector)

Constructional variation	Average daily distillate production per module [mL]	Maximum water temperature on tile [°C]	Average energy efficiency [%]
1—Concrete body	299.0	61.5	27.1
2—PS in base	343.0	62.5	31.0
3—PS in base and walls	369.0	64.0	33.3
4—SC, concrete body	369.0	66.5	33.5
5—SC, PS in base	424.0	69.0	38.5
6—SC, PS in base and walls	373.0	64.0	34.1

It can be seen that the applied polystyrene insulation proved to be effective, as maximum evaporation temperature and distillate yield rise proportionally with the amount of incorporated material. Also, the application of raw water preheating contributed to higher distillate yields, although in a lesser degree than expected—the daily yield difference between variation 3 and 6 is only 74 milliliters in average. Variation 5 exhibits the major distillate production of all modules which is reflected in the most elevated raw water temperature on the tile surface. Due to an inhomogeneous distribution of the copper pipes of the solar collector, the raw water output temperatures for the modules 4 to 6 differed, with module 5 constantly having the highest values. This observation emphasizes the importance of preheated raw water (preferably at a minimum level of 70 °C) for a high distillate yield. Concordantly, the energy efficiency is slightly higher for module 5 in comparison with the other variations. Although the obtained values are in the typical range for commercial solar stills—between 30 and 45% as described by McCracken and Gordes (1985)—the goal of raising the energy efficiency of the distillation process still has to be attained. The latter can easily be accomplished by the incorporation of more insulation material into the body of the modules and by lowering the raw water volume inside the still. Especially the water volume below the tile can be reduced, as only approximately 18% of the measured total daily distillate production originate from nocturnal distillation. Projecting the modular distillate yield to liters per square meter and day, the output lies between 2.7 and 3.9 L. Considering the exclusive use of standard, non-metallic construction materials, the results mark a solid base and are easily improvable by the implementation of the aforementioned measures.

16 Water Quality and Remineralization

Table 3 shows the results of the water quality analysis. In all determined parameters, a considerable reduction or complete elimination of the contaminants was achieved. This is especially important in the case of fluorides and lead,

Table 3 Chemical parameters of raw water and distillate (N.D. = not detected, N.E. = not established)

Parameter	Raw water	Distilled water	Official Mexican limits
pH	7.7	7.6	6.5–8.5
Color [Co/Pt units]	0.0	0.0	0.0–20.0
Odor	Odorless	Odorless	Odorless
Free residual chlorine [mg/L]	0.0	0.0	0.2–1.5
Total Hardness [mg $CaCO_3$/L]	60.0	16.0	500.0
Calcium hardness [mg $CaCO_3$/L]	53.0	12.5	N.E.
Magnesium hardness [mg $CaCO_3$/L]	7.0	3.5	N.E.
Chlorides [mg/L]	26.0	3.0	250.0
Fluorides [mg/L]	4.2	0.3	1.5
Nitrites [mg/L]	0.9	0.0	1.0
Nitrates [mg/L]	9.8	N.D.	10.0
Sulfates [mg/L]	56.0	3.0	400.0
Cadmium [μg/L]	0.1	N.D.	5.0
Lead [μg/L]	22.4	N.D.	10.0
Arsenic [μg/L]	14.0	N.D.	50.0
Total Dissolved Solids (TDS) [mg/L]	246.0	38.5	1000.0
Mesophilic aerobic microorganisms [CFU/100 mL]	28800.0	N.D.	N.E.
Total coliform bacteria [MPN/100 mL]	N.D.	N.D.	N.D.
Fecal coliform bacteria [MPN/100 mL]	N.D.	N.D.	N.D.

which exceeded the official limits in the raw water. Also, all mesophilic aerobic microorganisms were eliminated.

The minimal quantity of fluorides in the distillate is to be explained by the use of tap water for the preparation of the concrete mixture. The implementation of a distillate trough made of concrete proved to be a success in terms of remineralization: the parameters of calcium and magnesium hardness could be detected, together with low sulfate values, which indicate a positive influence of the concrete on the distilled water. For future implementations, the application of either limestone and/or dolomite material into the trough is seen as feasible and highly recommended. Due to the slow dynamics of the distillation process, a thorough remineralization can be expected.

17 Material Costs

Table 4 gives an overview of the material costs per solar still unit. Additional costs, such as work force or additional components for the construction of a system, are not considered. The calculated material cost is highly competitive in comparison to

Table 4 Material costs per solar still module

Position	Quantity	Cost (MXN)	Cost (USD)
Cement	13 kg	40.00	2.00
Gravel	18 kg	2.00	0.10
Sand	21 kg	5.00	0.25
Screws, nuts, wing nuts	3 each	10.00	0.50
Silicone caulk	28 mL	6.00	0.30
CPVC pipe ½″	30 cm	4.00	0.20
CPVC connector	3 pieces	60.00	3.00
Rubber gasket	132 cm	19.00	0.95
Cover glass	1	55.00	2.75
Mirror glass	1	30.00	1.50
Black tile	1	18.00	0.90
Metal frame	132 cm	50.00	2.50
Total		299.00	14.95

commercial solutions (as presented in *State of the art*), apart from facilitating self-construction.

The materials of the mold cost about 850.00 MXN (approximately 43.00 USD). Although the given costs have to be multiplied by the number of desired units and the cost for the additional infrastructure has to be added (which depends on the size and the conditions of installation), it is important to stress that these costs only arise once. The system itself does not have any continuous operating or maintenance costs. Therefore, the price per unit of water produced during the expected lifetime of the modules (approximately 20 years) decreases rapidly and gets near zero after the first two years of operation.

18 Conclusions

The presented concept of a low-cost and low-tech modular solar still was implemented successfully and the applied materials proved to be resistant during the experimental period. The use of concrete in combination with a casting mold for fabrication provides an easy and standardized way of mass production and dissemination of the technology.

In terms of distillate yield per square meter and day, the achieved quantity of 2.7–3.9 L is average for the technology. The achieved product water quality is very satisfactory; all of the analyzed contaminants are removed successfully and their values are either below the limit of detection or much below the values established in the Official Mexican Norm for Drinking Water. Furthermore, a basic remineralization with calcium and magnesium carbonate is achieved by using concrete for the distillate recollection trough. Therefore, the produced water is suitable for direct ingestion.

The material costs of approximately 300 MXN/15 USD per solar still module are competitive in comparison to commercial systems and can be lowered further by the production in small batch series. Especially in rural areas, where the access to adequately purified drinking water is difficult, the technology has the potential for being a reliable small scale solution for the provision of a sustainable water source.

Long-term experimentation, as well as field studies are planned to assure the correct functioning of the modules and the distillation system. In order to enhance the distillate yield, more insulation material will be incorporated into the modules. Also, pigments will be applied in the concrete mixture to selectively heat or cool certain areas of the still.

Acknowledgements I want to thank the Goethe Institute Mexico City for financing this project in the context of the binational year Mexico-Germany 2016/2017 (Matching Funds). Many thanks to Florian Schmitt from *Centro Cultural Aleman de San Luis Potosí, A.C.*, who promoted the project and helped in organizational matters. My gratitude to the Autonomous University of San Luis Potosí (UASLP), for providing the necessary installations, tools and workshops. In particular, to Dr. Yuri Nahmad Molinari from the institute of physics for the latter and his constant input of ideas. Also, a very heartfelt thanks to Mr. Enrique Martinez Hernández for his constant technical mentorship and participation in the construction of the molds. To Mario Alberto Betancourt Urbina and his students from the laboratory of construction materials (Faculty of Science) for their technical advice in concrete mixtures and casting, as well as for the provision of facilities for the latter and their practical help. To Dr. Catalina Alfaro de la Torre from the Faculty of Chemistry for her mentoring and the water quality measurements. To Dr. Roberto Bartali Marchetti from the Faculty of Science for his invaluable technical advice, cooperation and guidance.

References

Abdallah S, Badran O, Abu-Khader MM (2008) Performance evaluation of a modified design of a single slope solar still. Desalination 219:222–230. doi:10.1016/j.desal.2007.05.015

Alarcón-Herrera MT, Bundschuh J, Nath B et al (2013) Co-occurrence of arsenic and fluoride in groundwater of semi-arid regions in Latin America: genesis, mobility and remediation. J Hazard Mater 262:960–969. doi:10.1016/j.jhazmat.2012.08.005

Al-Hayeka I, Badran OO (2004) The effect of using different designs of solar stills on water distillation. Desalination 169:121–127. doi:10.1016/j.desal.2004.08.013

Badran OO (2007) Experimental study of the enhancement parameters on a single slope solar still productivity. Desalination 209:136–143. doi:10.1016/j.desal.2007.04.022

Badran OO, Al-Tahaineh HA (2005) The effect of coupling a flat-plate collector on the solar still productivity. Desalination 183:137–142. doi:10.1016/j.desal.2005.02.046

Biyoune MG, Atbir A, Bari H et al (2017) Remineralization of permeate water by calcite bed in the Daoura's plant (south of Morocco). Eur Phys J Spec Top 226:931–941. doi:10.1140/epjst/e2016-60181-6

Castellanos Coutiño J (2000) Modificación a la Norma Oficial Mexicana NOM-127-SSA1-1994, Salud ambiental. Agua para uso y consumo humano. Límites permisibles de calidad y tratamientos a que debe someterse el agua para su potabilización. http://www.salud.gob.mx/unidades/cdi/nom/m127ssa14.html. Accessed 17 Aug 2017

Derickson R, Bergsrud F, Seelig B (1992) Treatment systems for household water supplies—distillation. North Dakota State University

Foster R, Eby-Martin S (2001) Solar water purification for the border: solar distillation

Ghoneyem A, Ileri A (1997) Software to analyze solar stills and an experimental study on the effects of the cover. Desalination 114:37–44. doi:10.1016/S0011-9164(97)00152-5
Hanson A, Zachritz W, Stevens K et al (2004) Distillate water quality of a single-basin solar still: laboratory and field studies. Sol Energy 76:635–645. doi:10.1016/j.solener.2003.11.010
Kalidasa Murugavel K, Chockalingam KKSK, Srithar K (2008) Progresses in improving the effectiveness of the single basin passive solar still. Desalination 220:677–686. doi:10.1016/j.desal.2007.01.062
Keenan JD, Hegemann DA (1978) Chlorination and ozonation in water and wastewater treatment. Chemosphere 1:9–28
Kinzer J, Crane A (2006) UTC Solar Distiller—Appropedia: The sustainability wiki. http://www.appropedia.org/UTC_Solar_Distiller. Accessed 17 Aug 2017
Luptáková A, Derco J (2015) Improving drinking water quality by remineralisation. Acta Chim Slov 62:859–866. doi:10.17344/acsi.2015.1590
McCracken H, Gordes J (1985) Understanding solar stills. Volunteers in Technical Assistance, Alington, Virginia
Noble N (2014) Solar Distillation. Pract Action Inf Serv https://infohub.practicalaction.org/oknowledge/bitstream/11283/314193/1/4fec63f8-1398-4963-bc82-206c1661b3dc.pdf. Accessed 20 Aug 2017
Ojinaga Santana L, Foster R (2001) Experiencias en el uso de destiladores solares en San Luis Potosí y Chihuahua. In: XXV Semana Nacional de Energía Solar—ANES. San Luis Potosí, Mexico
Quaschning V (2005) Understanding renewable energy systems. Earthscan, London
Reyes-Gómez VM, Alarcón-Herrera MT, Gutiérrez M, López DN (2013) Fluoride and Arsenic in an Alluvial Aquifer System in Chihuahua, Mexico: Contaminant Levels, Potential Sources, and Co-occurrence. Water Air Soil Pollut doi: 10.1007/s11270-013-1433-4
Singh HN, Tiwari GN (2004) Monthly performance of passive and active solar stills for different Indian climatic conditions. Desalination 168:145–150. doi:10.1016/j.desal.2004.06.180
SolAqua LLC. (2017) Rainmaker™ 550 Solar Water Distiller. http://www.solaqua.com/solstils1.html. Accessed 5 Feb 2017
Tanaka H (2009) Experimental study of a basin type solar still with internal and external reflectors in winter. Desalination 249:130–134. doi:10.1016/j.desal.2009.02.057
Zieke G (2011) Development of a low-cost, high-efficiency solar distillation unit for small-scale use in rural communities

Author Biography

Gregor Zieke is a graduate engineer in Environmental Engineering and Renewable Energies (HTW Berlin, 2008), Master in Environmental Sciences/Technology and Resources Management in the Tropics and Subtropics (UASLP and Technical University of Cologne, 2011). Professor at the Center for Applied Investigation (CIAHab) of the Habitat Faculty (UASLP) between 2012 and 2013; lines of investigation: Renewable Energies and Technology for Rural Communities. Cooperations at Consultoría Ambiental Viveros de Coyoacán, S.C. and Organi-K Integradora de Servicios y Sistemas Ambientales, S.A. de C.V. between 2013 and 2015 in the area of Research of Sustainable Technologies. Cooperation in the joint project between UASLP and the Interamerican Development Bank *Solar thermal central for the Hospital Dr. Ignacio Morones Prieto* in San Luis Potosí during 2015 and 2016.

Vulnerability and Coping Capacity of the Population of Low Balsas to Extreme Environmental Hazards to a Sustainable Development: A Case Study of the Population of Cuitaz, Zirandaro, Guerrero

Liliana Aguilar-Armendáriz, Angel García-Pineda and Saray Bucio-Mendoza

Abstract On September 2013, hurricanes Ingrid and Manuel passed through the region of low Balsas causing great havoc. The region was also hit by the Chikungunya fever in 2015 and is considered a zone ruled by organized crime. Despite these prevailing circumstances, the present research was carried out. Following Beerman et al. (The role of institutions in the transformation of coping capacity to sustainable adaptive capacity, 2, 86–100, 2013), the present research was conducted trying to elucidate the coping strategies upon which the community had relied during the recovery from the 2013 flood and whether or not they had acquired adaptive capacities with special regard to the institutional role in it (Lebel et al. in Measuring vulnerability to natural hazards, United Nations University, Hong Kong, pp 359–379, 2006). The research highlighted social features and coping strategies of the Cuitaz, a community in Zirandaro, Guerrero which showed their strong dependence on external help to recover because of its marginalization. Results showed the lack of intervention of formal institutions to mitigate, prepare or prevent future similar scenarios. It was also found that during the reconstruction, the

L. Aguilar-Armendáriz (✉)
Faculty of Economics, Autonomous University of San Luis Potosí,
Av. Pintores S/N Col., Burócratas, SLP, C.P 78213 San Luis, Mexico
e-mail: liliana.aguilar.a@gmail.com

A. García-Pineda
Independent Researcher, Abasolo 374-2. Centro Histórico, C.P 58000
Morelia, Michoacán, Mexico
e-mail: gapi_angel@hotmail.com

S. Bucio-Mendoza
Center for Research in Environmental Geography, Autonomous University of Mexico,
Antigua Carretera A Pátzcuaro, Ex Hacienda San Jose de La Huerta 8701, C.P 58190 Morelia
Mich, Morelia, Michoacán, Mexico
e-mail: saraybucio@gmail.com

W. Leal Filho et al. (eds.), *Sustainable Development Research and Practice in Mexico and Selected Latin American Countries*, World Sustainability Series,
https://doi.org/10.1007/978-3-319-70560-6_8

government could have left the population in dire need out of the programs contrary to international provisions which Mexico has signed.

Keywords Vulnerability · Low balsas · Social structure · Formal institutions

1 Introduction

The vulnerability of a society to extreme environmental hazards is understood as the susceptibility to damage in the face of the emergence of events, such as floods, droughts or epidemics among others. Studies done by Cutter (2003), Cutter and Finch (2008), UNISDR (2004) have noted while working closely with populations, that some factors inherent to them exacerbate the damages caused by extreme events. Some of them have been related to the fact that the more vulnerable the population is, the more socio-economically unprotected. In this regard, the capacities of urban and rural societies to environmental hazards have also been approached (Beerman et al. 2012; Scheuer et al. 2011; Lebel et al. 2006; Billing and Madengruner 2006; Hutton and Hacque 2003) in order to complement the vulnerability assessment and/or to delineate better strategic planning.

Following Beerman et al. (2013), the present research tried to elucidate the coping strategies upon which the community had relied during the recovery from the 2013 flood and whether or not they had acquired adaptive capacities for future similar scenarios.

Understanding the coping strategies that a community can deploy in the event of a disaster can provide interesting insight into the role of social organization as well as the role of formal institutions during recovery. In this study the perspective of Lebel et al. (2006), regarding the institutional influence on vulnerability was followed. According to which, institutions can modify vulnerability and risks through several pathways such as: mitigation, preparedness, emergency and rehabilitation.

The Yokohama Strategy (UNISDR 1994), the Hyogo Framework for Action 2005–2015 (UNISDR 2005) and the Sendai Framework for Risk Reduction 2015–2030 (UNISDR 2015) have tried to delineate guidelines that pursue the decrease in vulnerability and better risk management around the world. By differentiating the needs between rural and urban areas they have emphasized the key role of formal institutions in such task. The Sendai Framework for Risk Reduction 2015–2030 establishes four priorities all pertinent actors should consider in order to diminish the vulnerability of the population: (1) The understanding of disaster risk in order to prevent, mitigate and respond in case of a disaster, (2) Strengthening the governance of disaster risk in order to promote collaboration among institutions to enforce mechanisms that reduce disaster risk, (3) Investing on disaster risk reduction that also promotes job creation, innovation and growth, (4) Enhancing preparedness in case of a disaster to be able to provide an effective response and a better recovery.

2 Understanding Vulnerability from Coping Capacities

According to Beerman et al. (2012), there was a bridge that needed to be crossed in order to understand how institutions accounted for the transformation of coping capacities into adaptive capacities which involved the preparation for future possible scenarios starting from a familiar one.

Coping capacities are defined as the manner in which people use available resources to face the adverse consequences of environmental disasters. Such resources must be seen from a wide perspective since local knowledge, social capital and immigrants are also part of them and are sometimes a very important one.

The research was conducted in two phases. The first one comprised a census, conducted throughout the entire community. The questionnaire consisted of 46 questions that were focused on six areas: Education, Income, Flood of 2013, Flood of 1967, Chikungunya Fever of 2015 and Organized Crime. Even though, the main theoretical questions posed in the present research weren't answered by the census, it served to outline main aspects of the community as a whole, and as a first approach to their reality.

The second phase was carried out with the intention of elucidating the coping capacities found in the social structure and whether they had been modified by past events through a contemporary case study (CSR). It was necessary to study the population in depth and to observe the different manifestations of vulnerability in such a small community from an anthropological perspective. This was possible due to the close ties that exist between the second author on the locality.[1]

The CRS is linear but iterative, the process comprises six stages that are to plan, design, prepare, collect, analyze and share, with permanent feedback between all stages.

Planning is the first stage and focuses on identifying the situation that is relevant to the application of this method. It implies a previous deep reflection on the characteristics of the phenomenon as well as an extensive investigation on the available information. The census allowed the initial diagnosis.

In the second stage corresponding to the design, the research questions, the theoretical propositions, the units of analysis, the logical linking of the data to the propositions and the criteria for the interpretation of the data were refined and delimited. At this stage, the theoretical development was also reviewed in order to properly locate the framework in which the study is located. Although it is a small population, in this part, it was decided to approach the phenomenon as a multiple case study, due to the different reactions to the phenomenon that were being found. The similar assimilation of risk after the recovery from people with different socio-economical background was defined as a theoretical proposition in this investigation.

---

[1]Angel Garcia Pineda grew up there. His family relations made the interviews with the population possible since organized crime usually does not allow strangers in it's territory.

There were defined other theoretical proposals on the relationship between the disaster of 2013 and the subsequent phenomenon of Chikungunya fever in 2015 from the perception of those affected; as well as the determining role of family solidarity in the reaction to the phenomenon and in the ability to overcome the processes.

Another theoretical proposition contrasted in the study was that prevention and early warning information to extreme event would lessen the damage. For the verification of these propositions, it was considered to make the approximation through several cases, in order to facilitate the comparison on the differences in how vulnerability manifests itself in the population.

The analysis was designed at different levels: (a) housing regarded as whether one or more families live in it; (b) the groups of households-families included in each case; (c) the entire Cuitaz community. These levels of analysis are due to the need to verify the theoretical propositions and make a comparison. The quality criteria regarding the validity of constructs, external validity and reliability of the research proposed by Yin (2014) were strictly taken care of.

On the third stage, preparation, the protocol of the case study was reviewed and it was given a final form with the final selection of the cases to be observed. Special reflection was made on the possible risks of the investigation and agreed on the preventive and training measures required to carry out the field work successfully. In the fourth stage, corresponding to the collection of evidence, the method admits multiple sources. However, by the nature of the study the main source in these cases was the face-to-face interview, which was contrasted with the available documentary information corresponding to the different units of analysis, databases to characterize the territory, and the researchers' own experience through participant observation.

In the census phase, the instrument applied contained 46 questions, including socio-economical background, land tenure and production, access to services, migration, as well as housing characteristics, the experience of the 1967 flood (for those who lived it) and 2013, Chikungunya fever and organized crime. In the second phase, the in-depth testimony of the heads of families selected for inclusion in the three cases was collected.

The analysis is the fifth stage, the research was qualitative of inductive type. It was revised and integrated the information carrying out an interpretation according to the technique of explanation building (Yin 2014). The sixth stage is the sharing, in which the report of the case study is made, which for this occasion was done following the linear-analytic structures (Yin 2014).

The main theoretical proposition was that there were different risk assimilation patterns during the recovery of people with a similar socio-economic background revealed by the coping and adaptive capacities of individuals. The role of public institutions on prevention, mitigation and rehabilitation is a key component in diminishing the vulnerability of the population and could be reinforced by applying the coping capacities of the clanship.

It included deep interviews to seven members of the clanship in order to harness their personal experiences during the disaster, as well as their motivations during the recovery. Cases were selected based on the decisions they made during the recuperation process, the first case corresponded to the families that relocated their homes. The second and third one were composed of families that did not change their residence, although some received help from government to rebuild their homes while others did not.

It was decided that CSR was the best strategy to address this research because of the nature of the questions, and the small number of identified populations that comprised one or a few families in each observable situation. Due to its qualitative nature, the present study allowed, on the one hand, a contrasting theory with respect to environmental factors influencing vulnerability, and on the other hand, the generation of new knowledge to the community learning about disasters and how to react to them.

The study uses the tools of qualitative research, combining different sources of information with personal experience as its key component as well as knowledge of the territory and its dynamics by some of the authors. The trust derived from its relationship with the community allowed the authors and those who collaborated in the information gathering process, to have little resistance from inhabitants to answering the questions and at the same time, guaranteed a high level of certainty regarding the information obtained face to face.

It is important to mention that although the authors are emotionally linked to the case study, the research was carried out with strict scientific rigor and without manipulation of the data, since the results were presented faithfully as they were expressed by the settlers. On the contrary, these characteristics facilitated the selection of participants and their taking part in each of the three cases presented.

Likewise, this confidence derived from closeness allowed the interviewees to feel comfortable telling their experiences and expressing their thoughts in response to questions, which was of great value and relevance to the investigation. This is particularly relevant in the region of Tierra Caliente, Michoacan and Guerrero, since it has been extremely difficult to carry out the fieldwork during the last decade. It is also relevant in the social sciences that involve high interaction with the local population (that distrusts outsiders) and are under the control of organized crime.

The described characteristics of the community being studied definitely make it a singular phenomenon. In addition, the data are more observable in the investigation than the actual number of respondents, so it is more of an in-depth study than a statistical one (Eisenhardt 1989). The main unit of analysis was the vulnerability to the flood of 2013, how it was lived, what the impacts on the population were and how the inhabitants assimilated the risk in order to recover. The 1967 flood, the Chikungunya Fever and organized crime were defined as complementary units of analysis that were also expressed in different layers of vulnerability that the population had faced.

The analysis made through CRS expresses the reality lived in a specific community and the results can not be extrapolated to other contexts which represents the main limitation of the research. However, it can serve to provide a deep insight reflections on the experience of those affected and the limitations that are found in the cases reviewed to achieve real sustainable development.

In order to understand the position of the formal institutions in the disaster, the Secretary of Civil Defense of the State of Guerrero was asked to provide information obtained in the field work such as the number of people displaced, the number of refugees, the amount of money invested on the recovery and the risk reduction strategies implemented ever since.

3 Characterization of El Cuitaz, Zirándaro, Guerrero

The National Council for the Evaluation of Social Development Policy (Coneval 2012) published that in 2010, 86.4% of the population of the municipality of Zirandaro lived in poverty, and 57.4% of it was considered to be living in extreme poverty.

El Cuitaz community is located in the municipality of Zirandaro, Guerrero and it is confined on the limit with the state of Michoacan in the Tierra Caliente[2] region. It is found by the Balsas River in a zone that is also known as *Low Balsas*. According to the National Institute for Geography and Statistics (Inegi 2010), there were 141 inhabitants in 2010. However, the census accounted only for 139 inhabitants in 44 households that only had pipe borne water and light as public services.

According to the Index of Marginalization by Locality that was published by the National Council of Population (Conapo 2012), Zirandaro was classified as a highly marginalized locality in 2010.[3]

The clan has a kindergarten, an elementary school and a junior high school that are also attended by children of other nearby communities. There is a vocational school in a neighboring locality on the state of Michoacan, where the young can continue their studies.

Fourteen inhabitants do not know how to read and write, 32 have completed their basic education and only 19 have completed vocational studies.

The community is part of the Ejido San Jeronimo del Cuitaz and therefore, some inhabitants have their own parcels of land. In total, 17 households of the Cuitaz have 140 ha of the Ejido San Jeronimo, with an average of 8.3 ha each. However, inequity is also visible in the distribution of the parcels, the person that has a bigger parcel holds 30 ha, while the person with the smallest parcel owns only 3 ha. Graphic 1 shows the manner in which the total number of hectares is divided among the holders. The crops traditionally grown are corn, sorghum and sesame. However, only 35% of communal land holders continue to grow their parcels annually with an approximate income of $217.00 (US dollars) per hectare. This is because the activity is dependent on rain. In a region of scarcity of rainfall, the low yields have

[2]The Tierra Caliente Region is named after its climate of high temperatures that range between 28 and 40 °C during the year. It can reach 50 °C in May.

[3]The latest index of marginalization by locality published by CONAPO is 2010. The index shows the lack of opportunities that a population has and considers three aspects: dwelling, health and income.

pushed the rest of the communal land holders (75%) to give their lands out on rent to a melon company at approximately $150.00 (US dollars) per hectare annually. The company also provides jobs for the people during periods of cultivation and harvest.

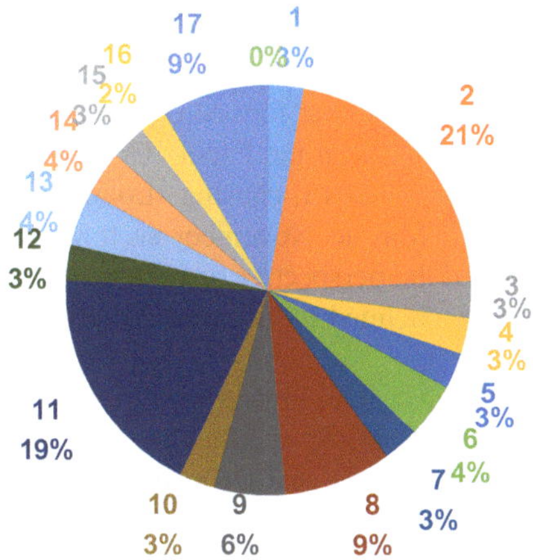

Graphic 1 Distribution of the common land parcels by households

Table 1 shows the main livelihood strategy adapted by individuals. It also shows the total number of individuals that depend on that activity, including housewives, children and elders that do not work.

Due to the low income received generally, kids start working from the early age of 12 and many of them grow up with the only option of migrating to the Unites States to live the *American dream*. It was found that 63% of families received income from their relatives in the US and 36% depended on it as their main revenue.

The census showed that eleven families have cattle, but in reduced land parcels they cannot have many, and with the constant increase in the prices of food for the livestock, their profitability level is also low. Only two families depended on cattle rearing as their main source of income, and they are also the ones that have a higher income in the community.

Table 1 Main livelihood strategy

Strategy	Activities	Average income per month	Number of households	Number of individuals
Agriculture	Rainfall-dependent agriculture	218 USD per hectare	22	80
Remittances	Remittances from family living in the US	170 USD per month	16	42
Livestock	Cattle rearing	500 USD per month	2	5

Twenty five families (56.8% of the population) received subsidies from the government in a bid to improve the income of the families in rural areas, as well as to stimulate children's education.

4 Coping Capacities After Hurricanes Ingrid and Manuel

Ingrid was a category 1 Hurricane that formed on the Gulf of Mexico and Manuel was a Tropical Storm that formed on the Pacific Ocean, they found each other on Mexican territory on September 2013, causing strong storms in the region. Both events occur naturally from May to November each year. However, their simultaneous appearance changed the perspective of some parts of the country. Guerrero, which is a state that is overwhelmed by inequity according to Coneval (2014) was one of the most affected.

According to the population interviewed using CRS, after three days of continuous storms the Balsas River started overflowing the riverbed. As a result, twenty four households were flooded (54%), and ten of them collapsed (23%). Seventy six inhabitants were displaced loosing most of their belongings in the process. Fortunately, no one died.

Nevertheless, the disaster had a huge impact on the population, some members had not fully recovered from the losses after three years and five months. The interviewees recalled that the light had been out for days therefore no media had been available to receive warning notices. The first generation of young educated adults living in Morelia, Michoacan, started organizing themselves through social media in order to help their community in the shortest possible time. Immigrants living in the United States followed their actions.

Table 2 shows the three coping strategies recalled in the CSR that were found. The first one accounted for the initial actions taken by the community. The second one was related to the formal institutional response to the recovery and the third one to the aid provided by the migrants living in Mexico and the United States.

Solidarity and cooperation were important assets when Ingrid and Manuel stroke the community. Affective bonds were also an important asset since migrants meant an essential source of income that meant provisions and in specific contexts the rebuilding of their homes and the recovery of their lost assets.

Migrants living in Mexico started collecting funds to deliver supplies to the suffering clan. They also visited the community to distribute the provisions and started organizing the community to look for aid and assistance from local institutions. Lime was immediately obtained in the municipal head, and the cleansing started in order to prevent diseases. Later on, the government also provided machinery to remove debris. They provided moral support to the community and helped them to overcome the psychological shock of the losses faster.

A few months after the disaster, only two households had relocated while most of the inhabitants went back to their previous homes (Graphic 2). When asked if they had received assistance during the disaster, almost 98% of the community claimed that

Table 2 Coping strategies in CSR

Strategy	Activity	Number of times pointed out (max 7)
Self-organization	• Gathering individuals belongings and taking them to shelters. • Organization of shelters for the displaced • Cleaning up the community	6
Institutional response	• Distribution of provisions • Rebuilding of houses	7
Aid from migrants	• Distribution of provisions • Help for self-organization • Moral support	7

they had received assistance from neighboring communities and relatives in other places. Only after being asked about it, did they recall the assistance of the government in cleaning the unpaved streets, removing debris and distributing provisions.

During the process of recovery, 98% of the population received help from relatives living in other regions of Mexico, whereas 81% of the population reported to have received help from their relatives in the United States and 6 households were rebuilt by the government.

Thirty three households recalled suffering from intestinal and eye diseases right after the event. None of them received medical assistance. Self-medication appeared to be a frequent practice and it was constant during the recovery phase due to the long travel distance needed to be covered before reaching a doctor as well as the bad conditions of the road at that time.[4]

4.1 Chikungunya Fever

It was interesting to note that for some people, the Chikungunya fever that struck in 2015 was directly related to the flood that occurred in 2013. 137 people in the community suffered from this disease in 2015, representing 98% of the total population.

None of the people surveyed had knowledge regarding the prevention nor the treatment of the chikungunya fever. Despite the fact that they all had access to public health services, the closest care clinic was 40 min away, which meant that their access involved travelling a considerable distance.

4.2 Flood in 1967

Fourteen households had at least one member of the community that lived through the flood back in 1967. According to their narrations, at that time, all households

[4]The closest doctor is located in Aratichanguio, a locality 40 min away from el Cuitaz.

were built with adobe, hence they all collapsed. Those who remembered the episode more clearly, recalled that the population did not receive any help during the disaster period nor during the recovery.

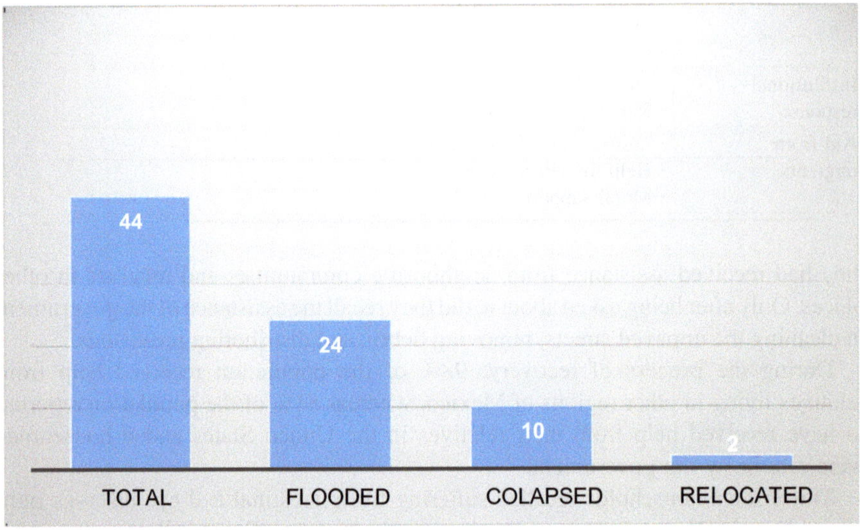

Graphic 2 Household after ingrid and manuel

Nonetheless, people returned to live in their previous location due to the proximity to the river. Back then, the community lacked access to pipe borne water, thus, inhabitants had to carry water to their homes in order to satisfy their needs. The closeness to the river was fundamental to their daily activities (Graphic 3).

4.3 Organized Crime

Police forces had no presence in the locality. The Comisario was in charge of keeping the order in the community. He is democratically elected every year in meetings with representatives of the municipality that imprint certainty and legality. However, he is left without means to enforce the law. Homicides can be arranged as vengeance without further consequences which probably makes El Cuitaz somewhat peaceful.

As a result of the low income labor in the region, organized crime activities have emerged. Young boys are easily drawn to a new way of making a living in such a precarious environment by: becoming assassins and saddled with the responsibility of protecting the chief of the ruling drug cartel in the region or watching over the drug crops for $60 US dollars per week. Future consequences, like being killed

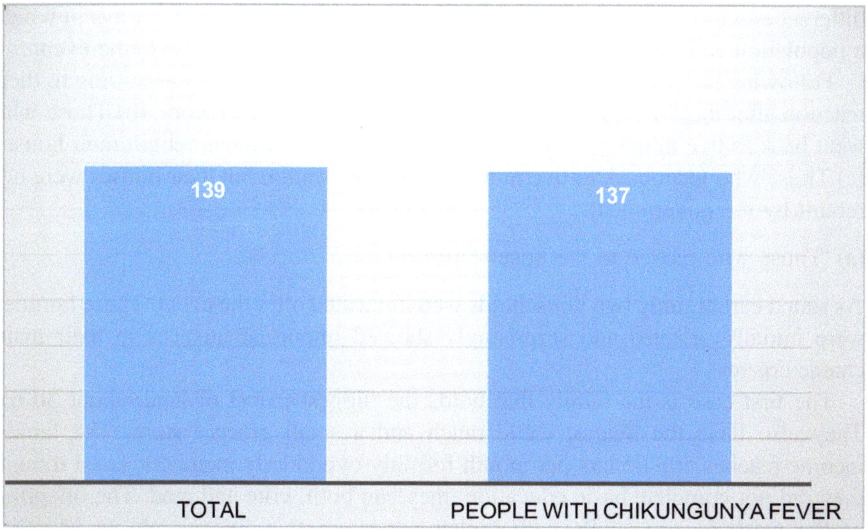

Graphic 3 Chikungunya fever

during a confrontation with other cartels or being arrested, are not put into consideration. After all, having access to fire guns may also increase their popularity and respect.

All of the people surveyed acknowledged the existence of an armed group in the community. Five of them reported to have had at least one conflict with them, but none of them explained why the conflict arose. None of the respondents admitted to having a member of their family in such groups therefore this activity could not be taken into account as a main livelihood for any household.

Two members of the community agreed to the presence of such groups within the community because they took into account the fact that they defended them from other armed groups in the region. On the contrary, the rest of the people surveyed answered that the presence of such group encouraged disputes with other groups. Furthermore, some pointed out that they sold drugs in the locality.

5 Multi-case Study: Vulnerability in El Cuitaz

Numerous studies have revealed that when a disaster occurs the vulnerable population in socioeconomic terms is also more affected. In this respect, Wisner (1998) stated that individuals that face an extreme event also cope with different constraints such as no access to financial credit, scarcity of income or assets, lack of access to information, among others.

In El Cuitaz, it was found that not all people were influenced in the same way by the flood that occurred in 2013. Therefore, special attention was paid to the differences within the community, in order to understand the diverse ways in which a population can become more or less vulnerable or resilient to extreme events.

Following the MCSR, the study was divided into three groups according to their reaction after the event: (a) Those who moved to the upper territory, (b) Those who went back to live in their previous location and the government rebuilt their houses (c) Those who went back to live in their previous location but their houses were not rebuilt by the government.

(a) Those who moved to the upper territory

As stated earlier, only two households were relocated after the event. These families were initially selected and surprisingly showed important nuances in their main characteristics.

The first case is the family that holds the biggest parcel of land, about 30 ha. They also have the biggest cattle ranch and a small grocery store. The family income reaches 500 dollars per month for only two elderly members. Even though they did not complete basic education, they can both write and read. The offspring of the marriage are adults with higher education that live mainly in Morelia, Michoacan and actively participated in the organization of the recovery aid.

They were able to seek refuge in their own store that is located in a higher territory. Shortly after the disaster and with the help of their sons, the couple started building a new home for the family next to the little shop. Two years after the event, the government built another house for them on the other side of the store.

The other case refers to a two-member family whose main income depends on their sons living in the United States and it is about $100 US dollars per month. They are both elderly and none of them can neither read nor write. Their previous home was among the first to be flooded and they lost most of their possessions due to the lack of warning information. Their home which was made of adobe collapsed. They took refuge with relatives living in higher territories. They had a small plot next to the highway which they fenced when they were young. There, they were able to build a house of sheets and cardboard where they lived for about one year and a half while the government built them a home.

In both cases, the relocation was due to several reasons, including the recognition of the possibility of occurrence, the thought that as they grew older they would have a diminished coping capacity, and the fact that they owned a lot in the upper territory. Gastrointestinal and eye diseases were present in both families as well.

In 2015, both families coped with the Chikungunya fever and opted for self-medication as a way of improving their condition. By the year 2016, the husband of the individual in the second case died, and it was believed to still be a consequence of such disease.

In both cases, they recognized the null prevention information they received from the institutions after the flood in 2013 nor during the epidemic they struggled with in 2015, plus the lack of infrastructure and the high cost of the medicine that hindered the access to health services.

(b) Those who went back to their previous location and the government rebuilt their homes

Only one household in this case was interested in participating in the study. The family is composed of an elderly person and her adult daughter. Even though they have 8 ha of land and a small cattle, their main income comes from some relatives in the United States, which is about 200 US dollars per month. One member of this family who has higher education and lives in Morelia, assisted in organizing the aid for the recovery.

When the disaster occurred, they did not envisage that the river would overflow its bed. When they realized that the flood was imminent, it was too late and they lost many of their belongings. Their relatives in the community helped them evacuate and they sought refuge with family members that were not affected by the flood in the same locality.

The previously mentioned house was built with adobe and tile roof, hence it collapsed. When they were able to return to their previous plot, they built a temporary home with sheets and cardboard, until the government rebuilt their homes. They also suffered gastrointestinal and respiratory illnesses after the flood.

When asked why they went back after the disaster, they mentioned that there weren't other lots available for sale in the community. They also stated that since their parcel of land had also been flooded, they couldn't consider building their homes there either.

They believed that the new house provided by the government did not really compensate for their losses because they deemed the material that was used as inappropriate. Furthermore, the roof made with metal sheet was unsuitable for such hot weather, so they couldn't stand the heat in the house during the day.

As a result of the Chikungunya fever, they both became infected and sought help in the public clinic. However, they didn't obtain prevention information during the flood of 2013 and the epidemic of 2015.

(c) Those who returned to their homes but did not obtain government's aid to rebuild their homes

The four interviews in this case were divided into two segments. The first one comprised of those who returned to their homes that had not collapsed and those who returned to a collapsed home and received not receive help for its reconstruction.

During the storms preceding the disaster, the interviewees dwelled on the fact that there was no warning information, thereby, provoking the dismissal of the event. It resulted in the loss of many of their material possessions. Two of the three families whose homes didn't collapse have an income of at least 300 US dollars per month. One of them has a family member who has undergone higher education and now teaches at the local elementary school. They were able to replace their lost goods within a few months. The third family interviewed is made up of two elders whose main income is dependent on their offspring in the United States. They took refuge at a relative's home and got help from the community to get their things out of their house, luckily for them, they did not lose much of their belongings.

These three families did not consider moving to upper territories because their homes represented a valuable asset. One of them mentioned that if her house had collapsed, her husband would have become ill because it meant everything to them. Their houses represented a major part of the stability achieved after many years of hard work and since they were built with brick and cement, they did not collapse.

They were also confident that the event would not re-occur soon. Some of them were present in the flood of 1967, but still believed that the events were isolated and would not happen soon.

For those who went back to their previous location which was a collapsed home and were unable to obtain help from the government for rebuilding, the story is different. Regarding this particular family, marginalization played a key role during the recovery. Due to the inability to prove that a little room was their dwelling, two young parents and their kids were left out of the government's reconstruction program.

Tied to the lowest income in the region's labor market, the possibility to rebuild a home for themselves had been futile by the time the research was conducted. They had been living in improvised homes made with metal sheets and cardboard since then. In this case, the government's aid was unable to bridge the gap of inequity, by widening it instead.

6 Conclusions

The inhabitants of El Cuitaz never expected a flood of the magnitude experienced in 2013. The incredulity combined with the total lack of warning information exacerbated the disaster by the loss of goods that the situation provoked.

Coping capacities are circumscribed to preexisting conditions related to income, health and education. Such preexisting conditions were the determinant factors in the decision to relocate or return to the old locations.

The research showed that special attention must be paid to the low income that has prevailed in the locality and that underlies vulnerability patterns, limiting the coping capacity and the post-disaster recovery.

On one hand, it was observed that those who had relatives in other cities were able to recover sooner than those who did not. Therefore, El Cuitaz is a community that depended on external aid as one of their main coping strategies.

Limited opportunities and low income have generated a breeding ground for organized crime that torments and narcotizes the population. The long distance needed to be covered in order to access medical services and the high costs of the medication, limited the coping capacity of the population to epidemics and post-disaster illnesses. Such situation have also pushed the inhabitants to indulge in self-medication as a common practice which can bring about serious health problems in the future. However, for many of them, it is the only possible way to relieve their diseases, thus, reinforcing vulnerability patterns.

In a community with a high degree of marginalization, inequities are also visible. The common land was distributed many years ago and no other alternatives to development have been stimulated in the region. This issue has enhanced the migration to the United States. Remittances, have therefore become a crucial income for many members of the community, especially among the elderly.

It can also be inferred that those who sent their offspring to bigger cities to pursue higher education saw the dividends of the invested effort in them since they were the first to offer help to the community in need. The aid consisted not only of provisions, but also of moral support and self-organization. Those descendants that have remained in the locality were unable to help their relatives overcome the situation.

Understanding risk was a main driver to relocate, however, the decision was taken based on collapsed homes. Those homes that didn't collapse were seen as solid structures that represented a lifetime effort that cannot be easily duplicated in a different location given their low income.

In a highly marginalized community such as El Cuitaz, it is imperative to propel development strategies that help people obtain better paid jobs and easier access to medical services. These would also help diminish disaster risk and limit the influence of organized crime in the clan.

The ones that had to go back to a collapsed home without any reconstruction aid are the ones in dire need. They were left out of the recovery plans and had no relatives in other cities, meaning that their recovery is not foreseeable in the near future.

The research showed that formal institutions in Mexico haven't been following the Sendai Framework for Disaster Risk Reduction 2015–2030 (UNISDR 2015) since the community didn't receive any warning information before the event or during the recovery to improve the response and preparedness for similar future scenarios. The reconstruction programs have worked for part of the community, but the aid has arrived too slowly and part of the population has been left out, thus, increasing their vulnerability. The decisions of whether to relocate their dwellings or not during the recovery were made by the actors, without taking into account institutional information on the risks they were facing.

A population without opportunities is partially blinded on its risk assimilation and therefore, condemned to reproduce vulnerability patterns. Public institutions have an important role to play in this respect. The Civil Defense Secretary in the state of Guerrero took two months to respond and provided scarce statistics, which encompassed the studied municipality and did not match the ones that were obtained in the field work. It may be possible that such information wasn't even generated or at least was not easily made available, thereby, contradicting international guidelines established by the United Nations.

The role of formal institutions is crucial for risk management and in this case, it was found that institutions owned by the government have not provided warning and recovery information that would help people prevent future disasters and manage risk in their daily decisions.

References

Beerman R, Quinn C, Paaavola J (2012) The role of institutions in the transformation of coping capacity to sustainable adaptive capacity. Environ Dev 2:86–100

Beerman R, Quinn C, Paaavola J (2013) Identifying drivers of household coping strategies to multiple climatic hazards in Western Uganda: implications for adapting to future climate change. Sustainable Research Institute Paper No. 51. University of Leeds

Billing P, Madengruner U (2006) Overcoming the black hole: outline for a quantitative model to compare coping capacities across countries. In Birkman (ed) Measuring vulnerability to natural hazards. United Nations University, Hong Kong, pp 403–414

Conapo (2012) Índice de Marginación por Localidad 2010. Consejo Nacional de Población. http://www.conapo.gob.mx/work/models/CONAPO/indices_margina/2010/documentoprincipal/Capitulo01.pdf. Accessed June 2016

Coneval (2014) Medición de la Pobreza por Entidades Federativas. Consejo Nacional de la Evaluación de Política de Desarrollo Social. Gobierno Federal. http://www.coneval.org.mx/InformesPublicaciones/Documents/Informe-pobreza-Mexico-2014.pdf. Accessed June 2016

Coneval (2012) Informe de Pobreza en México: el país, los estados y sus municipios, 2010. Gobierno Federal. Consejo Nacional de la Evaluación de Política de Desarrollo Social. http://www.coneval.org.mx/Informes/Coordinacion/INFORMES_Y_PUBLICACIONES_PDF/Informe_de_Pobreza_en_Mexico_2010.pdf. Accessed July 2016

Cutter S (2003) The vulnerability of science and the science of vulnerability. Annals Associ Am Geogr 93:1–12

Cutter S, Finch C (2008) Temporal spatial changes in social vulnerability to natural hazards. Proc National Acad Sci United States of America 105(7):2301–2306

Eisenhardt K (1989) Building theories from case study research. Acad Manag Rev 14:532–550

Hutton D, Hacque C (2003) Patterns of coping and adaptation among erosion-induced displaces in Bangladesh: implications for hazard analysis and mitigation. Nat Hazards 3:405–421

Inegi (2010) Censo de Población y Vivienda. Instituto Nacional de Estadística y Geografía. http://www.inegi.org.mx/sistemas/consulta_resultados/iter2010.aspx?c=27329&s=est. Accessed July 2016

Lebel L, Nikiitina E, Kotov V, Manuta J (2006) Assessing institutionalised capacities and practices to reduce the risks of flood disaster. In: Birkman (ed) Measuring vulnerability to natural hazards. United Nations University, Hong Kong, pp 359–379

Scheuer S, Haase D, Meyer V (2011) Exploring multicriteria flood vulnerability by integrating economic, social and ecological dimensions of flood risk and coping capacity: from a starting point view towards an end point view of vulnerability. Nat Hazards 58:731–751

UNISDR (1994) The Yokohama Strategy. United Nations Office for Disaster Risk Reduction. http://www.unisdr.org/files/8241_doc6841contenido1.pdf. Accessed Jan 2016

UNISDR (2004) Living with risk: a global review of disaster reduction initiatives. United Nations Office for Disaster Risk Reduction. http://www.unisdr.org/files/657_lwrsp.pdf. Accessed Jan 2016

UNISDR (2005) The Hyogo framework for Action 2005–2015. United Nations Office for Disaster Risk Reduction. http://www.unisdr.org/2005/wcdr/intergover/official-doc/L-docs/Hyogo-framework-for-action-english.pdf. Accessed Jan 2016

UNISDR (2015) The Sendai framework for risk reduction (2015–2030). United Nations Office for Disaster Risk Reduction. http://www.unisdr.org/files/43291_sendaiframeworkfordrren.pdf. Accessed Jan 2016

Wisner B (1998) Marginality and vulnerability. Why the homeless in Tokyo don't 'count' in disaster preparations. Appl Geogr 18:25–33

Yin R (2014) Case study research design and methods (Applied Social Research Methods Series). Sage Publications, London

Transference of Ecotechnology in Disadvantaged Regions of Mexico, Towards Sustainable Development

Lorena del Carmen Alvarez-Castañon, Daniel Tagle-Zamora and Maricruz Romero-Ugalde

Abstract The provision of ecological technology has been one of the strategies undertaken by the Government of Guanajuato, Mexico. Their ecotechnology programs include ecological artefacts that contribute to home improvement. The goals of program have been achieved (number of ecotechnology units installed); however, the impact has not been as high as expected due to high social rejection by the beneficiaries. Our hypothesis is that ecotechnology adoption failures are associated with the absence of a process that follows environmental education sustained in a process of transference of ecotechnology (TET) that would facilitate social adoption. The aim of this paper is to analyse the process of TET in two municipalities—*Penjamo* has a high rejection of ecotechnology and *Tierra Blanca* has a low rejection—and to determinate the factors that influence its social adoption. The results show that there are exogenous and endogenous factors that influence the social adoption of ecotechnologies. Hence, TET involves a social process of multilevel negotiation because its implementation depends on technical issues and on intervention strategies—these should consider social, cultural and political aspects. The challenge is to harmonize the vernacular knowledge of the region with the technical knowledge to improve socio-technical capabilities that promote development.

Keywords Transference of ecotechnology · Social adoption · Disadvantaged areas

---

L. C. Alvarez-Castañon (✉) · D. Tagle-Zamora · M. Romero-Ugalde
University of Guanajuato, Leon Campus, Leon, GT 37670, México
e-mail: lc.alvarez@ugto.mx

D. Tagle-Zamora
e-mail: da.tagle@ugto.mx

M. Romero-Ugalde
e-mail: mromero@ugto.mx

W. Leal Filho et al. (eds.), *Sustainable Development Research and Practice in Mexico and Selected Latin American Countries*, World Sustainability Series,
https://doi.org/10.1007/978-3-319-70560-6_9

1 Introduction

Over the last few years, the ways technology can, or should, contribute to social inclusion processes and the sustainable development, have been debated (Fressoli et al. 2013). Numerous studies argue that technology transfer (TT) is feasible in different regions where socio-environmental problems emerge (Burgos and Bocco 2016). The challenge is the social adoption of technology, above all in emerging areas, such as Mexico as regional imbalances are frequent in the coexistence of regions with strong social and economic lags, with innovative and dynamic regions (Burgos and Bocco 2016).

The government of the state of Guanajuato in Mexico has implemented the program "Impulse of social development" to improve the welfare of the vulnerable population (SEDESHU 2015). This program is pertinent to the reality of Guanajuato because 9.8% of its population report deficiencies in the quality and spaces of their homes and 14.9% lack of access to basic services in housing (CONEVAL 2010). The fourth aim of this program is to integrate ecotechnology in housing spaces that are not capable of satisfying a basic well-being line (SEDESHU 2015); these green technologies purport to mitigate environmental issues and provide basic services (UNDSD 2000). In the program, the installed ecotechnologies are: solar heaters, photovoltaic panels, rainwater harvesting, ecological stoves, concrete-iron cisterns, bio-digesters and dry baths (SEDESHU 2015).

The goal of the program has been accomplished, if you only measure the number of ecotechnology units installed. However, some indications show that there is resistance from the population in the use of these ecotechnologies, apparently, these could be generating local socio-environmental problems, even greater than the problems before installation. It is assumed that the failure is that the program is limited to the installation of the ecotechnology, without the execution of a technology transfer (TT) process that guarantees the social adoption of these devices. TT is part of a plan of adoption, assimilation, and technological learning (Hamidi and Benabdeljalil 2013; Jasso 1999). The TET is a matrix process based on the environmental education of the actors involved—users, local leaders, installers, among others—to achieve the effective assimilation of the technology, to build capabilities and new local knowledge; this is integrated into a series of transitional stages that favour the efficient use of the ecotechnology (Fressoli et al. 2013).

TET in dissimilar regions implies a participatory model in which the local actors deal with their own interests (Herrera 2006) and the requirements of the program that promotes ecotechnologies. TT is articulated through direct communication in a multidirectional learning process (Hamidi and Benabdeljalil 2013; Heijs et al. 2007). Therefore, the TET process must be designed in line with the heterogeneity of regional conditions where ecotechnology will be installed to reduce social, economic, and technological gaps. The TET will achieve the expected results only if the actors are fully involved in the process (Herrera 2006), and it is designed as a process that motivates the construction of technological capabilities per the characteristics and the profile of the region where it is implemented.

In these regions, it is necessary to reassess the role of all the social actors involved with a participatory approach of the TET process (Herrera 2006). Here, we must go beyond the classic approach of the TT that is limited to installing the ecotechnology and to train for its use (Fressoli et al. 2013). In this paper, we propose that TET process recognize the importance of cultural aspects to increase the odds of success (Lee et al. 2013). The purpose of this study is focused on the social adoption process of ecotechnology in disadvantaged regions of Mexico. We studied the municipalities of *Penjamo* and *Tierra Blanca*, Guanajuato; *Penjamo* reports a high rate of rejection and *Tierra Blanca* a low rate of rejection of the environmental technologies implemented because there is evidence of negative externalities that have been caused by the social rejection of the beneficiaries. According to the previous statement, we formulated the research question: which factors motivate the social adoption of ecotechnologies that help these municipalities have a sustainable development?

The paper is structured in five sections; in the first, we outline the reference framework of the research; in the second, we show the background of municipalities that we have studied and in the third, the methodological strategy is shown. In the fourth section, we present and discuss the results, and we close with the conclusions in the fifth section. Based on the results, we propose that TET should move towards a participatory and holistic process, that disposes spaces where knowledge is shared beyond the investment of government's programs in the ecotechnologies; in addition, the programs should be redesigned to encourage collective learning processes and sociotechnical partnerships between the different actors involved in the implementation and operation of ecotechnology.

2 Reference Framework

Technology transfer (TT) implies a transaction between the one who has the technology and the one who will use it (Fressoli et al. 2013), it is a planned displacement of the previously mentioned technology (Herrera 2006), in this case, it will be between the suppliers of the ecotechnology and the Secretary of Social and Human Development of Guanajuato (SEDESHU for Spanish initials) with the people benefited by the program. The classic view of TT considers that the key to success in technological adoption is the transmission of technical knowledge from one individual or organization to another for its application through a certain means of communication (Rogers et al. 2001); i.e., TT is a tacit and explicit knowledge transmission process between different actors involved in the process (Dominguez and Brown 2004).

According to Bozeman (2000), the TT process should consider five strategic dimensions: the actor who transfers, the transfer method, the transferred object, the transfer receiver, and the context or environment in which it takes place. In addition to this, according to Villavicencio (1994), the TT triggers technological learning to take place in four phases: when the technology is acquired, because the supplier transmits the necessary information for its use; when the technology is put into operation, because the user must be capable to make the technology work; when the technology needs maintenance or it must be repaired; when the user needs to modify, or improve the technology.

Pirela et al. (1991) explain that technological learning switches according to the characteristics of the technological culture of the actors and of the region where this transfer occurs. The authors sustain that TT generates four technological capabilities: the ability to complete their own knowledge as new technologies are incorporated; the capability of technological interdependence to systematize their knowledge and to accumulate a technological memory so the individual does not depend on the knowledge of the person who installs the technology; the prospective capability in which they can avoid negative externalities of technology in the short, medium, and long-term; the capability to adapt the technology change with the characteristics and profile of the region where the transfer takes place.

In this study, TET occurs in vulnerable households. Therefore, we identify the TT strategies per the characteristics of the studied regions. The "learning by doing" strategy is based on solving problems on the fly; it is assumed that the gradual use of technology leads to learning and efficient use of technology (Arrow 1974). The "learning by using" strategy defends that TT progressively accumulates skills and generates knowledge through the experience of using technology, over time it achieves its efficient use (Rosenberg 1979); this strategy is a hybrid, based on the interaction between "learning by doing" and "learning by using" (Lundvall 1988), where TT is generated by a collective process of technological learning (Villavicencio 1994).

The "learning to learn" strategy seeks to build specialized skills, since those involved in the TT process appropriate new knowledge and combine it with their own knowledge, it is possible to make efficient use of technology (Stiglitz 1987); this mix motivates the technology to solve social and environmental issues (Fressoli et al. 2013; Ilgin and Gupta 2009; Truffer and Coenen 2012). Hence, the relevance to integrate vernacular or local knowledge into the TET process as the technological culture is fundamental to achieve a successful process of TT based on the environmental education of the members of the community whose capabilities respond to different social needs.

A region with passive technological culture shows low levels of learning and poor technological memory; the technological culture is reactive when learning and technological memory is held by only a few settlers of the community. Therefore, the TT is vulnerable, fragile, and segmented. In the active technological culture, the community learn to coexist with the technology and TT is achieved with a backwards and forwards integration with the installers of the ecotechnology and within the community when the tacit knowledge is transmitted to the following generations (Lundvall 1988; Pirela et al. 1991; Stiglitz 1987).

Different studies in Mexico report that the institutionalization of "technological packages" in projects vertically designed by the government generate new local problems and the programs do not achieve the underlying objectives (Herrera 2006). Thereby, increasing the participatory processes of the TET is very important

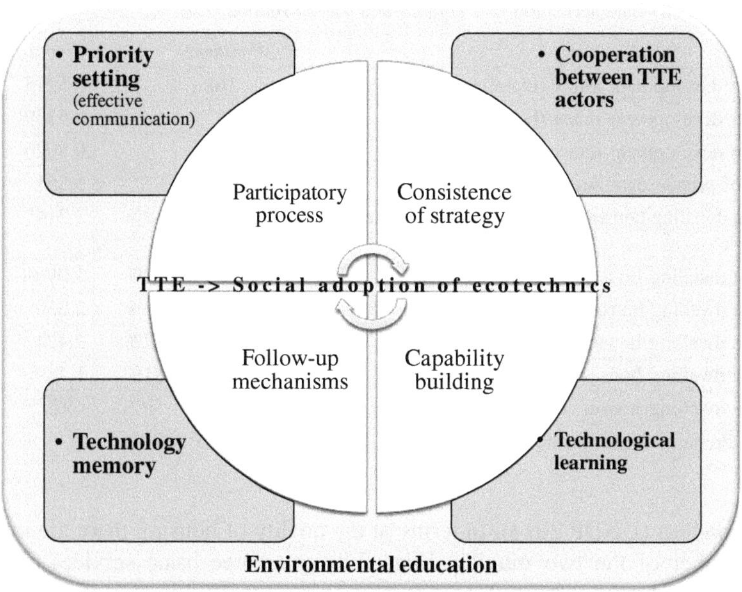

Fig. 1 TET model for dissimilar regions. *Source* Elaborated by the authors based on reference framework (Bozeman 2000; Dominguez and Brown 2004; Fressoli et al. 2013; Herrera 2006; Heijs et al. 2007; Lee et al. 2013; Lundvall 1988; Massa and Andersen 2000; Pirela et al. 1991; Rogers et al. 2001; Stiglitz 1987; Villavicencio 1994; among others)

to accelerate the construction of technological capabilities, to make technology work, and to promote its use (Massa and Andersen 2000).

Based on the literature review, Fig. 1 shows the theoretical model that explains TET in dissimilar regions such as Guanajuato. In this case, participative interaction is the basis of the TET process to motivate technological learning, which is necessary for the social adoption of these devices. The model may involve the beneficiaries in the installation of the ecotechnology, the installers in the monitoring of the use of technology, and higher education institutions in the design of environmental education programs, among others.

3 Background of *Penjamo and Tierra Blanca, Guanajuato, Mexico*

Despite the strong economic growth in the state of Guanajuato in Mexico (SDES 2016), it continues to have heavy debts in terms of social marginalization (CONAPO 2000, 2010). Specifically, *Penjamo* reports a human development index of 0.652 and *Tierra Blanca* of 0.633; however, when it is disaggregated, the index reveals greater vulnerability in the indicators of income and health for both

Table 1 Social characterization of *Penjamo* and *Tierra Blanca*

	Penjamo		*Tierra Blanca*	
Human development index (income)	0.6168		0.5914	
Human development index (health)	0.5351		0.5339	
Human development index (education)	0.8381		0.8026	
Total of private dwellings	35,786		3,861	
Private dwelling houses that have water from the public network	31,970	89%	3,214	83%
Private dwelling houses with drainage	29,494	82%	1,096	28%
Private dwelling houses with toilet or sanitary facilities	29,937	83%	2,350	60%
Private dwelling houses with electric power	34,881	97%	3,424	88%
Private dwelling houses with refrigerators	29,249	81%	1,789	46%
Private dwelling houses with washing machines	21,297	59%	692	17%

Source Prepared by the authors based on INEGI (2010) and UNDP (2014)

municipalities (UNDP 2014). In terms of the quality of housing there are important lags, neither of the two municipalities fully guarantee basic services of potable water, sewerage, and electricity to its populations. In addition, the problem of the lack of sanitary facilities and refrigerators in homes are strong (INEGI 2010), and it is aggravated the sanitation problems imposed by these absences (Table 1).

The municipality of *Penjamo* is integrated by 782 localities; its main economic activity is in the primary sector; 67.5% of the population is in poverty and 19.7% reaches levels of extreme poverty (INEGI 2010); the most worrying indicators of social deprivation are to health services, access to food, and basic housing services (CONEVAL 2010). *Tierra Blanca* is concentrated in 99 locations; its main economic activity is temporary agriculture; 70.7% of its population is in poverty and 28.1% in extreme poverty (INEGI 2010); the most worrying indicators of social deprivation are to basic housing services and income levels—the population with incomes below the welfare line is 38.2%—(CONEVAL 2010).

4 Methodology

The research described herein is explanatory and transversal. We used a qualitative methodological strategy through the ethnographic method to approach the object of study. We are interested in bringing out into open social trends, focusing on meaning and trying to understand social behaviour around ecotechnology. This research is a part of the project "*Transformación sociocultural, uso y aplicación de ecotecnias para el mejoramiento de la vivienda de las familias vulnerables de los municipios de Penjamo, Comonfort, Apaseo el alto, Tierra Blanca y San Felipe del estado de Guanajuato*" that it was funded by SEDESHU. As part of the methodological process, quantitative data was collected on the ecotechnologies installed in each municipality, which enriches the thick description of the interactions between the different actors (Clifford 2003) involved in the TET.

The purpose of this research is to study the TET in two municipalities—*Penjamo* has a high rejection of ecotechnology and *Tierra Blanca* has a low rejection—and to find the factors that facilitate social adoption of the technology in disadvantage areas of Guanajuato, Mexico. Insomuch as, this focusing is to analyse the conditions to help the TET and the social adoption of ecotechnologies for to achieve sustainable development of this regions, so the qualitative approach through the thick description allowed us to consider the subjective configurations of the social (Clifford 2003) was supported by the four tools of data collection, which facilitated and enriched the interpretation of the studied phenomenon.

During the project, these four implements were: semi-structured interview; non-structured interview; ethnographic letter; focus group in each municipality. The Table 2 shows the mechanisms of data collection and the interviewed; in parenthesis is indicated the number of people interviewed in each tool. For example, the focus groups were conducted and semi-structured interviews were applied to those responsible for the social development management of the municipalities; some beneficiaries and suppliers were also interviewed. Participant observation was another instrument of data collection that contributed to the meaning of the phenomena that occurs in *Penjamo* and *Tierra Blanca* for the reasons explained above.

The main limitation of this paper is the coverage of the study, since only two municipalities were studied. Although the qualitative strategy limits the generalization of results, the profile of the selected cases allows to assume that these are disadvantaged municipalities that could advance to sustainable development.

Table 2 Mechanisms of data collection in the project

Semi-structured interview	Director of social development in each municipality (1)
	Promoters of ecotecnias in each municipality (average = 8 in each municipality)
	Beneficiaries and candidates of the ecotechnology programs of SEDESHU (average = 40 in each municipality)
	Civil associations that promote ecotechnologies (3)
Non-structured interview	Research centres related to ecotechnology: UNAM. UAQ and UAM
	Officials of various instances of state government (5)
Ethnographic letter	Beneficiaries of the ecotechnology program, about their experience and perception on risks, barriers and benefits of eco-technology (average = 40 in each municipality)
Focus group in each municipality	Director of social development (1)
	Promoters of ecotecnias (8)
	Beneficiaries and candidates of the ecotechnology programs of SEDESHU (average = 55)

Source Prepared by the authors

This is important to confirm the findings of the investigation and to be able to identify additional categories in relation to the TET given this profile, i.e., due to the explanatory nature of the research, although it is not possible to infer about other municipalities, the results allow to generate a concrete an idea of how the transfer processes of ecotechnologies are executed and which are the factors that influence the social adoption of technology.

5 Results and Analyses

The use of ecotechnologies in the municipalities of Guanajuato can be traced from three decades ago; however, a strong ignorance about its use is also detected (Tagle 2016). From 2013 to 2015, the government program has installed more than 2,700 ecotechnology units—39.5% of these are solar water heaters; 36.0% bathrooms; 10.5% ecological stoves; 7.0% photovoltaic panels, and 7.0% dry toilets; with a budget of approximately four million dollars (SEDESHU 2015).

The process of installation in both municipalities was homogeneous. The general sequence is: the municipal government detects that there are resources in the program; the potential beneficiaries are identified and are selected according to the degree of marginalization of their locality; fieldwork is carried out to collect information on the houses that will be benefited, likewise, any necessary adjustments to the housing for ecotechnology installation are noted; the installation program of ecotechnology is defined between promoters and suppliers; installation is performed; the installed eco-technology is presented to the beneficiaries; the suppliers teach the beneficiaries to use and to maintain the technology; the beneficiaries sign an agreement of installation; photographic evidence is collected to demonstrate the installation.

Based on the fieldwork, the main exogenous factors that were identified as hurdles that impede the social appropriation of technology are:

1. The lack of diagnoses pertinent to the reality of the municipalities, without these diagnoses ecotechnologies are installed in houses that do not need them;
2. The weak or null participation of the beneficiaries during the process impedes the dialogue between vernacular and technical knowledge;
3. The absence of an effective and efficient communication process between the installer and the beneficiary, to avoid the generation of an environment of conflict and the lack of cooperation in the community;
4. The total absence of environmental education during the process of TET to guarantee the social adoption of the beneficiaries and socio-environmental contextualization of the community;
5. The scarce articulation and communication with the different governmental institutions involved in the installation of ecotechnologies;

6. The neglect of the gender perspective in the TET process, without preconceived preferences is assumed relevant since the main contact and recipients of the technologies are women.

In both municipalities, the beneficiaries do not know the word ecotechnology. These beneficiaries demand information and training for its use and its maintenance. About the main endogenous factors, aside from the evident economic insolvency and the undervaluation of the supports of the program, we identified:

1. The low level of education of the beneficiaries;
2. The disinterest and passivity of the community for the environment;
3. The disinterest of the domestic units regarding the use and management of the ecotechnologies;
4. The lack of commitment and of appraisement of their ecotechnologies.

In *Penjamo*, the most used ecotechnology is the solar heater. The septic system is the ecotechnology with the worst problems; there was a widespread carelessness and abandonment of the septic systems installed by the government program; the users changed the purpose and they used them as cellars or refuse disposal. The main cause was the unfeasibility of use by the type of soil of the communities of the basin; however, where its installation was feasible, success cases were identified in the use of this technology. Therefore, it was evidenced that the lack of knowledge increases the rejection level of ecotechnology.

In *Tierra Blanca*, we found that there are many houses with ecotechnologies and in some cases, more than one ecotechnology per dwelling. For example, concrete-iron cisterns are used to water the backyard orchard. The communities show a high interest in ecotechnologies. However, they expressed the lack of information, training, and monitoring provided by the government program. Although they recognize the benefits from the ecotechnologies, in some cases, they also abandoned them. In the case of the septic system, there is a strong uncertainty about the bio-filter. However, many beneficiaries modified it to function as a septic tank or to connect it to the drainage network; these changes affect the environmental function but not the social function of ecotechnology.

When comparing the social adoption of ecotechnologies in the studied municipalities, the results obtained allow us to assume that the TET in the framework of the program "Impulse of development" must be redesigned based on a participatory process. A TET process where the beneficiary of the ecotechnology temporarily share part of their daily space for the incorporation of technology is required. If the beneficiary of the environmental technology is involved in the process, then the mixture of knowledge would facilitate the social adoption of these devices (Fig. 2).

In Fig. 2, we show the participatory TET model. In this process, the beneficiary is the cornerstone and the agent of change, who achieves the continuous use of ecotechnology and guarantees the permanence and functionality of the installed technology. According to Herrera (2006), families promote the use of technology, beyond the promoter or the social program that finances them. During the TET process, capability building is achieved through environmental education,

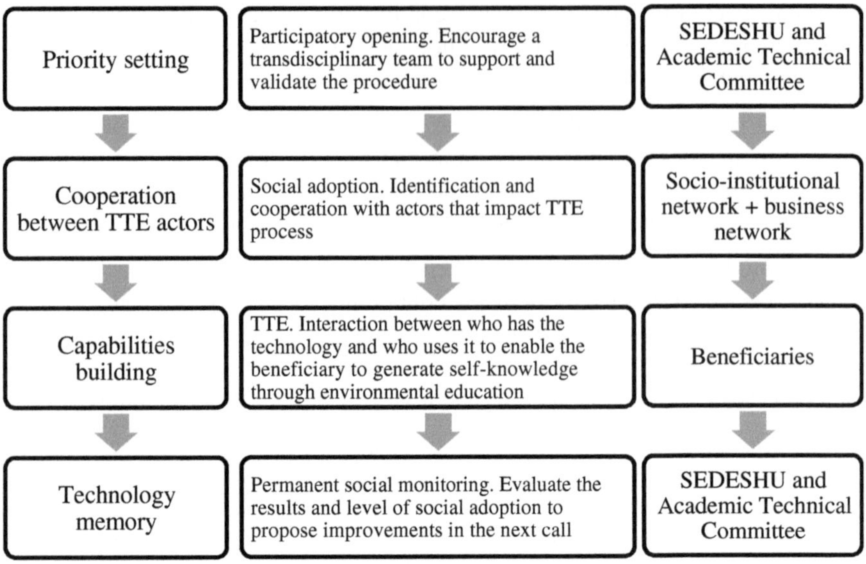

Fig. 2 Flowchart based on four participatory processes. *Source* Elaborated by the authors based on fieldwork

through both "formal" training given by the installer of ecotechnology and "non-formal" for the transmission of knowledge between the very inhabitants of the community and external actors, for example, educational institutions or NGOs, among others. According to Pieck (2011), environmental education outside the structures and the formal education system facilitate the adoption of technologies.

In this regard, there are innumerable proposals to train rural communities to build productive capabilities in their regions, which have been frequently implemented by Mexican social policy (Herrera 2006). However, the participative process of TET with environmental education must be multidirectional among groups or individuals so that it strengthens both knowledge and technical skills. Consequently, facilitating the social adoption of technology increase the probability of sustainable development in the municipalities. Training in rural regions is complex because of the nature of their social, economic, and environmental conditions (FAO 2012). In the case of vulnerable regions, the FAO (2012) considers three dimensions that condition or allow for training: the accumulation of knowledge and skills required to achieve the development of these regions; the formative process that would allow for capability building; the social context where the community develops.

The TET process must achieve the appropriate mix of objective environmental education strategies to foster capability building in communities. According to Herrera (2006), TET is a holistic, continuous and participatory process, it should be designed from the concrete reality of the communities where it is sought to stimulate the generation of knowledge and the construction of capabilities; it is through the systematization of experiences between the teacher and the student that it is possible to walk towards the achievement of actions that allows the initiation of a transformation of their reality (Pieck 2011). A participatory TET would hope to build capabilities to achieve social adoption of the environmental objects.

This is a counterproposal to traditional training processes in rural environments or rural schools; models that have been based on authoritarianism, vertical design, and have no link with reality (Mata 2013; Mata et al. 2007; Pieck 2011; Rendon et al. 2015). According to Pieck (2011), environmental education should promote technological learning on an integral, analytical, participatory and self-managing basis of their capabilities. In praxis, if we hope to achieve capability building in the region, then the TET should consider instruments such as: the transfer and mix of technical and vernacular knowledge; the preparation of technical manuals to explain the ecotechnology, thus, SEDESHU could follow the program at low cost; the technical assistance of technology providers for a considerable time after the installation of the technology; among others.

6 Conclusions

1. The main result was to identify that the factors of success and resistance are endogenous and exogenous to the beneficiaries; the mix of these factors weakens or strengthens the social appropriation of the ecotechnology.
2. In the case of *Penjamo*, a participatory society is identified, which could be the basis for generating positive synergies in the TET process; if and only if the TET is based on diagnoses pertinent to the reality of this municipality.
3. In the case of *Tierra Blanca*, there is a diversity of perceptions about ecotechnologies, although community members have "spread the word" that ecotechnologies bring economic benefits, thus, there is expectation about them. However, the most important problem is the abandonment of technology for lack of environmental education.
4. The results showed that TET involves a social process of multilevel negotiation because the functioning of ecotechnologies depends not only on technical issues of technology transfer, but also their intervention strategies. Intervention strategies should consider social, cultural, and political aspects of the region where the transfers take place.
5. The challenge is to harmonize the vernacular knowledge of the region and technical knowledge to improve socio-technical capabilities that promote their development; thus, the participative learning and the spaces of negotiation of knowledge between the TET actors would allow the empowerment of the users of the technology (Fressoli et al. 2013) in the TET process.
6. The TET must move towards a participatory process where the users of technology are considered active players in transfer processes.

Acknowledgements The authors wish to thank the Directorate for Research Support and Postgraduate Programs at the University of Guanajuato for their support.

References

Arrow K (1974) The limits of organization. Norton and Company Inc, USA, pp 13–80

Bozeman B (2000) Technology transfer and public policy: a review of research and theory. Res Policy 29(4–5):627–655

Burgos A, Bocco G (2016) Redes de conocimiento en contextos territoriales disímiles y contrastantes. In: Rodríguez J, Alvarez L, Tagle D, Coronado J (eds) Desarrollo desde lo local y dinámicas territoriales. Fontamara, Mexico, pp 61–85

Clifford G (2003) Interpretación de las culturas. GEDISA Editorial, Spain, pp 17–40

CONAPO (2000) Consejo Nacional de Población. Índice de Marginación por municipio, 2000. Mexico: CONAPO. http://www.conapo.gob.mx/work/models/CONAPO/indices_margina/indices/pdfs/AnexoB.xls. Accessed 31 Jan 2017

CONAPO (2010) Consejo Nacional de Población. Índice de Marginación por entidad federativa y Municipio. Mexico: CONAPO. http://www.conapo.gob.mx/work/models/CONAPO/indices_margina/mf2010/CapitulosPDF/Anexo%20A.pdf. Accessed 31 Jan 2017

CONEVAL (2010) Consejo Nacional de Evaluación de la Política de Desarrollo Social. Informe anual sobre la situación de pobreza y rezago social. Mexico: CONEVAL. https://www.gob.mx/cms/uploads/attachment/file/47984/Oaxaca_313.pdf. Accessed 31 Jan 2017

Dominguez L, Brown F (2004) Medición de las capacidades tecnológicas en la industria mexicana. Revista de la CEPAL 83:135–151

FAO (2012) Gestión social y emprendedurismo: 8 pasos para implementar un programa de capacitación con personas jóvenes rurales. Food and Agriculture Organization of the United Nations, Costa Rica, pp 3–31

Fressoli M, Garrido S, Picabea F, Lalouf A, Fenoglio V (2013) Cuando las transferencias tecnológicas fracasan: Aprendizajes y limitaciones en la construcción de Tecnologías para la Inclusión Social. Universitas humanística 76:73–95

Hamidi S, Benabdeljalil N (2013) National innovation systems: the Moroccan Case. Procedia Soc Behav Sci 75:119–128

Heijs J, Buesa M, Baumert T (2007) Sistemas nacionales de innovación: conceptos, perspectivas y desafíos. In: Buesa M, Heijs J (eds) Sistemas regionales de innovación: Nuevas formas de análisis y medición. Fundación de las Cajas de Ahorros, Spain, pp 17–64

Herrera F (2006) Innovaciones tecnológicas en la agricultura empresarial mexicana. Revista Gaceta Laboral 12(1):91–117

Ilgin M, Gupta S (2009) Environmentally conscious manufacturing and product recovery (ECMPRO): a review of the state of the art. J Environ Manage 91:563–591

INEGI (2010) Instituto Nacional de Estadística y Geografía. Censo de población y vivienda. Mexico: INEGI. http://www.beta.inegi.org.mx/proyectos/ccpv/2010/. Accessed 31 Jan 2017

Jasso J (1999) La madurez tecnológica mundial. Revista de la CEPAL 69:119–137

Lundvall B (1988) Innovation as an interactive process: from user-producer interaction to the national system of innovation. In: Dosi G et al (eds) Technical change and economic theory. Columbia University Press, USA, pp 349–369

Lee S, Trimi S, Kim C (2013) The impact of cultural differences on technology adoption. J World Bus 48(1):20–29

Massa I, Andersen M (2000) Special issue introduction: ecological modernization. J Environ Policy Plann 24:265–267

Mata B (2013) Nuestro caminar con las escuelas campesinas. Escuelas campesinas: 10 años en movimiento. Universidad Autónoma de Chapingo, México, pp 23–44

Mata B, López M, González S, Delgado V (2007) Escuelas campesinas en México: una visión desde los encuentros nacionales organizados por la UACh. Artículos y Ensayos de Sociología Rural 3:63–77
Pieck S (2011) Beyond post development: civic responses to regional integration in the Amazon. J Cult Geogr 28(1):179–202
Pirela A, Rengifo R, Arvanitis R, Mercado A (1991) Conducta empresarial y cultura tecnológica. Empresas y centros de investigación en Venezuela. CENDES, Venezuela, pp 1–98
Rendon R, Roldan E, Hernandez B, Cadena P (2015) Rural extension processes in Mexico. Revista Mexicana de Ciencias Agrícolas 6(1):151–161
Rogers E, Takegami S, Yin J (2001) Lessons learned about technology transfer. Technovation 21 (4):253–261
Rosenberg N (1979) Tecnología y economía. Gustavo Gili, Spain, pp 20–422
SDES (2016) Secretaria de Desarrollo Económico Sustentable. Actividad económica en Guanajuato. México: SDES. http://sde.guanajuato.gob.mx. Accessed 31 Jan 2017
SEDESHU (2015) Secretaria de Desarrollo Social y Humano. Manual Ciudadano. Guanajuato. México: SEDESHU. http://www.desarrollosocial.guanajuato.gob.mx/pdf/manual-ciudadano-2015.pdf. Accessed 31 Jan 2017
Stiglitz F (1987) Learning to learn: localized learning and technological progress. In: Dasgupta P, Stoneman P (eds) Economic policy and technological performance. Columbia University Press, USA, pp 125–153
Tagle D (2016) Reporte técnico final del proyecto Transformación sociocultural, uso y aplicación de ecotecnias para el mejoramiento de la vivienda de las familias vulnerables de los municipios de Penjamo, Comonfort, Apaseo el alto, Tierra Blanca y San Felipe del estado de Guanajuato. UG-SEDESHU, Mexico, pp 1–198
Truffer B, Coenen L (2012) Environmental innovation and sustainability transitions in regional studies. Reg Stud 46(1):1–21
UNDP (2014) United Nations development programme. municipal human development index in Mexico: new methodology. Mexico: UNDP. http://www.mx.undp.org/content/dam/mexico/docs/Publicaciones/PublicacionesReduccionPobreza/InformesDesarrolloHumano/UNDP-MX-PovRed-IDHmunicipalMexico-032014.pdf. Accessed 31 Jan 2017
UNDSD (2000) United Nations division for sustainable development. Agenda 21. USA: UNDSD. https://sustainabledevelopment.un.org/agenda21sp/index.htm. Accessed 31 Jan 2017
Villavicencio D (1994) La transferencia de tecnología: un problema de aprendizaje colectivo. Argumentos 10(11):7–18

Author Biographies

Lorena del Carmen Álvarez-Castañón is professor (*titular A*) at the University of Guanajuato. Member of the National System of Researchers (2013–nowadays). *Awards:* National Research Award for Scholars by the National Association of Colleges and Schools of Accounting and Administration (2012); Merit for academic excellence in the doctoral program; Merit for academic excellence by the National Association of Colleges and Schools of Engineering.

Daniel Tagle-Zamora is professor at the University of Guanajuato. Member of the National System of Researchers (2013–nowadays) and invited professor in the multidisciplinary program Environment and Resources Management (ENREM) at the University of Applied Sciences of Cologne (Germany) and the Autonomous University of San Luis Potosí (Mexico). He has participated in numerous research projects; He is leader for the academic group *Water, Energy and Climate Change.*

Maricruz Romero-Ugalde is professor at University of Guanajuato and, is the founder and general manager of the National Laboratory of teaching and research *Etnografía Audiovisual Interpretativa* (ETNOAI). Member of the National System of Researcher and PRODEP Profile. She studies the symbolic and visual anthropology and has been collaborating in institutions as ENAH, Film library of UNAM, INI, among others.

A Decalogue for Education for Sustainability Across Environmental Generic Competencies

Rosalba Thomas Muñoz

Abstract The results of this research bring the possibility to the universities and their professors to create a course based on a sustainability vision across environmental education and the notion of competencies. The research responds to the question: "What capacities should university graduates have with respect to the environmental and sustainability area, which allows them to understand and influence the world around them?" Three stages were carried out in the methodological design: (1) analysis of institutional, curricular and pedagogical conditions of the host institution; (2) interviews with university professors, environmental employers and student survey and (3) design of generic environmental competencies and a proposal for a sustainability action plan. This paper highlights the last two stages of a curriculum, describing how knowledge, skills, performance and contexts of action were identified, analyzed and integrated, according to the findings of the previous stages of the research. One of the main contributions of this paper is the possibility of having a Decalogue as a set of tools for higher education institutions interested in incorporating the sustainability perspective in a pre-graduate curriculum.

Keywords Environmental education · Interdisciplinary · Transversality Sustainability

1 Introduction

The historical conditions that were present in the emergence of the notion of disciplines at the beginning of the 21st century are changing fast. Boix (2004) point to some causes: the speed of communications, the production of knowledge and other important changes in the problems dealt with by academic and professional

R. T. Muñoz (✉)
Facultad de Ciencias, Universidad de Colima, José G. Alcaraz 551, El Diezmo 28019, Colima, Col., México
e-mail: rosalbathomas@gmail.com

W. Leal Filho et al. (eds.), *Sustainable Development Research and Practice in Mexico and Selected Latin American Countries*, World Sustainability Series,
https://doi.org/10.1007/978-3-319-70560-6_10

communities. As examples, these authors comment on the different forms of resolution of the professional problems present in the frontiers of knowledge:

- Research projects with experts from different disciplines. Interdisciplinary teams that combine engineers, scientists, writers, sociologists, entrepreneurs in products and corporate solutions.
- Health care teams—teachers, social workers, nurses, physical and behavioral therapists, computer programmers.
- Educational projects with psychologists, teachers, curriculum designers, artists, computer scientists and others, for the design of new learning technologies for the student (Boix 2004).

According to Follari (2007), "We are in a light, postmodern era, characterized by the loss of hard truths and the assumption of less strict and systematic modes of functioning. We now understand the world through paradigms that are constantly changing and transforming our society."

Education for sustainability is a consequence of this. It responds to complex, uncertain, unpredictable, unstable, self-organized, unfinished, unbalanced, and dependent but autonomous phenomena. Sustainability is not only achieved with technological solutions, political measures and economic resources, but also with changes in our way of thinking and acting. We need education that addresses challenges in a constructive, integrative and creative manner, while also contributing to the formation of more sustainable and resilient societies. Therefore, universities are a key sector due to their ability to integrate knowledge.

At the beginning, education for sustainable development should works in an academic field, with the construction of a model of thinking accord to this complex challenge, with the effort of the whole academic community (students, professors, families, etc.), in order to better know how can the impacts on environmental, social, political, economic be reduced. But, besides this, we need to know how to deal the challenge of the traditional education in México. In spite of this, universities are slowly advancing, and this must be so, because educational system is the basis of many structural changes in the paradigms of life and cultural systems. However, we must face it, by designing new educational models, focused on practice, in the development of skills and actions oriented by values of sustainability, that is, respect for the future.

This paper presents the findings of a research aimed at the design of an educational program for higher level learning that sought the multidisciplinary and transversal integration for the development of generic sustainability competences. Mexican teachers were interviewed to learn about their experiences in cross-cutting and interdisciplinary environmental issues, as well as graduates who worked in environmental contexts. Finally, we surveyed students to assess their knowledge and expectations about environmental training. For this research only the environmental variable is taken as an aspect to be highlighted in the Mexican curricula, given the gap that it presents with other social issues. The following is the way in which environmental components should be included as support for the teacher, regardless of the academic area in question.

This study is cataloged as an interpretative-comprehensive type and is oriented under the principles of qualitative research. It consists of the following stages:

(1) Analysis of official documents of the University of San Luis Potosí (UASLP), as an example of Mexican universities.
(2) Identification of key informants (interviews with 48 teachers and graduates, and a survey of 870 students).
(3) Identification of knowledge, skills and attitudes for environmental competencies and a final design of the program, vision, objectives and goals, as well as structure, contents, pedagogical and evaluation approach.

The methods used were:

The Atlas.ti software was used to analyze the results of the interviews. Atlas/ti®V 5.2.6 de Scientific Software Development (Visual Qualitative Data Analysis). It helps to establish patterns and conduct interpretation and analysis of information. The analysis of survey for descriptive statistical procedures was used through the Microsoft Office Excel 2003 program, with the aid of graphs and frequency tables combining, in some cases, two or more variables.

In this article we will describe only three stages, due to the time limit and to deepen into the content on environmental and sustainability competencies, as well as the possibility to be part of every step of the design and to offer professors some skills to incorporate into his classes this kind of concerns.

2 The Professors and Their Experiences on Environmental Issues[1]

The participation of teachers is undoubtedly fundamental in training processes. In this case they contribute their experiences and knowledge in two aspects: (1) incorporated environmental and sustainability issues in their classes and (2) with interdisciplinary exercises. For the analysis of the answers, two concepts were considered: transversality and interdisciplinarity.

After analyzing several definitions around transversality, we understand it as "a methodological strategy of curricular structuring that allows the interrelation of

[1]48 interviews were conducted with professors from the Faculties of Chemical Sciences, Engineering, Agronomy, Medicine, Economics, Accounting and Administration, Stomatology, Habitat, Nursing and Law. Institutes of Desert Zones, Geology and Metallurgy, Social Sciences and Humanities. Multidisciplinary Academic Units of the Media and Huasteca areas were able to cover careers such as Nursing, Biochemistry, Accounting, Law, Design and Architecture, Medicine, Stomatology, Economics, Agricultural Engineer, Plant Engineer, Agronomic Engineer, Chemical Engineering, Anthropology, Geography, History, Geological Engineer, Metallurgical Engineer, etc.

different degrees within a Higher Education Institution (HEI), articulating the learning of sustainability issues and promoting an understanding and analysis of the current challenges" (Thomas Muñoz 2011: 32). The teachers identified some examples:

- Specific activities incorporated or added individually and/or individually, without prior systematization, or articulation with the rest of the contents of the subject(s) it imparts.
- Crossed transversality when they deliberately incorporate environmental and sustainable learning into the curriculum, but without restructuring it, only adapting the learning to the form that the curriculum has.
- Issues elective or obligatory.
- Approaches or topics proposing a specific approach (such as meaningful or collaborative learning) or topics (the water problem, climate change, etc.) to provide a summary in their course.
- Projects, case studies, thesis or research projects between various areas, etc.
- Transversal thread, activities or materials, around which relationships are established with other contents or with the rest of the curriculum.
- Integrating nuclei restructuring the curriculum by integrating activities or materials that have been created specifically to include environmental and sustainability issues (research seminars, social service, etc.).
- Key issues, specific problems that incorporates activities across the curriculum (environmental culture, climate change, etc.).

In these partial findings, we observed that teachers reduce their experiences only with integrative projects, addressing concepts or topics such as sustainability or some management, implement a practical program of recycling, final disposal of hazardous waste, noise control, and water treatment, among others. However, transversality is much more than that when we integrate an interdisciplinarity perspective, as we will see forward.

Interdiscipline is understood as a way of:

> Building knowledge product of the communication and experiences between the different disciplines that converge in an Institution of Higher Education, different theories, methodologies and techniques of analysis oriented to solve a same problem. It is a process through which students establish relationships of cooperation, interaction and mutual help in the analysis and understanding of an environmental problem, in order to allow them to broaden their perspective and their capacity to understand the complexity of environmental problems (Thomas Muñoz 2011: 53).

Some definitions that guided teachers are:

- Multidiscipline: collaboration and help of various disciplines to solve a part of some environmental problem.
- Interdiscipline: several disciplines collaborating to jointly carry out a specific environmental project.
- Transdiscipline: several disciplines building common knowledge in an environmental project.

The results of the interviews show that natural sciences experiences are focused on having classes with examples of environmental problems, practice, and social service. In engineering and technology, these integrative activities are carried out with projects between various disciplines, thesis, integrated work among teachers, and workshops of environmental and sustainable issues. Health sciences include multidisciplinary science on postgraduate programs, projects on environmental impacts, and sustainable development commissions with government, social and academic participation in various subjects and summers of sciences. For social sciences, activities are carried out through common trunk materials. In the agronomic sciences, courses of agroecology are carried out with expert talks and case studies and finally, in the administrative sciences research projects are carried out with the students.

In spite of performing these exercises on their own initiative, teachers have important difficulties when working with disciplinary integration exercises, such as: lack of communication among students, cultural issues, bureaucratic problems, complexity of disciplines, technical terms, procedures and language of each area, saturation in materials, disparate schedules, lack of availability of other teachers and coordination between faculties, apathy of students, absence of ability to understand nuances in problems, need to have a common understanding of ideas, deficiency of economic resources, and trained personnel, among other things.

During the stage three, these results help to anticipate complications of interdisciplinary exercises, and to consider the operability, the institutional procedures, and the complexities of the disciplines involved (methodological and technical).

3 Environmental Education from the Students

We seek to diagnose which knowledge; skills and attitudes can serve as a basis for the formation of environmental and sustainable generic competencies in students, as well as identify expectations around these issues. A survey was applied to 802 students (447 women and 355 men) from 28 UASLP degrees, with an average age between 20 and 24 years. The results show that students do have knowledge of basic ecology and global, regional and local issues, but in matters related to a more advanced knowledge of sustainability, they no longer feel capable.

Students consider these next topics as missing for their professional training:

- Basic concepts of ecology.
- Know how to characterize an environmental and sustainable problem.
- Recognize the systemic relationships of environmental and sustainable problems.
- Collaborate with other professions to solve an environmental and sustainable problem.
- Collaborate with teams to make decisions.

- Know how to organize communities to solve environmental and sustainable problems.
- Conduct original and interdisciplinary research on environmental issues.

For students, interdisciplinary teaching was expressed through their understanding of specific topics, from ecological and biological to political and sociological. The students also identified strategies to learn environmental issues, highlighting the use of media and talks with teachers. However, they propose some themes to address interdisciplinarity:

- Respect and care for the genetic diversity of ecosystems and species.
- Values and behaviors between humans and nature.
- Forms of organization of civil society and citizen participation.
- Forms of distribution of wealth.
- Legal frameworks that guarantee respect for people and the environment.
- Air quality, water, soil, climate and energy.
- Equity between genders.

On the expectations and needs from the students about environmental and sustainable issues the findings show next needs for training:

- Recognize social, environmental, economic, political and cultural impacts of human activities.
- Identify major environmental, economic, political and cultural problems in Mexico and the planet.
- Know and understand basic concepts about environmental, economic, political and cultural problems related to their professions.

This finding helps us to understand that, for students the idea about the complexity of sustainability, must go beyond an ecological and reductionist vision about contemporary world to a complex, uncertainly and chaotic system in which they live. In this sense, all kind of programs, degrees or workshops which wants to concern about sustainability education and future, should be able to achieve this purpose.

4 Perspective from Professionals to Define Environmental Competences

We understand the notion of competencies as

> The repertoire or set of knowledge, skills and attitudes that are articulated in a specific way [it is not the simple sum] in a context of determined action to allow a person to perform a task or to fulfill a goal (Notion constructed by Nieto-Caraveo 2005, from Díaz Villa 2004, Catalano et al. 2004 and Novick 1997, cited in Thomas Muñoz 2011, p. 60).

Therefore, a student will be competent not only when he has knowledge or skills about sustainability, but also when he has demonstrated his competence through certain performances. From the professional's perspective we define a group of environmental and sustainable generic competencies (ESGC), identifying knowledge, skills, performance and attitudes related to the sustainability, common in all (generic) degrees. This task is not only a list of skills, but also contexts of action in which sustainability learning could be observed.

In this context, to identify this knowledge we were guided by the question: what information is considered indispensable for a professional who performs in a sustainability related work? The main findings were: environmental impact of each profession and alternative solutions, legislation, regulations, audits, annual operating documents, ISO instruments (14-000 and 14-001) and ecological regulations, among others. Besides, concepts of ecology and sustainability were conceived as key knowledge.

Know-how relates to the skills to recognize procedures, mechanisms and methodologies, therefore the findings included: know how to search and analyze information, evaluate and diagnose a problem, propose interdisciplinary solutions, develop projects and communicate to community applicable measures. Skills were also manifested by practical achievements that make them visible and serve to assess the level of learning acquired. Although, how to apply these performances skills varies depending on the discipline and performance context. Some of them performances were:

- Manage information.
- Analyze and report relevant information.
- Argue and debate ideas.
- Understand texts on legislation.
- Develop projects.
- Identify elements of an environmental problem.
- Develop problem solving alternatives.
- Manage projects.
- Make decisions.
- Solve problematic situations; and,
- Reach agreements with other professionals.

But, how should students and professionals working on sustainable issues are related? We identify necessary traits in the behavior and attitudes that motivate their congruence with a concern for the sustainable world. It is worth mentioning that this was one of the most ambiguous components, due to the difficulty of professionals to differentiate behavioral and attitudes, or to find differences between "having conscience," "worrying" or "do concrete actions." For all of them, were almost the same attitudes, which are put at stake to solve a sustainable problem:

- Be aware of sustainability challenges (global, regional or local).
- Show responsibility for your actions and decisions.
- Sensitivity to the environment.

Contexts of action identify moments and circumstances in which these competences can be implemented. Taking into account that these are not static, but evolve and redefine, it is important always to consider the changes and implications that have the places where we put into practice these competencies, as well as the disciplines involved. For example, while in health sciences a competence is achieved with laboratory practices, in social sciences and agronomy they are made on field work.

Once the basic components of ESGC's have been identified, their integration should not be linear, that is, we should not only list the components that interest us the most, but also articulate them in context. To complete the work on ESGC, is necessary to know institutional framework of a university, add the experiences of transversality interdisciplinary including the interest of teachers and students, the normative possibilities (institutional regulations), didactic (to have innovative equipment or materials) and transversal study plans, among others.

5 Toolbox to Design Environmental and Sustainable Generic Competencies

The competence approach in higher education seeks to motivate students to practice their profession from the earliest formative levels, allowing them to know and develop early those tasks that will necessarily arise in practice. However, in addition to this disciplinary knowledge, students should be able to be resourceful (human, technological, financial and material), to solve emerging problems and to meet the demands of social, political, environmental and cultural dimension's working in a multidisciplinary way. In this sense, we will describe some environmental and sustainable generic competencies that can be used to put into the practice the findings of this research, taking further teachers' desire to promote an integrative, meaningful and collaborative sustainable education, to an better understanding about the impacts of their profession so that as graduates they will be responsible citizens of the world in which they live.

5.1 Environmental and Sustainable Generic Competencies

Competence 1. The student will be able to argue, expose or debate (their) ideas on an environmental and sustainable problem to support the decision making of an interdisciplinary team. Context of action: business, urban or rural populations and governmental dependencies (Table 1).

But, as we see, a competence is not complete without specific skills and performances. The findings of research propose the following (Table 2).

Table 1 Elements that compose competence 1

Knowledge	Evidence	Criteria for evaluating
Basics concepts to understand ecology and his interactions with human life	Bibliographical sheets with key notions	Information and language must be consistent, coherent and argumentative and integrate reports with results
Basics concepts to understand sustainability and his interactions between environmental, political, economic, social and cultural dimensions	Presentations in front of the group to discuss their ideas, positions and answer questions	The data must be correct and well used. References should be reliable and well cited. Interventions must be precise, coherent and purposeful
Basics concepts to introduce environmental impact and specific problematic	Practice to visit a community. Cite the references used. Prepare reports with agreements	The community visited should be supported by material (photographs, interviews, etc.). Final products should contain the original approaches and agreements as a result of discussions
Identify main environmental management instruments according to the case (legislation, standards, certificates, ISO, etc.);		

Table 2 Necessary skills and performance to this competence

Skills	Performances
Investigate information	Present in front of a group with mastery of the subject
Public speaking	Identify reliable information for analysis
Argue their posture	Argue their position in team discussions
Make decisions in interdisciplinary teams	Show evidence of performance in accordance with the needs of their professional activities (projects, conclusions, etc.)

Table 3 Elements that compose competence 2

Knowledge	Evidence	Criteria for evaluating
Basics concepts to understand sustainability and his interactions between environmental, political, economic, social and cultural dimensions	Conceptual maps on specialized readings	The maps must contain the basic information from the author consulted, be clear and schematic
Information about methodology for sustainability projects	Bibliographic records with the main methodologies used for the project in question	The bibliographic records must show the author's ideas in a clear and synthetic way
Community features	Log of the field visit to the chosen population or community	The field visit should be supported by a field log and additional material (photographs, interviews, etc.)

Table 4 Necessary skills and performance to competence 2

Skills	Performances
Analyze and synthesize information	The student highlights the main ideas of a specialized reading
Communicate with people	The student must have a conversation with the population (individual or in groups)
Develop an intervention project	The student must prepare an intervention project and make known the alternative solution
Make known alternatives to the problem	

Within the scope and limitations of this competence, performances should be reinforced periodically, for example, the argumentative ability is developed only with practice. Decision making requires several stages to develop fully and the search for information must continue and evolve depending on sources, materials and learning resources.

Competence 2. The student will be able to elaborate projects of intervention with conceptual bases of the environmental sciences to present alternatives of sustainability in urban or rural populations. Context of action and realization: urban or rural localities (Table 3).

To complete the elements of competence 2, is necessary define skills and performance. In this case, is important that professors have some information about methodology on systems theory and complexity theory, because sustainable problems are not solved with only one perspective but also an integrated and holistic approach (Table 4).

Besides, this competence requires the prior management of economic resources to move into the population or communities.

Competence 3. The student will be able to identify, understand and apply the environmental management instrument(s) appropriate to the sustainability problems of an organization. Context of action and realization: companies with environmental or sustainable focus (Table 5).

Skills and performance recommended use while developing this competence (Table 6).

It is important to remark that it is not necessary for students to understand all the management tools that exist, but to identify what they are used for and to recognize

Table 5 Elements that compose competence 3

Knowledge	Evidence	Criteria for evaluating
Environmental management tools and knowledge on regional and local sustainable issues	Reports of information about experiences of environmental or sustainable problems in an organization and presents it to the group	Coordinate teamwork; Select the management tools relevant to the case
External and internal regulations of the selected organization		Have enough information for the analysis

Table 6 Necessary skills and performance to this competence

Skills	Performances
Identify the characteristics of the environmental problem within the organization	Make a diagnosis of the problem and an evaluation of the identified impacts
Select management tools and methodology applicable to the case	Apply the methodology of the selected management tool
	Conduct an interview or survey to compare and support the information previously obtained

the environmental and sustainability problems that are solved with each one. The limitation of this competence lies in the management capacity of students has to convince the organization to collaborate with them. In this case, professors may help connecting personal of organization to make sure the collaboration with students.

All this findings from professors, students and professionals join with documental information from the context of universities allow us to reconstruct a sustainable action plan that will serve the universities to get in the way of incorporating these perspectives into the curriculum. In next we will describes main guidelines to construct an action plan for sustainable education.

6 10 Guidelines for a Sustainable Action Plan

The purpose of this guideline is to contribute to sustainable training experiences as an integral strategy, establishing a relationship with professional environmental fields, and allowing the students to argue, expose and debate their ideas to carry out their profession with a sustainability vision, providing arguments for responsible decision making with society and environment.

On the bases of the findings of this research, we recommend the following support guide:

(1) To offer an optative course to all degrees students, with the possibility to obtaining optional and binding credits.
(2) The overall strategy of this plan is carried out by modules in which 7 stages can be addressed during the semester:

   a. Introduction to sustainability topics and course objectives.
   b. Organization of the group in working teams to understand a case study.
   c. Carry out a field practice to know the real conditions of the problem.
   d. Systematize the information obtained during the visit.
   e. Carry out a second field practice to gather missing information.
   f. Organize in group for the decision making around the case study.

g. Present the final results to the group and teacher.

(3) Organize the work in face-to-face, additional hours and field practice, through weekly sessions of 2 h each.
(4) Teaching and learning strategies should be based on three possibilities:

a. Case-based learning, this means to provide a problem that represents real-life conflict situations.
b. Collaborative learning: Students develop the ability to learn in collaboration with their peers, the tasks can be diverse; the goal is to learn from their peers and to generate interdisciplinary learning experiences.
c. To the previous approaches of teaching and learning is added the one of the significant learning that emphasizes the utility of the knowledge for the student, the application of his disciplinary training to the solution of problems that he will face in his work context. These approaches represent new cognitive challenges for students.

(5) It is necessary to create links with some department or address in the university that is related to environmental issues, groups of teachers, environmental areas or a degrees with this same approach, which will help in the coordination of teachers' proposals, web platform work, materials, facilities, etc.
(6) It is also recommended to select topics that can be coordinated by the teacher according to their experience and their relevance according to the local context and interest of the student. Some examples of topics in Mexico can be:

a. Risks health due to the environmental impact of productive activities.
b. Environmental contingencies for hazardous waste.
c. Urban development and environmental impact of productive activities.
d. Availability and quality of water.
e. Remediation and restoration of contaminated sites.
f. Risk communication for contaminated populations.

(7) The teacher responsible for the course must at first to create the case to be analyzed (preferably local problems), and establish previous contact with the community, to agree on possible dates and timetables; considering that there are only 5 weeks of work per module, this is the best to take advantage of time.
(8) You must have all the work material previously organized for the student in digital format, leaving the first week to familiarize yourself and the students with the web platform, personal and group email account of the course.
(9) Evaluation strategies should consider theoretical and practical aspects. Therefore, it cannot be based only on finished products or written exams, but be based on a triangulation of evidence which allows the teacher to complete the evaluation:

a. Evidence of process: compare information that students had at the beginning of the course and at the conclusion.

b. Evidence of knowledge acquired: which was the knowledge acquired.
c. Evidence of the products made: what products the student made to confirm his learning.

(10) It is recommended to have in general bibliography materials and resources of support that orient the student on elaboration of oral presentations, conceptual maps, work reports, etc. This material should be part of the resources contained in the support package that should be given to the student at the beginning of the course.

7 Conclusions

Currently, environmental education is still conceived by a part of the population as an education that prepares people to conserve nature, is thought of as those activities that promote awareness about environmental problems and also promotes a change in the behavior of people. Unfortunately, these approaches are not enough for this educational program, or to address sustainability issues that include environmental education as one of its most effective tools.

Education for sustainability goes beyond considering only environmental problems; this approach is aimed at seeking the integration of these with the social, political, cultural and economic, as this presents the opportunity to promote a true integral and professional environmental training. Environmental education can no more be conceived as education that teaches only about nature, since the problems of civilization today should be part of an environmental problem, but also social, economic and political, and should be understood and resolved as a whole.

Based on the findings of this research, the methodology used helped us to identify the institutional support structure that a university should offer to lay the foundations of an education plan for sustainability (that section is not seen in this article because of space issues); we deepen the identification of experiences of interdisciplinarity and transversality of university professors, as part of an exercise for the design of generic environmental competences, for which they also made an approach to professional graduates from 28 university degrees, who commented on the knowledge, skills and attitudes necessary for a job related to environmental issues. The answers were very interesting, and we raised the need to incorporate some thematic and practical that must be taught generically during the passage through the university.

Besides, through these findings we now are able to know that it is possible to design other environmental generic competences from the previous ones. For this, it will necessary to select knowledge of interest, to combine skills or contexts of action that allow adapting the same competence in different moments. However, this exercise is conceived as a support tool to create other activities that foster concern and responsibility of students and teachers around the environment, as well as identify environmental impacts of their degrees. It is suggested to investigate other components of generic environmental competences, according to their context, (university, subject, educational level and environmental problems present in their locality). In this way, there will be interest in the whole educational community and it will have better results.

Identifying a large number of elements that make up the sustainable and environmental generic competencies, it is clear there is difficulty developing more than one in each semester, not only because of the limitations of time, space and resources, but also because the acquisition of an advanced level of competence is needed in practice and in a full-time context. For this reason it is recommended to design and work with students on a single competence per course.

The concepts of competencies, transversality and interdisciplinarity give educational program innovative features, not only because the competences allow detailing features of a relevant and flexible profile, but also because the support of the host institution is required. This allows for combined curricula, obtaining of credits by election, freedom of professorship and, above all, the real interest to improve society, for which discourse is required, in addition to effective programs, with resources and personnel assigned to follow up.

References

Boix M (2004) Assessing student work at disciplinary crossroads [Electronic Journal] January/February 2005, 37(1). http://www.heldref.org/change.php. Accessed 2 July 2007

Follari R (2007) Interdiscipline in teaching. [Electronic Journal] POLIS, J Bolivarian University 5 (16). http://www.revistapolis.cl/16/diezyseis.htm p 12. Accessed 2 July 2007

Thomas Muñoz R (2011) Education for sustainability driven by change and innovation in higher education. University of Colima, México, p 178

Author Biography

Rosalba Thomas Muñoz is graduated in Political Sciences (1999), master's degree on environmental sciences (2007), and Ph.D. in Social Sciences (2012). She is currently Associate Professor in the University of Colima working at environmental sciences postgraduate program. She has been member of the National Academy of Environmental Education since 2009. As a young researcher, she received an honorable mention from the Potosino Council of Science and Technology in 2007, in the scientific research category. She received the State Ecology Award in 2011 and the State Youth Award in 2006. She has published three books: Education for Sustainability Driven by Change and Innovation in Higher Education (2011), Sustainability Stories for Reading without silverware (2016), "Is the discourse of sustainability in Mexico a media construction?" (In press). Edited by the University of Colima, Mexico. His areas of research are discourse studies, environmental education and regional sustainability projects.

Oil Exploitation in Yasuni Biosphere Reserve. Impact on Ecuador's Commitment with Sustainability

Alicia Anahí Cisneros Vidales and Víctor Mauricio Barriga Albuja

Abstract The purpose of this research is to identify how oil exploitation inside the Yasuni Biosphere Reserve, one of the most biodiverse places on Earth, affects the Ecuadorian government's commitment to its sustainable development. The research is based on the bibliographic review of the reserve's management plans, legislation regarding environmental protection and land management; as well as previous works on documenting key factors and their impacts in the reserve, focusing on oil exploitation. For this analysis, Yasuni has been considered as a complex system, in which key factors determine changes in four dimensions and their interactions, defined to represent sustainable development: social, environmental, political and economical. In the proposed analysis, components and actors are identified for each dimension, and their relationship is described at the different scale of their impact. By understanding these roles and their significance, it is determined which dimension is the most influential, therefore establishing if it supports sustainable development or not. In Yasuni, it is clearly established that the economical and political dimensions determine the future of the reserve, which is considered highly unsustainable and depicts an important dependence of the country and Yasuni on oil extraction in order to get financial resources. The identification of key factors

A. A. Cisneros Vidales
Calle 16 725 Col. Industrial Aviación, CP 78140 San Luis Potosí, Mexico
e-mail: anahicisvi@gmail.com

V. M. Barriga Albuja (✉)
Agustin Mentoso N47-184, CP 170510 Quito, Ecuador
e-mail: mauricebar@yahoo.com

A. A. Cisneros Vidales · V. M. Barriga Albuja
Programas Multidisciplinarios de Posgrado En Ciencias Ambientales,
Universidad Autónoma de San Luis Potosí, San Luis Potosí, Mexico

A. A. Cisneros Vidales · V. M. Barriga Albuja
ITT, TH Köln, Cologne, Germany

W. Leal Filho et al. (eds.), *Sustainable Development Research and Practice in Mexico and Selected Latin American Countries*, World Sustainability Series,
https://doi.org/10.1007/978-3-319-70560-6_11

and actors however, brings the possibility of promoting a more egalitarian distribution of power among these, in decision making, enabling a path towards sustainability and engaging the Ecuadorian government with it.

Keywords Yasuni · Sustainable development · Oil exploitation

1 Introduction

Yasuni Biosphere Reserve (YBR) is located in the Ecuadorian Amazon, in the provinces of Orellana and Pastaza. The reserve comprehends an area within which the Yasuni National Park (YNP) is contained, which constitutes about 60% of the reserve (Albacete et al. 2004). Due to availability reasons, some of the data presented here is specific to the national park, but it remains relevant to the analysis of the whole biosphere reserve.

To determine how the oil exploitation inside this biosphere reserve affects Ecuador's commitment with the principles of Sustainable Development, a Sustainability Assessment of this area has been conducted. This complex system consists of four dimensions that classify the nature and interests of specific actors that interact to determine the present and future of people, resources and biodiversity here encompassed.

These four dimensions include the three commonly depicted in the origins of sustainable development by the Brundtland report (Brundtland et al. 1987): economic, social and environmental, plus a political dimension.

To strengthen the structure of this complex system, local, national and global scales of incidence are established. The period included by the analysis starts at the early beginnings of oil exploitation in the Ecuadorian Amazon (circa 1972) to the present.

2 Precedents and History

Yasuni Biosphere Reserve has 1,682,000.00 ha, of which YNP contributes 982,000.00 ha (Albacete et al. 2004). It is the largest natural protected area in continental Ecuador and it is of great importance ecologically and ethnically for the country and the world (Fig. 1).

Yasuni is known for its outstanding biological diversity (Romo 2010; Bass et al. 2010; Ministerio del Ambiente del Ecuador 2011) but the reserve is also inhabited by the last voluntarily-isolated indigenous communities in Ecuador and some of the last such communities in the world. They are known as *Indigenous People in Voluntary Isolation,* or PIA (Pueblos Indígenas Aislados) in Spanish. The two PIA communities found are Tagaeri and Taromenane, but previously, the now contacted Waorani community was also in self-isolation.

Fig. 1 Location of Yasuni National Park. *Source: Google maps*

The territory where the PIA communities live, the Waorani Ethnic Reserve and what is now the national park was declared Biosphere Reserve by the United Nations Educational, Scientific, and Cultural Organization (UNESCO) (Ministerio del Ambiente 2011). This declaration occurred in 1989 with the goal to protect the unique cultural and biological diversity found in these areas. To protect its unique cultural and biological diversity, in 1989 the territory in which the PIA communities live, the Waorani Ethnic Reserve and what is now the national park, was declared a Biosphere Reserve by the United Nations Educational, Scientific and Cultural Organization (UNESCO), (Ministerio del Ambiente 2011). Later, in 1999 in order to protect the PIA communities, it was established—by executive order 552 and published in the Official Registry, Supplement 121 on February 2nd 1999—a dedicated zone inside the biosphere reserve to guarantee PIA isolation and access to natural resources, it is called the "Zona Intangible Tagaeri-Taromenane" or ZITT (Ministerio del Ambiente 2016) (Fig. 2).

In the context of this analysis, it is very important to understand what the designation of a Biosphere Reserve is. As declared by UNESCO, biosphere reserves are "'*Science for Sustainability support sites'—special places for testing interdisciplinary approaches to understanding and managing changes and interactions between social and ecological systems, including conflict prevention and management of biodiversity*", (UNESCO 2016). Biosphere reserves are divided into three interrelated zones that perform complementary functions. Those functions are: 1) *core,* which is a strictly protected area dedicated to conservation. In the case of Yasuni, the area of the National Park is considered the core as defined by the National Environmental Authority's Strategic Management Plan (Ministerio del Ambiente 2011) *Buffer,* that surrounds the core area and allows activities destined to reinforce scientific knowledge, training, education and are "*compatible with*

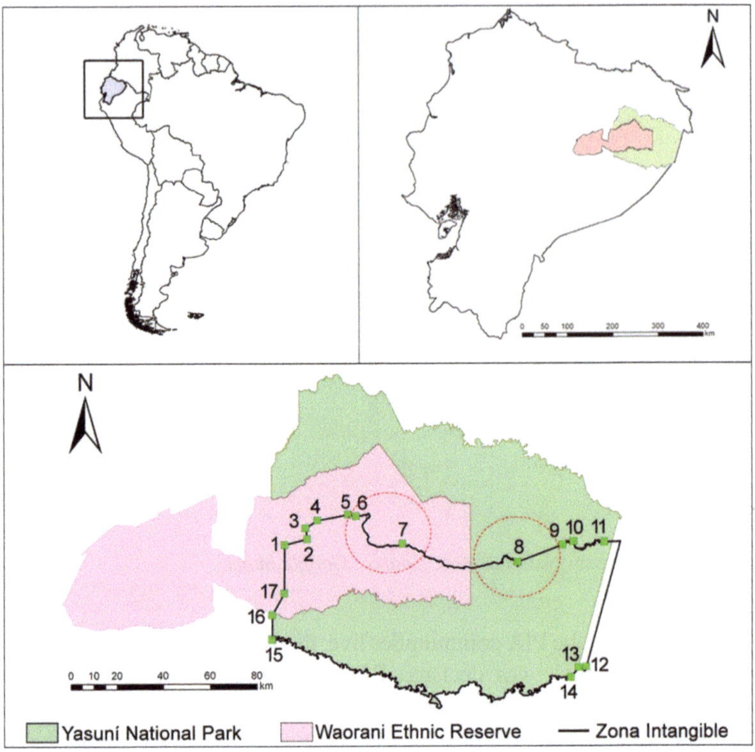

Fig. 2 The biosphere reserve: Yasuni National Park, Waorani Ethnic Reserve and Zona Intangible Tagaeri Taromenane (ZITT). *Source:* Pappalardo et al. (2013)

sound ecological practices" (UNESCO 2016). *Transition* area is the one allowing the greatest activity and it promotes economic development in accordance with sustainable development.

Oil exploration and exploitation in the Amazon region of Ecuador began before the establishment of the biosphere reserve. Although, after the national park's creation in 1979, and despite its legal national protection category, oil concessions have been granted within park limits.

The latest conflicts have been related to the cancelation of a government proposal (initially envisioned by civil society groups, launched in 2007 as a government-adopted initiative and cancelled in 2013), popularized around the world as pioneer in its nature, which would preserve untapped all the oil contained in two blocks (31 and 43) in Yasuni National Park if, in exchange, the International Community would grant Ecuador half the income that it would otherwise get by exploiting these resources (i.e. 3.6 billion USD). The proposal was called the Yasuni-ITT Initiative, after the Ishpingo-Tiputini-Tambococha (ITT) locations where the blocks are (UNDP 2013).

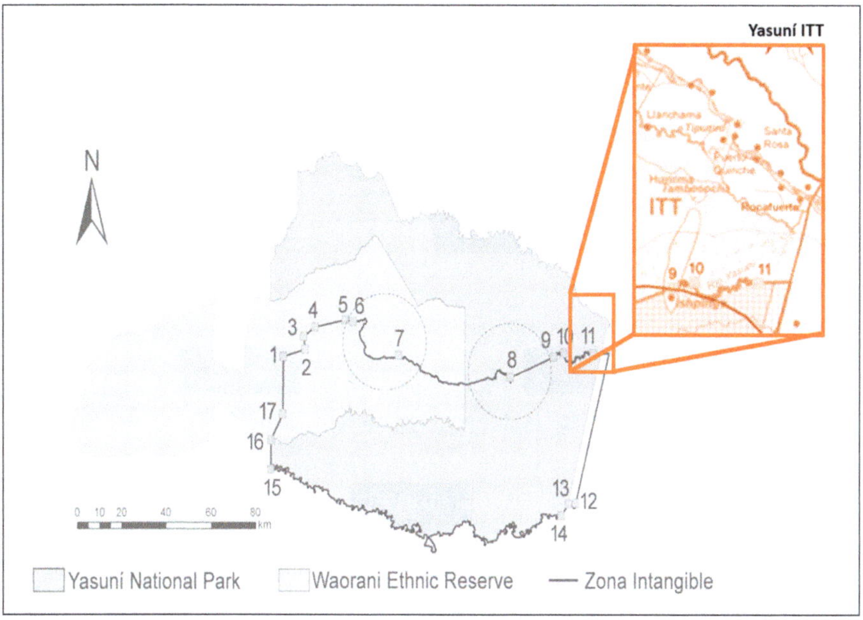

Fig. 3 Adapted from: location of the current oil extraction conflicts. Yasuni Biosphere Reserve and ZITT (Zona Intangible). *Source:* Pappalardo et al. (2013)

The conflict that was created not only has strong environmental questions but also has social issues, since the ITT exploitation proposal overlaps the ZITT area, reserved for the PIA communities, where non-renewable resource extraction is prohibited by Ecuador's Constitution (Art. 407) (Figs. 3 and 4).

3 Sustainability Dimensions

As previously explained we understand sustainability as an interaction of processes and actors from four dimensions: social, economical, political and environmental. This is an essential step in analyzing YBR as a complex system. The principal components that play a role in the reserve are identified in each dimension; this will be useful to later define specific actors for each dimension (Fig. 5).

4 Environmental Dimension

The YNP is the largest continental protected area in Ecuador. As it is part of the tropical rainforest Amazon, it preserves a great biodiversity of flora and fauna according to research during the last decade (Ministerio del Ambiente 2015).

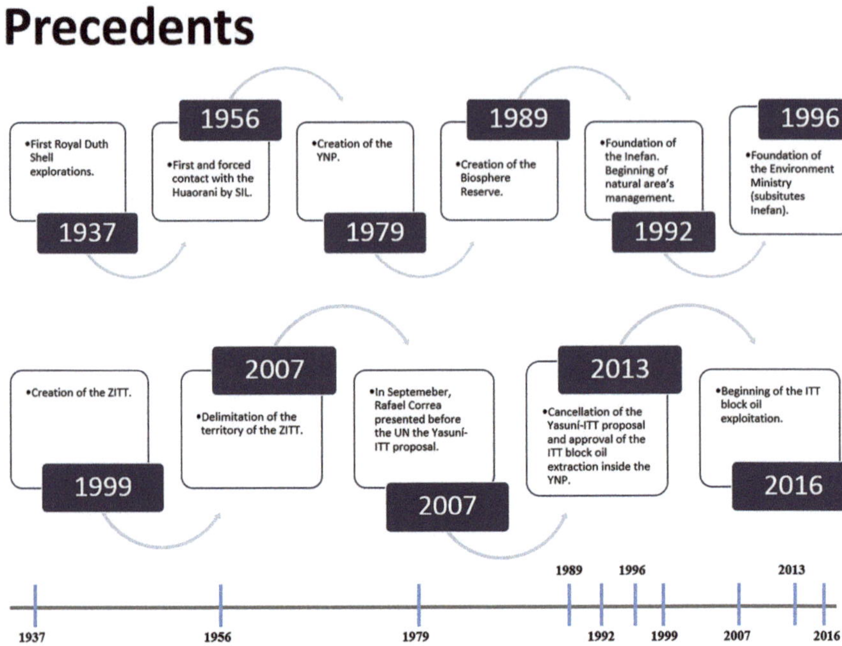

Fig. 4 Precedents timeline for Yasuni Biosphere Reserve

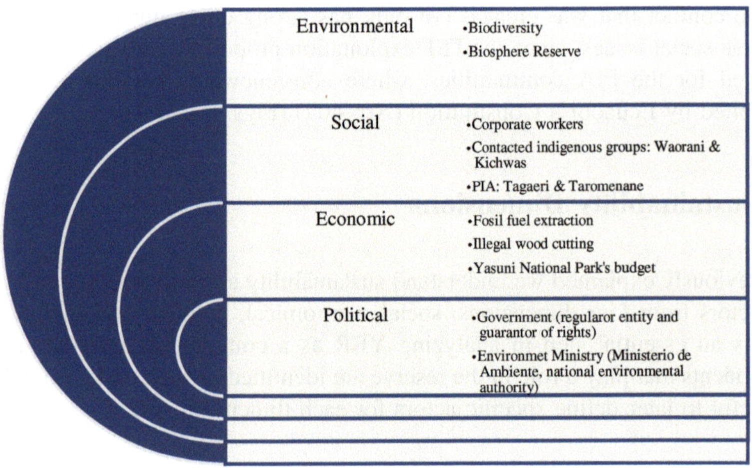

Fig. 5 Sustainability in Yasuní National Park. Components under the social, economical, political and environmental dimensions

According to the current Management Plan of the YNP (Ministerio del Ambiente del Ecuador 2011), there are 2274 species of trees and bushes; 204 species of mammals; 610 species of birds; 121 species of reptiles; 139 species of amphibians; more than 268 species of fish; hundreds of thousands of species of insects. In addition, it is also considered to have the highest number of vertebrates in the world (Vallejo et al. 2011). Species that can be found here are the jaguar (*Panthera onca*), the anaconda (*Eunectes murinus*), the harpy eagle (*Harpia harpyja*), the pygmy marmoset (*Callithrix pygmea*). The great diversity within this park enables it to be considered as one of the most biodiverse places in the world. According to the Ministry of the Environment, in a single hectare of YNP, 650 species of trees are found, which is more than occur throughout North America.

In-between the Napo and Curaray rivers there is a complex hydric network composed by the Tivacuno, Tiputini, Yasuní, Nashiño, Cononaco and Tigüino rivers, which form the lowest basin of the Napo. More than one million hectares of rainforest occur where this network of rivers intersects. YNP's Management Plan indicates that the entral and western part of the park has wide plains that are flooded periodically during the rainy season. These floods nourishe the earth with organic materials. The plains are surrounded by small hills that never get flooded. As a result the species composition in the unflooded region differ from those in the flooded area and different forests can grow. Near the border with Peru the soil is poorly permeable, permiting the rainwater to accumulate and to form extensive marshes. Certainly, YNP's topography is a determing factor that has promoted the development of its outstanding diversity.

Forests cover most of the YNP, including the communities Añangu, Nueva Providencia, Indillana, Llanchana or Mandaripanga, which are located on the banks of the Napo River (Sistema Nacional de Áreas Protegidas del Ecuador 2014).

As this review suggests, the conservation of YNP is vital not only for Ecuadorians but for the whole world, as it is home of thousands of species and has a large cultural wealth.

5 Social Dimension

Between the Napo and the Curaray rivers, groups of ancestral indigenous people occur, among them the Waorani, who used to live isolated. Some of their activities to survive include hunting, fishing and fruit harvesting.

The first forced contact with the Waorani was in 1956 (Larrea 2010). This first contact was by the controversial "Summer Language Institute (SIL)", which was founded in 1934, and which had as headquarters the Oklahoma State University since 1941 (SIL International 2015).

The objective of the SIL was to translate the New Testament, Christian hymns and Scriptures into the ancestral languages of the tribes in isolation. This project took place in Latin America, Africa, the East Indies, the Pacific islands and Russia (A. Del Valls 1978). The SIL practiced the Christian Evangelical religion and its purpose was to convert the natives to this religion (Cannell 2006). Since no one knew the language of the Waorani, at first it was impossible for the SLI to communicate with the Waorani. Local settlers abducted a Waorani girl named Dayuma, who would serve as an interpreter between the SIL and its people (Vera 2007). In 1956, the SIL started to relocate the Waorani community, from their forests to 16,000 hectares in the so-called "Huaroani Protectorate", where many of them live isolated from their ancestral practices and at the same time marginalized from the western society (Larrea 2010).

Before the SIL's arrival, the Waorani community was divided into two groups because of a power struggle. One of the groups fled and hid in the jungle and was never contacted. They formed the community called Tagaeri, but in the density of the jungle there was also another community, which the Waorani did not know: the Taromenane (Vera 2007).

As the first oil companies settled in the Waorani's territory in 1937, two new communities of *mestizo* settlers who came from other parts of Ecuador to work for the oil exploitation companies settled, most likely on Waorani territory. Since then, a complex relationship exists between the traditional indigenous communities and the oil companies. The indigenous communities gained access to *westerner's* resources, which were traded by the oil companies for access to their natural resources. These interactions have caused several changes to the traditional culture of the indigenous communities and their ancestral life-style. Nowadays, territories, boundaries and limits are not controlled and patrolled. There is a low presence of authorities, which allows logging and other illegal activities to take place. These activities compromise the sustainability of the YNP.

6 Economic Dimension

Although YNP represents the largest continental portion of the National Protected Areas System of Ecuador, receives one of the smallest budgets. According to YNP's management plan (Ministerio del Ambiente 2011) the cost of park maintenance amounts to 340,000 USD per year, in case of including an integral management plan even more than 490,000 USD. However, the POA 2011 (Operative Annual Plan) states that current requirements are between 300,000 USD and 800,000 USD per year. Based on these estimates, projected five years budget, considering an integral management, is in Table 1.

The financing alternatives for YNP depend on national or local income. Tourism operator patents, visitor entrances, research and film & photography permits, visitor centers, handicrafts make the local income. The Ministerio del Ambiente (2016), states that the total income generated by these economic activities would be approximately 50,000 USD per year.

On the other hand, national income comes from tax revenues, capitalized funds, National Environmental Fund (FAN) and oil benefits. According to Ministerio del Ambiente (2016), YNP is funded by a fiduciary fund created by Ecuador's National

Table 1 Estimation for five years of the average general expense per program for Yasuni National Park

Expense category	Year 1	Year 2[a]	Year 3	Year 4	Year 5
Conservation of the natural and cultural heritage	150,000	155,535	1,612,742	1,672,253	1,733,959
Control and surveillance	200,000	207,380	2,150,323	222,967	2,311,945
Environmental participation, communication and education	150,000	155,535	1,612,742	1,672,253	1,733,959
Research	150,000	155,535	1,612,742	1,672,253	1,733,959
Tourism	150,000	155,535	1,612,742	1,672,253	1,733,959
Total expense	800,000	829,520	8,601,293	8,918,681	924,778

[a]Projections were estimated with an inflation rate of 3.69% per year

Source: Ministerio del Ambiente (2016). Plan de Manejo del Parque Nacional Yasuní. Quito, Ecuador

Financial Corporation (CFN) which allocated 4 million USD to guarantee the park's management over the next few years. The oil benefits in the park come from blocks 14, 15, 16, 31, and the recently exploited ITT.

In 2011 these blocks were managed by Petro Oriental (Block 14), REPSOL YPF (Block 16) and Petroamazonas (Blocks 15 and 31), whose contracts end in 2018. Although most of the oil extraction comes from the core and buffer areas of the biosphere reserve there was no legal instrument that allowed non-park areas to receive funding from oil benefits before the Inter-ministry agreement 003-2011, signed between the National Secretary of Planning and Development (SENPLADES) and the Ministry of Finance of Ecuador. Articles 1 and 2 of this agreement established that the economic surplus generated by oil companies will receive the 12% to finance and execute programs and investment projects in Ecuador.

YNP's Management Plan, together with the development programs constitute the technical basis to request funding from SENPLADES and the Ministry of Finance. According to data from the Internal Tax Service (SRI), it is estimated that the 12% of oil taxes for REPSOL YPF and Petro Oriental for year 2010 were approximately 20 million USD. According to the country's legislation, between 0.6 and 0.75% of that 12% surplus can be requested for maintenance of the National Park. This resources should be part of the fiduciary fund that should guarantee the park's maintenance even after oil extraction is finished.

Finally since 2002, the YNP receives financing from the Fund for Protected Areas (FAP). In 2010 the funds received reached 57,557 USD, the highest amount received during the last 9 years (Ministerio del Ambiente 2016).

As this information shows, management plans and financial projections are only available for YNP, which is the core area of the biosphere reserve. This clearly gives the idea that budget for maintenance, with all its challenges, might be only available to the National Park but not available for the rest of the biosphere reserve.

From the principal components found in the economic dimension it is clear that oil exploitation is the most important and impactful economic activity and the main source of financing for the park's maintenance budget. However, logging (mostly selective) is another economic activity. Logging is illegal within the YNP's borders given its protected area legal status, and has an important negative impact on ecosystems. It is conducted by some indigenous communities found in association with oil exploitation camps (Vera 2007). More recently, illegal loggin has occurred by invasive loggers from Peru (Hill 2017).

7 Political Dimension

This dimension involves the historical presence of conflicts, generated by the exclusion of the communities and indigenous groups traditionally living in the biosphere reserve in the decision-making processes. These patterns have been greatly worsened by the cancellation of the Yasuni-ITT project in 2013 to leave oil untapped forever.

A paragraph from Albacete et al. (2004) in a *ParksWatch* evaluation, in collaboration with Wildlife Conservation Society (WCS), already in 2004 states: "*Decision-makers have warranted priority to oil extraction over conservation in the area, as was clearly demonstrated in 1990 when the Ecuadorian Government changed Yasuní National Park's borders in order to allow oil extraction in block 16* (*Villaverde* et al. 2004)". This is a clear example of how the political and economic dimensions are tied in the decision-making, probably since the first oil exploitations.

Stidsen (2009) prepared a timeline of oil companies' presence in Waorani and ZITT territory based on CONAIE (2006):

- Royal Duth Shell (1937).
- TEXACO (1964–1989) ESSO Hispanoil Block8, during the 70s.
- Petrocanadá Block9.
- BRASPETRO (later PETROBRAS) Block17.
- ARCO Block10, today AGIP.
- CONOCO (February 1986–marzo 1989) Block16.
- MAXUS (1990–1995) Block1.
- YPF/MAXUS 1997 Block16.
- Repsol/YPF 1997 today in Block16: Block31 Petrobrás.
- ENCANA Block14 and 17.
- ORYX Block21.
- Perez Companc Block31.

Inevitably, the political and economical dimensions share similarities from the point that oil became a significant part of Ecuador's national income. This is evident since Ecuador government's need for income, contradicts the fact that it is also the authority in charge of providing permission for explorations. Oil revenues have set the environmental and social dimensions aside, to the point that the existence of non-contacted indigenous communities have been ignored when permission for oil companies to operate have been granted.

As Stidsen (2009) mentions, in the Amazon, oil companies have been free to act without the government's control concerning their interactions with indigenous communities. "Community relations", as the relationships of dependence on the oil companies are called, have interfered systematically to diminish the communities' empowerment and have corrupted, persecuted and frightened indigenous leaders.

As mentioned above, the cancellation of the Yasuni-ITT initiative in 2013 brought still more pressure on the relation between the social/environment and the political/economic dimensions. It is important to note, that according to Ecuadorian Constitution, non-renewable resources are banned for exploitation in protected areas, unless the President, without any other consultation, determines it is a national priority. Therefore, enormous responsibility and power of decision-making, falls on one person only.

After President's Correa announcement, an important activist group in Ecuador, called Yasunidos, started collecting signatures to run a plebiscite to ask the Ecuadorian public if they agreed to proceed with the exploitation of Yasuni. The activists collected 757,623 signatures, well beyond the 500,000 minimum required. However, and with very questionable arguments, the Electoral National Committee (CNE) rejected approximately 380,000 signatures. A complete report on the chronology of events is exposed by Yasunidos (2016).

8 Sustainability in Scales

8.1 Local Scale

A depletion of eco-systemic qualities and disarrangement in the local social structures exist in YNP, which is caused by several complex interactions that condition social class structure, manpower organization, interchange systems, State's activities, ideologies, beliefs and social organization (Bunker 1988: 21). The Kichwas, Waorani, Tagaeri and Taromenane ethnic groups who face the effects of resources extractions in YNP. These are communities that share objective and subjective characteristics, which allow them to share a sense of belonging to the community.

According to Stavenhagen (2000), two basic components that serve the identification of the ethnic groups are social organization and territory, as these are directly related with the processes of resources management. On one hand, social organization refers to the intrinsic network of social relationships of the individuals. On the other hand, territory is considered as a symbol of power, a political instrument. For the different indigenous communities that live in YNP and YBR territory. This means a space under their control, where they can develop and reproduce social and cultural activities to subsist and that provide them a sense of belonging (Castro Lucic 1998). However, the indigenous communities allow the oil exploitation and logging in order to have technological benefits or access to otherwise scares economic resources. In addition, the YNP's management department, the local government of Orellana and other institutions, that used to be in charge of the park, are also important actors in this resources extraction. Despite the fact that the YNP was created in 1979, it was until 1992 that government started to plan a management of the area. On this context, the former INEFAN was crated and started a process to delimitate the area in order to stablish management units that could allow to preserve the nature and the culture of the YNP (Paúl Cisneros 2008). In the same year, the INEFAN got a resolution that permitted the oil company Maxus to settle in the YNP and extract oil, in change of funding natural resources research. Nowadays, there are seven exploited oil blocks that overlap with indigenous territories of the YNP.

In 1996, the Ministry of the Environment (MAE) was founded and the INEFAN was eliminated. This new entity defined the borders of the communities and in 2001, with the support of the CONFENIAE (Confederación de Nacionalidades Indígenas de la Amazonía Ecuatoriana), they created an agreement, which aim was to guarantee that the benefits of the use of the natural resources in their territories will be for the indigenous communities (Cisneros 2008). Moreover, in November 2001 riparian communities were allowed to extend their territory to the north of the Tiputini River. In parallel, the FCUNAE and the non-riparian communities started to search founding and allies to develop economical activities in the YNP; and in 2001 they achieved the permission to develop ecotourism activities.

Historically, the oil exploitation within the territory of the YNP and the YBR has generated changes in the area management that enabled to full-fill oil industry's necessities and exigencies. The aggravating circumstance is that the profits generated with the exploitation of oil resources in these areas, represent a large portion of Ecuador's fiscal budget, therefore there is little hope for conservation. However, the potential biological and cultural heritage loses in the YNP and the YBR will be a catastrophe, from local to planetary scale.

8.2 National Scale

At the national level the YBR represents the struggle to preserve not only biodiversity but also the last communities that live directly in nature and from nature, in self-isolation, as our ancestors did before the colonization. According to Paúl Cisneros (2008) for more than 350 years the Ecuadorian Amazon has suffered of the extraction of its resources, which has also caused the breakup between human communities and the environment, which is in turn, according to Bunker (1988) one of the main causes for chronic impoverishment.

Oil revenues from resources located in the biosphere reserve on the other hand represent one of the main sources of economic income for the country, which can only look forward to exploiting more oil in search of financial stability. However, for the reserve's maintenance, the budget is always short and it requires hard work to distribute the oil benefits.

Another important fact is that the contacted indigenous communities, specially the ones living in the north part, also depend now directly from the resources that oil companies provide, and which have constantly depicted a relationship of control and power over the indigenous communities by means of assistencialism.

In the territory of the YBR, this dependency exist since the beginnings of the inclusion of the Waorani as beneficiaries of some minimum infrastructure provided by oil companies in the 80s, to the current more strategic plans run by the different oil companies to negotiate and facilitate the extraction in the region.

The so called "communitarian plans" represent a relationship model between oil companies and the Waorani, and despite having some positive changes through the years, still focus on the maximum profit of oil companies, and establish a relationship in which, the Waorani, as beneficiaries of some goods and services provided by oil companies, are forced to consume and support this model (Rivas Toledo and Lara Ponce 2001).

Previously, the strategies in the communitarian plans reduced Waorani "protectorates", but these days, they are offered different types of jobs at the multi-nationals, starting a complex dependence on money and westerner services (Rivas Toledo and Lara Ponce 2001).

At the same time, the increasing pressure on this region and the communities living here has given birth to several environmental movements and voluntary associations that work hard in the pursue of conservation. Among these we have Yasunidos and Yasuni Oro Verde.

Also, we have to consider the sovereignty factor, as the YBR suffers of illegal logging that benefits from its geographic location, precisely in the border with Peru in the East. That is why the control organisms like Police and Military play another important role here, having to patrol and develop strategies together with the different government agencies and the Park management.

8.3 Global Scale

The Amazon rainforest provides important ecological services not only to Ecuador but to the world. Some of these services are: carbon storage in the biomass, regulation of the hydric cycle and the climate, rain generation at regional level, soil conservation, biodiversity support (Wirth et al. 2009). Global deforestation and pollution of rainforests have severe consequences for climate change.

Seen as systems of energy flux, Paúl Cisneros (2008) states that the economic practices in the Amazon, protected by the law and the Government, allow goods to be exported in an extraction of matter and energy model, that in turn simplifies the natural environment to the resources obtained. This process of appropriation affects the environment making it every time less capable of saving energy and take part of the semi-closed flux of matter and energy.

Not to say the least, being close to the headwaters, Yasuni and its possible environmental degradation may have important impacts to the whole Amazon river basin, which is one of the most important pluvial systems in South America.

In fact, one of the strongest arguments that the Yasuni-ITT project presented to the world was the possibility to avoid the emission of more than 492 million tons of CO_2 to the atmosphere, which would happen with the oil extraction.

Yasuni's importance to the world goes far beyond Ecuador; several environmental movements and organizations have placed their eyes here and support campaigns to protect it. In several around the world, the news about Yasuni are followed and have impact in their strategies. Not to forget the importance of the Yasuni-ITT proposal, which however cancelled, still remains as a valid and exemplifying strategy to finance the State, clearly demonstrating that alternative economic models can be developed to diversify the income sources.

9 Conclusions

After the analysis of the existing relationships in the YNP, we conclude its sustainability is highly at risk, which has several social and environmental negative irreversible consequences. The main factor that compromises the sustainability of the YNP is the influence of Ecuador's State policies, which are a result of the historical management and use of lands that integrate Ecuador's Amazon. In addition, the increasingly dependence on economic and political factors in the processes of decision-making, the political annulment of citizens' participation, the lack of transparency in the land's adjudication processes, and the lack of PIA's protection, threat the future of this Biosphere's Reserve.

Ecuador is a country which although its economic and development policies have advanced, reality has remained unchanged to a great extent since the decade of the 70s when oil started to be exploited. Although Ecuador's discourse is that it seeks to overcome its oil dependency, in reality there has been little to no investments in other economic alternative activities, and if it wasn't for international assumed commitments regarding climate change, there wouldn't be many efforts towards the diversification of the economic activities or interest in energy efficiency. Oil dependency is predicted to increase as a source of energy, which threats directly the natural zones with oil reserves of the YNP and the whole Amazon region.

The system of the YNP is complex and requires numerous actions and decisions, in order to ensure a sustainable development and eliminate the unsustainable practices that are currently taking place on these lands. Unfortunately, the complexity of the system is related to several stakeholders, some even contradictory, who pursue their own interest and who are incapable to reach common ground.

Vital factors to be modified urgently are citizen's involvement and contacted indigenous groups participation, as they can ensure a transparent execution of the protection laws of the PIA. Impunity, in addition to a lack of transparency in the decision-making processes can also be reduced and eliminated through social involvement in legal processes. It is necessary to mention that both, impunity and lack transparency, are currently some of the reasons why the various stakeholders remain divided and uncommunicated.

To achieve the sustainability of the YNP, structural changes must be generated at the local scale, even though regional and global factors can also impact its stability at a large scale. In order to strength this weakness, the government must define an economy's diversification plan and reduce the dependency on natural non-renewable resources. We conclude that social participation is an improvement opportunity, since non-empowered population cannot participate in the decision-making of resources' extraction of the YNP, which is one of the main causes for non-sustainability.

References

Albacete C, Espinosa P, Prado W (2004) Rapid evaluation of the Gran Yasuní Napo. ParksWatch, Durham, NC

Bass MS, Finer M, Jenkins CN, Kreft H, Cisneros-Heredia DF, McCracken SF, Di Fiore A (2010) Global conservation significance of Ecuador's Yasuní National Park. PLoS ONE 5(1):e8767

Brundtland G, Khalid M, Agnelli S, Al-Athel S, Chidzero B, Fadika L, … de Botero MM (1987) Our common future (\'brundtland report\')

Bunker S (1988) Underdeveloping the Amazon: extraction unequal exchange and the failure of the modern state. University of Chicago Press, Chicago, p 279

Cannell F (2006) The anthropology of Christianity. Duke University Press

Castro-Lucic M (1998) La cuestión indígena: ¿Hacia territorios como base cultural o identidad sin territorio? Tercer Congreso Chileno de Antropología. Universidad de Chile. 11 p

Cisneros P (2008) Gobernanza ambiental, conservación y conflicto en el Parque Nacional Yasuní, en: La guerra del fuego. Políticas petroleras y crisis energética en América Latina, Guillaume Fontaine y Alicia Puyana, (Coordinadores), FLACSO Ministerio de Cultura

Coello F, Nations J (1989) "Plan preliminar de manejo del parque nacional Yasuní Reserva de la Biósfera". Quito: MAG. 121 p

Contraloría General de la República de Costa Rica (2015) Indicadores para la medición del impacto de la participación ciudadana en las EFS de la OLACEFS. Retrieved from http://www.olacefs.com/wp-content/uploads/2016/01/Indicadores-de-PC-OLACEFS.pdf

Del Valls TA (1978) El Instituto Lingüístico de verano, instrumento del Imperialismo. Nueva Antropología III(9):117–142

Faisal T (2012) Forecasting and analysis of the potential renewable resources for ecuador's power sector (2012–2032). University of Applied Sciences Cologne, Cologne, Germany

Gallopín G, Hammond A, Raskin P, Swart R (1997) Branch points: global scenarios and human choice. Stockholm Environment Institute, Estocolmo, Suecia. PoleStar Series. Report No. 7. Ver http://www.gsg.org

Hill D (2017) Ecuadorians denounce foreign loggers in Yasuni national park. The guardian. Retrieved from https://www.theguardian.com/environment/andes-to-the-amazon/2017/jun/16/ecuadorians-denounce-foreign-loggers-in-yasuni-national-park?CMP=twt_gu

Jorgenson JP, Coppolillo PP Análisis de amenazas como una herramienta para la conservación del Yasuní en Ecuador. Retrieved from http://manejofaunasilvestre.org/DesktopModules/Bring2mind/DMX/Download.aspx?Command=Core_Download&EntryId=5311&language=es-ES&PortalId=86&TabId=3469

Larrea C (2010) Yasuní-ITT: una iniciativa para cambiar la historia: Quito, Gobierno Nacional de la República del Ecuador
Martínez-Alier J (1992) "El ecologismo de los pobres". Envío Digital, 125. Disponible [02/09/2006] en: http://www.envio.org.ni/articulo/
McDonald T (1989) "Respuesta indígena a una frontera en expansión: Conversión económica de la selva Kichwa en hacienda ganadera", en Whitten N (ed), Amazonía Ecuatoriana La otra cara del progreso. 3ra edición. Abya Yala, Quito, pp 171–199
Meadows (2009) Thinking in systems. A primer. Earthscan, London, 239 p. 718
Ministerio del Ambiente (2015) Recuperado de "Sistema Nacional de áreas Protegidas", el 29 de octubre de 2016, de: http://areasprotegidas.ambiente.gob.ec/info-snap
Ministerio del Ambiente (2016) Recuperado de "Minsiterio del Ambiente", el 29 de octubre de 2016, de: http://www.ambiente.gob.ec/
Ministerio del Ambiente del Ecuador (2005) Análisis de las necesidades de financiamiento del Sistema Nacional de Áreas Protegidas del Ecuador. MAE, Quito, p 142
Ministerio del Ambiente del Ecuador (2011) Plan de Manejo del Parque Nacional Yasuní
Pappalardo SE, De Marchi M, Ferrarese F (2013) Uncontacted huaorani in the Yasuní biosphere reserve: geographical validation of the zona intangible Tagaeri-Taromenane (ZITT). PLoS ONE 8(6):e66293. doi:10.1371/journal.pone.0066293
Raskin P, Banuri T, Gallopín G, Gutman P, Hammond A, Kates R, Swart R (2002) Great transition: the promise and lure of the times ahead. Stockholm Environment Institute, Global Scenario Group, Stockholm Environment Institute, Boston
Rivas Toledo A, Lara Ponce R (2001) Conservación y petróleo en la Amazonía ecuatoriana: Un acercamiento al caso huaorani. Abya-Yala, EcoCiencia
Romo D (2010) Lo bueno, lo malo y lo feo de la propuesta Yasuní-ITT. Polémika 5(1)
SIL International (2015) A Brief History of SIL International. Retrieved from: http://www-01.sil.org/sil/history.htm?_ga=GA1.2.670113866.1475638218
Stavenhagen R (2000) Conflictos étnicos y Estado nacional. México: UNRISD—Siglo XXI:1–46
Stidsen S (2009) El Mundo Indigena 2007. IWGIA
UNDP (2013) Ecuador Yasuni ITT Trust Fund. Retrieved from http://mptf.undp.org/yasuni
UNESCO (2016) Reservas de biósfera. Retrieved from http://www.unesco.org/new/es/natural-sciences/environment/ecological-sciences/biosphere-reserves/
UNICEF (2004) Nacionalidades y pueblos indígenas, y políticas interculturales en Ecuador. Una mirada desde la educación, Quito-Ecuador
Vallejo MC, Larrea C, Burbano R, Falconí F (2011) La Iniciativa Yasuní-ITT desde una perspectiva multicriterial. Programa Conjunto para la Conservación y Manejo Sostenible del Patrimonio Natural y Cultural de la Reserva de Biosfera Yasuní, Ecuador
Vera CA (Writer) (2007) Taromenani. El exterminio de los pueblos ocultos.: Cámara oscura
Villaverde XF, Ormaza V, Marcial YJ, Jorgenson (2004) Parque nacional y reserva de la Biosfera Yasuní: historia problemas y perspectivas. Ediciones ABYA-YALA, Quito, Ecuador
Wirth C, Gleixner G, Heimann M (2009) Old-growth forests: function, fate and value–an overview. In Old-growth forests. Springer, Heidelberg, pp 3–10
Yasunidos (2016) Yasunidos. Retrieved from http://sitio.yasunidos.org/

Author Biographies

Alicia Anahí Cisneros Vidales is Mexican, with a bachelor in Architecture by *UASLP* in Mexico. Studies on Bioclimatic and Sustainable Architecture at the *UNAM* in Mexico increased her interest in sustainable and resilient development. Nowadays she pursues a double Master degree in Environmental Sciences and Natural Resources Management at the *UASLP* in Mexico and the *TH Köln* in Germany.

Victor Mauricio Barriga Albuja is Ecuadorian, with a bachelor in Electronics Engineering by *ESPE* in Ecuador and a diploma in Sustainable Development by *UTN* in Argentina. His deep interest in environment conservation and human-nature relationship has been nourished by multiple professional and traveling experiences in several countries around the world, in multi-cultural environments and his passion, which is Photography. Currently pursuing a double degree Master's Program in Environmental Sciences between *UASLP* in Mexico and *TH Köln* in Germany, he devotes his time to better understand how humanity can develop in harmony with nature and provide sustainable solutions to our needs.

Individual Resilience and the Environmental Education for Sustainability as a Base of Community Resilience. A Case Study with High School Teachers

Erick Cajigal, Ana Lucía Maldonado and Edgar González-Gaudiano

Abstract This text describes the methodology and first results of a case study on the community resilience of cities that are recurrently affected by floods in the state of Veracruz, Mexico. A problem arising from the changes in the frequency and intensity of tropical cyclones resulting from climate change which mainly affects populations in poverty and those settled in high risk areas. It has been documented that populations resist, recover and even transform themselves in the face of adversity, which has been called community resilience. Community resilience is based on individual resilience and the latter can be developed through education. For this, a significant figure is needed, among other things, to provide relevant models of resolution as well as leadership. The case study confirms that young people can play a determining role during the floods; therefore the significant figure can be originated with theirs teachers. So, teachers with resilient characteristics are necessary to strengthen resilience in their students. Then, students could share individual resilience with their friends, family and neighbors, to create a social belt of transmission of resilience permeated by environmental education for sustainability. For this it is necessary to have resilient teachers. Therefore, this research identifies the individual and community resilience of high school teachers to provide a diagnosis to be considered in actions that improve their capacities of prevention, recovery and transformation to floods.

E. Cajigal (✉)
Universidad Veracruzana, Campus Sur. Paseo 112, Lote 2, Sección 2A, Edificio B, 3er Piso, Col. Nuevo Xalapa, Xalapa, Veracruz, Mexico
e-mail: cajigal1983@hotmail.com

A. L. Maldonado · E. González-Gaudiano
Instituto de Investigaciones en Educación, Universidad Veracruzana, Campus Sur. Paseo 112, Lote 2, Sección 2A, Edificio B, 3er Piso, Col. Nuevo Xalapa, Xalapa, Veracruz, Mexico
e-mail: maldonado.analucia@gmail.com

E. González-Gaudiano
e-mail: egonzalezgaudiano@gmail.com

W. Leal Filho et al. (eds.), *Sustainable Development Research and Practice in Mexico and Selected Latin American Countries*, World Sustainability Series,
https://doi.org/10.1007/978-3-319-70560-6_12

Keywords Individual and community resilience · Climate change · Teachers

1 Introduction

Research on community resilience is currently in a significant position compared to other studies, due to increased knowledge and growing concern about climate change (CC). Climate change is defined as a phenomenon associated to an increase of environmental deterioration, as well as to a series of disasters in different populations on a global level never seen before (Gutiérrez and González Gaudiano 2010).

CC is the result of the current model of developing countries. Considering the environmental situation and its increasingly serious nature; many cases leads to irreversible changes in nature. So in theory, it is not sustainable to continue to maintain the rhythm of exploitation of natural resources. Social inequality and global problems brought about by the devastation of natural resources (Colín 2003).

Economic growth has been the basis of development, without regard to social and environmental consequences (Maldonado 2008). The term development itself has a great myth within which phenomena such as industrialization, urbanization, concentration and accumulation of capital, as well as destruction of nature are concealed (Gutiérrez and González Gaudiano 2010).

There is a relationship between the model of growth, CC and disasters. That is, the type of current development is associated to progress and welfare, material growth and a high consumption of goods and services. In this regard, based on our lifestyles and model of growth, Riechmann (2005), argues that tropical cyclones (TCs) are born from the exhaust pipes of our cars and the chimneys of our nuclear power plants.

Some disasters are processes that are getting socially developed (with the deforestation, pollution, consumerism, inefficiency, corruption, among others), that harshly reveal themselves during episodes caused by extreme natural events, such as a tropical cyclone (Ruiz 2005). In other words, disasters are constructed in and by society, so they can be considered as latent phenomena with the expectation that contingencies may occurred.

In sum, the dominant model of development contributes to make worse diverse environmental problems such as CC. The intensity and frequency of some natural phenomena are related to CC, such as tropical cyclones. Nevertheless, this causal relationship is not always evident.

Therefore, CC is complex considering that it is difficult to see all its effects; with consideration that available scientific knowledge itself does not have all the answers about it. Some natural events have conditions that could be associated to science as an effect of CC. In recent years, for example, tropical cyclones such as Katrina (2005), Dean (2007), Irene (2011), Ingrid and Manuel (2013) have been registered around the world, and at the same time, global average temperature has increased (Riechmann 2005; IPCC 2015).

Variations of climate exist as a consequence of the global average temperature. In addition, some phenomena such as frost, storms, droughts and floods affect entire populations because of the intensity in which they occur. In this sense, communities in marginal, economic and social conditions are highly vulnerable because they are more susceptible to suffer damages (Chardon and González 2002).

In 2014, hydrometeorological disasters were the phenomena with most impact, accounting for 58.7% of the total fatality of all disasters reported (Guha-Sapir et al. 2015). The effects of CC are global and harshly hit developing countries, especially those with poor communities settled in high-risk areas (González Gaudiano 2007). Coastal areas and communities that are close to rivers are most prone to flood disasters.

Mexico is located within the intertropical region in the middle of two large oceans: the Atlantic and the Pacific. TCs occur on both coasts and on average, 25 TCs are recorded per year. Around four or five of them usually enter in the territory and cause severe damages because of heavy rains with their consequential floods (CENAPRED 2001).

In recent years, in the coastal state of Veracruz, on the Gulf of Mexico, there is a large area that has been impacted by several TCs (Karl in 2010, Ernesto in 2012 and Barry, Fernando and Ingrid in 2013), which have resulted in disasters.

The highest level of precipitation during TCs has increased in the state of Veracruz. This is consistent with the observations made around the world, which indicates a significant statistically increase in rainfall intensity, although in a contradictory way, the annual average of precipitation in the state of Veracruz as a subtropical area has decreased (CNIAN 2012). The increases of the precipitation level of TCs, the vulnerability of towns of Veracruz and its geographical location have caused floods with human and economic losses. Communities respond in different ways to these types of disasters.

On the other hand, the IPCC (2015), defines resilience as the ability to recover of social, economic and environmental systems in order to confront, respond, or reorganise in the face of a hazardous event. In this way, the systems mentioned preserve their essential function, identity and structure while maintaining the capacity for adaptation, learning and transformation. Being resilient makes it possible to confront adversities by learning and transforming to reduce or eliminate damages. Individuals can develop resilience (Melillo 2001), and if many people combine their individual resilience, they can create community resilience (Gamarra 2010), which allows populations to adapt, learn and even transform in the face of adversities.

In this context, where the model of development, as well as CC and its effects frequently impact the populations of the state of Veracruz causing disasters, it is necessary to study the factors that favour the development of individual resilience in significant people who exert a leadership in the community. In our study, we have chosen high school teachers and the idea is to consider our results in order to construct strategies aimed at developing abilities that allow local residents to resist, recover and/or transform themselves in the face of floods.

This research considers high school teachers as significant adults. Therefore, they are relevant figures in the creation of resilience (Munist et al. 1998; Melillo 2001). Teachers can provide resolution models, as well as leadership in young people, who can share this information to their families, neighbours, and friends to create a social driven belt, permeated by Environmental Education for Sustainability (EES).

2 The Social Driven Belt of Resilience and Environmental Education for Sustainability

The social driven belt is an analogy of the transmission belt related to mechanics. It is taken to the social field to illustrate the process and the elements that contribute to the development of resilient abilities in the face of floods. The transmission belts have the function of synchronously move two or more pulleys, so that they can generate power for a common purpose. In this process there are two types of pulleys: the drive pulleys and the driven pulleys. The drive pulleys are the ones that begin the movement. These pulleys rotate while transmit energy to the driven pulleys through the transmission belt that unites them.

In the social field, as in the mechanical process, there are elements that begin the movement, to whom we will name drive actors. These drive actors transmit information, motivation and knowledge to the driven actors through this social driven belt. These driven actors will subsequently become in drive actors, starting in this way a cyclical process.

A social driven belt gets established as the intangible element with properties of cohesion and coordination that involve the process of activation of drive and driven actors for the good of the community. These properties are continuously promoted and provided by all actors. The social driven belt unlike mechanics, considers that the elements in the process are not static. That is why actors will appear in different moments like drive or driven actors taking into account their abilities.

Teacher's work must transcend the classroom. It is inconceivable that the role of a teacher cannot be thought as static, nor it can be reduced to manage his students' learnings without interventions and without the contexts in which the educational process is developed. Teachers must consider the attention to social and environmental needs in their didactic strategies (Zabalza 1993), which means that they must extend their work beyond the classroom. The teacher would begin the movement (Fig. 1).

Although the social driven belt involves teachers, their own students, who are in their adolescent period, would be the immediate actors moved by the social driven

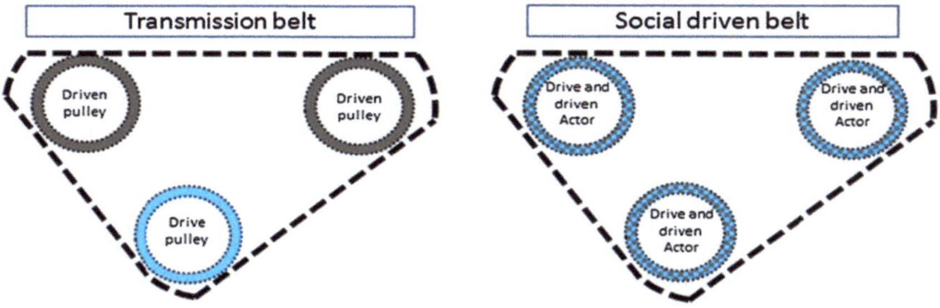

Fig. 1 Transmission belts: mechanics and social. *Source* Author's own elaboration

belt. It is in the adolescent period when skills are acquired, and behaviours are established, and the most important point is that project of life starts to be developed. In this period, individuals already have a type of formal thinking that allows them to confront, reflect, debate, analyse and draw their own conclusions that are connected to their realities (Alchourrón et al. 2001).

This belt will be permeated by the EES. This type of education appears as a viable proposal that contributes to confront environmental problems and their consequences (González Gaudiano and Arias 2009), which are greatly derived from the model of development, the CC and its effects, as we have been saying. The EES is understood as a liberating education that seeks the social transformation towards a living in order to better coexist between us and nature (Terrón 2010). It is characterized by a complex approach that allows the understanding that environmental problems are not isolated but are interrelated with other problems.

Social change from schools and promoted by EES can contribute to build a critical and independent thinking in students, families and communities when more teachers become involved in taking the challenge of developing the capacity to understand coherently the reality and the meaning of the world (González Gaudiano 2007). Therefore, the social driven belt permeated by the EES pretends a social change, where high school teachers and their students initially participate, followed by the rest of the community, creating in this way the conditions for building community resilience, in order to understand and to deal with environmental issues and their effects, such as floods.

The study is located in three towns in the central area of the state of Veracruz: Tlacotalpan, Cotaxtla and La Antigua. These towns have been affected several times by floods, having tropical cyclones Karl and Matthew as the most severe in 2010 (Tejeda 2011; Méndez 2014). This research considers four high schools. The town of La Antigua and Cotaxtla have one state high school each and Tlacotalpan has two state high schools.

3 Resilience: Concept and Study

In the social field, the term of resilience allows to explain, identify, evaluate and reinforce systems, institutions and people's abilities (Maldonado and González Gaudiano 2013). The term can be defined from three approaches (Uriarte 2013): stability, recovery and transformation. In this research, we agree with the process approach towards transformation. This approach emphasises people's abilities to adapt to changes, the ability to learn, creativity, future orientation, strengths and opportunities rather than hazards and weaknesses. In some cases, disasters can be constituted in opportunities in order to change or improve the inefficient living conditions of those affected (Olabegoya, 2006 in Uriarte 2013).

According to Melillo (2001), there are two interpretative paths of resilience. The first one is the geneticist interpretation that confirms the existence of individual who possess resilient innate characteristics. The second interpretative path is that resilient characteristics can be developed in certain situations throughout the growth of people. We based on this second path in which we associate the figure of significant adults or teachers, who can develop resilient capacities in their students so these last ones also will help the rest of the community to develop those capacities.

Munist et al. (1998) and Melillo (2001) have documented that throughout the growth of people, there is interaction with individuals, such as parents, educators and carers, who make sure that resilient capacities get developed. These individuals have a high social status before their children, students and people with whom they coexist, which is very appropriate in order to provide models of resolution and leadership.

In addition to the *significant adult figure*, there are other factors for the development of resilience. The first factor is belonging to *cohesive groups such as families*. Salamanca (2009), studied the experience of disasters lived by poor and middle-class families in Bolivia, making a comparison between the results obtained. It is highlighted among his results that disasters impact the stability of the family, so it is likely to find disruption and rupture in family relationships. However, beyond the socioeconomic level, the impact on family stability is in function of family cohesion prior to disaster (Salamanca 2009). Along those lines, Henderson (2001), has identified that resilience and socioeconomic status are not related because poverty does not make impossible the development of family cohesion, so the economic situation of a family is not a condition for it to be resilient or not, although the bond between members does it.

The second factor is the *Around the "I"* (Henderson 2001). It consists of three elements: (a) I have (b) I am, and (c) I can. These factors are characterized by some issues that are important when is time to confront adversities such as the attitude, self-confidence, character and support from close and beloved people.

In the case of teachers, the third factor is *work environment*. This factor gathers some elements such as prosocial bonds, life skills, affection and support, opportunities and high expectations and clear and firm boundaries (Henderson and

Milstein 2003). These six elements consider different actors with whom the teacher daily interacts, such as directors, students, parents and other teachers.

The fourth and last factor is *health and physical condition*. Healthy conditions without weight problems provide advantages compared opposite conditions when we found ourselves in front of adversity (Richardson et al. 1990).

In sum, the factors of development of resilience in teachers are: *cohesive groups or families, the Around the "I" factor, work environment, and health and physical condition.* The development of resilience in teachers is possible due to the strengthening of these factors. What is more, community resilience could be produced if people's individual resilience develop, coordinate and interact to each other in the same territory (Gamarra 2010).

In regard to community resilience there are other factors involved. Suarez Ojeda (2001) has identified four factors. The first factor is *collective self-esteem*, which refers to pride and satisfaction that a person has for feeling part of the community and the place where he/she lives. The second factor is *cultural identity*, which refers to a person's involvement in local traditions and acceptance and practice of his/her neighbours' ideologies related to entertainment, politics and religion. The third factor is *collective honesty*, which appreciate local authorities and neighbours' honesty. The fourth factor is *social humour*, which refers to the ability of kindly express a gloomy fact, so that through the storytelling, it is possible to cause a tranquillity effect before some tragedy.

In addition to these four factors, the authors of this article suggest two more factors derived from experiences in towns in Veracruz. These suggestions also come from researches made by Gamarra (2010). These other factors are *significant institutions* and *reflection from experience*. The *significant institutions* have a decisive role that directly impact on communities' tasks. In Gamarra's research (2010) made in Peru, it is illustrated how two institutions, one educational and the other a religious one worked together for the strengthening of local leaderships to put up with an armed conflict. The work of these two institutions got documented. They developed individual resilience until turn it into community resilience, which helped them to resist the armed conflict without suffering displacement.

Finally, the factor *reflection from experience* refers to the reflection on the damage experienced individually against the damage experienced by other most affected people. To think of oneself as fortunate to receive less damage than others, provides a feeling of optimism that activates the strengths to confront, overcome and even learn from adversity.

As we have pointed out, this research aims at providing elements that can help as basis in order to develop individual and community resilience in towns affected by floods in the state of Veracruz. This would be accomplished by high school teachers as significant adults through social driven belt permeated by the EES.

4 Methodology to Identify Individual and Community Resilience

The methodology used to study the factors involved in the development of resilience in teachers, as well as the recognition of community resilience from the teacher's perspective, are due to techniques and procedures inherent to quantitative research. The categories of this research are related to the previously identified factors. Five categories were considered for individual resilience, each with their own variables:

1. *Work environment*: social bonds; understanding, involvement and acceptance of school rules and policies; peer interaction's; opportunities for training that favour their educational practices; recognition and appreciation of their work in the community; recognition of places for initiatives within the school.
2. *Around the "I"*: (a) (I have) I have people with whom I can trust to and appreciate me; I have people with whom I can count on before an adverse situation; I have people from whom I can learn correct behaviours. (b) (I am) I am a person for whom others feel appreciation and affection; I feel happy when I do something good for others; I respect myself and respect others; I am sure that everything will be fine; I take responsibilities for my actions. (c) (I can) I talk about things that trouble me; I seek the way to solve my problems; I constantly reject the activities that could be dangerous; I look for help if I need it.
3. *Family*: levels of cohesion, communication, organisation and cooperation for daily activities and adverse situations.
4. *Significant adult figure* (self-recognition): to model behaviour; to support when there are problems; to create an atmosphere of respect and collaboration; acceptance by students and family.
5. *Health and physical conditions:* chronic diseases; frequency of common diseases; type of food; participation in sport activities; identification of overweight.

In regard to the identification of the community resilience of three towns mentioned before, whose perspective comes from high school teachers, six categories were considered with their own variables, which are related to the factors mentioned before:

1. *Collective self-esteem*: there is pride for being part of the community; satisfaction for the place where someone lives.
2. *Cultural identity*: the individual participates in local traditions; he/she celebrates local festivals; he/she shares the ideologies of majority of his/her neighbours (religion, politics, and entertainment).
3. *Collective honesty*: Appreciation of his/her neighbours and local authorities' honesty.
4. *Reflection from experience of others*: there is a reflection on the degree of damage suffered by one person compared to the one suffered by others most affected.

5. *Social humour*: Appreciation for sense of humour.
6. *Significant institutions*: level of student's meaning about school; level of community's meaning about school; identification of significant institutions in the community; participation in significant institutions.

With all these parameters using a range of 48 items, a survey was integrated. The 90% of these items are dichotomous and are related to individual and community resilience, although social aspects are also investigated in order to identify the participants. In this sense, the sample of participating teachers was established by the confidence of the study that is 90% with a 10% margin of error.

The instrument was subjected to a pilot test with high school teachers of the town of Juchique de Ferrer in Veracruz, a place that has also been affected by tropical cyclones. The collected data was considered for Kuder Richardson reliability test with a value of 0.81, which means that it is a reliable instrument (Campo and Oviedo 2008). The application of the instrument in the field work did not have complications, which allowed involving more subjects than the established in the sample. As well as, the collected data was submitted to Kuder Richardson test, resulting in reliable values in all cases.

The study of individual resilience was applied to a representative sample of teachers from the four schools mentioned in Table 1. However, the identification of community resilience was only made from the perspective of teachers since it was not possible for lack of time and resources to expand research to representative samples of populations in each municipality. Knowing of the first source the community resilience of the municipalities may be of interest for future research.

With the purpose of firstly generate frequency tables that allowed the description of the results, the collected data were introduced in the SPSS program. The progress of this research is currently in this point. Afterwards, combinations of results will be made to find correlations, in order to define better the findings and conclusions.

Table 1 Field work

Town	High school's name	Total population of teachers in high school	Sample established	Participating teachers	Kuder Richardson result
La Antigua	Agustín Yáñez	42 teachers	27 teachers	28 participants	0.86
Tlacotalpan	Avelino Bolaños Palacios	21 teachers	17 teachers	18 participants	0.82
Tlacotalpan	CBTIS 35	26 teachers	19 teachers	20 participants	0.80
Cotaxtla	Telebachillerato Cotaxtla	9 teachers	9 teachers	9 participants	0.82
	Total	98 teachers	72 teachers	75 participants	

Source Author's own elaboration based on field work

For this analysis, descriptive statistics was used with the procedures: counting and percentages. Each item was designed to indicate whether or not it contributed to the development of resilience. The answers provided elements to know the state of each one of the variables and the sum of the answers allowed identifying the individual resilience of the teachers and we approached to know the community resilience of the towns.

5 Outcomes. First Approaches to Individual and Community Resilience

The results of individual resilience indicate that there are categories with poor participation in their development, such as: *work environment* and *health and physical conditions*. Despite the fact that teachers have the opportunity given by directors for training and improve their educational practice, in all cases it is evident that *work environment* is not satisfactory; teachers also point out that their work is not recognised inside and outside the school. In addition, with divided opinions, teachers affirm that high school's rules do not apply equally to all teachers and besides this; there are no places for making work suggestions that could complement the school curriculum.

In regard to *health and physical conditions* category, the results are divided. In high schools from Cotaxtla and La Antigua, three of five items were positive, although in two high schools from Tlacotalpan, three of five items were negative. Let us remember that overweight is one of the items investigated in this research, where all the cases were negative, finding this problem as crucial to other health and physical problems. Hence this category weakly contributes to resilience in two high schools (Cotaxtla and La Antigua) and in the other two high schools this category does not contribute to the development of resilience.

Health and physical conditions represent individuality what gives it advantages over other categories. The improvement of health and physical conditions, placing overweight as a crucial subject, will greatly depend on the teacher himself, this is the reason, and so the attention goes directly to his/her work.

The categories *around the "I"*, *family* and *significant adult figure* were positive in all cases, with high percentages in most of the items. In other words, the combination of teacher's personality, attitude, support and character with the belonging to cohesive families and his/her self-recognition before his/her students contribute to the development of resilience. For example, the category *around the "I"* registered in all cases, higher percentages to 77.8%; in this regard, the *family* category averages 79.9% with positive answers among all cases; and the *significant adult figure* obtained in all the cases percentages above 80%. This information points out that the three categories mentioned here are highly strengthened and strategies could be reliably supported in them.

In regard to teachers' perception of community resilience, the *social humour* category is considered to have a high contribution to community resilience; in all cases it was positive with values above 88%. In different moments of the field work when interacting with teachers, we noticed that they have sense of humour, which is a characteristic that can transmit tranquillity effects and it helps to activate strengths before a tragedy. Being a capacity that allows defusing and overcoming misfortunes and calamities, it emphasizes strengths and helps to overcome unpredictable, illogical, unusual, and contradictory situations.

Similarly, the category *cultural identity* obtained positive results. This category places Tlacotalpan as the town with the best contribution because all items were positive in Avelino Bolaños and CBTIS 25 high schools having as result five of six positive items. The town of Cotaxtla was rated positive with five of six items and the results of La Antigua were also positive to a lesser extend having as result four of six items.

The results of *cultural identity* indicate that there are items where all cases coincide. For example, the involvement in entertainment activities, parties and local traditions are located in positive results in all cases, however, most people negate their participation in political activities (in elections). At first, the absent of participation in political activities was considered as a negative result, nevertheless the result was inverted when analysing it, because of the non-participation in political activities is a sign of cultural identity since most people shares the rejection to these activities.

The *collective honesty* category is negative in two high schools: one is located in Tlacotalpan (CBTIS 35) and La Antigua. In the other two high schools (Tlacotalpan, Avelino Bolaños and Cotaxtla) the values are divided: one of two items was positive. There is an item rated as negative in all cases which is the local authorities' honesty. In general, this category does not contribute to community resilience.

The results of the *reflection from experience* category were positive for Tlacotalpan and La Antigua. This means that teachers perceive that their community is not as affected as others. On the contrary, teachers from Cotaxtla point out that their community has been the most affected. In a section of the *resilience survey*, it is inquired whether the survey respondent teacher lives in that community or if he/she moves from one community to another; The results obtained indicate that eight of nine teachers do not live in Cotaxtla and the one who does live in the community considers that there were other most affected communities.

The previous result about the feeling of belonging to the community does not allow to describing it as the most affected. In other words, when teachers point out that their towns were not so damaged like others, it is implicit some competitiveness between communities derived from the feeling of belonging that suppress them from seeing themselves as vulnerable.

The following category is *collective self-esteem* that identified pride and satisfaction towards the place where teachers work. In all cases, the result was positive, although the percentage in Cotaxtla was the lowest what suggests that there is a correlation between collective self-esteem and reflection from experience. Eight of nine teachers do not live in Cotaxtla but they move regularly and the feeling of belonging to the place where they live is greater than the one of the place where they work.

Finally, the *significant institutions* category indicates that teachers perceive their educational institutions as significant in their community. In this category, other significant institutions from different levels such as churches were identified. The results agree with the study made by Gamarra (2010) that indicates that these two institutions are significant for communities in Peru.

To sum up, the categories that contribute to the development of individual resilience are: *around the "I"*, *family*, and *significant adult figure*. On the other hand, the categories that required to be considered as areas of opportunity are: *health and physical conditions* and *work environment*. The results of community resilience indicate that: *social humour, cultural identity, reflection from experience* and *collective self-esteem* categories, contribute in a positive way; and the category that requires attention for being considered in the development of community resilience is *collective honesty*. In addition, the identification of *significant institutions* allows visualising links for the development of resilience. In the towns studied in this research, the significant institutions are the educational and religious.

6 Conclusions

The effects of climate change are global, however, the impacts are different in each population because of their different capacities to resist, recover and transform. The floods caused by tropical cyclones represent the greatest human and economic losses in the world and besides, some projections indicate that these phenomena will become even more intense, what represents more extreme environmental contingencies with their consequential damages.

The populations of this research have been severely damaged by floods, however they have demonstrated to have a high level of resilience. According to this study, community resilience can be strengthened by identifying and organising significant leaderships such as high school teachers. Although the results are preliminary, they help us to recognise and approximate to the factors that contribute the most to individual resilience, like: *around the "I"*, *family*, and *significant adult figure* and community resilience, like: *social humour, cultural identity, reflection from experience* and *collective self-esteem.* Also, the study allows us to identify the variables that difficult the development of individual resilience, like: *health and physical conditions* and *work environment,* and community resilience *collective honesty.*

Therefore, it is important to analyse the social components of resilience, as well as educational strategies appropriate to sociocultural characteristics of each town in order to improve its resilient capacities. The data collected will be considered in a future research in order to developing abilities that allow local residents to resist, recover and/or transform in the face of floods.

References

Alchourrón M, Daverio P, Moreno E, Piatinni J (2001) Promoción de la resiliencia en adolescentes de una escuela semirrural. En Melillo A, Suárez Ojeda EN (Compiladores). Resiliencia descubriendo las propias fortalezas. Paidós, Argentina, pp 169–184

Campo A, Oviedo H (2008) Propiedades Psicométricas de una Escala: la Consistencia Interna. Revista Salud Pública 10(5):831–839

CENAPRED (2001) *Diagnóstico de Peligros e Identificación de Riesgos de Desastres en México*. Secretaría de Gobernación, México

Chardon AC, González JL (2002) Amenaza, Vulnerabilidad, Riesgo, Desastre, Mitigación, Prevención. Primer acercamiento a conceptos, características y metodologías de análisis y evaluación. IDEA UN, pp 1–38

CNIAN (2012) Cambio climático. Evidencias, impactos y opciones. Respuestas a preguntas comunes sobre la ciencia del cambio climático. Academia Nacional de Ciencias, Estados Unidos

Colín L (septiembre, 2003). Deterioro Ambiental vs. Desarrollo económico y social. Revista Boletín IIE 27(3), 103–108

Gamarra J (2010) *Resiliencia social y cambio en comunidades campesinas afectadas por conflicto armado interno: el caso de las comunidades de Incaraccay y Tanquihua en la provincia de Cangallo, Ayacucho*. Instituto Peruano de Educación en Derechos Humanos y la Paz, Perú

González Gaudiano EJ (2007) Educación y cambio climático: un desafío inexorable. Trayectorias 25:33–44

González Gaudiano EJ, Arias MA (2009) La educación ambiental institucionalizada: actos fallidos y horizontes de posibilidad. Perfiles Educativos 31(124):58–68

Guha-Sapir D, Hoyois P, Below R (2015) *Annual Disaster Statistical Review 2014 The numbers and trends*. Bélgica, Centre for Research on the Epidemiology of Disasters (CRED)

Gutiérrez E, González Gaudiano E (2010) *De las teorías del desarrollo al desarrollo sustentable: construcción de un enfoque multidisciplinario*. Universidad Autónoma de Nuevo León, México

Henderson N (2001) Introducción. Nuevas tendencias en resiliencia. En Melillo A, Suárez Ojeda EN(Compiladores). Resiliencia descubriendo las propias fortalezas. Paidós, Argentina, pp 19–30

Henderson N, Milstein M (2003) *Resiliencia en la escuela*. Editorial Paidós, Argentina

IPCC (2015) Synthesis Report. Contribution of Working Groups I, II and III to the Fifth Assessment Report of the Intergovernmental Panel on Climate Change. Grupo Intergubernamental de Expertos sobre el Cambio Climático, Suiza

Maldonado AL (2008) *La contribution de l'organisation communautaire et des participants à un mode de gouvernance civique en environnement dans trois régions administratives du Québec*. Laval University, Canada

Maldonado AL, González Gaudiano EJ (2013) De la resiliencia comunitaria a la ciudadanía ambiental. El caso de tres localidades en Veracruz, México. Revista Íntegra Educativa 6(3): 13–28

Melillo AC (2001) Prefacio. En Melillo A, Suarez Ojeda E (Compiladores). Resiliencia descubriendo las propias fortalezas. Paidós, Argentina, pp 15–18

Méndez LM (2014) *Criterios para la formulación de estrategias de gestión del riesgo ante eventos climáticos extremos en la localidad de Cotaxtla, Veracruz*. Universidad Veracruzana, México

Munist M, Santos H, Kotliarenco M, Suárez Ojeda E, Infante F, Grobert E (1998) *Manual de identificación y promoción de la resiliencia en los niños y adolescentes*. Organización Panamericana de la Salud, Estados Unidos

Richardson GE, Neiger BL, Jensen S, Kumpfer KL (1990) The Resiliency Model. Health and Education 21(6):33–39

Riechmann J (2005) Un mundo vulnerable. Ensayos sobre ecología, ética y tecnociencia. Catarata, España

Ruiz JC (2005) De la construcción social del riesgo a la manifestación del desastre. Reflexiones en torno al imperio de la vulnerabilidad. Revista Desacatos 19:99–110

Salamanca LA (2009) Estudio de resiliencia en desastres naturales en seis barrios de la ciudad de La Paz, Bolivia. En Ramírez F (Coordinador), Estrategia Internacional para la Reducción de los Desastres, La gestión del riesgo urbano en América Latina: recopilación de artículos. pp 205–222

Suárez Ojeda EN (2001) Una concepción latinoamericana: la resiliencia comunitaria. En Melillo A, Suarez Ojeda EN (Compiladores), Resiliencia descubriendo las propias fortalezas. Paidós, Argentina, pp 67–82

Tejeda A (2011) Introducción. Apuntes corográficos de las inundaciones en el estado de Veracruz. En Tejeda A (ed), Las inundaciones de 2010 en Veracruz. Memoria social y medio físico. Consejo Veracruzano de Investigación Científica y Desarrollo Tecnológico, México, pp 23–42

Terrón E (2010) *Educación ambiental. Representaciones sociales y sus implicaciones educativas*. Universidad Pedagógica Nacional y Horizontes educativos, México

Uriarte J (2013) La perspectiva comunitaria de la resiliencia. Psicología Política 47:7–18

Zabalza M (1993) El ambiente desde una perspectiva curricular. En Caride JA (Coord.), Educación Ambiental: realidades y perspectivas. Universidad Santiago de Compostela, España, pp 243–296

Assessment of Competencies for Sustainability in Secondary Education in Mexico

Jorge Gustavo Rodríguez-Aboytes and Luz María Nieto-Caraveo

Abstract Given a conceptual model of competencies for sustainability, such as the generic competency for sustainability established in the secondary education latest reform in Mexico, an assessment of competencies for sustainability was conducted in order to determine, in an exploratory way, the level of achievement of those competencies in the students of the main high school institution in the city of Matehuala. Within the assessment process, an analysis of the institution curriculum was required, and interactions with and teachers through interviews, focus groups, and workshops. An assessment framework was constructed in order to design an instrument that generates the learning evidence regarding those sustainability competencies and, thus, to infer their levels of achievement. The instrument consisted of a performance task, and a knowledge and attitudes questionnaire. It was applied to 60 students, divided in 3 groups according to their level of instruction. Despite the small differences among the three groups, the results showed that all students have good attitudes towards sustainability and understand the basic terms related to it; however, their performance in systemic thinking, prospective thinking and the ability to propose strategies are not developed well enough to face the challenges that the environmental and civilization crisis demand nowadays.

Keywords Sustainability competencies · Assessment · Secondary education

J. G. Rodríguez-Aboytes (✉)
Programas Multidisciplinarios de Posgrado en Ciencias Ambientales, Universidad Autónoma de San Luis Potosí, Ave. Manuel Nava 201 2do Piso Zona Universitaria, C.P. 78210 San Luis Potosí S. L. P, Mexico
e-mail: iq_gustavo91@hotmail.com

L. M. Nieto-Caraveo
Facultad de Ingeniería, Universidad Autónoma de San Luis Potosí, Dr. Manuel Nava no. 8 Col. Zona Universitaria Poniente, C.P. 78290 San Luis Potosí S. L. P, Mexico
e-mail: lmnieto@uaslp.mx

W. Leal Filho et al. (eds.), *Sustainable Development Research and Practice in Mexico and Selected Latin American Countries*, World Sustainability Series,
https://doi.org/10.1007/978-3-319-70560-6_13

1 Introduction

Humankind faces the biggest challenge in all of its existence: The environmental and civilization crises, which have marked a new chapter in human evolution and history (Leff 2007). In order to face and to solve these crises, sustainability and Sustainable Development (SD) have established the basis to initiate the change in the direction of meeting present and future generation needs WCED (1987), without exceeding Earth's natural capacities of resources regeneration and pollution cushioning. Despite the variety of concepts, terms and discourses about SD (e.g. Amador and Padrel 2013; Baker 2016; Gutierrez and González 2010; Hopwood et al. 2005), this type of development has risen as a paradigm shift in which previous ways of understanding and influencing reality were not good enough (Barth 2015).

Sustainability entails changes in people's behavior and mindsets; thus, Education for Sustainable Development (ESD) plays a crucial role in shaping people's and social group's capacities to contribute to the transformation and building of a better world through the development of competencies for sustainability. In this way, the research problem is that given a conceptual model of competencies for sustainability, what is the level of achievement of such competencies? The purpose of this research was to know the levels of competencies development that the students in secondary education have achieved in order to improve those levels. This research was conducted within the normative and conceptual framework of the latest secondary education reform in Mexico (SER) and the scholar space of Matehuala High School (MHS).

2 Methodology

Figure 1 shows the general steps that were taken to do this research. It is divided in three general parts (Rodríguez-Aboytes 2017):

1. *Conceptual step:* This consisted in a general literature review about competencies for sustainability models and the description of this type of competencies

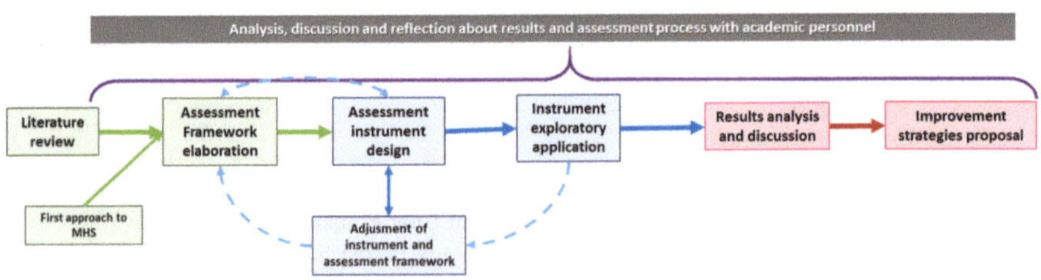

Fig. 1 General steps to elaborate the research. *Source* Rodríguez-Aboytes (2017)

in the SER in Mexico. This step also included a curriculum analysis and a first field approach to MHS in order to explore the procedures and mechanisms that this scholar institution was implementing in competencies for sustainability development and assessment through open interviews, focus groups, and workshops to academic personnel and students. All of this previous work allowed constructing an assessment framework that establishes the levels of achievement of these competencies and the evidences to infer those levels.

2. *Operative step:* With the assessment framework constructed, an assessment instrument was elaborated in order to generate the evidence to determine the levels of achievement of these competencies for sustainability. An exploratory application was conducted and the results permitted to calibrate and adjust the assessment framework and the assessment instrument itself.
3. *Discussion and analysis step:* After the operative step, an analysis of the assessment results was done in order to discuss them in terms of pedagogic strategies, curriculum pertinence, and sustainability competencies models. Strategies of improvement to MHS academic personnel and the SER were proposed.

3 Research Orientation

This research combines different types of research because it belongs to many scientific fields: social, educational and environmental. Therefore, it can be said that this work is part of the emergent body of knowledge: Sustainability Science (Kajikawa et al. 2007; Spangenberg 2011).

There is foregoing research in assessment of competencies for sustainability, such as the work of Remington Doucette and Musgrove (2015), which is one of the most representative in the field of ESD. Moreover, another type of work is the research of Shephard et al. (2014) who did an assessment of environmental literacy in Otago University, focusing on environmental attitudes and knowledge. Despite these works, this research can be considered exploratory in the sense that assessment of competencies for sustainability is not a common and systematized practice.

The main issue in this type of research is of epistemological nature, being that there is no unique way to know the development of competencies for sustainability; contrary to the technique and administrative orientation that approaches this kind of assessment in elementary, secondary and higher education in Mexico. So this research did not yield results that can be extrapolated to all secondary education population in Mexico, not even to all MHS students, because it would be necessary to have solved the epistemological issue. Therefore, this research is an exploratory study that goes from the epistemological to the technical nature (Rodríguez-Aboytes 2017).

4 Competencies for Sustainability

What should students do in order to contribute to SD? Which are the competencies that they need to accomplish that? These questions are being solved through the proposal of models and frameworks of competencies for sustainability. These frameworks entail changes in curriculum, methods and pedagogical practices, and even changes in scholar institutions management.

The environmental and civilization crises, as well as the sustainability challenge, are seen as problems to be resolved in which informed and conscious action is required. That is why one of the models is the action competence proposed by Mogensen and Schnack (2010), as an elemental issue in the future citizen and stakeholders education. Other models start off from the so called environmental literacy, which is the general capacity to reflect about the actual state of society and natural world and to make decisions in favor of sustainability (Ansari and Stibbe 2009; Parkin et al. 2004; Stibbe and Luna 2009).

One of the competencies for sustainability framework is the *Gestaltungskompetenz* model or shaping competency, which is the capacity to act and solve problems about society according to SD principles (de Haan 2006, 2010). Nonetheless, the most convincing and structured framework is the work of Wiek et al. (2011), who collected, through a literature review, a model that systematizes previous models and proposals such as those mentioned above. Their framework establishes five key sustainability related competencies that are based on the capacity of problem solving: systems-thinking competence, anticipatory competence, normative competence, strategic competence and interpersonal competence (Wiek et al. 2011).

The competencies for sustainability domain are given by all the frameworks and models presented above (see Table 1). It is not about getting a unique model of competencies, since each one of them must be considered in the context and education need of the region in which they are embedded. Each model or framework has different characteristics, so they must be discussed and analyzed regularly in order to improve them.

With this brief description and analysis, it can be said that there is not a single competency model for sustainability. On the contrary, when the topic of competencies for SD rises, it is about many models or frameworks well organized through a strong conceptual basis. It is important, as Wiek et al. (2011) and Barth (2015) emphasize, that competencies for SD are not a 'laundry list'. Moreover, one has to take into consideration the theoretical model which the terms of competence and competencies are based on, since these are object of continuous debate, controversies and recontextualizations (Díaz Villa 2010).

Therefore, what are the competencies that future citizens and stakeholders have to develop? The Wiek et al. (2011) framework provides the best answer so far with competencies that they established. Despite these models described, there are few frameworks that make the competencies operative in their development and assessment. Two works in targeting this are Remington Doucette and Musgrove (2015) and Wiek et al. (2015) papers.

Table 1 Competencies for sustainability conceptual frameworks

Author	Framework or proposal	Characteristics
Mogensen and Schnack (2010)	Action competence	Action oriented competencies founded on the *Bildung* concept
Parkin et al. (2004)	Sustainability literacy: Learning and skills for SD	To understand the need for change to a sustainable way of doing things To have sufficient knowledge and skills to decide and act in a way that favors SD To be able to recognize and reward other people's decisions and actions that favor SD
Stibbe and Luna (2009)	Handbook of sustainability literacy	General expectations highlighting systemic-thinking abilities
Ansari and Stibbe (2009)	Skills for sustainability literacy	Capacities and abilities to deal with the sustainability challenge: social problems understanding, working in groups, citizenship, wise use of technology, self-reflection, and multidimensional research
de Haan (2006, 2010)	Gestaltungskompetenz	A framework integrated by: Future thinking competence Interdisciplinary work competence Transcultural understanding competence Participatory abilities competence Planning competence Empathy, compassion, and solidarity capacities Etc
Brundiers et al. (2010)	Key competencies in Sustainability	Strategic knowledge cluster Practical knowledge cluster Collaborative cluster
Wiek et al. (2011)	Key competencies in Sustainability	Systems-thinking competence Anticipatory competence Normative competence Strategic competence Interpersonal competence
de Kraker et al. (2007)	Transboundary competence	Ability to think, communicate, learn, and collaborate across the boundaries that divide perspectives in crucial transition processes
Hidalgo and Fuentes (2013)	Basic Competencies for Sustainability in Higher Education	Cognitive competencies Methodological competencies Attitudinal competencies
Cebrián and Junyent (2015)	Competencies in Education for Sustainable Development	Four competence dimensions: Practical skills Ethical values Attitudes Emotions

Source Rodríguez-Aboytes (2017)

4.1 Competencies for Sustainability in Mexico's Secondary Education

The specific scope of this research is the SER in Mexico and the scholar space of MHS. The SER is formed by many federal and official statutes that were enacted in 2008 (DOF 2008a, b, 2009, 2012). With the objective of articulating all the secondary education system, the SER established a set of generic, disciplinary and professional competencies as part of Mexico's secondary education graduate profile. The SER 444 official statute set eleven generic competencies in which a competency for sustainability was included (DOF 2008a).

The competency for sustainability and its attributes says (DOF 2008a, p. 5, author's own translation), "It contributes to sustainable development, critically, with responsible actions".

Attributes:

1. "Assume an attitude that favors environmental problem solutions in local, national, and international contexts".
2. "Recognize and comprehend biological, economic, political and social implications of environmental damage in an interdependent global context".
3. "Contribute to reaching equilibrium between short and longtime interest in relation to the environment".

This competency is actually a set of three competencies for sustainability that are the attributes described. These competencies have a mixed focus, ambiguous and poorly clear. Analyzing the previous competency statement, the term 'critically" and 'responsible actions' alludes an empowerment approach of ESD (Barth 2015) or critical-social (Caride and Meira 2000). Also, those terms entail a humanist-critical discourse of SD (Barth 2015), transformation position (Hopwood et al. 2005) or sustainability ideal model (Baker 2016).

Despite the approach of these competencies for sustainability, they are embedded within a larger model that is organized in a 'laundry list' structure. The consequence of this is that the eleven generic competencies are fragmented among them originating a competition in the curriculum of secondary education institutions. This has greater implications in practical pedagogies of development and assessment of competencies: teachers and academic managers have serious difficulties in assessing the competencies, turning their concern and efforts into technical and administrative aspects leaving alone the conceptual and epistemological ones.

4.2 Competencies for Sustainability in MHS Curriculum

Matehuala city is located in the north of the state of San Luis Potosí, Mexico. Matehuala High School (MHS) is the main secondary institution in the city and is

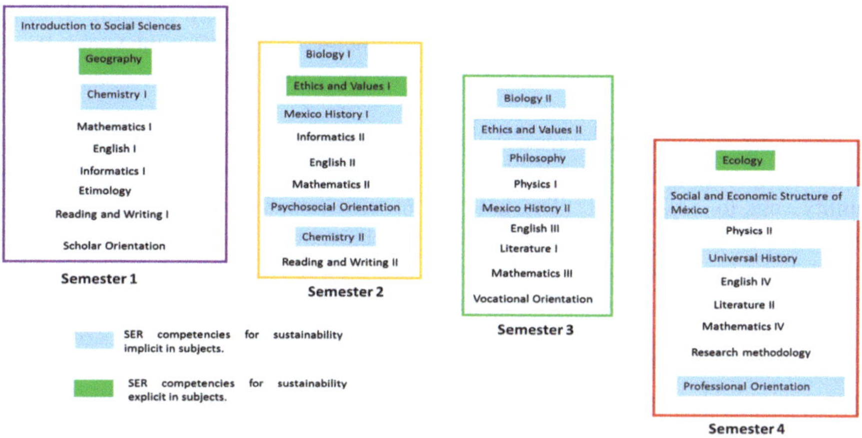

Fig. 2 SER competencies for sustainability detected in MHS curriculum. *Source* UASLP (2015, cited in Rodríguez-Aboytes 2017)

the only high school institution of San Luis Potosí Autonomous University. The study plan is structured in four semesters (two years) that must be completed by the students in order to graduate from high school and enter higher education.

A curriculum analysis of MHS was conducted in order to understand the incorporation of the SER competencies for sustainability. This required an in-depth analysis of each study program to see the extent to which these competencies are being developed and assessed. In Fig. 2, the study plan of MHS is presented with the competencies detected in each syllabus.

It is shown that the SER competencies for sustainability are explicit (green ones) only in Geography (semester 1), Ethics and Values I (semester 2) and Ecology (4 semesters). Therefore, only in these syllabuses, the development and assessment of competencies for sustainability are done. Nevertheless, in the specific analysis of each syllabus (blue ones), it was detected that the competencies, in an implicit way, are also being developed.

This analysis reinforces the discussion in the previous section: there is a fragmentation of all the eleven generic SER competencies, which causes a competition among them. MHS, as well as many secondary education institutions, already have methods, mechanisms, and assessment procedures, but they all are oriented to technical aspects. This exercise allows seeing that within MHS curriculum there is a basis for the development of the SER competencies for sustainability.

5 Assessment Framework

The assessment of competencies for sustainability is a type of learning assessment. This assessment is the inquiry about the learning of students (Maki 2004). It is not information gathered, it is more an inquiry process that seeks to answer if the students really accomplish the learning outcomes at the end of a teaching-learning period and how the scholar institution contributes to this (Palomba and Banta 1999). In this case, the learning outcomes are the competencies for sustainability levels of achievement.

However, the assessment process requires a reference to which the performance or learning can be compared. Students have to develop competencies for sustainability, but to what level? The assessment is essentially a comparison and an inference activity, so knowing the levels of achievement of competencies for sustainability requires evidence from which these levels can be inferred. The SER neither sets the levels of achievement of such competencies nor the evidence needed, so an assessment framework is crucial in order to do the evaluation. Without this framework, any evaluation or assessment is tremendously difficult and meaningless. That is why it was argued before that the first problem in this type of assessment is more epistemological than technical. Figure 3 shows the gap between the reference (the levels of achievement) and the evidence needed.

Hence, the assessment framework is the bridge between conceptual domain that is going to be assessed and the evidence that is required. In other words, it is a conceptual model that establishes the concrete and well detailed references, and the evidences that an assessment instrument will generate (see Fig. 4).

Examples of these assessment frameworks can be found in PISA Draft Science Framework (OECD 2013); the National Environmental Literacy Project (Hollweg et al. 2011), as well as in the recent paper of Wiek et al. (2015), in which a framework for operationalizing competencies for SD was proposed.

Subsequently, an assessment framework was built in this research. The framework is structured in seven steps as shown in Fig. 5. The first two steps correspond to the general statement of the SER regarding the competency for sustainability and its attributes. The next five steps were created from the literate review of

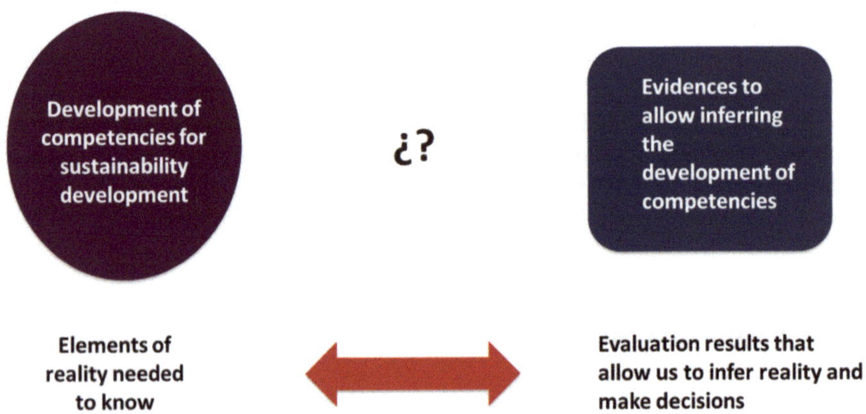

Fig. 3 Representation of the epistemological problem regarding the assessment of competencies for sustainability. *Source* Rodríguez-Aboytes (2017)

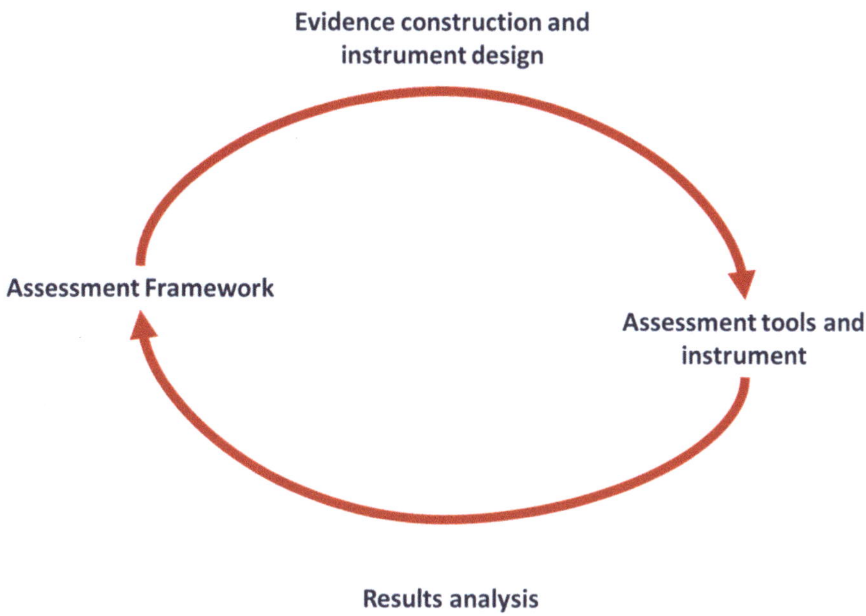

Fig. 4 Assessment framework dynamic and function. *Source* Rodríguez-Aboytes (2017)

competencies for SD models and the analysis of MHS curriculum and syllabuses (Rodríguez-Aboytes 2017).

The attributes are actually the competencies for sustainability, so they are the first level of the assessment framework. These competencies, as mentioned before, are not enough to constitute the levels of achievement. Therefore, these three competencies were broken down in constitutive elements defined by the authors.

For each constitutive element, performance indicators were defined and established. These indicators are the specific learning outcomes. Each performance indicator was divided into three levels of performance. These are the ultimate criteria from which the levels of achievement will be inferred. With all of this established, a group of possible evidences was generated in order to construct an instrument or set of assessment tools.

The performance indicators levels do not mean anything by themselves with regard to the competencies levels of achievement. It is important to differentiate between them since the first are specific for each performance indicators, whereas the second are the qualitative integration of the assessment results, valued through the performance indicators (see Fig. 6).

The six steps of the assessment framework are described in the following three tables; each table refers to one competency: Competency 1 (see Table 2), Competency 2 (see Table 3) and Competency 3 (see Table 4).

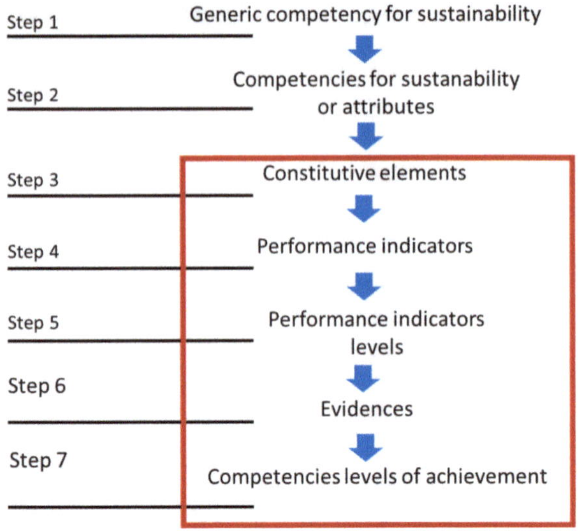

Fig. 5 Assessment framework structure. *Source* Rodríguez-Aboytes (2017)

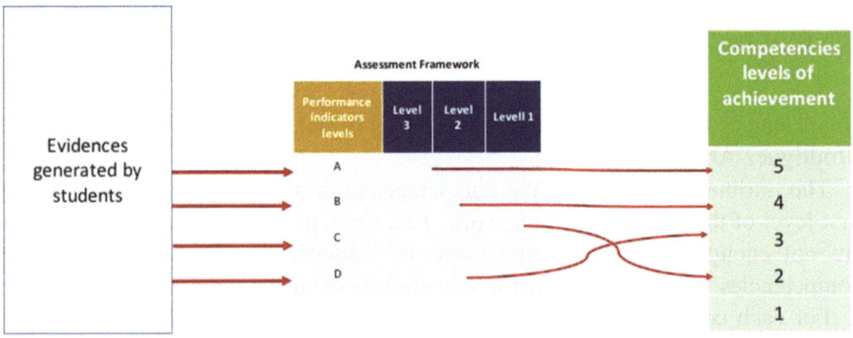

Fig. 6 Dynamic between evidences, performance indicators levels and competence levels of achievement. *Source* Rodríguez-Aboytes (2017)

6 Specific Assessment Tools and Instrument

To assess the SER competencies for sustainability development an assessment instrument was necessary. This instrument is a measurement artifact that, through the student's responses, generates the evidence required that is compared to the references established in the assessment framework. The results of this comparison occur with the performance indicators levels which are integrated in a qualitative way to infer the levels of achievement. In other words, the instrument was designed from the assessment framework described above.

Table 2 Performance indicators for competency 1

General competence	"Contributes to sustainable development, critically, with responsible actions"		
Competencies	1. "Assume an attitude that favors environmental problems solutions in local, national and international contexts"		
Constitutive elements	Environmental basic dispositions	Respect for diversity and plurality	Commitment and responsibility for sustainability
Performance indicators (The student is able to…)	1. Express feelings of concern 2. Show dispositions to know 3. Show dispositions to act 4. Trust his/her actions contribution	1. Respect different ideas and opinions 2. Value ethnic, cultural and gender differences	1. Assume a commitment to contribute to sustainability 2. Respond and reflect over his/her own action

Source Rodríguez-Aboytes (2017)

Inasmuch as it is pretended to infer the levels of achievement of sustainability competencies, the correct way to do so is through the observation of the student's performance. This performance is the answer, activity and the evidence that the student will show during a situation or problem that needs to be solved; in other words, a situation in which the student has to mobilize and articulate their knowledge, attitudes, and skills. That is why the instrument consisted in a performance task (Chun 2010) complemented by a knowledge and attitudes questionnaire (see Fig. 7). The performance task consisted of open constructed questions whereas the others instruments consisted of discrete or closed questions (Rodríguez-Aboytes 2017). The instrument considered context variables such as level of studies, gender, age, and scholar shift.

The attitudes and knowledge questionnaire involves the performance indicators that require a low demand of cognitive resources; the performance task involves the indicators that require a major cognitive resources demand such as systemic-thinking and making decisions.

The instrument was applied without any previous technical calibration of the items because the purpose was to verify if the assessment framework really worked. In other words, the results were directed to covering the epistemological gap.

The instrument was applied to 40 students of the MHS and 20 students who had recently graduated from this school. In total, only 57 answered. It was applied in three different groups according to their level of study: the first year students (A1), the second year students (A2), and the recently graduated students (A3). It is necessary to make it clear that a statistical sample was not determined because the purpose of the research was not to extrapolate the assessment results to the MHS

Table 3 Performance indicators for competency 2

General Competence	"Contributes to sustainable development, critically, with responsible actions"			
Competencies	2. "Recognize and comprehend biological, economic, politic and social implications of environmental damage in an interdependent global context"			
Constitutive elements	Basic Sustainability Issue Comprehension	Problematization	Critical information use	Systemic analysis
Performance indicators (The student is able to...)	1. Recognize basic terms and concepts 2. Comprehend basic notions, concepts and terms	1. Formulate the main problems that appear in a given situation 2. Identify and describe the role of main stakeholders	1. Seek information from a diversity of sources to investigate a problem 2. Select and discriminate information according to the source origin and intentionality 3. Identify the main ideas in the information 4. Formulate coherent answers based on the information analyzed	1. Comprehend interrelations between dimensions 2. Comprehend interrelations between geopolitical scales 3. Analyze the complexity of a given problem 4. Analyze the historical context of a problem 5. Articulate knowledge from different scientific fields

Source Rodríguez-Aboytes (2017)

student population. The instrument was designed to be answered in two hours. Paper and pencil were used in the application format. Each student answered the instrument individually.

The main results of the assessment, the levels of achievement, are ordinal measurement type, so the statistical analysis used was non-parametric (Zorrilla 1988). The achievement levels are presented in terms of frequencies. The principal factor associated to the variability of the levels of achievement is the level of studies, but also an analysis between the other context variables was conducted. The statistical software used in this research was Minitab 16.

Table 4 Performance indicators for competency 3

General competence	"Contributes to sustainable development, critically, with responsible actions"		
Competencies	3. "Contributes to reaching equilibrium between short and longtime interest in relation to the environment"		
Constitutive elements	Future thinking	Establishment of strategies	Communication, organization and leadership skills
Performance indicators (The student is able to…)	1. Construct future scenarios 2. Use future related concepts 3. Explain the contribution of their future professional career regarding to a sustainability problem	1. Set action plans to solve a problem 2. Assess the advantages and limitations of strategies that propose 3. Make decisions on a founded basis	1. Set communication channels to construct group solutions 2. Recognize his own limitations and bias regarding the solution of a problem 3. Mobilize and organize people for solving problems

Source Rodríguez-Aboytes (2017)

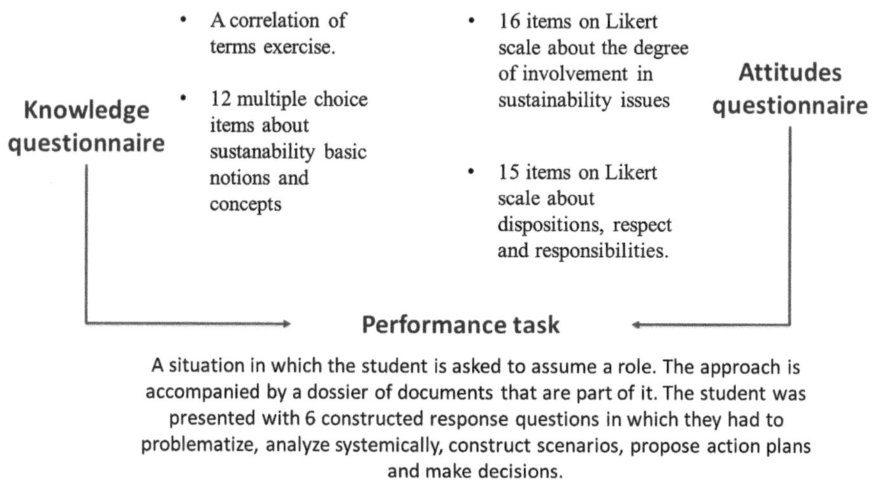

Fig. 7 Assessment instrument design. *Source* Rodríguez-Aboytes (2017)

7 Results and Discussion

The resulting achievement levels of the first competency were high, reaching the maximum levels, but, for the other competencies, the resulting achievement levels were too low (see Fig. 8). There were no differences between frequency distributions according to all the context variables, using parameters such as Spearman's rho, Kendall's Tau-b and test for concordance. This means that the students show good attitudes to sustainability but not to systemic-thinking, strategic, and future thinking capacities (Rodríguez-Aboytes 2017).

The statement that the higher level of studies, the higher levels of achievement of the competencies for sustainability would be, was not confirmed by this research. The curriculum establishes that there is a progression in the competence development when students advance in their level or grade of studies. In the case of MHS, the A1 and A2 groups have not yet completed the Ecology Syllabus, whereas the A3 group has already completed it. The study conducted by Remington Doucette and Musgrove (2015) showed that students who have studied modules related to sustainability have a better performance with respect to sustainability competencies. However, there was no difference found in this research.

From the assessment results, certain types of sustainability discourses (Hopwood et al. 2005) were observed as well as others ways of thinking. The performance tasks open answers allowed to analyze that: individual and isolated efforts, the high priority given to recycle solid waste individually, or the behaviorism tendency of punishment and reward.

The disparity between the resulting levels of achievement of the first competency and the other two originate a discussion around the mind gap between the environmental knowledge and concern and pro-environmental behavior (Kollmuss and Agyeman 2002).

The analysis of the incorporation of the generic competency for sustainability in MHS curriculum expressed that a fragmented conception of curriculum blocks the transversality in the trajectories of competencies for sustainability development in secondary education. The proposal strategies to improve the competencies for sustainability levels of achievement consisted in (Rodríguez-Aboytes 2017):

1. Promoting and offering refresher courses to the MHS academic personnel, in topics about sustainability in a systemic perspective way.

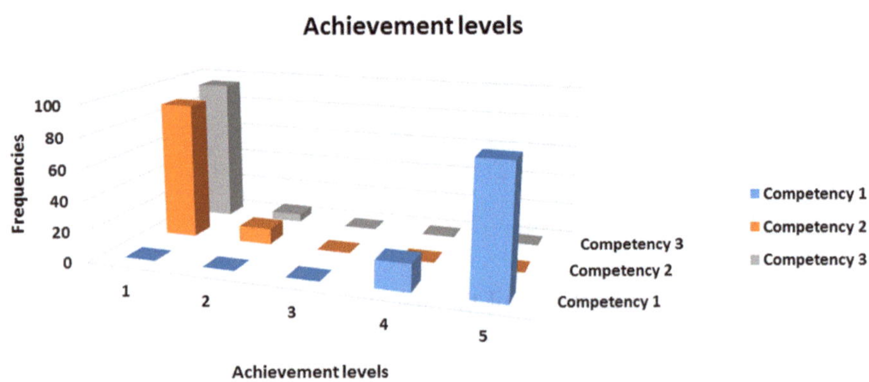

Fig. 8 Resulting competencies for sustainability levels of achievement. *Source* Rodríguez-Aboytes (2017)

2. Emphasizing the critical use of information within teaching and learning practices.
3. Carrying out curricular and extracurricular activities which foster the creation of future scenarios and the proposal of strategies to solving problem.
4. Boosting workshops for the development and assessment of performance tasks with MSH teachers.
5. Retaking the assessment and evaluation competencies process as a collective project which involves all academic stakeholders of the MHS.
6. Proposing and presenting modifications to the normative and conceptual SER framework generic competencies.

8 Conclusions

The assessment framework proposed in this work can be considered as the beginning of the construction of a solid and coherent assessment system. Despite the potential improvement and modifications to this framework, this demonstrated to be robust and powerful to infer the competencies for sustainability levels of achievement. Furthermore, the assessment framework allows observing the epistemological interwoven in which this type of assessment is embedded.

The performance task demonstrated to be a powerful instrument to collect evidences from a high cognitive processes such as systemic, future and strategic thinking. Nonetheless, the performance task is an instrument rarely used in the context of this research. The MHS that responded this instrument was not familiarized with the performance task, so this was a factor that influenced on the assessment results because students found it as an activity which obligated them to answer thoroughly. The performance task can even be used as a pedagogical learning resource to be implemented along a course.

Beyond the results, this question cannot be answered in a simplistic way. First of all, the three competencies cannot be taken separated from each other because all of them jointly correspond to the generic sustainability competency. Even though students showed favorable attitudes towards sustainability, it is clear that they are not able to contribute to its construction.

It should be noted that this should not lead mechanically to negative judgments about the teaching-learning processes in MHS, the curriculum and the SER, since the situation, as it can be seen throughout the research, has a high degree of complexity. It is not known, because that was not the purpose of this research, what results were obtained from the causes, although some hypotheses for future research may be considered.

On the one hand, the general approach to sustainability competencies in the SER establishes a high level of demand and does not mark the limits that would be necessary for the secondary education level. It is very clear that "contributing to sustainable development, critically, with responsible actions" is a high-level purpose, which would be expected of any adult, citizen of this planet and that can acquire multiple expressions.

On the other hand, it should be researched to what extent the high fragmentation of the competences proposed by the SER makes it difficult to integrate them into the specific curricula of the subsystems of the Mexican secondary education system and into the teaching and learning processes. As discussed earlier, the issue is not only about contents in the syllabus nor the use of more or less active teaching techniques, as would seem to be understood from a superficial reading of this subject.

Finally, as noted earlier, the lack of guidance in the SER on assessment approaches, methods and techniques could be adding more difficulty to the challenge of competencies development, especially because of the crucial importance of formative evaluation approaches.

Thus, the assertion that the students who responded to this assessment do not meet the achievement levels on the competencies for sustainability has to be seen as a challenge to deepen the aforementioned structural causes and to establish better conditions for its development.

There was a great challenge and conceptual and epistemological difficulty in the realization of this project. Actually, the field of evaluation is one of the most misrepresented and co-opted by external discourses and agendas unrelated to the purposes of evaluation. Therefore, it requires a good theoretical basis on which to anchor to navigate around three fields of ultimate difficulty: sustainability, competencies, and evaluation. Part of the contribution of this research is that it works jointly with these three fields and gives them a sense and orientation.

Further research can be developed from what is proposed in this work. First of all, it is necessary to continue this evaluation process by improving the reference framework and the assessment instrument. Another future work is to assess whether there are actually differences in achievement levels according to contextual variables such as age, degree of studies, gender, environment in which one lives, etc.

Acknowledgements We would like to thank to Escuela Preparatoria de Matehuala for opening their scholar space and access to academic personnel and students; and to Secretaría Académica de la Universidad Autónoma de San Luis Potosí for their support in this research project. This paper is product of a M.Sc. Thesis work financed by the Mexican National Council of Science and Technology through the scholarship No. 334036 given to the main author.

References

Amador F, Padrel C (2013) Integrating sustainability into the University: past, present and future. In: Caeiro S, Filho W, Jabbour C, Azeiteiro U (eds) Sustainability assessment tools in higher education institutions: mapping trends and good practices around the world. Springer International Publishing, Switzerland, pp 65–78

Ansari WEl, Stibbe A (2009) Public health and the environment: what skills for sustainability literacy—and why? Sustainability 1(3):425–440. http://www.mdpi.com/2071-1050/1/3/425/

Baker S (2016) Sustainable development 2nd edn. Routledge, New York

Barth M (2015) Implementing sustainability in higher education: learning in an age of transformation 1st edn. Routledge, New York

Brundiers K, Wiek A, Redman CL (2010) Real-world learning opportunities in sustainability: from classroom into the real world. Int J Sustain High Educ 11(4):308–324. doi:10.1108/14676371011077540

Caride JA, Meira PÁ (2000) La Construcción Paradigmática de la Educación Ambiental. Educación Ambiental y Desarrrollo Humano. España, Ariel Educación, pp 189–248

Cebrían G, Junyent M (2015) Competencies in education for sustainable development: exploring teacher's views. Sustainability 7(3):2768–2786

Chun M (2010) Taking teaching to (Performance) task: linking pedagogical and assessment practices. Change, (April), 22–29. doi:10.1080/00091381003590795

Díaz Villa M (2010) Introducción al estudio de la competencia. Educación y competencia. In: Flexibilidad y competencias profesionales en las universidades Iberoamericanas. México: Ediciones Pomares, p 29–58

DOF (2008a) Acuerdo 444 por el que se establecen las competencias que constituyen el marco curricular común del Sistema Nacional de Bachillerato. Diario Oficial de la Federación, México

DOF (2008b) Acuerdo número 442 por el que se establece el Sistema Nacioanl de Bachillerato en un marco de diversidad., México: Diario Oficial de la Federación

DOF (2009) Acuerdo número 486 por el que se establecen las competencias disciplinares extendidas del Bachillerato General. Diario Oficial de la Federación, México

DOF (2012) Acuerdo número 656 por el que se reforma y adiciona el Acuerdo número 444 por el que se establecen competencias que constituyen el marco curricular común del Sistema Nacional de Bachillerato, y se adiciona el diverso número 486 por el que es establecen la. Diario Oficial de la Federación, México

de Haan G (2006) The BLK '22' programme in Germany: a 'Gestaltungskompetenz' based model for Education for Sustainable Development. Environ Educ Res 12(1):19–32. doi:10.1080/13504620500526362

de Haan G (2010) The development of ESD-related competencies in supportive institutional frameworks. Int Rev Educ 56(2):315–328. doi:10.1007/s11159-010-9157-9

De Kraker J, Lansu A, Van Dam Mieras R (2007) Competences and competence-based learning for sustainable development. In: Crossing boundaries. Innovative learning for sustainable development in higher education. Frankfurt, p 103–114

Gutierrez E, González É (2010) De las teorías del desarrollo al desarrollo sustentable 1st ed., México: Siglo XXI Editores. Universidad Autónoma de Nuevo León

Hidalgo LA, Fuentes MA (2013) The development of basic competencies for sustainability in higher education: an educational model. US-China Educ Rev 3(6):447–458. Available at SSRN: https://ssrn.com/abstract=2486082

Hollweg KS, Taylor JR, Bybee RW, Marcinkowski TJ, McBeth WC, Zoido P (2011) Developing a framework for assessment environmental literacy. Washington, DC: North American Association for Environmental Education. Available at http://www.naaee.net

Hopwood B, Mellor M, O'Brien G (2005) Sustainable development: mapping different approaches. Sustain Dev 13(1):38–52. doi:10.1002/sd.244

Kajikawa Y, Ohno J, Takeda Y, Matsushima K, Komiyama H (2007) Creating an academic landscape of sustainability science: an analysis of the citation network. Sustain Sci 2(2):221–231. doi:10.1007/s11625-007-0027-8

Kollmuss A, Agyeman J (2002) Mind the gap: why do people act environmentally and what are the barriers to pro-environmental behavior? Environ Educ Res 8(3):37–41. doi:10.1080/13504620220145401

Leff E (2007) La complejidad ambiental. Polis [En línea], 16. URL: http://polis.revues.org/4605

Maki PL (2004) Assessing of learning: building a sustainable commitment across the institution 1st edn. American Association for Higher Education, ed., Stylus Publishing, Virginia, United States

Mogensen F, Schnack K (2010) The action competence approach and the new discourses of education for sustainable development. Environ Educ Res 16(1):59–74. doi:10.1080/13504620903504032

OECD (2013) PISA 2015 draft science Framework., pp 1–54. Available at: https://www.oecd.org/pisa/pisaproducts/Draft PISA 2015 Science Framework.pdf. Accessed November 28, 2015

Palomba CA, Banta TW (1999) Assessment essentials: planning, implementing and improving assessment in higher education. Jossey- Bass, San Francisco

Parkin S, Johnson A, Buckland H, White E (2004) Learning and skills for sustainable developmetn: developing a sustainability literate society. Forum for the future, London

Remington Doucette S, Musgrove S (2015) Variation in sustainability competency development according to age, gender, and disciplinary affiliation. Int J Sustain High Educ 16(4):537–575. doi:10.1108/IJSHE-01-2013-0005

Rodríguez-Aboytes, JG (2017) Evaluación de competencias para la sustentabilidad en educación media superior. Master's Thesis. Universidad Autónoma de San Luis Potosí. San Luis Potosí. México. Available at: http://ninive.uaslp.mx/jspui/bitstream/i/4210/3/MCA1RAJ201701.pdf

Shephard K, Harraway J, Lovelock B, Skeaff S, Slooten L, Strack M, Furnary M, Jowett T (2014) Is the environmental literacy of university students measurable? Environ Educ Res 20(4):475–495. doi:10.1080/13504622.2013.816268

Spangenberg J (2011) Sustainability science: a review, an analysis and some empirical lessons. Environ Conserv 38(3):275–287. doi:10.1017/S0376892911000270

Stibbe A, Luna H (2009) Introduction. The handbook of sustainability literacy: skills for a. Totnes, Green Book, pp 1–6

UASLP (2015) Plan de estudios de la Escuela Preparatoria de Matehuala. Available at: http://www.preparatoria.uaslp.mx/Paginas/Programas.aspx. Accessed November 1, 2015

World Comission of Environment and Development (WCED) (1987) Our common future. Oxford University Press, Oxford

Wiek A, Bernstein M, Foley RW, Cohen M, Forrest N, Kuzdas C, Kay B, Keeler LW (2015) Operationalising competencies in higher education for sustainable development. In: Hand of higher education for sustainable development. Routledge, New York, NY, pp 297–317

Wiek A, Withycombe Keeler L, Redman CL (2011) Key competencies in sustainability: a reference framework for academic program development. Sustain Sci 6(2):203–218. doi:10.1007/s11625-011-0132-6

Zorrilla S (1988) *Introducción a la metodología de la investigación* 6th ed., México, D.F: Ediciones Océano

Methodological Proposal to Evaluate Touristic Activity with Local Sustainability Criteria in the Hydrographic Sub-Basins of the Huasteca Potosina, Mexico

Salvador Luna Vargas, María de la Luz Valderrábano Almegua, Irma Suárez Rodríguez and Larisa Alcérreca Molina

Abstract This work presents a methodological proposal based on a system of sustainability indicators through the Framework for the Evaluation of Natural Resource Management Systems Incorporating Sustainability Indicators, from which a comparative analysis of the *Huasteca Potosina* tourism systems was carried out, with the purpose of establishing interregional comparison standards. The indicators allowed the monitoring and evaluation of the degree of sustainability of the sub-basins by means of initiatives of citizen participation in the processes of planning of the main tourist destinations in the regional scope. In this way, it was possible to identify the conditions that limit and strengthen the development of

---

Translator/Reviser: Martha Elva Martinez Carrillo: interpreter, language trainer and translator.

Translator: BAE. Alfredo Lara Álvarez: professor and translator at the Autonomous University of San Luis Potosí (UASLP).

S. L. Vargas (✉) · L. A. Molina
Section of Graduate Studies and Research, School of Tourism, National Polytechnic Institute (Instituto Politécnico Nacional), Miguel Bernard 39, La Escalera, 07630 Mexico, Mexico
e-mail: salvador_luna14@yahoo.com.mx

L. A. Molina
e-mail: larisa_alcerrecam@hotmail.com

M. de la Luz Valderrábano Almegua
Interdisciplinary Center for Research and Studies on Environment and Development, National Polytechnic Institute (Instituto Politécnico Nacional), 30 de Junio de 1520 S/N, La Laguna Ticomán, 07340 Mexico, Mexico
e-mail: mvalderrabano@ipn.mx

I. S. Rodríguez
Multidisciplinary Academic Unit Huasteca Area, Autonomous University of San Luis Potosí (Universidad Autónoma de San Luis Potosí), Romualdo Del Campo 501, Rafael Curiel, 79060 Ciudad Valles, San Luis Potosí, Mexico
e-mail: irma.suarez@uaslp.mx

W. Leal Filho et al. (eds.), *Sustainable Development Research and Practice in Mexico and Selected Latin American Countries*, World Sustainability Series,
https://doi.org/10.1007/978-3-319-70560-6_14

sustainability in tourism, in order to foster a reflection on the commitments that can be adopted by tourism service providers to carry out actions that promote equity and improve the living conditions of the inhabitants of the local communities through the adequate conservation and restoration of the ecosystems from an ecological and social context that allows to develop an alternative model of sustainable tourism. The results generated criteria to formulate recommendations that contribute to the sustainable development of the region through the decision-making of the local population and the public administration; this is why this article is essential for the creation of measurement tools used to determine the evaluation methods for tourism activity in rural areas.

Keywords Sustainable tourism · Water resources management
Rural sustainability · Sustainability assessment · Touristic resource utilization

1 Introduction

Tourism, like other activities, depends primarily on hydrological resources, and involves transforming the landscape in a way that takes into consideration the natural, sociocultural and economic context of the communities that are tourist destinations in order to develop recreational spaces through the creation and re-creation of environments geared toward visitors (Anton 2009). For this to occur, transportation, lodging, dining and recreational services must be in existence; nevertheless these entail the creation of negative effects such as: generation and inadequate management of wastes, pollution of bodies of water and overexploitation of natural resources, all of which alter habitats and cause displacement and replacement of flora and fauna species (Maldonado 2006; WTO 2009; Monterroso 2011; Ivanova 2012; Álvarez 2015); in addition to socioeconomic impacts like displacement of people, migration and poorly compensated jobs (Ávila 2013; Conde 2013), economic dependence (García et al. 2013), cheap labor (Sartorello 2013), loss of identity in the population, loss of the mother tongue and the sociocultural space (Pereiro 2015).

Thus, it becomes necessary to reinterpret resource management from a multidisciplinary perspective in order to achieve a comprehensive vision of the land and tourism system of each region. As such, it is convenient to address the tourism phenomenon by using a sub-basin approach, as this allows for understanding of the interrelationship among natural resources (climate-topography-soil-vegetation) as well as their appropriation and use in addition to the impact on water quantity, quality and seasonality in specific geographic spaces (Cotler and Priego 2007; Burgos et al. 2015). This approach enables assessment that uses multi-criteria decision analysis techniques in order to promote decision making by organizations and institutions involved in tourism activity.

The importance of sustainability analysis in tourist destinations lies in designing operational frameworks which allow for tangible evaluation of plans, programs and projects aimed at fostering more equitable and environmentally healthy social development in rural communities through actions and strategies that facilitate decision making when facing challenges related to the land, political relations, civic engagement, public-private partnerships and the legal and normative frameworks applied to the territory, carried out collectively and individually, allowing for the comprehensive understanding of the limiting factors and opportunities for sustainable development. For this reason, this research proposes the creation of tools that allow an improved decision making process in regards to policies and actions to be implemented, with the purpose of providing methodologies for sustainability analysis and the recovery of natural resources to be applied from within the community space as a self-management, autonomy and independence exercise, since most methodologies applied have used academic, technical and external approaches (Masera et al. 2000; Pensado-Leglise 2010), since the communities are who can provide a diagnosis, outline and execute alternative projects, as well as correct errors, through a solid social foundation that allows for compatible development (Torres-Carral 2015). Therefore, this paper presents the results that allowed for the implementation of measurement tools used to determine tourism activity evaluation methods in rural areas, in order to promote the inhabitants' decision-making and the planning of projects aimed at improving the social metabolism in rural areas.

2 Tourism Sustainability Assessment

On an international scale, sustainability assessments have become a determining factor in the evaluation of environmental processes as well as socioeconomic phenomenon, however, a majority of said assessments have been designed for analysis in the biophysical arena, excluding the social and cultural aspects, which renders these assessments unsupported and not all inclusive; which originate from the concept of sustainable development (World Commission on Environment and Development 1987), that intends to sustain the physical medium over a period of time long enough to maintain global social well-being (Macedo 2005), however, these types of concepts make no reference to changes in usage, advantage and management of nature and are simply limited to quantitative changes (Torres-Carral 2009); and although "*sostenible*" is the most exact translation of "*sustainable,*[1]" the issue continues to be a topic of debate among researchers in the comparison between "*sostenible*" and "*sustentable*" since, for some researchers, the differences are based on the phase of the process and not on the limits of development. In terms of official Mexican discourse the option chosen is "*sustentable,*" nevertheless, some authors and institutions still use both concepts interchangeably.

[1]Researchers such as Torres-Carral (2016), argue that the differences between "*sostenible*" and "*sustentable*" are only important in the Spanish language, since the first concept refers to maintaining a process in time, while the second refers to nurturing that process.

Tools for measuring and managing the economic, sociocultural and environmental impact related to tourism were developed in order to establish a tourism system in the geographic spaces, in addition to monitoring the impact resulting from tourism activities, with the objective of identifying a model for sustainable development, in other words, the objective is to have a model that is economically viable, socially acceptable and ecologically rational, as mentioned by The United Nations (2002). The latter requires analysis of the threats to tourism activity, as is the case with climate change, because weather is a determining factor as it pertains to the duration and quality of the tourism experience, as it may attract or dissuade visitors, since tourism is highly sensitive to weather. Thus the development of the tourism industry is highly dependent on hydrometeorological conditions and, as such, is a highly vulnerable tourism destination activity; nonetheless, it is also necessary to identify the impact that may contribute to climate change and to look for alternatives that may decrease those effects, as tourism is one of the sectors that most contributes to greenhouse gas emissions. According to data from the Ministry of Tourism (SECTUR 2014), tourism's contribution to greenhouse gas emissions is approximated to be between 5.2% and 12.5%, 39.6% of which is due to air transportation, 35.8% is due to other forms of transportation, while 21% is due to lodging, and other activities comprise 3.4%.

After examining tourism's dynamic we can mention that, in terms of policy and management, there have been global efforts to develop indicators that contribute to decreasing the problems and difficulties that could be found in any destination, primarily coastal areas, small islands, and in ecotourism and cultural tourism destinations. The first signs of these efforts were by the World Tourism Organization in 1993. On the other hand, proposals have been developed at the international level for analysis and indicators of sustainability in the tourism activity, such as those designed for monitoring (Macário et al. 2013), the systemic (López-Ridaura et al. 2002; Salinas and La 2006), urban tourism (Hiu-Kwan and Hon-Wan 2012; Zheng-Xin and Linglin 2014), cultural tourism (Hernández and Vaquero 2012; Lozano-Oyola et al. 2012); rural tourism (Blancas et al. 2011), Unesco World Heritage Sites (Van-Oers and Pereira-Roders. 2012), marketing sustainable tourism (Dolnicar and Leisch 2008; Hultman and Säwe 2016) and the frameworks for assessing sustainability in tourism (Gyou-Ko 2005; Klein-Vielhauer 2009); in addition to the analysis of sustainability in the quality of the water and its water footprint in tourism (Yang et al. 2011; Mansir et al. 2015).

3 Methodology

The objective of this research is to propose a set of indicators to assess the impacts on the overall health and stability of the *Huasteca Potosina* hydrographic sub-basins in Mexico, for the case study where their natural resources are used for tourist purposes.

The general purpose of the assessment is to identify the variables and factors that contribute to promote the development of the rural communities whose inhabitants use the natural resources for their daily lives and other purposes such as to provide leisure and recreation services for the tourists. And in doing so, also identify the possible damages to the environment and the sustainability of the existing resources.

With such an assessment at hand, it would be possible for the rural communities' inhabitants to design strategies and plans to improve their living conditions by means of the appropriate ecosystem restoring and conservation activities.

With the above purposes, we propose to use the concepts and Framework for the Evaluation of Natural Resource Management Systems Incorporating Sustainability Indicators (Known in Spanish as MESMIS). In this framework the indicators are not an instrument merely qualifier of options, but rather a promoter of the operationalization of the concept of sustainability "in the pursuit of a more equitable social development and environmentally healthy for rural communities" (Masera et al. 2000).

Although the framework has been used in the last few decades to evaluate systems for agricultural production, forestry and/or livestock, Ortiz-Ávila (2008) mention that it can also be used in activities related to the management of natural resources, such as fisheries, aquaculture, management of flora and fauna, and ecotourism.

For that reason we choose to address the tourist systems from a sub-basin approach, where the water resource is the social and economic relevant axis for the rural and urban community activities of the *Huasteca Potosina*, and plays a key role in the "social metabolism" (Toledo 2013).

It is important to mention that the methodology used in this study has already been proposed as an operational-methodological framework for tourism activity (Arriola 2003), and has been implemented by some authors in the form of experimental trials (Zizumbo et al. 2006; Arévalo et al. 2015). Nevertheless, none of these proposals were operationalized using a sub-basin approach, nor were they derived from local sustainability criteria, in other words, these were simply based on observational criteria and statistics published by government entities, without resorting to primary sources. This is why this research was executed as outlined in the following sections.

4 Definition of the Object Being Evaluated

The "*Huasteca*" Region has been considered by several researchers as a region rich in biodiversity and culture, a result of its geological history, volcanic phenomena, the physiography, climates, and its landscapes[2] (Puig 1991; Rzedowski 1992;

[2] Cultural diversity in the *Huasteca Potosina* is clearly, so only in this region are three distinct ethnolinguistic groups of affiliations *Náhuatl*, which belongs to the Uto-Aztecan family

Ariel de Vidas 2003; Hudson 2004; Puig and Lacaze 2004; Ruvalcaba 2004; Stresser-Péan 2008; Palomo 2010), which allowed various ethnic groups held demonstrations to give cultural identity and reason of being of the ancient Mesoamerican peoples of the region, among which are the *Nahuas*, *Huasteco* and *Pames*. The ancient traditions of these cultures were modified during the colonial era; however, a large part of their vision of the world was preserved in an effort of syncretism between Catholicism and indigenous thought. It is as well as the cultures of the *Huasteca* are an amalgam between the worldview of the ancient world and colonial elements brought by the Spanish people and the slaves from Africa, thus forming the region called the *Huasteca* region, which has experienced in its evolution, the cultural transformation of the landscape. This gave rise to a series of events that projected to this region as a natural destination for travelers seeking an alternative to sun and beach tourism, above all, those who had as their main reason for adventure tourism. In this way, the tourist activity was added to the agricultural activities such as the cultivation of sugar cane of colonial introduction in the plains, the coffee in the highlands and the cultivation of citrus fruits which replaced the large portions of forest (Barthas 1996), In addition to the livestock introduced since the arrival of Europeans. In such a way that the nature tourism began to grow in the region after the 1980´s decade, sporadically, and it was not until 2001 that the Ministry of Tourism (SECTUR 2006), elaborated proposals for the development of the segment of nature in the country, which allowed the society will seek viable alternatives for the communities in rural areas with the greatest natural attractions, slowing down the expansion of agriculture and livestock which reduced the productivity of the jungles and forests of the region, however, in the growth of the tourist flow of the *Huasteca Potosina*, it was not generated a responsible participation and coordinated with the communities, governmental institutions and visitors, as well as not designed a strategic planning to decrease or eliminate the environmental, social and economic impacts. This led to the commodification of cultural[3] and natural heritage, under the interests of individuals and with zero benefits for the social groups, most of which are evicted from their territories (San Juan 2009; Vargas 2016) And deprived of the benefits that allowed them to take advantage of the natural and cultural resources of their own communities, creating social conflict and forced migration (Luna and Muñoz-Güemes 2016), while those who stay in the same place "become part of the staff of the services of these companies and earn wages that keep them in poverty" (Ávila 2013), in addition to the low social participation due to several factors such as the disinterest, indifference and/or apathy (Ferney et al. 2015). This problem is compounded by the

(uto-Aztec), Huastec (*teenek*) to the family mayanse and pame (*xi'oi*) to the family of the Otomanguean stock.

[3] The concept of commodification makes reference to the fact that the producer of the culture offers for sale or change their culture in order to meet the needs of the tourist, losing the true value of the knowledge of the community for the community, running the risk of transforming these knowledge for the sake of tourism, to make it more attractive to tourists. See: (Calleja and González 2016).

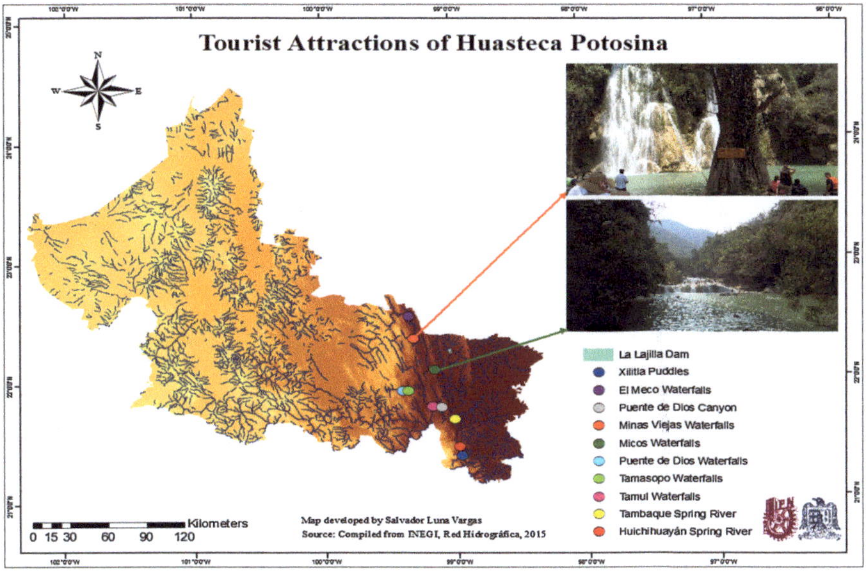

Fig. 1 Tourist attractions along water bodies and streams

process of massification of nature tourism in the *Huasteca Potosina*, which continues increasing but with no management plan and administration of tourism.

Currently there are 14 sites with the greatest concentration of tourists in the *Huasteca* region; however, the 78.57% of those sites belong to a body or stream of water (Fig. 1), where the "*Río Gallinas*", "*River Los Naranjos*", "*Río Tampaon*" and "*Río Axtla*" are the main engine of tourist activity in the region. Within these sub-basins of the aquifer extends *Huasteca* region, which has a total volume of extraction of 22.2 million m^3 of water per year, of which 12.9 million m^3 (57.9%) are destined for agricultural use, 0.8 million m^3 (3.4%) for domestic use, 1.1 million m^3 (4.9%) industrial use, 0.1 million m^3 (0.3%) for urban public use, 6.1 million m^3 (27.6%) for use in livestock, and finally the increase in services uses 1.3 million m^3/year (5.9%) of the total volume (CONAGUA 2016).

This study took place in an area within the *Huasteca Potosina* region that comprises the Northern Gulf Hydrological-Administrative Region (*HAR*) IX, which is subdivided into the *Río Tamuín* hydrological region 26, in the *Río De los Naranjos* sub-basin. The sub-basin encompasses an area of 2015.39 km (Fig. 2). The specific sites being analyzed are the *Minas Viejas ejido*[4] tourist destination,

[4] An *ejido* is a juridical entity with legal personality and an estate comprised by a collection of assets and rights, called *ejidal* property, which protects the integrity of the land for human settlement and regulates the advantage of forests and communal water sources in order to elevate the quality of life for its inhabitants, and adopt measures that will be most beneficial in terms of the use of productive resources. See: (Orozco 2010).

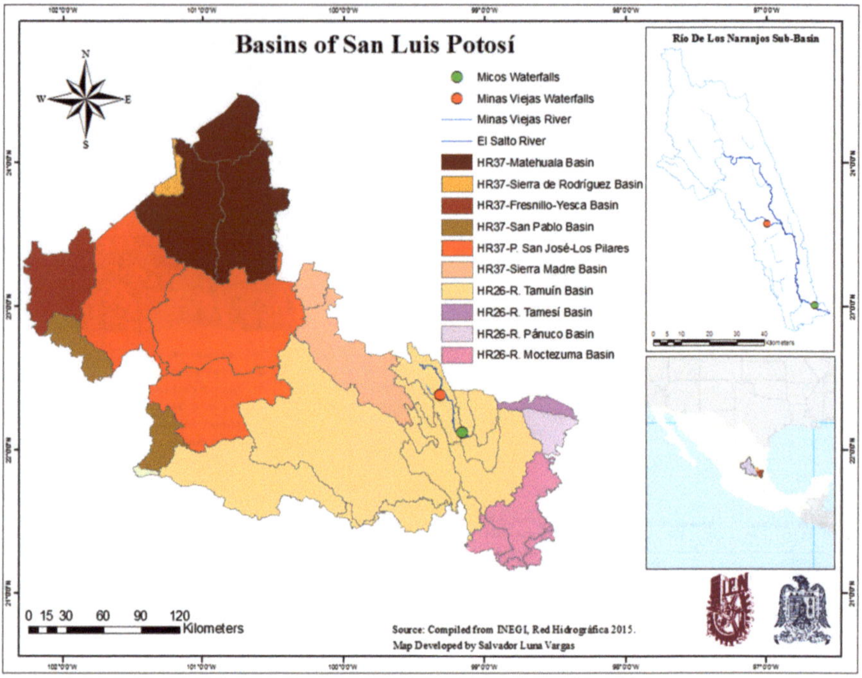

Fig. 2 Basins of San Luis Potosí

located in the municipality of *El Naranjo*, and the *Micos* Waterfalls that are located in the *ejido* of *El Platanito* which is part of the municipality of *Ciudad Valles*, *San Luis Potosí*. The physiographic province that comprises the study area occupies the greater part of the *Sierra Madre Oriental* mountain range and, in the case of *Micos*, its eastern edge is adjacent to the Gulf Coastal Plain. With regard to the physiographic subprovinces, *Micos* and *Minas Viejas* are on the border between the *Carso Huasteco* and the *Gran Sierra Plegada* mountainous regions. Given the conditions within the site, and in order to identify economic benefits, a comparative analysis of the two locations that share and use the bodies of water for recreational purposes is proposed. Therefore, the analysis will generate criteria to help the key agents make informed decisions, in addition to fostering agreements with local and municipal authorities in order to contribute to sustainable tourism development in the region.

5 Diagnostic Criteria and Indicators

According to the proposed methodology, the evaluation of the sustainability should be carried out through the comparison of one or more systems with a reference system in such a way to avoid absolute claims, "it is impossible to obtain an absolute measure of sustainability" (Masera et al. 2000). For that reason, we developed a characterization of the tourism management system that includes the following concepts: (1) biophysical components; (2) the inputs and products required to perform the activity; (3) the practices that involve each system, and (4) the main socio-economic characteristics of the providers of tourist services, as well as the levels and types of organization (Table 1).

These components allow for the formulation of diagnostic criteria for the design of joint sustainability indicators in order to describe processes specific to tourism activities, and which depend on a specific problem's characteristics and data availability. In this case, data was generated from primary sources; values were assigned to qualitative data collected from the members of the tourism committee,[5] and were compared to the data provided by the town's key agents who provided a more objective point of view to the research.

6 Measurement and Monitoring of Indicators

After analyzing the management systems for the object of study, it is necessary to identify potential and critical points that should be evaluated, and which could limit or strengthen the systems' abilities to be sustained over time (Table 2). The points that should be identified could be environmental, technical, social and economic factors or processes, that individually or collectively, could have an effect on the permanence of the tourist destination, however, because the evaluation is an interactive process, it may not take specific aspects into consideration; nevertheless, these points will allow for the creation of a link between attributes, diagnostic criteria and indicators. Under this premise, the town's key agents were consulted about the critical points in order to enable selection of diagnostic criteria based on the attributes outlined in the methodology.

Subsequently, surveys were designed based on the diagnostic criteria, and included 3 items per each diagnostic criterion and 5 responses per each item. The responses were arranged using a Likert-type scale, or summative evaluation method, in order to use host variation in attitudes in regards to their locations' sustainability of tourism activity as a measure of host perception. What resulted from this were 33 indicators that range from favorable to most unfavorable and include a neutral or midpoint. Values were assigned to each response in order to categorize each item according to criteria used by the International Union for Conservation of Nature and the United Nations Development Program (Guijt and Moiseev 2001). In this manner, the indicators developed using an ordinal scale are

[5]These committees are replaced every three years and the state's Office of Tourism proposed their formation. The committees have at least four members and include a president, a treasurer, a secretary and an aide. These members are elected by the community and their work is volunteer work, although they do receive economic remuneration by way of revenues generated by tourists a majority of the time.

Table 1 Tourism management systems of reference

Determinants of tourism system	Tourist attraction: *Micos* Waterfalls	Tourist attraction: *Minas Viejas* Waterfalls
	Reference management system	Comparative management system
Original byophysics	Climate: Humid semi-warm type with abundant rainfall during the summer, with a winter percentage lower than 5% (A) C (m) (w), and a precipitation of 1500–3000 mm (Cantoral 2004). The original vegetation consists of remnants of low deciduous forest with trees ranging in size from 7 to 8 m, with two shrub layers: one tall, 2–4 m and the other low, from 0.50 to 1.50 m. Although the dominant tree species is *Pithecellobium flexicaule*, species perennifolia (Puig 1991). This author concludes that this forest is present as a phase of regressive evolution of tropical deciduous tropical forest. It is worth mentioning that large portions of original forest have been replaced by monoculture sugar cane. In addition the tourist site *Pago Pago* (*Micos*) has a gallery forest dominated by *Taxodium mucronatum*. The dominant soils are Leptosols and Vertisols (INEGI 2010). In this locality were found 58 species of diatoms, result of a great variety and heterogeneity of microenvironments. For this reason there is a greater probability of presence of diverse species in an area with a wide range of substrates, illumination and current velocity (Cantoral 2004)	Climate: Humid semi-warm type with abundant rainfall during the summer, with a winter percentage of less than 5% (A) C (m) (w). Precipitation of 1500–3000 mm (Cantoral 2004). The original vegetation is composed ofremnants of Tropical Forest Low Deciduous with an arboreal stratum, whose average height is 8–12 m. The shrub stratum measures 3–6 m, while the herbaceous stratum is poorly developed. Among the dominant species are *Acacia coulteri*, *Bursera simaruba*, *Phoebe tampicensis*, *Guazuma ulmifolia* and *Casimiroa pringlei* (Puig 1991). In this locality, the forest has also been largely replaced by sugar cane monoculture, while along the banks of the Minas Viejas river there is a gallery forest of *Taxodium mucronatum*. The dominant soils are Luvisols and Leptosols (INEGI 2010). It is possible to highlight the great richness in diatom species diversity in the localities of *El Salto*, *El Nacimiento de El Salto* and *El Meco*, located in the upper basin, highlighting the need to carry out more phytophoric and diatom studies in the *Huasteca potosina*, with the aim of generating indicators of the ecological quality of water
	Water bodies: Due to its karstic nature, the *Huasteca* aquifers are more susceptible to pollution than those on other regions, due to the relative ease with which water is filtered (Montejano 2004)	

(continued)

Table 1 (continued)

Determinants of tourism system			Tourist attraction: *Micos* Waterfalls Reference management system	Tourist attraction: *Minas Viejas* Waterfalls Comparative management system
Management technologies	Tourism segment		Nature Tourism	Nature Tourism
	Activities		Contact with nature, camping, picnic, hiking, waterfall jumping, kayaking, boat ride	Contact with nature, camping, picnic, abseiling
	Services		Rent of life jackets, boat ride and waterfall jump guides	Rent of life jackets, surveillance and lifeguard
	Infraestructure		Signeage, restrooms, camping, paved trail, paved roads, retaining wall, lodging rooms, booth, groceries store and parking	Palm shades, restrooms, picnic areas, booth, trail and parking
	Tecnology used		Manual	Manual
	Labor employed		The committee is comprised of 39 persons, which includes lifejacket renters, boaters, turistic guides, housekeeping, restrooms and parking, and people who sell food. The *ejido* has a population of 167 inhabitants and a high marginalization grade (SEDESOL 2010)	The tourism committee is made up of 15 persons, plus the additional persons during the vacation season. The services offered are lifejacket rental, security, booths, restrooms, and food sale. The *ejido* has a population of 758 inhabitants and a high marginalization grade (SEDESOL 2010)
	Waste management	Wastewater treatment type	Septic tank/ sedementary tanks	Septic tank/ sedementary tanks
	Conservation practices	Waste management	Municipality garbage collection	Municipality garbage collection
		Soil conservation practices	Ditches, agriculture partnership and reforestation	No
		Water conservation practices	No	No
Socioeconomic and cultural	Alternative activities		Seasonal sugarcane harvest and labor on private parcels	Seasonal sugarcane harvest and labor on private parcels
	Constitution of the turistic organization		Community organization	Local Tourism Committee

(continued)

Table 1 (continued)

Determinants of tourism system		Tourist attraction: *Micos* Waterfalls	Tourist attraction: *Minas Viejas* Waterfalls
		Reference management system	Comparative management system
	Service providers characteristics	Community organization throughout the year	Community organization during rest days and mainly holiday season
	Service purpose	Obtain monetary income	Obtain monetary income for reinvestment in infrastructure
	Tourist organization characterisctics	The organization has been formed for more than 5 years	Change of committee every 3 years

Table 2 Potential critical points for the sustainability of the tourism system

Characteristic	Critical factors
Productivity	Unemployment
	Low yields
	Poor quality of the product
Stability	Groups not organized
	Conflict of interest
	Permanence of tourism committees
Reliability and resiliency	Deforestation
	Soil and water resources contamination
	Soil loss or degradation
Adaptability	Seasonality of tourist activity
	Poor diversification of tourism product
	Untrained tourism service providers
Equity	Migration
	Null participation of women
	Unfair distribution of benefits
Self-management	Deterioration of community organization
	Poor linkage with academic institutions
	Poor linkage with governments

Table 3 Categorization criteria for indicators, subcomponents and components

Result (score)	Classification
0.80–1.00	High
0.60–0.79	Good
0.40–0.59	Average
0.20–0.39	Poor
0–0.19	Bad

converted into numeric variables that range from sustainable to undesirable or unsustainable (Table 3). It is worth mentioning that the indicators were designed to be easily measured and interpreted by rural community residents, and were also compared with other analyses proposed by some of the investigators previously mentioned.

Once data is collected by using the tool, the next step is to obtain the average value of each item answered by an absolute majority (half plus one) of the tourism committee members, this will subsequently be averaged by the total number of respondents in the system of reference (21 in *Micos*) and in the comparison system (8 in *Minas Viejas*), respectively. The average value of each component or subindex is calculated for this and involves determining its average value (Ibáñez 2014):

$$\bar{x} = \frac{1}{n}\sum_{i=l}^{n} a_1 = \frac{a_1 + a_2 + a_3 \ldots\ldots a_n}{n}$$

where:

x = average value of the subcomponent or component,

n = number of indicators for the subcomponent or component,

ai = individual value of the indicators for the subcomponent, from a_1, a_2, …, a_n

As mentioned before, the indicators were designed according to the diagnostic criteria that were previously conceived from the potential critical points derived from the vernacular knowledge of the locations' key agents[6] though the attributes already established by the MESMIS methodology, results which can be found in (Table 4).

Once the indicator values are obtained, the results are presented in graphical form in order to simplify the process of comparative analysis of the sustainability of the tourism management systems in the locations that use bodies of water for recreational purposes.

[6]Vernacular knowledge refers to popular knowledge, traditional skills and the cultural experience of living space.

Table 4 Diagnostic criteria and sustainability indicators for evaluation of tourism activity

Attribute	Diagnostic criteria	Indicators	Area of evaluation	Micos	Average	Minas viejas	Average	Total Micos	Total mims viejas
Productivity	Efficiency	Employment opportunities Tourism infrastructure coverage Faiitrade Income	SEc Ec SEc	18.2 17 17.4	0 87 0.81 0 83	7.2 6.2 6	0.90 0.78 0.75		
		Total		17.53	0.83	6.47	0.81	0.83	0.81
Stability	Risks sharing	Abandonment of traditional activities due to tourism organisational consolidation Length of stay	S S SEc	116 17.4 17.8	0.55 0.83 0.85	3.6 6.4 7.2	0.45 0.80 0.63		
		Total		15.60	0.74	5.73	0.63	0.74	0.63
Reliability & resiliency	Diversity	Biological importance of the locality Geological significance of the region Use of local knowledge and skills	Em Em SEm	19.4 18.6 19.4	0.92 0.89 0.92	7.2 7.4 7.6	0.90 0.93 0.95		
		Total		19.13	0.91	7.40	0.93		
	Resources conservation	Contamination of water bodies Deforestation Sod conservation	Em Em Em	16.4 16.8 14 4	0.78 0.80 0.69	7.6 5 3.6	0.95 0.63 0.45		
		Total		15.87	0.76	5.40	0.68		
	Fragility system	Ability to return to equdibnum after a weather phenomenon Economic progress Insecunty Impact	SEm SEc S	16.8 11.4 12	0 80 0.54 0.57	4.4 4 5	0.55 0.50 0.63		
		Total		13.40	0.64	4.47	0.56	0.77	0.72

(continued)

Table 4 (continued)

Attribute	Diagnostic criteria	Indicators	Area of evaluation	Micos	Average	Minas viejas	Average	Total Micos	Total mims viejas
Adaptability	Learning proccess strengthening	Tourism service provider trammg	S	19.2	0.91	5.8	0.73		
		Quality of courses and workshops	S	18.8	0.90	6.8	0.85		
		Environmental education	SEm	15.8	0.75	5.4	0.68		
		Total		17.93	0.85	6.00	0.75	0.76	0.69
	Capacity for change and innovation	Seasonality	SEcEm	18.4	0.88	6.2	0.78		
		Diversification of tourist product	S	9.8	0.47	3.2	0.40		
		OrganEation with other organsed groups	S	13.4	0.64	5.6	0.70		
		Total		13.87	0.66	5.00	0.63	0.76	0.69
Equity	Quality of life	Garbage collection	SEm	19.2	0.91	8	1.00		
		Water access sources	SEm	15.2	0.72	4.6	0.58		
		Migration	S	17.4	0.83	3.2	0.40		
		Total		17.27	0.82	5.27	0.66		
	Distribution of costs and benefits	Marginality of the local population due to lack of access to the economic benefits of tourism Fair distribution of benefits intergenerationally Participation of women in tourism	S Ec S S s s	15.4 15 16.8	0.73 0.71 0.80	4.2 6.2 7.2	0.53 0.78 0.90		
		Total		15.73	0.75	5.87	0.73	0.79	0.70

(continued)

Table 4 (continued)

Attribute	Diagnostic criteria	Indicators	Area of evaluation	Micos	Average	Minas viejas	Average	Total Micos	Total mims viejas
Autodependey (selfmanagement)	Self-sufficiency	Dependence on public programs	S Ec	11	0 52	5.2	0.65		
		Dependence on pnvate sector	S Ec	13	0 62	3.4	0.43		
		Dependence on external inputs	S Ec	13.8	0.66	6.2	0.78		
		Total		12.60	0.60	4.93	0.62		
	Organization	Support from academic institutions	S	11.6	0.55	4	0.50		
		Social organization among service provider	S	14.8	0.70	4.6	0.58		
		Coordination with the municipal authorities	S	15.8	0.75	6	0.75		
		Total		14.07	0.67	4.87	0.61	0.63	0.61

Evaluation areas: Social **S**, Environment **En** and Economic **Ec**

7 Integration of Results

This stage requires integration of the results obtained through the monitoring of indicators, as well as, synthesis of the information in order to subsequently issue value judgments about the tourism management systems analyzed, in order to provide the necessary tools for decision making about the changes required to improve the evaluated management systems. This can be summarized in five important ways: (1) Combining data in a matrix of indicators (Table 4); (2) Determining thresholds or values for indicators based on the perceptions of the host community (Table 2); (3) Developing indices for indicators (Table 3); (4) Presenting all results in graphical or table format (Fig. 3), using multi-criteria decision analysis techniques; and (5) Examining relationships, including the effects of feedback mechanisms among indicators. Once numerical indices are developed, the systems evaluated can be compared by considering indicators with mixed techniques that combine graphical representations with numerical information. What was used in this case was an amoeba diagram, which requires drawing a radar chart in which each indicator is represented on a separate axis, with the external radial axis representing the optimal value or threshold. This allows for qualitative representation of the desired objective's degree of coverage and also allows for a simple, graphical and comprehensive comparison of limiting factors and opportunities for sustainability, and serves as a useful tool for planning and incorporating

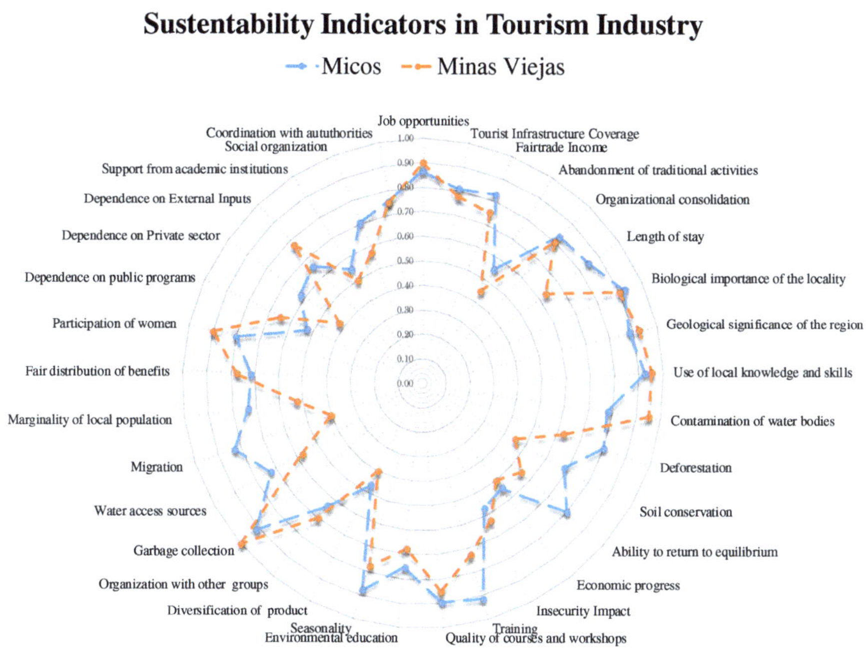

Fig. 3 Amoeba diagram evaluation of two tourism systems

strategies that strengthen the social, environmental and economic profile of the systems under evaluation (Masera et al. 2000).

Once results are incorporated, what follows is analysis of the results obtained from the tourism management systems and, through discussion of the principal elements, involves evaluation of the factors, in terms of political, socioeconomic or environmental constraints within the system that allow or impede the tourism systems to improve sustainability. The sustainability assessment for the two tourist sites is what follows:

As shown in (Fig. 3), with regard to employment, the survey respondents think that tourism has been an employment generator, however, the seasonal nature of the *Huasteca Potosina* makes that employment temporary, causing able-bodied men to migrate to the big cities in search of an income that will insure their families' well-being. *Minas Viejas* is the town with the highest migration rates, and the cities of *Matamoros* and *Monterrey* are the most frequent destinations in the search for employment. The *ejido* of *El Platanito* (*Micos*) has a lower migration rate due to better income from tourism activity, which allows residents to alternate between providing tourism services and their daily activities; as is the case with subsistence agriculture involving corn and beans, as well as, sugar cane; however, emigration, primarily to the United States, is still prevalent. It is worth mentioning, that organized tourism activity has taken place in *Micos* since the 1990s, whereas, in *Minas Viejas*, this activity is more recent. Although this town has been around for longer than 5 years, it has not managed to establish itself because its tourism committees change every three years, whereas membership in the *Micos* community organization is more stable. Tourism infrastructure in *Micos* is more extensive as the locality has some rooms for lodging, campgrounds, bathrooms and dressing rooms, as well as, dining establishments. In the case of *Minas Viejas*, the tourism infrastructure is insufficient even though the locality possesses the necessary elements for tourism activities. The roadway infrastructure is in poor condition as a result of the commercial trucks that transport sugar cane. However, Minas Viejas is the farthest town from *Ciudad Valles*, a vital hub in the *Huasteca Potosina*.

In regard to the biological importance that the residents assign to their towns, it can be said that they consider this, and geological diversity, important factors in the development of the communities, however, there still exists a lack of awareness about the communities' geological assets. Nevertheless, the majority of the survey respondents acknowledged that local skills and knowledge, primarily in the areas of artisanal crafts, are still being practiced, as well as, traditional agriculture, indigenous languages and herbal medicine, which demonstrates that a great sense of belonging still exists within these towns. This can help strengthen management and protection of the natural resources in these territories, where logging is still a big problem (mainly in *Minas Viejas*) resulting in soil erosion. Meanwhile, in *Micos*, the pollution of bodies of water is a problem that is increasingly more evident, due to residuals released by the sugar mill, fertilizers and tourism to a lesser degree; these things decrease the ecosystems' abilities to recover from disturbances, such as wildfires and flooding. From this stems the importance of having the necessary environmental education and courses and workshops in the tourism field, as well as, in the management and conservation of resources, emphasizing the pressing need to create a link between academic institutions and local communities for the development of tools that allow comprehensive management of resources through a diversified and sustainable tourism product, in such a way that helps promote practices to achieve economic and social development, without significantly compromising the environmental conditions of the territory, through the adequate conservation and restoration of natural resources, promoting diversity and cultural pluralism, that promotes the necessary goods and services for society, guaranteeing acceptable and well paid employment for present and future generations, so that they can define their priorities, identities and values.

8 Conclusions

In recent years, the increase in the number of tourists to the region has become a determining factor in the deterioration of the environment, due to inadequate planning of tourism activity and processes that do not include the local communities, which is why it has not manifested directly in commercial dynamism nor in the development of new tourism-based employment; additionally, the visitors' low spending and the widespread increase of nature-based tourism in the *Huasteca potosina*'s bodies of water have provided questionable benefits to the rural communities who use, exploit and utilize natural resources designated for visitors' leisure and recreation.

In the future, the work completed in this paper will be useful in researching the tourism phenomenon with a hydrographic-sub-basin-as-territorial-unit approach, which intends for the indicators to not be solely a tool for evaluation of options, but rather a proponent for the operationalization of the sustainability concept, "in the search for a more equitable and environmentally sound social development for rural communities" (Masera et al. 2000), in addition to its application of transdisciplinary principles (Lang et al. 2012; Betancourt 2016), this approach builds bridges among researchers, key agents and facilitators; thus becoming a new option for evaluating tourism activity according to local sustainability criteria, given that, the methodology used in this research has previously been used to evaluate agriculture, forestry and/or livestock (primary sector) production systems, this approach aims to innovate through the use of evaluation methods of tertiary activities, such as the tourism's situation, but addressing the primary and secondary sectors as a whole; which transforms this work as a new way to evaluate tourism sustainability in rural spaces, offering the possibility of replicating the measurement tools by transferring the knowledge into the social and scientific practice of other rural space regions. In this way, the objective is to formulate recommendations that will gear tourism practices toward a more sustainable tourism model, and as such, it becomes necessary to outline future research that allows for the design of alternative activities that promote equitable endogenous development, in order to reactivate economic dynamics and increase the standard of living for rural inhabitants, thereby strengthening their culture and identity processes without compromising natural resources in the medium term or long term.

It is worth mentioning that financial resources can be a limiting factor for this type of methodology, since costly equipment would be required if there is a desire to delve into analysis of the physical-chemical and biological properties of the water, as well as more study time and transportation resources, taking into consideration the risks in the study area. However, this proposal presents strengths in the understanding of landholding, division and location of employment, participation, and access to natural resources.

In this way, we highlight the importance of creating a link between academic institutions and local communities for development.

Acknowledgements We wish to thank the members of the tourism committees for their unconditional support. We want to especially thank Mr. Alejandro Hernández Hernández for his collaboration in the administration of the surveys that were an essential part of this research.

References

Álvarez L (2015) El arrecife mesoamericano, un paraíso que peligra por su belleza. La Jorn ecológica 199:3–4

Anton S (2009) Los paisajes temáticos. In: Busquets J, Cortina A (eds) Gestión del paisaje. Manual de protección, gestión y ordenación del paisaje. Barcelona: Ariel

Arévalo G, Guerrero H, Ortíz C (2015). Evaluación de la sustentabilidad del turismo de salud en Michoacán, México. Paper presented at the XX Encuentro Nacional sobre Desarrollo Regional en México, MuCIC, Morelos, 17–20 Nov 2015

Ariel de Vidas A (2003) El trueno ya no vive aquí. Representación de la marginalidad y construcción de la identidad teenek (Huasteca veracruzana, México). D.F.: CIESAS, COLSAN, CEMCA, IRD

Arriola O (2003) Evaluación de la sustentabilidad turística. Notas para una propuesta operativo-metodológica. In: Palafox A (ed) Desarrollo Sustentable del Turismo. Contribuciones Académicas del 1er Congreso Internacional. Chetumal, Mexico: Universidad de Quintana Roo

Ávila A (2013) Turismo y pueblos indígenas de México: despojo y veredas de apropiación comunitaria. In: Carámbula M, Ávila L (eds) Patrimonio biocultural, territorio y sociedades afroindoamericanas en movimiento. Buenos Aires, CLACSO

Barthas B (1996) De la selva al naranjal (Transformaciones de la agricultura indígena en la Huasteca potosina). In: Bovin P (ed) El campo mexicano: una modernización a marchas forzadas D.F.: CEMCA and ORSTON

Betancourt A (2016) Experiencias históricas de diálogo intercientífico: Resiliencia, historicidad e inteligibilidad de ciencias subalternas. In: Delgado F, Rist S (eds) Ciencias, diálogo y saberes y transdisciplinariedad. Aportes teórico metodológicos para la sustentabilidad alimentaria y del desarrollo. La Paz: Agroecología Universidad Cochabamba

Blancas F, Lozano-Oyola M, González M, Guerrero F, Caballero R (2011) How to use sustainability indicators for tourism planning: The case of rural tourism in Andalusia (Spain). Sci Total Environ 412–413:28–45. https://doi.org/10.1016/j.scitotenv.2011.09.066

Burgos A, Bocco G, Sosa J (eds) (2015) Dimensiones sociales en el manejo de cuencas. CIGA-UNAM, D.F

Calleja C, González A (2016) Mercantilización de la cultura en aras de ofrecer una experiencia turística estandarizada. Reflexiones desde el caso de Cozumel. Rev Iberoamericana de Turismo-RITUR 6:82–95
Cantoral E (2004) Diatomeas de ríos en la cuenca baja del Río Pánuco. In: Ruvalcaba J, Pérez-Zevallos J, Herrera O (eds) La Huasteca, un recorrido por su diversidad. CIESAS, COLSAN, El Colegio de Tamaulipas, D.F
CONAGUA (2016) Acuerdo por el que se da a conocer el resultado de los estudios técnicos de las aguas nacionales subterráneas del acuífero Huasteca Potosina, clave 2418. In: el Estado de San Luis Potosí, Región Hidrológico-Administrativa Golfo Norte. Final Report. San Luis Potosí, Mexico: SEMARNAT
Conde N (2013) Diagnóstico del sistema turístico mexicano. Un enfoque hermenéutico. Instituto Politécnico Nacional, D.F
Cotler H, Priego Á (2007) El análisis del paisaje como base para el manejo integrado de cuencas: el caso de la cuenca Lerma-Chapala. In: Cotler H (ed) El manejo integral de cuencas en México, 2nd edn. D.F.: SEMARNAT, INE
Dolnicar S, Leisch F (2008) Selective marketing for environmentally sustainable tourism. Tourism Manag 29:672–680. https://doi.org/10.1016/j.tourman.2007.07.010
Ferney H, Aguilar M, Medellín P (2015) Gestión participativa en la Cuenca del Río Valles, Oriente de México. In: Burgos A, Bocco G, Sosa-Ramírez J (eds) Dimensiones sociales en el manejo de cuencas. D.F, CIGA-UNAM
García I, Becerra R, Pérez G (2013) Uso, aprovechamiento social y conservación de las plantas medicinales en México. In: Carámbula M, Ávila L (eds) Patrimonio biocultural, territorio y sociedades afroindoamericanas en movimiento. Buenos Aires, CLACSO
Guijt I, Moiseev A (2001) Conjunto de Herramientas para La Evaluación de la Sostenibilidad. UICN, Gland and Cambridge
Gyou-Ko T (2005) Development of tourism sustainability assessment procedure: a conceptual approach. Tourism Manag 26:431–445. https://doi.org/10.1016/j.tourman.2003.12.003
Hernández M, Vaquero M (2012) Load capacity in large cultural tourism resources. Anales de Geografía de la Universidad Complutense 32:253–274. https://doi.org/10.5209/rev_AGUC.2012.v32.n2.39720
Hiu-Kwan E, Hon-Wan E (2012) Critical social sustainability factors in urban conservation. The case of the central police station compound in Hong Kong. Facilities 30(9/10):396–416. https://doi.org/10.1108/02632771211235224
Hudson PF (2004) Geomorphic context of the prehistoric Huastec floodplain environments: lower Pánuco basin. Mexico J Archaeol Sci 31:653–668. https://doi.org/10.1016/j.jas.2003.06.002
Hultman J, Säwe F (2016) Absence and presence of social complexity in the marketization of sustainable tourism. J Cleaner Prod 111:327–335. https://doi.org/10.1016/j.jclepro.2014.11.053
Ibáñez R (2014) Adaptación de metodologías globales a casos locales. Propuesta de modelo para el análisis de sustentabilidad en pequeñas localidades costeras con actividad turística en México. In: Monterrubio J, López Á (eds) De la dimensión teórica al abordaje empírico del turismo en México. Perspectivas multidisciplinarias. D.F.: UNAM, UAEM
INEGI (2010) Compendio de Información Geográfica Municipal 2010. Instituto Nacional de Estadística y Geografía, Aguascalientes, Mexico
INEGI (2015) Red Hidrográfica escala 1: 50 000 edición 2.0, Mexico. Retrieved from http://www.inegi.org.mx/geo/contenidos/topografia/regiones_hidrograficas.aspx
Ivanova A (2012) El cambio climático y el turismo: impactos, adaptación y mitigación. In: Ivanova A, Ibáñez R (eds) Medio ambiente y política turística en México. SEMARNAT, INE, AMIT, UABCS, D.F
Klein-Vielhauer S (2009) Framework model to assess leisure and tourism sustainability. J Cleaner Prod 17:447–454. https://doi.org/10.1016/j.jclepro.2008.07.006
Lang DJ et al (2012) Transdisciplinary research in sustainability science: practice, principles, and challenges. Sustain Sci 7:25–43. https://doi.org/10.1007/s11625-011-0149-x

López-Ridaura S, Masera O, Astier M (2002) Evaluating the sustainability of complex socio-environmental systems. The MESMIS framework. Ecol Ind 2:135–148. https://doi.org/10.1016/S1470-160X(02)00043-2

Lozano-Oyola M, Blancas J, González M, Caballero R (2012) Sustainable tourism indicators as planning tools in cultural destinations. Ecol Ind 18:659–675. https://doi.org/10.1016/j.ecolind.2012.01.014

Luna S, Muñoz-Güemes A (2016) Relaciones Interétnicas y la Competencia por el Aprovechamiento Turístico del Agua en la Huasteca potosina, México. Gestión de la Innovación como agente determinante del cambio. Hacia un emprendimiento sostenible. Tijuana, Mexico: ILCSA

Macário V, Pasa C, Ataíde G (2013) Indicadores de Sustentabilidad para la Actividad Turística. Una propuesta de monitoreo usando criterios de análisis. Estud y Perspectivas en Turismo 22(2):177–197

Macedo B (2005) El concepto de sostenibilidad. UNESCO, Santiago

Maldonado C (2006) Turismo y comunidades indígenas: Impactos pautas para autoevaluación y códigos de conducta. OIT, Geneva

Mansir A, Abdul-Nassir M, Khamaruzaman W, Amirhossein M, Rosilawati B (2015) A GIS-based water quality model for sustainable tourism planning of Bertam River in Cameron Highlands. Malaysia Environ Earth Sci 73:6525–6537. https://doi.org/10.1007/s12665-014-3873-6

Masera O, Astier M, López-Ridaura S (2000) Sustentabilidad y manejo de recursos naturales: El marco de evaluación MESMIS. MundiPrensa-GIRA-UNAM, D.F

Montejano G (2004) Proyecto ficoflora de la Huasteca: avances y perspectivas. In: Ruvalcaba J, Pérez-Zevallos J, Octavio H (eds) La Huasteca, un recorrido por su diversidad. CIESAS, COLSAN, El Colegio de Tamaulipas, D.F

Monterroso N (2011) La insustentabilidad del turismo. In: Arroyo J, Corvera I (eds) Desarrollo insostenible. Gobernanza, agua y turismo. Zapopan, Mexico: Universidad de Guadalajara, UCLA Program on Mexico, PROFMEX-WORLD, Juan Pablos Editor

Orozco P (2010) Naturaleza del ejido de la propiedad ejidal: Características y limitaciones. Rev Mexicana de Derecho 12:163–193

Ortiz-Ávila T (2008) Caracterización de sistemas de manejo de recursos naturales. In: Astier M, Masera O, Galván-Miyoshi Y (eds) Evaluación de sustentabilidad. Un enfoque dinámico y multidimensional. Valencia: SEAE, CIGA, ECOSUR, CIEco, UNAM, GIRA, Mundiprensa, Fundación Instituto de Agricultura Ecológica y Sustentable

Palomo N (2010) La gestion des plantes médicinales chez les communautés autochtones Nahuas de la Huasteca Potosina, Mexique. Université de Montréal, Montréal

Pensado-Leglise M (2010) Referentes de indicadores en la evaluación económica con criterios ambientales para núcleos agrarios. In: Cantú R (ed) Los desafíos ambientales y el desarrollo en México. Plaza y Valdés, Instituto Politécnico Nacional, LASUR, Madrid

Pereiro X (2015) Reflexión antropológica sobre el turismo indígena. *Desacatos*. Rev de Antropología Soc, D 47:18–35

Puig H (1991) Vegetación de la Huasteca. México. Estudio fitogeográfico y ecológico. D.F, ORSTOM, INECOL, CEMCA

Puig H, Lacaze D (2004) Huasteca y biodiversidad. In: Ruvalcaba J, Pérez-Zevallos J, Herrera O (eds) La Huasteca un recorrido por su diversidad. CIESAS, COLSAN, El Colegio de Tamaulipas, D.F

Ruvalcaba J (2004) La agricultura de roza en la Huasteca, ¿suicidio o tesoro colectivo? In: Ruvalcaba J, Pérez-Zevallos J, Herrera O (eds) La Huasteca un recorrido por su diversidad. CIESAS, COLSAN, El Colegio de Tamaulipas, D.F

Rzedowski J (1992) Diversidad y orígenes de la flora fanerogámica de México. In: Trueba C (ed) Rev Cienc UNAM 6:3–21

Salinas E, La OJ (2006) Turismo y Sustentabilidad: de la teoría a la práctica en Cuba. Cuad de turismo 17:201–221

San Juan R (2009) "Un Territorio Partido por la Mitad" Interpretaciones, Prácticas y Contiendas alrededor del Nacimiento Huehuetlán y Nacimiento Xilitla, S.L.P. San Luis Potosí, Mexico: COLSAN

Sartorello S (2013) Educar para el arraigo sociocultural. El perfil de egresado de alumn@s indígenas en una propuesta educativa intercultural y bilingüe en Chiapas. In: Carámbula M, Ávila L (eds) Patrimonio biocultural, territorio y sociedades afroindoamericanas en movimiento. Buenos Aires: CLACSO

SECTUR (Secretaría de Turismo) (2006) El turismo de naturaleza: retos y oportunidades. SECTUR, D.F. Retrieved from https://manuelmiroglio.files.wordpress.com/2011/05/el-turismo-de-naturaleza-en-mexico.pdf

SECTUR (Secretaría de Turismo) (2014) Guía Local de Acciones de Alto impacto en Materia de Mitigación y Adaptación al Cambio Climático en Destinos Turísticos Mexicanos. D.F.: SECTUR

SEDESOL (Secretaría de Desarrollo Social) (2010) Catálogo de Localidades. Retrieved from http://www.microrregiones.gob.mx/catloc/Default.aspx?tipo=clave&campo=mun&valor=24

Stresser-Péan G (2008) Viaje a la Huasteca con Guy Stresser-Péan. FCE, CEMCA, D.F

Toledo V (2013) El metabolismo social: una nueva teoría socioecológica. Relaciones, XXXIV 136:41–71

Torres-Carral G (2009) El desarrollo sustentable en México (Visión crítica hacia un desarrollo compatible). Universidad Autónoma de Chapingo, Plaza y Valdés, D.F

Torres-Carral G (2015) Sustentabilidad y compatibilidad. Universidad Autónoma de Chapingo, Texcoco, Mexico

Torres-Carral G (2016) Desarrollo Social Sustentable. Paper presented at the V Coloquio de Estudiantes del CIIEMAD, Centro Interdisciplinario de Investigaciones y Estudios Sobre Medio Ambiente y Desarrollo, Mexico City, 17–18 Nov 2016

UN (2002) Plan de Aplicación de las Decisiones de la Cumbre Mundial sobre el Desarrollo Sostenible. Cumbre Mundial sobre el Desarrollo Sostenible. Johannesburg: UN

Van-Oers R, Pereira-Roders (2012) Historic cities as model of sustainability. J Cult Heritage Manag Sustain Dev (2):4–14. doi: 10.1108/20441261211223298

Vargas B (2016) El turismo y la mujer indígena: el caso de Tancanhuitz en la huasteca potosina. In: Suárez I, Gutierrez A (eds) Turismo, sociedad y cultura. Visiones interdisciplinarias para el desarrollo. Corral, Chile: RECINATUR

World Commission on Environment and Development (1987) Report of the world commission on environment and development "our common future". UN, Oslo

WTO (2009) De Davos a Conpenhague y más allá: el avance de la respuesta del turismo al cambio climático. Documento de referencia de la OMT. Copenhagen, WTO

Yang M, Hens L, De Wulf R, Ou X (2011) Measuring tourist's water footprint in a mountain destination of Northwest Yunnan. China J Mt Sci 8:682–693. https://doi.org/10.1007/s11629-011-2062-2

Zheng-Xin W, Linglin P (2014) A systems thinking-based grey model for sustainability evaluation of urban tourism. Kybernetes 43:462–479. https://doi.org/10.1108/K-07-2013-0137

Zizumbo L, Monterroso N, Chaisatit N (2006) Turismo y desarrollo forestal en la comunidad de San Pedro Atlapulco, Estado de México. Paper presented at the V Congreso Internacional y XI Nacional de Ciencias Ambientales, Universidad Autónoma del Estado de Morelos, Oaxtepec, Mexico, 7–9 June 2006

Author Biographies

Salvador Luna Vargas has a bachelor's degree in Sustainable Tourism from the Autonomous University of San Luis Potosí (*UASLP*, as per its Spanish acronym) and a Master's Degree in

Tourism Administration and Innovation, from the National Polytechnic Institute (*EST-IPN*, as per its Spanish acronym). He has worked in the generation and application of knowledge in the area of: Society, Sustainability, and Tourism. He is a member of *Red de Investigadores y Centros de Investigación en Turismo, (RICIT*, as per its Spanish acronym, Tourism Researchers and Research Centers Network), member of the *Red Mexicana de Cuencas Hidrográficas*, Remexcu, (Mexican Hydrographic Basins Network) and part of the team of Mexican researchers for the *Red Iberoamericana de Ciencia, Naturaleza y Turismo,* (*RECINATUR*, as per its Spanish acronym, Ibero-American Network of Science, Nature and Tourism).

María de la Luz Valderrábano Almegua is a research professor of the Interdisciplinary Center for Research and Studies on Environment and Development (*CIIEMAD-IPN*, as per its Spanish acronym). She holds a master's degree in Science Methodology and Chemical and Industrial Engineering. She has managed projects and participated in institutional projects on the analysis of Mexican environmental policy, management of decent work, evaluation of public policy methodologies, mobbing and emotional intelligence, environmental pollution and education. She is currently a Professor and Dean.

Irma Suárez Rodríguez has a bachelor's degree in Hotel and Restaurant Tourism Business Management, a Master's degree in Tourism Business Marketing, and is a full-time research professor in the Sustainable Tourism bachelor's degree program, and Planning Secretary of the Multidisciplinary Academic Unit in the *Huasteca* Area (*UAMZH-UASLP*, as per its Spanish acronym). She is a representative of the Mexican Chapter of the *Red Iberoamericana de Ciencia, Naturaleza y Turismo*, (*RECINATUR*, as per its Spanish acronym, Ibero-American Network of Science, Nature and Tourism). She is currently the leader of the project for the inclusion of the *Huasteca Potosina* in UNESCO's Global Geoparks Network.

Larisa Alcérreca Molina has a doctorate in Architecture from the Polytechnic University of Cataluña, a Masters in Urban Environment and Sustainability, and a bachelor's degree in Architectural Planning and Design. She is currently a full-time research professor in the Graduate Studies and Research Department of the National Polytechnic Institute's School of Tourism, (*EST-IPN* as per its Spanish acronym). She is part of the *IPN's Red de Medio Ambiente* (National Polytechnic Institute Environment Network). She is currently a leader of the Graduate Studies and Research Department.

Mercury Concentration in Hair Due to Environment on Two Populations in Mexico

Isela Martínez Fuentes and Rocío García Martínez

Abstract The study of mercury pollution has been intensified through the last decades due to the high toxicity of this heavy metal and its increasing availability in the environment; since mercury is produced by both natural and anthropogenic processes. Mercury is an element naturally present in air, water and soil, leading to its accumulation in all living beings without being essential for any biological process. The measurement of the corporal mercury load in humans is made through the use of different biological markers such as nails, teeth, bones, saliva, urine, blood and hair. Our objective was to quantify total mercury in hair of two populations: one rural population (miners of the region of San Joaquin, Queretaro, Mexico) and one urban population of the Metropolitan Area of Mexico, and to compare the results of a population potentially exposed by the explotation of mercury in mines and a community not affected by mercury emissions. Each participant provided a hair sample and completed a questionnaire assessing potential exposures and health outcomes. We found average mercury concentrations of 32.07 $\mu g\ g^{-1}$ and 2.62 $\mu g\ g^{-1}$ in the rural and urban population, respectively. The great difference between these values is probably due to a difference in the time of exposure for each population. In both cases, the populations studied exceeded the maximum allowable limit established in standards and by national and international agencies, mainly due to the direct exposure of mercury vapors in miners and by anthropogenic sources in the urban population.

Keywords Mercury · Hair · Toxicity · Analysis · Bioaccumulation

---

I. M. Fuentes (✉)
Posgrado en Ingeniería Ambiental, UNAM-IMTA. Blvd, Paseo Cuauhnáhuac 8532, Progreso, 62550 Jiutepec, Morelos, Mexico
e-mail: ise_la19@yahoo.com.mx

R. G. Martínez
Centro de Ciencias de la Atmósfera, UNAM. Cd. Universitaria, Circuito Exterior s/n, 04510 Mexico, CDMX, Mexico
e-mail: gmrocio@atmosfera.unam.mx

W. Leal Filho et al. (eds.), *Sustainable Development Research and Practice in Mexico and Selected Latin American Countries*, World Sustainability Series,
https://doi.org/10.1007/978-3-319-70560-6_15

1 Introduction

Mercury is an element that cannot be destroyed and is a global threat to human and environmental health. In Latin America and the Caribbean, artisanal and small-scale gold mining represents the main source of mercury emissions, releases, and consumption. However, another source of concern is the primary production of mercury. In the case of Mexico, in the past 2 years the informal production of mercury mining has increased 10-fold. Considering this scenario, an intervention program was initiated to reduce health risks in the mining communities. The program's final goal is to introduce different alternatives in line to stop the mining of mercury, but introducing at the same time, a community-based development program. This increased concern about the health of persons exposed to very low environmental mercury concentrations is because mercury causes subclinical effects at low concentrations.

In 2003, after having performed a global assessment, the United Nations Environment Program (UNEP) found that there was sufficient evidence of significant global adverse effects from mercury and its compounds to warrant further international action to reduce the risks to human health and the environment from the release of mercury and its compounds to the environment. In January 2013, an intergovernmental negotiating committee agreed on the text of the Convention on Mercury, and in October 2013 the Convention was signed in Minamata, Japan, by 128 countries. In its first article, the Minamata Convention states that its objective is to protect the human health and the environment from anthropogenic emissions and releases of mercury and mercury compounds (UNEP 2013). Mercury (Hg) is a toxic chemical element widely used by humans and produced consequently in vast quantities as a result of industrial processes, mining, and international trading.

Hg possesses several physical and chemical states (elemental/inorganic/organic), each with its own intrinsic toxic properties. In term of toxicology, organic Hg and specially methylmercury (CH_3Hg^+) has a higher toxicity than elemental Hg (Hg^0) and inorganic compounds (HgS, HgO, $HgCl_2$, among others). Even a low quantity of Hg may cause serious health problems: Hg exposure leads to diseases of the nervous, digestive and immune systems as well as damage to the lungs, kidneys, skin and eyes (Ramírez 2008).

The distribution and toxicity of mercury depend on various factors, including the hydrodynamic characteristics and the physico-chemical conditions of the area, the assimilation of mercury by living organisms and its interactions with other constituents of the environment which can modify its form in the water, soil, air or within the bodies of living organisms. Hg occurs naturally in a wide variety of organic and inorganic compounds not only in the solid state, but also dissolved in water and, exceptionally, in the atmosphere due to its high vapor pressure (Fig. 1).

The analytical diagnostic of Hg poisoning is difficult: urinary excretion of this element is not an assessable index since organic Hg is stably absorbed, and in the highest concentration, in red blood cells. Blood tests thus indicate the mineral concentration at the cellular level. In both cases, it only reflects Hg concentration in

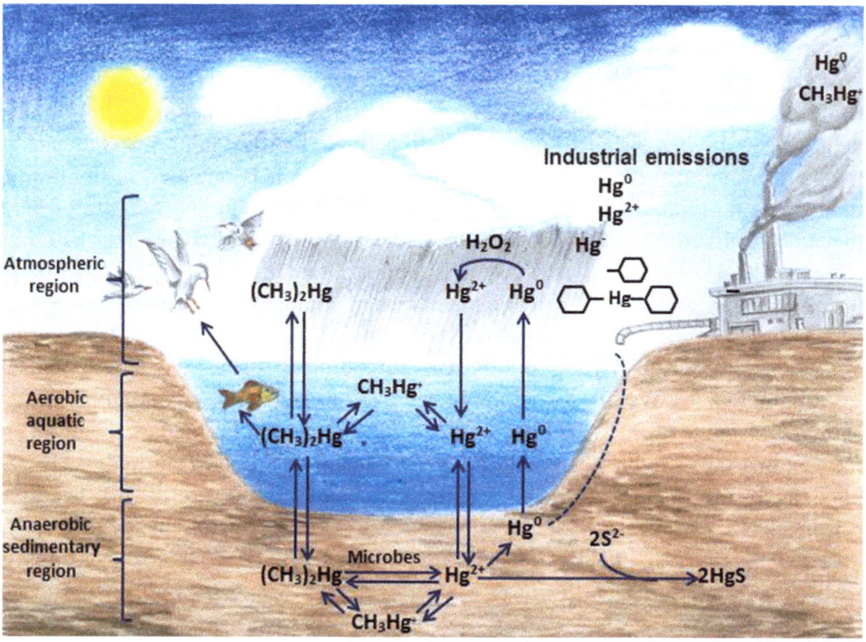

Fig. 1 Biogeochemical mercury cycle. Realized by Daniel Ramírez

the body at the moment the sample is taken. It is thus necessary to uncover other biomarkers that will allow early detection of mercury contamination, particularly in high-risk populations.

Hair is the best indicator of Hg exposure because it is affected by the diet, the environment and the profession of the individual. Analyses of hair samples enable to precisely evaluate the mineral concentration inside the body, since it accumulates in proteins rich in sulfhydric groups. Therefore, hair works as a dosimeter and the detected quantity is proportional to the concentration of mercury in the organism (Echegaray and Gómez 1984). It also allows the evaluation of the levels of pollution through time. It is possible to tell how much mercury a person has been exposed to by testing their hair, blood and urine. According to The World Health Organization, Dental amalgam has been identified as the largest single source of continuous Hg exposure for members of the general population who possess amalgam fillings (WHO 2000).

The principal advantages of hair as a biomarker are the following: (1) it is a non-invasive method of analysis, (2) the sample is stable at room temperature, easy to transport and to store for long periods of time and (3) it does not require a specialized training to collect. However, it is important to consider that hair doesn't grow evenly and the substances remain according its growth at an approximated rate of 1.20–2.0 cm/month, exposed to external pollution.

Hg exposure in non-exposed populations often happens through dental amalgams, polluted food (mainly fish and seafood) and the use of personal hygiene products containing this metal (hair dyes, creams and clarifying soaps). Except some outstanding cases, human health is not in danger because of Hg present in nature. The higher risk to human health derived by the natural abundance of mercury is due to the occupational exposure, because it is obtained in the mining industry as a principal product or byproduct in the extraction and refining of other metals and minerals, as well as in iron and steel industry and some other activities (Cespón 2001). For these reasons, it is important to develop exact and precise techniques that will provide reliable results when applied in a complex matrix such as hair.

Research suggests that Hg and its derived chemical species enter the organism by different pathways, as for example, inhalation, ingestion, absorption through skin and placenta (Fig. 2). The toxic effects of organic and inorganic Hg are due to the fact that they bind to the cellular organic constituents rich in sulfhydryl groups, affecting various metabolic and enzymatic systems of the cell and its wall. The toxic action of Hg on enzyme systems occurs because it precipitates the proteins synthesized by the cell, mainly the neurons, and because it inhibits the groups of several essential enzymes (Ramírez 2008).

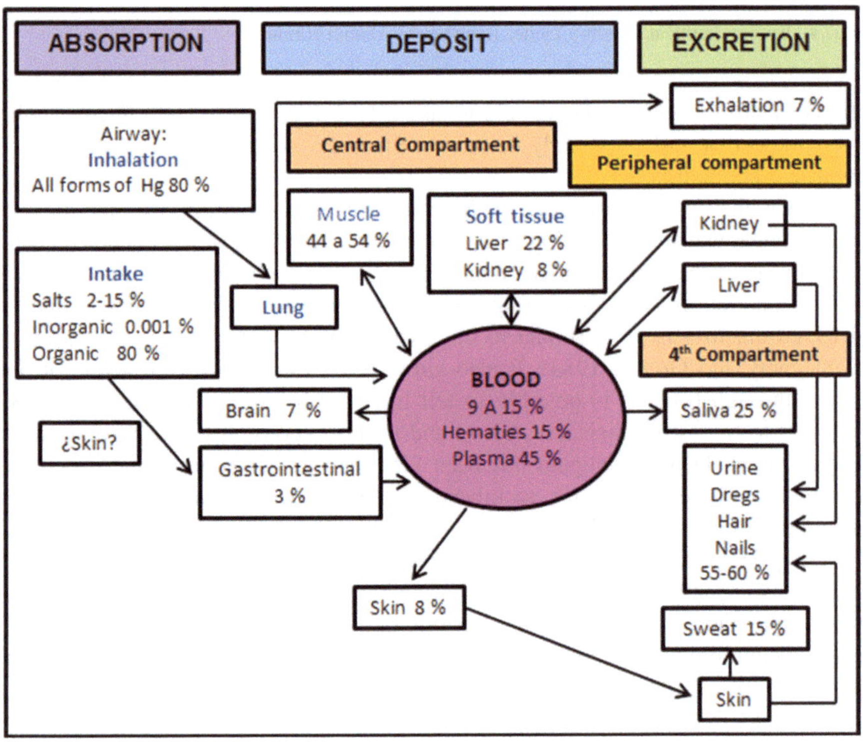

Fig. 2 Toxicokinetic model of inorganic mercury (Ramírez 2008)

International Agencies like the United Nations Environment Programme (UNEP), the Food and Drug Administration (FDA), the World Health Organization (WHO) and the United States Environmental Protection Agency (USEPA) stablish the maximum allowable limit of Hg for non-exposed people's hair as 1–2 $\mu g\ g^{-1}$, for people that have a diet rich in contaminated fish as 10 $\mu g\ g^{-1}$ and for occasionally exposed people during odontological activities as 7 $\mu g\ g^{-1}$ (UNEP 2002).

Because the importance of its effects, Hg is a chemical product of global awareness, especially for its transport at long distance in the atmosphere, its persistence in the environment after being anthropogenically introduced, its capability of bioaccumulation in ecosystems and its effects against human health and the environment (Carretero and Pozo 2007) Minamata Convention on Mercury was created to take actions on the environmental storing of Hg and waste control, to set expiry dates and stablish manufacture, importation and exportation protocols of Hg for 2020 to utilize economically feasible techniques and to produce new products that lack this metal in their composition. Some exceptions have been made for some products without any reliable substitution and also, the Hg content has been limited for some applications like low-energy lamps. The Convention committed to devise new strategies to identify and evaluate Hg polluted locations for and appropriate management, focusing in human health (MEJ 2013).

In Mexico, there are few studies, and even fewer regulations, that indicate the maximum limits of exposure to this metal. A few references can be found about exposure by odontological activities (analyses of hair and urine), but none about Hg exposure in high polluted areas. In this study, we performed hair analyses on two different populations: a rural, high-risk population and an urban one.

The first population is located in the mining region of San Joaquín, Querétaro. This is considered a partially exposed population because of the explotation of inorganic Hg mines (HgS, cinnabar), in which the informal Hg exploiting represents the principal economic income for the inhabitants. The importance of the study in this area relies in the Hg exposure by different pathways (absorption through skin and inhalation) that workers must endure because of the lack of the necessary safety equipment, affecting severely their health through continuous exploitation and handling of HgS. Other consequence of this industry is the production of hazardous waste that continuously damages the ecosystems and people in the area, due to the capacity of mercury to remain in the environment for decades, centuries and even thousands of years.

The second population of interest is located in the urban area (Metropolitan Area of Mexico City) in which the principal sources of mercury pollution are anthropogenically produced by industrialization, the use of products with high content of Hg (hair dyes, clarifying soaps and creams) and the possible consumption of contaminated seafood.

In large cities like the Metropolitan Zone of Mexico City, air, water and soil composition has changed because of pollution with chemical products and powder from excessive use of cars, factories and lack of vegetation, asphalt of other kind of coatings on soils, as well as a variety of activities that use hazardous solvents and toxic substances as Hg. Because of these actions, common use products and even food have been polluted affecting human beings through several pathways engaging their health (ME 2012).

In Mexico there has not been a detailed evaluation about production, emission and spreading of Hg. In this work, we propose an assessment of the degree of Hg exposure in a population dedicated to Hg mining, which has affected for several years to the Terrestrial Systems (water, soil, air and plants).

2 Metodology

This study was performed to two different Mexican populations: one of them potentially exposed to Hg (miners) and living in the mining area of San Joaquín, Querétaro; and the other one in the southern part of the Metropolitan Zone of Mexico City (Ciudad Universitaria), an area that is not exposed to Hg by mining activity.

The present study is organized in two parts. The first stage corresponded to the diagnostic, the organization and the scheduling of the performed tasks. This was achieved by an activity-time survey answered by the participants, including personal details (age, profession, and geographic localization), diet, personal hygiene, use of cosmetic treatments on the hair and oral health.

The second stage corresponded to the recollection of the samples: hair samples of 24 miners that currently work in the mercury mines of la Poza, Rosario, Otatal, Atenea, la Fortaleza, la Maravilla and la Barranca, located in the region of de San Joaquín; and hair samples from 36 students of Ciudad Universitaria in Mexico City. This last group consisted of 18 men and 18 women around 21 to 25 years-old.

Hg has great affinity for sulfhydryl groups of keratin and other hair proteins. Hair is one of the favorite matrices used for Hg determination because it is a non-invasive procedure, provides a simple sample with an exposure profile inclusive over time and that can remain with little concentration changes for years. Furthermore, the main advantage of hair analysis is that once the Hg has been incorporated, it cannot return into the bloodstream.

The back of the head is the most highly recommended part of the scalp for sampling because it is the area the less sensitive to external pollution. 2–3 locks of hair from 1–2 cm long were taken from the back of the head of the volunteers (Cespón 2001).

The samples were then placed in polyethylene tubes, previously washed with a nitric acid solution and sterilized in order to avoid microbial contamination and possible pH variations.

After the determination of the elements of endogenous absorption, the washing procedure is the next fundamental part of the analysis. An ideal washing must eliminate only the external pollutants, leaving the endogenous elements untouched.

The International Atomic Energy Agency (IAEA) has examined different procedures of washing and has determined all the variables associated. As the incomplete eradication of exogenous pollutants and the partial elimination of endogenous elements is hard to control, the IAEA method suggests a sequential washing process of the hair samples with a mixture of water and acetone, with constant stirring, removing powder particles, sweat, fat and other external substances (López 2013).

After the hair samples were washed, they were decanted and dried at room temperature. Then the acid digestion of the samples was performed in a microwave oven. Each sample of dried Taraxacum officinale were weighed and placed in liner tubes. 10 mL of nitric acid and 5 mL of hydrogen peroxide were added. A control (blank) was performed for each digestion cycle, consisting of the mixture of reagents without hair reference material. Prior to chemical analysis, acid digestion of the samples was done with a MARS 5 analytical microwave oven, using 15 mL of 3 M nitric acid and 5 mL of hydrogen peroxide at 110 psi, for 45 min; after digestion, the solution was poured into a 25 ml volumetric flask, then deionized water was added to the mark and kept at 4 °C until the analyses were carried out.

The samples were analyzed by graphite furnace atomic absorption spectrometry (GFAAS) with a 932AA double beam GBC device, coupled with a 3000 graphite furnace accessory system, which consists on a GF3000 graphite power supply and a PAL3000 furnace auto sampler, both computer controlled.

Calibration was made using certified standards that were prepared within the expected concentration range for traceable samples according with the National Institute of Samples and Technology (NIST). Calibration curves with sample concentrations of 1, 2, 5, 8, 10 and 15 $\mu g\ L^{-1}$ and their respective blanks were used. Concentration of Hg was determined in a wavelength of 253.7 nm.

3 Results and Discussion

According to the Mexican Norm (NMX-EC-17025-IMNC-2006, General Requirements for the Competence of Test and Calibration Laboratories), the validation of the analytical methods used is a fundamental requirement before realizing an analytical measurement, because analytical method performance vary between laboratories. Furthemore, analytical methods are specific of the matrix with which they were developed, and then validated, in the first place. They cannot be used to measure in any given matrix.

In this study, the analyses were performed by Spectrophotometric Tests by Atomic Absorption, in which the performance parameters are: linear range, detection limit, limit of quantification, sensitivity and percentage of recovery.

Linearity. To ensure quality control/quality assurance during sampling and analysis only plastic and glassware were used. Mercury National Institute of Standards and Technology standard (NIST; $Hg(NO_3)_2$ in 2 mol L^{-1} HNO_3, 1000 mg L^{-1} Hg was used to obtain calibration curves and for validation, 10 calibration curves were realized, observing at which point these curves lose linearity. Quality controls (QC) were evaluated from the calibration curves (Table 1), where the equipment recorded a concentration of Hg concentration and absorbance for the blank of 0 μg L^{-1}. In Graph 1, the correlation was (R^2) de 0.9993, the good line were 0 a 15 μg L^{-1} (Table 1).

The numerical value of the detection limit (LD) and the quantification limit (LC) were calculated from the absorbance reading value of the reactive targets of the 10 calibration curves performed.

Detection Limit. The limit of detection (LD) is defined as the lowest concentration at which the analityc can be detected, but not necessarily quantified under the established experimental conditions (García 2010).

$$\textit{Limit detection}(LD) = 2.1\ \mu g\ L^{-1}$$

Quantification Limit (QL). The limit of quantification is defined as the lowest concentration at which the analytic can be quantified with acceptable established experimental conditions (García 2010).

$$\textit{Quantification Limit}(QL) = 3.1\ \mu g\ L^{-1}$$

Sensibility. Sensibility is understood as the change in instrument response in relation to a change in analytic concentration, as visualized by the slope of the average curve obtained (Graph 1). The sensitivity is 0.016. To evaluate the confidence interval we applied a t-Student distribution test for 95% confidence with n = 10, following Eq. 1.

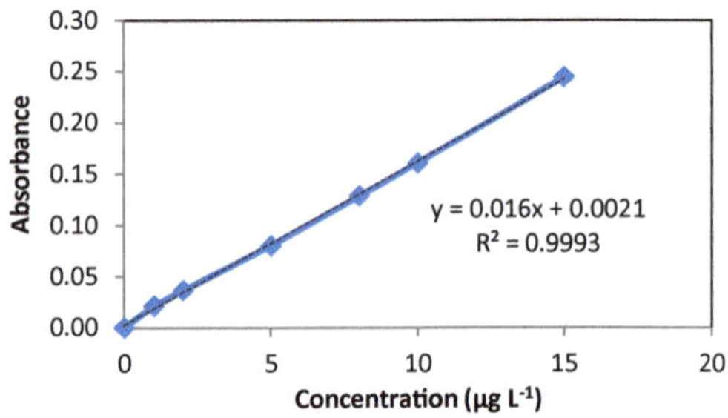

Graph 1 Average calibration curve

Table 1 Average calibration curve

Sample	Conc (μg L^{-1})	Abs
Blank	0	0.0000
Standard 1	1	0.0211
Standard 2	2	0.0362
Standard 3	5	0.0800
Standard 4	8	0.1285
Standard 5	10	0.1604
Standard 6	15	0.2441

$$\mathbf{Inter.\ de\ confianza} = \mathbf{t_9} * \frac{\mathbf{typicalerror}}{\sqrt{10}}$$

Where:
$t_9 = 54.64$
Typical error = 0.0002

$$\textit{Sensibility} = (0.016 \pm 0.0035)L\ \mu g^{-1} \quad (1)$$

Recovery rate: the recovery is defined as the analytic fraction added to a test sample (added or fortified sample) prior to the analysis, whose concentration is effectively determined by the method.

According to Rangel (2015), the percentage of recovery obtained in the validation of the method is 98.41%.

3.1 Non-exposed Population by Profession (Student Population in the Southern Area of Mexico City)

This population is comprised of 36 people, including 18 men and 18 women. The identification code of the samples was assigned according to the gender (H: for men and M: for women), followed by a number from 1 to 18, corresponding to the number of the participant (Table 2).

Table 2 Concentration of Hg according to gender

Men				Women			
Code	Conc ($\mu g\ g^{-1}$)	Code	Conc ($\mu g\ g^{-1}$)	Code	Conc ($\mu g\ g^{-1}$)	Code	Conc ($\mu g\ g^{-1}$)
H1	2.20	H10	1.99	M1	2.64	M10	1.78
H2	0.83	H11	1.39	M2	2.31	M11	3.21
H3	1.49	H12	0.94	M3	6.16	M12	1.70
H4	3.41	H13	1.39	M4	1.15	M13	2.22
H5	1.09	H14	7.76	M5	0.99	M14	1.72
H6	1.25	H15	7.96	M6	2.27	M15	2.09
H7	8.50	H16	1.72	M7	4.40	M16	1.99
H8	2.24	H17	2.00	M8	2.87	M17	0.61
H9	5.43	H18	0.33	M9	2.87	M18	1.67

The statistical parameters (mean, maximum, minimum and standard deviation) obtained from the samples of this population are shown in Table 3.

The obtained data were statistically analyzed with central tendency measures and variability degree in order to observe the data location, its dispersion with respect to the average and the possibility to present a normal distribution. Correlation tests (Spearman) were performed between the main variables (gender and age).

An increased dispersion of the obtained data is observed for male population, mainly from 23 years-old, followed by 22 and 21 years-old, whereas the rest does not show dispersion because of the number of participants of the same age. For female population, dispersion is only observed in 22 years-old volunteers (Fig. 3). Spearman correlation coefficient was used as statistical tool to verify the independence between total Hg concentrations in the whole sampling campaign and to identify the independence between two variables (age and Hg concentration). The principal objective of this technique is to recognize if there is a common source of gender and if there exist any correlation.

Scatter plot and Hg variability (Fig. 4) show that the average values and the standard deviation are above the allowed values dictated by the WHE, EPA and other national and international agencies. These results indicate that random errors

Table 3 Statistic parameters of Hg concentration by gender (n = 36)

Gender	Average conc ($\mu g\ g^{-1}$)	Max conc ($\mu g\ g^{-1}$)	Min conc ($\mu g\ g^{-1}$)	D.E. ($\mu g\ g^{-1}$)
Men	2.88	8.50	0.33	2.64
Women	2.37	6.16	0.61	1.29

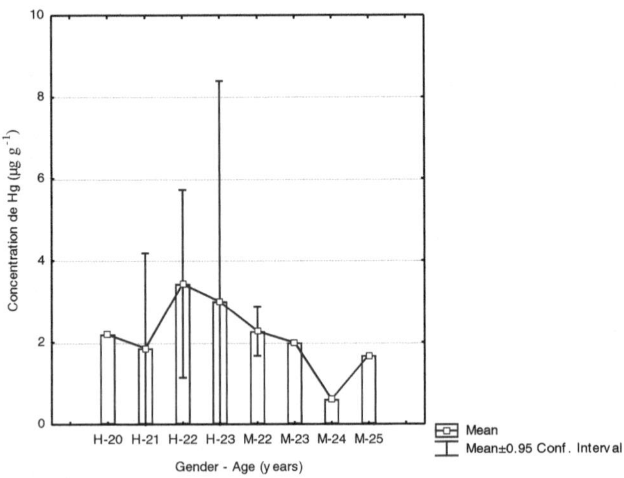

Fig. 3 Histogram of frequency of men and women with respect to age

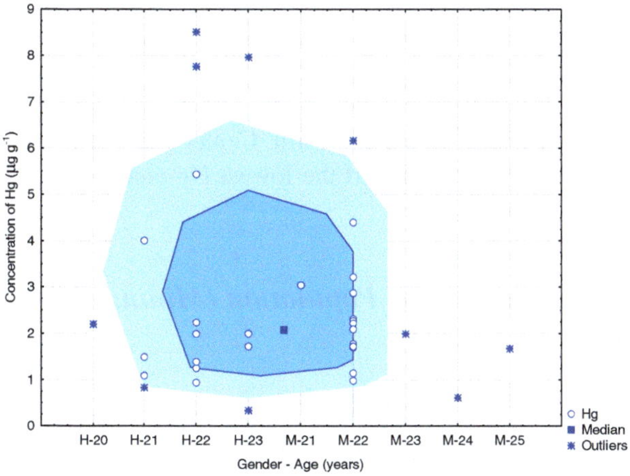

Fig. 4 Scatter plot and Hg variability for men and women with respect to age

Table 4 Hg concentration with respect to the frequency and percentage of participant distribution classified by geographical position (n = 36)

Location	Frequency	% frequency	Max conc ($\mu g\ g^{-1}$)	Min conc ($\mu g\ g^{-1}$)	Aver Conc ($\mu g\ g^{-1}$)	D.E. ($\mu g\ g^{-1}$)
Center	6	16.67	7.76	0.99	2.89	2.45
North	8	22.22	7.96	0.61	2.86	2.38
South	2	5.55	1.99	1.49	1.74	0.35
East	14	38.89	8.5	0.33	2.46	2.24
West	6	16.67	5.43	1.39	2.73	1.52
Total	36	100				

of age and gender are under control of mobility, inhabitation and food variables that are non-controlled variables.

The graphs also display that 86% of the population of interest is two times above the permissible limit (1 $\mu g\ g^{-1}$), considering that it is not a community with high consumption of seafood. The excess of Hg is due to environmental exposition and the use of hair products like hair dyes because at least two of the participants with the highest Hg concentration use to dye their hair in a regular basis, besides; the man with the highest amount of Hg is the only one that uses this type of products.

The excess of Hg observed must be linked to environmental exposition, such as the use of hair products like hair dyes for example, since at least two of the participants that exhibited the highest Hg concentration declare dyeing their hair on a regular basis. Furthermore, in the male group, the highest Hg concentration value correlates with the only individual making use of this type of products.

The Hg concentration measured was thus compared to the individual geographical position of each participant (Table 4), with Mexico City as the reference point and assigning coordinates as Center, North, South, East and West. The relations found by this analysis, with frequency and percentage of distribution, are summarized in Table 4.

The highest Hg concentration was found in participants living in the East of Mexico City, followed by those in the North, Center and West. The participants of the southern part of the city exhibited the lowest Hg concentrations.

3.2 Exposed by Profession Population (Mining Region, San Joaquin)

This population was comprised of 24 miners, who were all active at the moment of the analyses in one of the 6 mines that were included in this study in the mining region of San Joaquín.

The samples identification codes were assigned with the first letter of the name of the mine in which the participant worked (P: Pozas, R: El Rosario, O: El Otatal, A: Atenea, M: La Maravilla, F: Fortaleza y C: Carpintería), followed by a sequential number (1, 2, …,) depending on the number of volunteers for each of the location studied. The data collected and the statistical parameters for this population are shown in Table 5 and Table 6, respectively.

The information gathered from the time-activity survey applied to this population indicate in general terms that the volunteers do not eat seafood and that their diet consists in balanced meals prepared at home. They also eat 3 fruits or vegetables daily and drink mainly milk as a detoxifying beverage.

In the first place, an exploratory analysis was performed in order to detect odd values caused by a wide data dispersion (Fig. 5), there are two atypical values are observed, among them the minimum value (0.80 $\mu g\ g^{-1}$) obtained, since this is too

Table 5 Hg concentration in miners (n = 24)

Code	Conc. ($\mu g\ g^{-1}$)	Code	Conc. ($\mu g\ g^{-1}$)	Code	Conc. ($\mu g\ g^{-1}$)
P1	16.40	R5	11.05	A1	5.79
P2	16.57	R6	3.90	A2	28.20
P3	115.10	R7	121.60	M1	2.44
P4	65.29	R8	5.21	F1	4.00
R1	68.57	O1	40.38	F2	6.62
R2	2.88	O2	7.13	F3	3.15
R3	10.08	O3	142.56	F4	13.21
R4	62.24	O4	16.39	C1	0.80

Table 6 Statistic parameters of Hg concentration (n = 24)

Average conc ($\mu g\ g^{-1}$)	Max conc ($\mu g\ g^{-1}$)	Min conc ($\mu g\ g^{-1}$)	D.E. ($\mu g\ g^{-1}$)
32.07	142.56	0.80	41.89

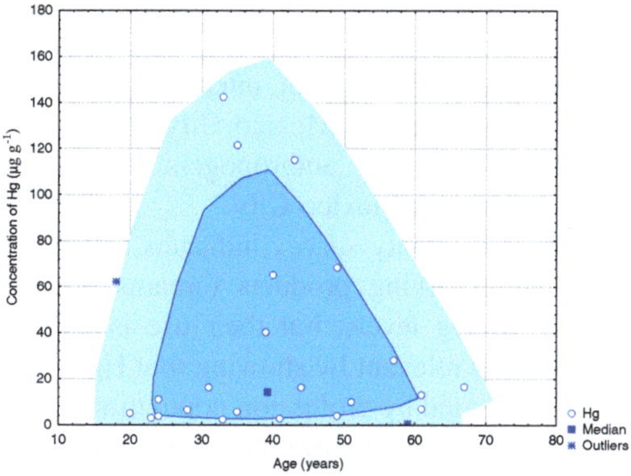

Fig. 5 Scatter plot and Hg variability with respect to age

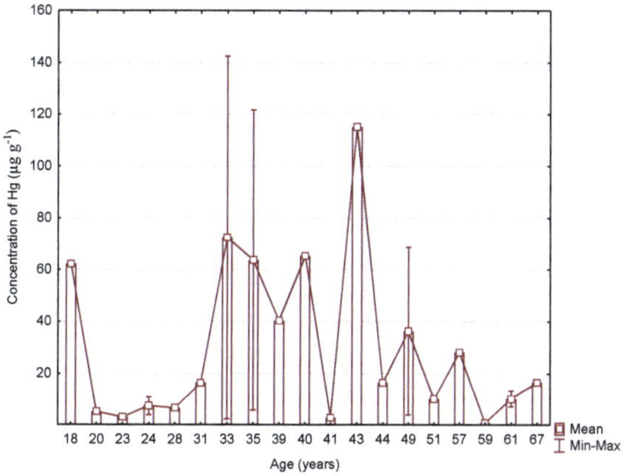

Fig. 6 Histogram of frequency with respect to age

far away from the subsequent value, which is 2.44 μg g^{-1}, which may possibly be related to the participant's age (18 years).

The frequency distribution of Hg histogram (Fig. 6) shows that more that 60% of the Hg concentrations are in the range of 0–20 μg g^{-1} with a frequency of 16; dismissing values of Hg concentration lower to a 10 μg g^{-1} 54% of the population is above the permissible value reported by UN, PAHO and others (10 μg g^{-1}) as the limit value for people that are exposed by profession, considering an odontological exposure because there isn't a reported value for mining activities.

4 Conclusions

According to the results of this research, Hg concentration on both populations is above the permissible limits established by National and International Agencies (USEPA, WHO, PAHO) that symbolizes a health risk which thus represents a real risk for human health.

The average values of Hg concentration in the non-exposed population are 1–2 times above the established limit, having a considerable dispersion of the obtained data for the population of interest. Hair Hg concentration of the population of Metropolitan Zone of Mexico City also shows important geographical variations probably because of anthropogenic local sources of mercury pollution in the northern areas of Mexico City.

The time-activity survey indicates that the utilization of hair treatments as hair dyes and clarifying products (creams and soaps) are important factors in the increase of Hg levels, but their use is independent of the gender and age. The research was relevant by showing that Hg levels are higher in men tan in women in the non-exposed by profession population which is attributed to diet habits, because men eat seafood in a more regular basis than women, who eat frequently fruits and vegetables. The factors that affect mercury levels are diet, occupation, age and sex; all have some effect on mercury levels in the hair, as mentioned above.

The advantages of hair Hg assessment include the fact that hair collection is non-invasive, and get good response rates in all population subgroups, where obtaining blood samples can sometimes be difficult from some of the subgroups, such as women for example. Additionally, hair is a time record marker of MeHg exposure in individuals and can be used to estimate Hg exposure over extended periods of time such as fetal exposure during gestation.

Even if both populations exceeded the permissible limits, it is important to note that there is a large gap the values of the concentration of Hg obtained since the student community (control population) is not rigorously exposed to contamination by Hg, the highest concentrations of Hg measured are 8.50 $\mu g\ g^{-1}$ in men and 6.16 $\mu g\ g^{-1}$ in in women, respectively. The lowest concentrations detected were 0.33 $\mu g\ g^{-1}$ and 0.61 $\mu g\ g^{-1}$. On the contrary, the highest concentration was 142.56 $\mu g\ g^{-1}$ and the lowest was 0.80 $\mu g\ g^{-1}$ in the population of the mining region. The wide interval observed between these last two values can be explained by the amount of time each miner has been working in a mine, which would influence the amount of time each individual would be exposed to the contaminant.

References

Carretero I, Pozo M (2007) Applied mineralogy: health and environment. Thomson, Spain, Madrid

Cespón M (2001) Development of automatic analytical methods for the determination of metals in the working environment. USC, Department of Analytical Chemistry, Nutrition and Bromatology, Barcelona

Echegaray R, Gómez M (1984) Determination of mercury in hair as an expression of mercury exposure. Public Health 6(31):87–92

García C (2010) Determination of mercury by cold vapor generation and detection via absorption and atomic fluorescence. Thesis of Degree in Chemistry. Sartenejas, Simon Bolivar University

López P (2013) Concentration of mercury in hair of a sample of the adult population of Yucatan, descriptive study. Thesis of master in Science. Merida Yucatan, Mexico. Unit Mérida, Department of Human Ecology. Cinvestav, IPN

Mexican Norm NMX-EC-17025-IMNC-2006 General requirements for the competence of the testing and calibration laboratories. Mexican Institute of Normalization and Certification

Ministry of the Environment (ME) (2012) Inventory of pollutant and greenhouse emissions. Metropolitan Zone of the Valley of Mexico. http://www.aire.cdmx.gob.mx/descargas/publicaciones/flippingbook/inventario-emisioneszmvm2012/#p=20. Accessed on 29 Nov 2014

Ministry of the Environment of Japan (MEJ) (2013) Teachings of Minamata disease and the management of mercury in Japan. Division of Environmental Health and Safety, Department of Environmental Health. https://www.env.go.jp/chemi/tmms/pr-m/mat01/es_full.pdf. Accessed on 10 Feb 2015

Ramírez A (2008) Occupational mercury poisoning. An Fac Med (69)1:46–51, Perú

Rangel M (2015) Quantification of Hg in water for human consumption from wells in the mining region of Taxco de Alarcón, Guerrero. Thesis of Degree in Chemistry. Mexico. Faculty of Chemistry, UNAM

United Nations Environment Program (UNEP) (2002) Global assessment on mercury. Chemical products. Geneva, Switzerland. https://noharm.org/sites/default/files/lib/downloads/espanol/Evaluacion_Mundial_Mercurio.pdf. Accessed 22 on May 2015

United Nations Environment Program (UNEP) (2013) Minamata convention on mercury. Geneva, Switzerland

World Health Organization (WHO) (2000) Environment Health Criteria 214, Human Exposure Assessment. International Programme on Chemical Safety (IPCS). Geneva, Switzerland

Author Biographies

Isela Martínez Fuentes, Bachelor B.S., graduated from the Faculty of Chemistry at the National Autonomous University of Mexico (UNAM), She worked at the Center of Atmosphere Sciences, UNAM in the quantitative analysis of heavy metals in environmental and biological matrices. Her research is focused in environment distribution of mercury, its interaction and its effects on urban and rural areas using Hydride Generation Atomic Absorption Spectrophotometry. She is currently working as a student in the Master in Environmental Engineering in the Mexican Institute of Water Technology in the wastewater treatment with nanoparticles and magnetic separation.

Rocío García Martínez, Earth Sciences Ph.D., graduated from the Earth Sciences doctoral program at UNAM. She is a Level I member of the National System of Researchers and a Level C member of the Incentive Program for Academic Performance at UNAM. She is also a member of the CONACYT list of Accredited Assessors in Physics, Math and Earth Sciences. Her research has been dedicated to: the isotopic characterization of rain and superficial water to identify sources of aquifer pollution recovery; the study of heavy metals and organic compounds over environmental bioindicators as atmospheric pollution indicators and the study of atmospheric mercury in rainwater and atmospheric sprays in urban, suburban and rural areas.

Organization and Community Management of Water for Domestic Use in a Rural Community in Nicaragua

Juana María Zavala-Figueroa and Juan Alberto Velázquez-Zapata

Abstract This work aims to analyze the experience of a Nicaraguan rural community in the management of water for domestic use through the conformation of community committees, and the relationships that such committees have with the government and other relevant organizations. The research is conducted through the qualitative paradigm of social sciences; hence, the methodology is based on a case study where empirical data were collected through semi-structured interviews, focus group discussions and participant observation. The community organization for local management has emerged in order to meet a specific need; in this case, the water supply for domestic use. According to the collective action theory, this work demonstrates that the internal rules and regulations have favored the economic sustainability, the empowerment and the active participation of the community organization in order to obtain autonomy in the management of its resources. This research is important for it shows that the state's recognition of the community organizations has favored the cooperative effort between communities, government and non-governmental organizations, facilitating the management process and the access to financial, technical and training resources.

Keywords Community · Organization · Community management
Water supply · Nicaragua

J. M. Zavala-Figueroa (✉)
El Colegio de San Luis, Parque de Macul No. 155,
78294 San Luis Potosí, S.L.P., Mexico
e-mail: Jmaryzavala24@gmail.com

J. A. Velázquez-Zapata
Consejo Nacional de Ciencia y Tecnología, El Colegio de San Luis,
Parque de Macul no. 155, 78294 San Luis Potosí, S.L.P., Mexico
e-mail: javelazquezza@conacyt.mx

W. Leal Filho et al. (eds.), *Sustainable Development Research and Practice in Mexico and Selected Latin American Countries*, World Sustainability Series,
https://doi.org/10.1007/978-3-319-70560-6_16

1 Introduction

Water management in Latin America can be classified in three major classes: public, private and social. In public management, the state is responsible for the water supply. On the other hand, in the private management model, the state delegates its responsibility to companies which are interested in an economic profit. In contrast, social management aims to satisfy a local need; hence, this management focuses on environmental and economic sustainability (Sandoval and Günther 2013).

Alternatively, community water management is a type of cooperative management (between community, government and non-governmental organizations) that has been implemented in rural Nicaraguan communities. This management is based on Community Water and Sanitation Committees (CWSCs), which are local organizations recognized by the Nicaraguan Law.

Several authors have studied the CWSCs in Nicaragua; for example, Kreimann (2014) analyses the community management of water resources in the country and claims that only 520 CWSCs work properly, since the committees do not have enough members. Other studies have focused on the forming process of a given committee; for instance, the study of Reyes et al. (2008) depicts the problems that a rural community faced in order to establish a CWSC. The author found that the organization structure in the community plays a key role in the establishment of a community water committee. However, the study of Lorío (2004) evaluates community management of water resources in a drought context, and the author argues that the rules established by the communities are more severe in water scarcity conditions.

This study is important because analyses the co-management model in a rural community in Nicaragua. The research aims to answer the question: is an organized community more prone to reach sustainability in managing its water resources? The study focuses on three organization experiences in the community of Ducualí, which is located in the municipality of Palacagüina (northern Nicaragua). The manuscript is organized as follows:

Firstly, the methodology is presented and the concepts of organization and community are reviewed; secondly, a brief historical review on the community management of water resources in Nicaragua is given; thirdly, the experience of the three CWSCs that operate in the community of Ducualí is discussed; and, finally, concluding remarks close the manuscript.

2 Methods

The research is conducted from the qualitative paradigm of social sciences; therefore, the methodology is based on a case study where empirical data were collected through semi-structured interviews, focus group discussions and participant observation. The methodology consisted of the following steps: first, the users (one member per family) of each CWSC were identified and listed; and second, the people to be interviewed were selected with the random sample technique. In addition, focus group discussions were conducted for each CWSC in order to obtain information about the committees' background.

3 Concepts of Community and Organization

In literature, the discussion about the concepts of community and organization is addressed from the point of view of different disciplines. In social sciences, the discussion focuses on the social interactions based on culture, values, traditions and beliefs developed in a given space. Nevertheless, from the point of view of administration sciences, the discussion focuses on the relationships between stakeholders and the decision processes in order to satisfy a need and develop skills (Bernal et al. 2014).

In particular, the concept of community in this research is understood as "a social actor with its own identity and a common voice arising from collective action as a complex process of decision" (Bernal et al. 2014). The people in an organized community share values and objectives leading to a common identity and to the development of social action in order to search for mutual interests and satisfy necessities.

Regarding the concept of organization, this research takes the point of view of Antonio et al. (2010), which argues that, in an organization, a social group coordinates its activities in order to achieve common goals by delegating functions through a hierarchical order; however, Olson (1992) claims that organizations are often used to accomplish individual benefits. Therefore, an organization requires commitment, cooperation and trust from its members to delegate tasks. The social organizations, in rural and urban areas, are a good example of people's ability to coordinate efforts to respond to a specific need.

4 Community Water Management in Nicaragua

In Nicaragua, the water and sanitation policy has been adjusted to the economic model of the moment (OPS-OMS 2004). In that context, the first Community Water and Sanitation Committees emerged in the decades of 1960 and 1970 (Alianza por el Agua 2015). At the same time, several social movements appeared in Latin America with the purpose of providing water supply to rural communities. Over the years, such movements showed that they are robust social structures that provide solutions to the lack of basic services (CARE International and Avina 2012).

In the 80s, a partnership model for water resources management between the state and the community was adopted in Nicaragua. This model took into consideration cooperative principles which favored the organization of community committees and the implementation of projects towards social welfare in the rural regions. In the 90s, the Community Water and Sanitation Committees were accepted by the state, and they were promoted as the organizations in charge of the administration and maintenance of the community water supply systems. Between 2007 and 2014, the number of legalized Community Water and Sanitation Committees in Nicaragua increased as a result of the implementation of Act 260 (Waters General Act) and Act 722 (Special Act on Water Committees). The CWSCs legalized in that period represent 55% of the total committees in the country.

However, the programs designed to provide water supply in rural communities in Nicaragua had been influenced by nongovernmental agencies that worked in collaboration with the official Water and Sewerage Enterprise (ENACAL). The SIASAR (Spanish acronym for Information System on Rural Water and Sanitation) system (2016) reports 6836 rural communities in Nicaragua, out of which 3670 communities have some kind of community water management. From this group of communities, 93.7% have legalized CWSCs, 2.4% have no legalized committees, and in the last 3.9% the water supply is administrated by community cooperatives in collaboration with ENACAL or the municipality.

The CWSCs were designed with the intended purpose of contributing to the social and economic development of Nicaragua by creating conditions to ensure access to drinking water (Gaceta 2010). The CWSCs are based on five guiding principles (Gaceta 2010):

1. Volunteerism of the community members in the community management of the water resources.
2. Equity, meaning all community members shall participate on equal footing.
3. Plurality, as recognition of the different values, opinions and practices in the community by the local, regional and national authorities.
4. Solidarity in the community management of water resources, by seeking common good and preventing individual interests.
5. Respect and defense of the community autonomy in the frame of the rights and obligations established by laws.

The guiding principles are based on the collective action theory, and take into account successful experiences in community management of water resources in Latin America. In particular, the Nicaraguan Water Policy recognizes and promotes the collective action in the management of water resources of common use. However, there is not a single model of community management, for it accommodates to the needs of a given country, region and locality.

Community water management does not imply the government's abandonment of its responsibility in water supply. It is a different model that the authorities implement in order to guarantee the sustainability of the service in long term (Aguilar 2011). This means community water management is a joint venture between community organizations, non-profit organizations and the Nicaraguan state to provide water supply to rural communities.

Regarding the collective action theory, Ostrom (2011) claims that "humans are able to organize by themselves and collaborate in order to generate initiatives to survive for long periods". For the author, a self-organized group with clear rules has better opportunities to survive; therefore, the agreements, limits, sanctions and the conflict resolution procedures are the key factors for a successful self-government organization. In this research, Ostrom theory is followed by the analysis of the community water management in a rural community located in Nicaragua.

5 Organizations Involved in Water Management in the Rural Community of Ducualí

The case study is the community of Ducualí, which is located in the municipality of Palacagüina, Department of Madriz (Fig. 1). The community is situated in the dry corridor of Central America, which has a smaller precipitation than the rest of the region. In Palacagüina, the precipitation ranges from 650 to 850 mm/yr. The rainy season is from May to November and the dry season is from December to April. In the dry season, the water from rivers and streams decreases (Pérez and Zelaya 2013); as a consequence, the shallow aquifer also decreases and the levels of the wells are reduced. This situation generally causes water scarcity as 95% of the water supply in Nicaragua relies on groundwater (FAO and CEPAL 2009). Regarding the demography, the annual population growth is 0.11%, and the main economic activities are the cultivation of flowers and tobacco; construction and services.

In Nicaragua, a community water organization usually begins with a request from the community to the municipal authorities for a water supply system. For their part, the authorities seek the financial resources to support the project with the national government and international agencies (Alianza por el Agua 2015). The first water managing projects in the municipality of Palacagüina were implemented with international agencies' funds. Some of these agencies were founded in the 70s; for example, the Alliance for Progress, which was an American program to establish cooperation between the U.S. and Latin America; and in the 80s, the Swiss Agency for Development and Cooperation (SDC). Nowadays, the main international agencies that work in the municipality are the Spanish Cooperation Agency for International Development (AECID) and the Rotary Club.

Figure 2 shows the Venn diagram that represents the relationship between the community and the main external organizations for water supply in Ducualí. In this figure, the community is represented by the largest circle, while the community organizations are located inside and the external organizations outside the main circle. The size of the circles represents the influence that the organizations have for water supply in the community.

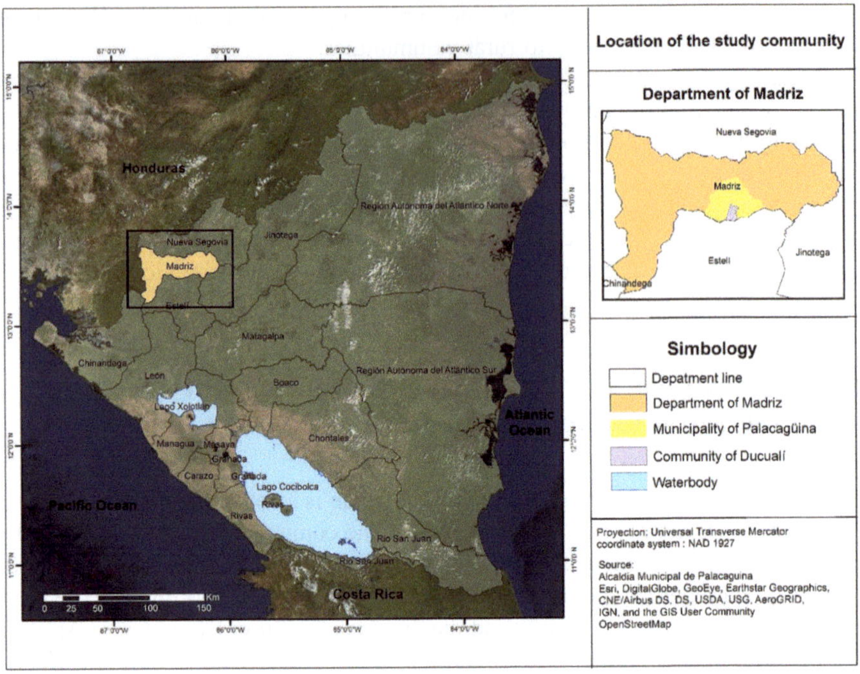

Fig. 1 Location of the study community. *Source* Own elaboration

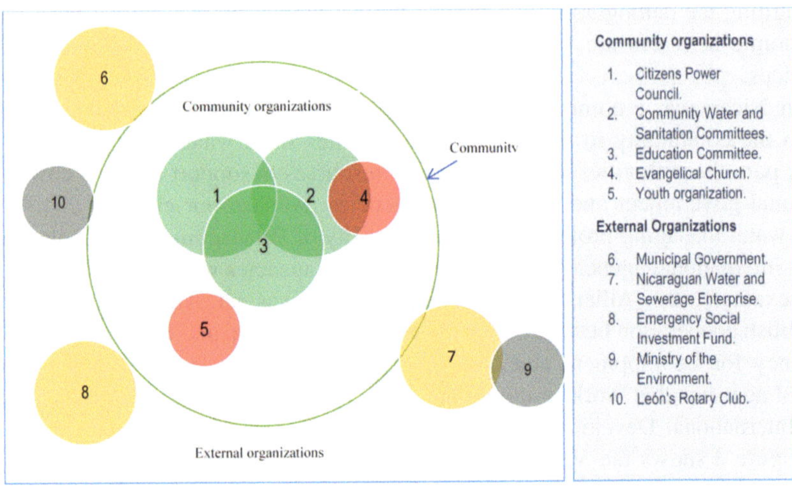

Fig. 2 Organizations involved in water management in the rural community of Ducualí. *Source* Own elaboration

From Fig. 2, it can be depicted that the most important community organizations are the Citizen Power Council (CPC), which are official organizations created in order to identify the beneficiaries of government social programs (Bay 2010), the CWSCs and the Education Committee. In fact, some citizens are often members of both the CPC and the CWSCs. Additionally, the Evangelical Church is the main religious institution and it has an important role in the promotion of cooperation activities in the community. For instance, the Evangelic church encourages solidarity between people; therefore, if a family does not have access to water supply, other members of their church donate their water to help.

The most important external organizations are the local government, ENACAL and the Emergency Social Investment Fund. Those organizations have provided equipment such as pumping and pipeline systems. Added to that, the Ministry of Environment trained the CWSC board on accounting principles and water rates through the Program for Hydrographic Watershed Management. Figure 2 evidences the relationships that different organizations built on the basis of the beliefs and traditions of the community, with the purpose of providing water supply to Ducualí.

6 Community Water and Sanitation Committees in Ducualí

In the 70s, the water supply in Ducualí was based on raw water from the Estelí River and the Pueblo Nuevo River, which are tributaries of the Coco River. However, some environmental problems emerged through the years, which motivated the organization of water community committees. Among the environmental problems, we found river pollution, droughts and hurricanes (such as Mitch in 1998). Therefore, in order to improve the access to water supply, the first water committees were organized to build new infrastructure.

Ducualí has four administrative sectors; three of them have water-supply networks which are administrated by two legalized CWSCs. The other sector has community wells and no pipeline system, forcing the population to collect water from hand-dug wells near rivers.

The distribution systems that operate in Ducualí are based on electric pumping systems installed in wells; first, the water is stored in tanks and, then, distributed by gravity through the mini-aqueducts to each house. In addition, the houses are provided with a water meter. The water from hand-dug wells is collected through hand pumps and this water is generally carried by women and children.

In Nicaragua, the CWSCs are formally organized (in most cases after distribution system are built) in order to guarantee the management and the sustainability of the infrastructure in the long term. The basic organization structure of a CWSC in Nicaragua is integrated by: General assembly, board and support commissions (Alianza por el agua 2015).

As mentioned before, there are two legalized CWSCs in Ducualí; nevertheless, most of the communities in the municipality of Palacagüina have only one committee. Each CWSC has a board formed by a president, a treasurer, and a monitor, and other commissioners who are democratically elected every two years in a General Assembly. It should be noted that CWCS are independent from each other.

The two CWSCs, namely Carlos Fonseca Amador and Union y Progreso, were formed and legalized after the enactment of the Special Law on Community Water and Sanitation Committees in 2010, and they are a good example of economical and organizational sustainability that guaranteed the water supply. In that aspect, the internal rules and regulations have favored the active participation of the community in order to obtain autonomy in the management of their resources in long term. Among the internal rules in those CWSCs we found:

1. Accountability of administrative processes.
2. Minimum water rate for a monthly consumption of 10 m^3. In case of larger consumptions, the water rate increases.
3. In order to have access to the water supply system, households should obtain the authorization of the board and should pay the connection.
4. Water supply from mini-aqueducts is only for domestic use.
5. Penalties to users with two-month late payments.
6. The board members of the CWSCs should be permanently settled in the community.

The internal rules in each CWSC are essential to the management process because they assign responsibility to community members. In that aspect, the collective action is achieved through the allocation of rights and obligations among the community members, and the collaborative work between users and board members. In addition, there is also joint cooperation between the community and the government. Furthermore, other external organizations have favored the empowerment and the active participation of the community citizens in order to obtain autonomy in the management of their resources.

The water committee of Ducualí Abajo is a community water organization formed in 1994 in order to manage the wells in the sector of the same name. Nowadays, the water committee is not in operation as the citizens have not established organizational structures. The community does not have rules, and the assembly is only called when a problem arises. Once the problem is solved, the community organization is inactive. Therefore, it is evident that citizen participation, organizational structure and committee legalization are necessary for an adequate water supply system.

Indeed, the legalized CWSCs have a better access to financial, technical and training resources. Such resources allow the strengthening of community capacities. According to some CWSC board members, the recognition of the community organizations improved the social relationship between users and the committee members and, in consequence, the access to water supply in the community.

7 Conclusions

In Nicaragua, about 50% of the rural communities have legalized Community Water and Sanitation Committees. That is, the Special Law on CWSCs in the country does not guarantee the success of water community management at the local level. Local water management depends on several factors and social actors with a key role in obtaining rights for the community and in enforcement of the citizen's participation. For this reason, the social interactions, values, beliefs and traditions in the community, as well as the leadership of the board members, are fundamental aspects in the collective action process. In addition, the active participation of water users, institutions and non-profit organizations in the water supply process allows the organization of the community to preserve its autonomy.

The study focuses on three organization experiences in the community of Ducualí, which is located northern Nicaragua. Three water committees in Ducualí were organized to find a solution to some environmental problems that made difficult the water supply in the community, such as river pollution, droughts and hurricanes. In this research, we showed that two of the CWSCs have taken advantage of the co-management model for the sustainable water management at local level in a rural community in Nicaragua. On the other hand, the third water committee is not organized, so it does not have rules and water supply problems are still common. Our study indicates that the organizational structure and committee legalization are necessary for an adequate water supply system. The rules, sanctions and water rates established by the communities are essential in building awareness among the local population on the rational management of the water resources.

All in all, the water community management empowered the population in decision-making; consequently, the community is more prone to defend its rights. This research was conducted in one rural community in Nicaragua. Future work should include several cases studies in Latin America in order to compare community water management policies in the region.

References

Aguilar E (2011) Gestión comunitaria de los servicios de agua y saneamiento. Su aplicación en México. Naciones Unidas. Retrieved from http://repositorio.cepal.org/bitstream/handle/11362/26079/S2011150_es.pdf?sequence=1&isAllowed=y

Alianza por el Agua (2015) La gestión de los comités de agua y saneamiento en Nicaragua. Retrieved from http://alianzaporelagua.org/documentos/Gestion_Comunitaria_Agua/Nicaragua.pdf

Antonio L, Tlaxcala B, Xocua M (2010) Participación y organización de actores sociales para el uso del agua y el manejo de los manantiales de Zacamilola, Atlahuilco, Veracruz. Edition digital. Retrieved from http://www.uv.mx/bbuvi/Parcela%2017.pdf

Bay K (2010) The return of the left in nicaragua: citizen power councils, pro-poor social services and regime consolidation. In: The American Political Science Association Annual Meeting, Washington, DC, 5 Sept 2010

Bernal A, Rivas L, Peña P (2014) Propuesta de un modelo de co-gestión para los Pequeños Abastos Comunitarios de Agua en Colombia. Perfiles latinoamericanos 22(43):159–184
CARE Internacional-Avina (2012) Programa Unificado de Fortalecimiento de Capacidades. Módulo 3: Organización comunitaria Cuenca, Ecuador. Retrived from http://www.avina.net/avina//wp-content/uploads/2013/03/MODULO-3-OK.pdf
Gaceta (2010) Ley especial de comités de agua potable y saneamiento. Retrieved from http://www.lagaceta.gob.ni/gaceta-111-lunes-14-de-junio-2010/
FAO & CEPAL (2009) Los recursos hídricos de Nicaragua. Retrieved from http://coin.fao.org/coin_static/cms/media/5/12820625348650/fao_nic_recursoshidricos_cepal.pdf
Kreimann R (2014) Los Comités de Agua Potable y Saneamiento y la gestión social de un bien común en Nicaragua. Los casos de los CAPS de El Edén y Chompipe. Agua y Territorio 2:34–47
Lorío A (2004) Procesos organizativos, regulación y tecnologías para el manejo y conservación del recurso hídrico y mitigación de la sequía, subcuenca del Río Aguas Calientes, Nicaragua. Dissertation, Centro Agronómico Tropical de Investigación y Enseñanza
Olson M (1992) La lógica de la acción colectiva. Bienes públicos y la teoría de grupos. Primera, edición edn. Limusa, México
OPS-OMS, ENACAL, RASNIC & Gobierno de Nicaragua (2004) Análisis sectorial de agua potable y saneamiento en Nicaragua. Retrieved from http://www.bvsde.paho.org/bvsacd/cd51/nicaragua/intro.pdf
Ostrom E (2011) El gobierno de los bienes comunes. La evolución de las instituciones de acción colectiva. Segunda edición. México D.F. Fondo de Cultura Económica
Pérez C, Zelaya C (2013) Informe final de la consultoría. PNUD. Retrieved from http://www.farem.unan.edu.ni/redcambioclimatico/docs/PNUDResumen%20Ejecutivo%20Escenarios%20de%20CC.pdf
Reyes B, Paiz C, Lira M, López N, Gómez S (2008) Creación y funcionamiento del Comité de Cuenca Bimunicipal Aguas Calientes. Sistematización de experiencias. Turrialba, Costa Rica, CATIE. Serie Técnica. Informe Técnico, (369), 36
Sandoval A, Günther M (2013) La gestión comunitaria del agua en México y Ecuador: otros acercamientos a la sustentabilidad. Ra Ximhai 9(2)
SIASAR (Information System on Rural Water and Sanitation) (2016) Línea de base por comunidad del Sistema de Información de Agua y Saneamiento. Retrieved from http://www.siasar.org/descargas/descargas_bd.php

Author Biographies

Juana María Zavala-Figueroa graduated in Environmental Sciences from the National Autonomous University of Nicaragua. Currently, she is a student at El Colegio de San Luis' Master's Degree in Sustainable Water Management program. She has worked in training programs for youths in communities located in northern Nicaragua.

Juan Alberto Velázquez-Zapata received a Civil Engineering degree and an Environmental Engineering Master's degree from the Autonomous University of San Luis Potosí, Mexico. He received his Ph.D. in Civil Engineering from Laval University at Québec City, Canada. Currently, he is a CONACYT (National Council of Science and Technology) researcher at El Colegio de San Luis.

Towards a Participative Environmental Management Education Model Using Information and Communication Technologies

Gloria Peza-Hernández, Ana Laura Barrera-González and Ana Ilse Benavides-Lahnstein

Abstract This text studies beliefs, attitudes and knowledge of environmental management and environmental education that were expressed by participants during a diploma course called Environmental Management and Community Participation. Overall, the purpose of the study was to evaluate the outcomes of the diploma course through the participants' inputs and enrich the understanding on this type of educational programmes. The investigation focuses on learning evidence from four participants and presents data from the initial and final questionnaires. In addition, this study reflects upon the implications of the participants' critical thinking and knowledge in relation to environmental management and environmental education, as well as the role of Information and Communication Technologies (ICT) as a support for their professional performance. On one hand, the findings revealed that the participants used their professional background and notions of community participation during the planning of environmental management projects, which was enriched by the activities of the course. On the other hand, the findings show that student perceptions on community participation lacked specificity and critical analysis. This suggested that, in a second edition of the diploma course, the learning experience could be improved by exploring further about the professional background and practices of the participants during the development of the programme. It is argued that the latter would allow identifying substantial needs in the planning of environmental management projects (an activity in the diploma course) and assisting in the articulation of sustainable development principles in community participation.

G. Peza-Hernández (✉) · A. L. Barrera-González
Institute of Investigation, Innovation, and Postgraduate Studies in Education (IIIEPE by Its Acronym in Spanish), Calle Nicolás Bravo, S/N. Centro, C.P. 64000 Monterrey, N.L., México
e-mail: gloria.peza@iiiepe.edu.mx

A. L. Barrera-González
e-mail: analaura.barrera@iiiepe.edu.mx

A. I. Benavides-Lahnstein
School of Education, University of Leeds, Hillary Place, Leeds LS2 9JT, West Yorkshire, UK
e-mail: edaibl@leeds.ac.uk

W. Leal Filho et al. (eds.), *Sustainable Development Research and Practice in Mexico and Selected Latin American Countries*, World Sustainability Series,
https://doi.org/10.1007/978-3-319-70560-6_17

Keywords Teacher training · Pre-service teachers · Sustainable development
Environmental management · Community participation

1 Introduction

The purpose of this study was learning about the environment-related perspectives of four participants in a diploma course named Environmental Management and Community Participation. These perspectives include the study of beliefs, knowledge, and attitudes about the participants' environmental awareness, environmental management, and community participation. Assisted by professionals in the field of education, the Environmental Management and Community Participation diploma course motivated the development of environmental management projects at a local level (Nuevo León, Mexico). The analysis of the participants' perspectives served as a reflection to evaluate the efficacy and limitations of the course and offer recommendations to similar educational endeavours.

This study is divided into four sections. The first chapter gives contextual information in relation to the diploma course. The second one provides the theoretical framework consulted for conducting the study. The third one specifies the methodology used in the study, which has a descriptive approach that elucidates how the learning activities of the course were used to reflect on their professional practice (Lucca and Berríos 2003). The fourth chapter presents the analysis and discussion of the beliefs, attitudes, and knowledge, as well as the areas of opportunity detected in the participants' work and forum discussions. The conclusions and recommendations give advice for subsequent developments of this and similar programmes. The aim is to use strategic planning and actions to create a process of collective actions that would foster socially responsible and sustainable participation.

2 Contextual Framework

In Nuevo León, recent public policies, which are socially constructed and conducted by state bodies (Tamayo 1997, p. 281), have promoted community participation in processes such as environmental training of teachers and the creation of Citizen Councils for, amongst others, pertinent environmental conservation and resource management. Accordingly, the Environmental Management and Citizen Participation diploma course aimed at supporting its participants (people from public and private sectors) to develop viable, situated, and sustainable project proposals by making use of their knowledge and experience in environmental management.

The main challenge faced—in this first edition of the diploma course—was to generate, from a higher education institute, a cross-sectional and interinstitutional program aimed at professionals with no apparent relation to teaching. Therefore, as a first achievement the diploma course encouraged the collaboration of experts, tutors, and students. A total of 22 women and 18 men participated in the course. Most of them had a university degree or certificate related to environmental protection and a strong interest in community development.

The diploma course was characterised by Information and Communication Technologies (ICT) support, a continuous follow-up of the learning activities, and academic coordination between the various tutors involved. The course comprises five blended or mix-learning modules—a technique consulted in Fidalgo (2007)—that were developed through a virtual education platform.

The virtual platform was a space, mainly, for forum discussions and homework submission. Classroom sessions were dedicated to presentations given by the tutors, sharing experiences, and targeting any doubts regarding their work in the course. The diploma course not only encouraged communication between different institutions and sectors to develop community projects for sustainability, but it also succeeded at producing reflections on ethical, methodological, and conceptual principles of environmental management.

3 Theoretical Framework

The framework of environmental management and community participation, the United Nations Educational, Scientific and Cultural Organization (UNESCO) suggested the need for responsible institutional management and supporting the professionalization and importance of educators (Navarro-Hoyos 2010). In this context, the formation of human resources as a process, seeks to re-dimension an environmental consciousness, so that the human being, harmonizes with its social and natural environment. In response to this, intervention policies suggested by the United Nations (UN) (2015a, b) and UNESCO (2015) support the need to promote joint actions between societies and governments to achieve a better quality of life.

To promote innovation and technology in favour of community participation, social inclusion and sustainable development (UNESCO 2012), the diploma course opted for the blended learning method (Fidalgo 2007). This education model consists of a combination of classroom sessions and independent work. ICT was incorporated in the structure of the diploma course through face-to-face and online pedagogical support, as well as resources, such as videos, online lectures, tests, and more.

According to Cacheiro (2011), ICT facilitates the creation of collaboration networks in students' individual learning and work. Such collaboration is achieved by a virtual exchange of learnings between students and tutors, which contributes to achieving the cognitive, procedural, and attitudinal processes that the established learning objectives aim for. Furthermore, the use of virtual spaces is inclusive and appropriate for working professionals because it is not constricted by the geographical location or time limits.

3.1 Views About Environmental Management

Environmental management has been considered a tool (Robinson 2012) and a demand of sustainable development (UNESCO 2015). Barrow (2006 pp. 163–164) describes environmental management as a process of decision making about the allocation of natural and artificial resources. According to him, environmental management is meant to satisfy, in the first place, basic human needs, and, in the second place, improve environmental quality (ibid.). Hence, environmental management in coordination with increasing environmental awareness can contribute towards reaching balance and integration of the social and natural worlds.

3.2 Dimensions of Environmental Awareness

The colloquial use of terms like *environmental awareness* makes them difficult to define (Amérigo 2006). According to academic work published on this topic (Ocaña et al. 2013; Febles 2004 in Alea 2006; Moyano and Jiménez 2005), environmental awareness includes mental representations, such as knowledge, attitudes, and beliefs. Within the environmental framework, these exist in a system of values and experience which creates a proactive relation between the individual and his or her environment. From this perspective, environmental awareness, as defined by Chuliá (1995, in Gomera-Martínez et al. 2012), has four dimensions:

- Cognitive (degree of information and knowledge).
- Affective (emotions, beliefs, and feelings towards environment).
- Conative (attitude and willingness to participate).
- Active (carrying out practical work and pro-active action).

In this study, beliefs are a part of subjective knowledge, generated at everyone's individual level. Moreno and Azcárate emphasize that an individual's beliefs explain and justify many personal and professional decisions and actions in his or her life (2003, p. 267). On the other hand, Ocaña et al. (2013) argue that beliefs are a decisive factor in forming personal judgements and opinions. Attitudes, however, are viewed as behavioural tendencies by these authors.

4 Research Design

The research design involved exploring a framework (theory above) which determined a series of questions (epistemology) that were examined systematically (methodology, analysis) (Denzin y Lincoln 1994, p. 11). The researchers employed the qualitative methodology and the structure of case studies (Denzin y Lincoln 1994; Creswell 2013). This study was conducted through case studies (Yin 2014) to research the phenomenon in question from different participants' perspectives. The data collection and analysis of this research gathered and defined relevant aspects of the participants' knowledge, attitudes, and beliefs on environmental management, citizen participation and education.

4.1 Applied Methods and Techniques

The data collection techniques used were documentation (mainly online forum discussions and the final project proposals), and initial and final questionnaires. These are common techniques recommended for case studies (Simons 2009). The investigators also used the non-participative observation method and took field notes (Lucca y Berríos 2003) which were used to amplify the context of outcomes and conclusions presented in the study. The purpose of the initial and final questionnaires was to gather information about participants' knowledge before and after the diploma course. The first questionnaire included 15 items divided between seven sections, containing conative, cognitive, and affective type items (more information in Sect. 3 of the theoretical framework). The second one comprised 25 items divided between seven sections to assess how useful the diploma course was in relation to each participant's needs. Questionnaires contained open- and close-ended alternative questions according to Likert scale, as well as multiple-choice questions. The forum discussions online ranged from environmental management initiatives to environmental education programmes.

4.2 Description of the Cases

Four participants were chosen as units of analysis for the case study—two men and two women, representing different areas of professional development. Two out of the four were referred to the program and the other two participated out of their own interest. The data collected from these units of study stand out for being more complete, and display more consistency in following the recommended rubrics for learning of the diploma course. Table 1 shows information about the participants and their initial motivation to join the course.

4.3 Analysis and Interpretation of the Findings

The data presented in this study refers to knowledge, attitudes and beliefs related to three dimensions of environmental awareness: cognitive, affective, and conative (Chuliá 1995 in Gomera-Martínez et al. 2012). Each dimension was analysed based on a scale of beliefs (Ocaña et al. 2013), as well as in the context of environmental management, citizen participation, and the use of ICT. The main source of evidence

Table 1 Summary of the units of analysis

Unit of analysis	Work field	Purpose of participation	Name of the project
Participant 1 (P1)	Academic	Interested in getting involved with environmental and community management	Implementation of environmental education workshops for the sustainability of communities at Cumbre National Park in Monterrey
Participant 2 (P2)	Employee at a municipal level office	Unsure of her reason to participate in the diploma course. Not sure of the relation between her work (at a municipal office that promotes tourism) and environmental protection	Promoting the region of Sierra de Santiago and the Álamo Canyon as destinations for enjoying natural areas sustainably
Participant 3 (P3)	Private sector employee	Interested in launching a campaign to make better use of resources at his workplace	Promoting ecotourism by encouraging people to care for resources without altering environmental balance through the activities of ecotourism
Participant 4 (P4)	Member of a civil organisation	I wish to turn the public square into a communal area for the families of the neighbourhood	Offering a different perspective of the local park: from a relaxation area to an oxygen-generating garden that supports family life

in this analysis is information from the participants' verbal utterances on their understandings and opinions, as salient reflection of designing their environmental management proposals.

4.4 Knowledge—A Key Element in Environmental Management

The cognitive dimension comprises the level or grade of environmental information. The participants' knowledge was identified through information gathered which talks about environmental management, environmental problems, and/or environmental education. The level or grade of environmental information is understood by the cognitive dimension.

One of the forum discussions was about involving communities and public service to improve protection, conservation, and exploitation of the country's natural resources. A universal concern expressed by participants, specialists, and tutors was how to reach environmental justice through effective, efficient, clear, and transparent application of and in compliance with federal environmental legislation in force. This would need to be done through attention to popular complaint, and through actions of inspection, verification, vigilance, and use of voluntary instruments. Furthermore, the evidence shows environmental awareness was an essential part of the participant's knowledge of environmental management and environmental education.

In the forum discussion, the participants P2 and P3 agreed on the role that federal bodies play in targeting environmental problems and the management service they provide. P2 suggested that the importance of environmental management carried out by federal bodies also depends on the community's level of involvement:

> All bodies play an important role, all trying to reach the same objective of protecting the environment in one way or another […] it is not only the Secretariat of Environment and Natural Resources (SEMARNAT by its acronym in Spanish) that is obliged to solve the problem [air pollution issue]. All the local people share this responsibility. We are still unaware of the dimensions of the problem. In Mexico City, which has an air pollution issue, the Federal Attorney for Environmental Protection (PROFEPA by its acronym in Spanish) is in charge of imposing anti-pollution regulations on companies to reduce the high pollution rates, but well, as we have seen, air pollution is what has been making the front pages in the past few months. If we all did our part, we could make a great team.

Almost up to the halfway point of the programme, the participants agreed on the need to promote connections between higher education and scientific research institutions, and different social sectors in Nuevo León. The following opinion was expressed by P4:

> [T]here is definitely communication between Mexican institutions, however, it is insufficient, considering that the programmes they implement are standardised for the entire country, whilst environmental problems are different in each part of Mexico. I argue that it is important to restructure the actions of environmental protection based on the needs of each region, build stronger links with universities to develop suitable programmes and actions for each region, and complement each programme by developing the public's abilities in relation to environmental education.

The latter coincides with the analysis done by P3 who reflected on his experience during the elaboration of the project he carried out as the end product of the diploma course:

> Throughout this project, we learnt that in order to analyse ecotourism projects, it is important to carry out a thorough analysis of their possible versions, in order to offer socially responsible services that provide a global vision focused on the care and preservation or natural resources native to the region, to attain sustainable development in the social, economic, and environmental context.
>
> It is important to keep promoting the culture of each environment, so as to ensure that these natural resources prevail, whilst tourism continues in all regions.

Environmental education appeared to be pertinent among the participants to communicate the value of conserving cultural heritage and biodiversity, which promotes interrelation between culture, tourism, and environmental protection. P1 added the following thought in the forum discussion about environmental education in basic education:

> There is no doubt that schools play an important role in sustainable development, yet we are far from launching a curriculum that includes Environmental Education for Sustainability.

> I argue that the Secretariat of Public Education (SEP by its acronym Spanish) should modify its plans of study and include environmental education – not as an elective, but give it the importance it deserves.

Participants were unaware of the role of environmental education in school programmes launched by the SEP. This is also rather common among many basic education teachers. Therefore, it was beneficial for the participants to have online forum discussions about the current situation of this topic.

Regarding the importance of context, most participants working in the public sector highlighted the importance of recognising the characteristics of each region in the context of environmental management. For example, P3 commented on the situation in Mexico City:

> [T]here is a specific objective in Mexico City right now, to reduce the emission levels of pollutants (though Mexico City is not the most polluted city in the country, it is the one that has received the most publicity, and one that has acknowledged the issue), and both the local and the federal government are (for now) focused on this.

This was also the case for participants working in the private sector. P1, a private sector academic, said in the forum discussion that it was important to adapt national programmes to the specific regional needs, which proves a necessary condition for environmental management. The perceptions of the four participants coincided in the idea that actions of intervention between different sectors must always be aimed at managing and solving environmental problems in the 'general-to-specific' method. When asked about this, they said "it is because we always need to do it in stages and only if the manager gives his permission".

There are similarities between all four cases regarding their knowledge of actions taken by governmental authorities in charge of environmental care, and the daily practices that they supported. In general, P2, like other employed in the public sector, showed a lack of accurate knowledge about the institutional responsibilities and function of their government in regards to environmental management, as so, they qualified the term sustainability as inapplicable in their praxis. Nevertheless, it appeared that participants adhered neither to the basic, nor the complementary reading that was supplied as part of the diploma course, due to the little use of the ICT that they made during the course. The other cases, working at citizen organizations and the private sector, recognised that general prescriptions for environmental management are inefficient, because environmental management and community actions depend on the specific conditions of each community.

5 Participants' Attitudes About Environmental Management

The conceptualized conative dimension implies the personal efficacy perceived in the environmental action and the behavioural disposition towards diverse environmental behaviours. The participants not only displayed their own attitudes, but often identified and discussed the importance of pro-environmental attitudes of communities. P1 favoured proactive attitudes towards environmental management in one of the forums:

> I think it is crucial to adapt educational programmes and include environmental education programmes in them […] also put it into practice through actions, such as recycling, reforestation, etc. […] the Autonomous University of Nuevo León (UANL by its acronym in Spanish) has a community development program, where students must also engage in theoretical environmental education courses such as reforestation projects.

Some beliefs about attitudes were oriented to specific areas of environmental management and support. For instance, in a section of the initial questionnaire where conservation of environment and actions are prompted, P2 stated "The conservation of wild flora will be one of our priorities". Other attitudes were about how and what methods are best to achieve environmental management, and the impact of our lifestyles in the environment. P2, reflected on the importance of pro-environmental attitudes and action in the civil society:

> The actions we take day by day necessarily have an impact on the environment. While human beings have some resistance to "change", in today's times we cannot but flow and understand that our way of life as we know it today is harmful, not only for the Earth, for us as well. To create a path to sustainability, we need "green" actions. Actions so simple that practically anyone can do, and besides, far from being an economic expense, these actions involve great saving of resources. Sharing the car: It is only necessary to agree with your neighbours, family and friends to go to work together, to the supermarket, to pick up the children to school.

The excerpt above shows the attitude from the citizens towards participation is believed to be important. In this way, regardless of their level of involvement, is pertinent to promote training on environmental management of resources in addition to participation and intervention techniques, so that citizens can participate in the creation and evaluation of plans, projects and programs. Although, citizen participation was often considered limited because environmental education implies an ongoing and life-long learning process. In this regard, P3 added:

> Education is a process, so a course that lasts a year or a few months is not enough to bring out the changes in us that we need or want. It does, however, lay the bases to create them, as long as one is convinced they want to make it happen, they want to create, and (most importantly) take care of something like a vegetable patch, or engage in hydroponics.

Regarding attitudes towards the *use of ICT*, the findings of the final questionnaire revealed that the four participants increased their comprehensive approach to the local environmental issues they focused on. The comprehensive character of their vision was demonstrated through their perceptions of the excessive use of energy, and consequently through the actions aimed at sustainability and energy efficiency that they mentioned:

- *I make use of daylight hours to do things around the house.*
- *I use energy-saving light bulbs (at home, at the office, at school), and switch off unnecessary lights.*
- *I switch off the electrical appliances that are not in use.*

In this field of exploring the local environmental situation from a holistic perspective to define the measure of communal responsibility, participants always demonstrated solidarity and good teamwork. They also recognised the mostly governmental limitations that apply to their proposals due to the specific context. It is interesting to note here that most group members built a relationship, and later started collaboration which lasts to this day.

6 Environmental Beliefs and Citizen Participation

The opinions expressed in the above sections show the presence of environmental awareness of environmental management at a national level, as well as a comprehensive perspective of local issues. The expression of beliefs that involve citizen participation among the four participants studied where varied; from beliefs about institutional exchanges (which distils in some of the quotations above) to local and individual scenarios in the targeted communities of the participants' projects.

It was argued during the diploma course that projects whose viability is only determined by technical terms sometimes end up failing because locals do not accept or come to terms with them (Peza and González 2011). If, however, the topics of a training programme are relevant to the participants' real needs and specific problems, such as in citizen participation, it is feasible to discuss possible solutions together with the interested parties, as supported by P4:

> Offering a different perspective of the local park: from a relaxation area to an oxygen-generating garden that inspires the community and invites people to exercise, contributing to the wellbeing and lifestyle of families, and raising awareness of how important it is to care for and conserve it as part of our environment and existence.

Other beliefs where identified whilst the participants made observations about the actions of others towards the environment. The participants showed generally less interest in exploring consequences of environmental problems, such as acid rain, and focused on well-known pollution issues—even though Nuevo León, in general, has a multiplicity of pressing environmental problems. For instance, results from the initial questionnaire exposed that participants' beliefs were limited to locating water pollution issues. In their view, people have a central role in protecting the environment through different actions like saving water (to benefit everyone). The actions mentioned were:

- Collecting and reusing water;
- Closing the tap whilst soaping the hands and brushing the teeth;
- Collecting rain water for domestic use.

In one of the forum discussions, P2 discussed pollution as a general issue in the targeted neighbourhood and identified citizen participation as a solution, which was also a common topic during the sessions of the diploma course:

> […] the cleanliness of our river is a great worry in our municipality, since it is one of our tourist attractions. We put our mind to it, get rubbish bags, even invite our guests, and all go out to clean up together. I think we should go to kindergartens and primary schools to educate them on how one should not just throw rubbish anywhere, teach them how to look after the environment, and save water […] and as such, we would make a small contribution to our environment.

There were some specific work-related situations that a few participants mentioned, in line with fieldnotes, and a possibility for the exchange of knowledge in community engagement of citizen participation, as expressed by P1:

> Through environmental education, we prepare ourselves to understand and appreciate the relations that are formed between people, their culture, and the environment that surrounds us.

P1 addressed the relation between the interdependence of sustainable development and environmental education:

> We cannot talk about sustainable development without the participation of all social actors – we all pollute, but few are those who contribute to caring for the environment […] Needless to say, we need more relations of interdependence to look after the environment. Private companies need to collaborate with state institutions and the civil society at taking actions that promote a responsible use of natural resources.

Finally, beliefs on *the use of ICT* that participants expressed in their final projects, in the context of the perception of environmental importance at the local level, showed a higher level of environmental awareness, i.e. a critical perception and acknowledgement of the need for improvement, using innovative ecotourism resources, as explained by P4:

> Promoting ecotourism by inviting walkers to actively participate in caring for resources without altering the environmental balance, using activities that offer a variety of foci and options to foster a better knowledge and use of the nature.

7 Conclusions

The four dimensions of the environmental awareness model used in this study were identified across the cases. On the one hand, during the development of the diploma course, the participants often displayed pro-environmental values and attitudes, showing aspects of their environmental awareness that are relevant to the affective and conative dimensions. On the other hand, despite the professional experience of the participants in environmental management, their knowledge on the topics studied, their decision-making and the actions taken in their projects are characterised by a lack of depth and critical analysis. Therefore, the cognitive and active dimensions of the participants' environmental awareness revealed areas of improvement, which here is believed are connected to their general performance in environmental management and community participation. During the diploma course, the researchers identified the lower emphasis in the cognitive and active dimensions involve:

- Lack of training in environmental education, environmental management, and community participation;
- Little use of the course's literature and ICT tools;
- Lack of support (financial and moral) in their workplaces and communities.

For example, the participants' comments involving environmental education and community participation included, mostly, claims about involving communities in every day pro-environmental activities, yet failing to suggest how to operationalize these claims in their context. The diploma course offered literature and contact with experts on these matters, but there was little engagement of the participants with it. Each participant faced significant challenges to involve their communities in projects of environmental management and environmental education; hence, they identified a need to delve into decision-making and the inclusion of local communities involved in environmental management programmes. However, those in leading positions of community-level environmental management admitted that their training in the social and/or environmental field was still insufficient. Overall, these issues, especially related with the cognitive dimension, represented obstacles to develop comprehensive and successful project proposals in the diploma course, since it was constantly necessary to address 'gaps' in the participants' knowledge and skills.

It was also identified that the use of ICT, which was little valued in the first five classroom sessions, supported cooperative work and contributed towards reflections on situated professional practice (Wenger 1998). The participants' attitudes favoured an exchange of knowledge between municipal and private sector employees, recognising the strengths of combining the support from public and academic sectors. The ICT tools also helped creating links between the participants' professional practices, as demonstrated by the improvement of two specific skills: Increased feeling of belonging and local identity; and, recognition of natural, cultural, economic, social, and political conditions that characterise the place where they live. Therefore, the ICT tools in the diploma course were a positive and helpful aspect of the programme, yet insufficient in the face of other challenges from the cognitive and active dimensions of environmental awareness.

8 Limitations

It is acknowledged here that these conclusions are limited to the small sample of participants, the context and methods involved in this study; however, small scale studies are useful to corroborate previous research results and contribute to understanding and improving future research studies.

9 Recommendations

In subsequent editions of the diploma course, learning tasks will be strengthened by actions, techniques and management processes, as well as areas for further reflection and analysis of the actions planned for the projects. Regarding the development of the diploma course, a future recommendation is to cut down on the hours of the programme. A 120-h course proved difficult for the participants, albeit blended and supported by ICT, wherefore it is preferable to offer three short consecutive courses. In addition, the experience of using digital tools in educational platforms should also be supported, and appropriate support offered if needs be. Furthermore, the authors consider that in this type of education, future programmes would benefit from considering the relevance of the following topics:

- Analysis of individual and collective participation;
- Reflection on social, cultural, and economic surroundings;
- Appreciation of natural diversity;
- Appreciation of social and cultural diversity;
- Situated learning.

It is fundamental for the institutions that offer environmental education programmes to have an interdisciplinary team with experience in teaching, environmental management, investigation and innovation, environmental communication, assessment, and monitoring. This team should make consistent use of medium and long term strategies. Some objectives might be the following:

- An alternative to adapting to climate change in the region studied;
- Guidance to lifestyle changes, aiming for responsible consumption;
- Creation of a state-level information system for environmental education;
- Providing training, refresher courses, and certification to environmental teachers.

We recommend making new programmes of environmental education simpler when it comes to communication between institutions and social sectors (fostering cooperation between educational and environmental institutions at federal, local, and municipal levels). They should also facilitate the improvement of participants' critical purpose, theoretical and practical footing, demonstrate cooperation between different sectors and organisations, and aim to use relevant available resources that are adequate to participants' areas of specialisation. Future studies could address ways of project development strategies in the fields of planning, monitoring, and assessment. At the same time, it is imperative that environmental education projects, such as the diploma course discussed in this study, help people understand the global interactions on our planet, and encourage people to commit to direct environmental actions. Since becoming aware of issues that afflict the world, such as poverty, inequality and inequity, is the key to solving problems that compromise the sustainability of the world.

References

Alea A (2006) Diagnóstico y potenciación de la educación ambiental en jóvenes universitarios. Odiseo, Revista electrónica de Pedagogía 3(6). Retrieved from: http://www.odiseo.com.mx/2006/01/print/alea-diagnostico.pdf. Accessed 4 Apr 2016

Amérigo M (2006) La investigación en España sobre actitudes proambientales y comportamiento ecológico. Medio Ambiente y comportamiento Humano. 7(2):45–71

Barrow C (2006) Environmental management for sustainable development, 2nd edn. Routledge, New York

Cacheiro ML (2011) Recursos educativos tic de información, colaboración y aprendizaje. UNED. Madrid: Facultad de Educación. Departamento de Didáctica, Organización Escolar y Didácticas Especiales. Retrieved from: http://acdc.sav.us.es/pixelbit/images/stories/p39/06.pdf. Accessed 5 Feb 2016

Creswell J (2013) Qualitative inquiry and research design. SAGE Publications, London

Denzin N, Lincoln Y (1994) The sage handbook of qualitative research, 3rd edn. SAGE Publications, Thousand Oaks

Fidalgo A (2007) ¿Qué es b-learning? Innovación Educativa. Retrieved from https://innovacioneducativa.wordpress.com/2007/03/10/%C2%BFque-es-b-learning/. Accessed 10 Mar 2007

Gomera-Martínez A, Villamandos-de la Torre F, Vaquero-Abellán M (2012) Medición y categorización de la conciencia ambiental del alumnado universitario: contribución de la universidad a su fortalecimiento. Profesorado. Revista de Currículum y Formación de Profesorado 16(2):214–228

Lucca N, Berríos R (2003) Investigación Cualitativa en Educación y Ciencias Sociales. Publicaciones Puertorriqueñas, Puerto Rico

Moreno-Moreno M, Azcárate-Giménez C (2003) Concepciones y creencias de los profesores universitarios de matemáticas acerca de la enseñanza de las educaciones diferenciales. Enseñanza de las Ciencias. Investigación Didáctica 21(2):265–280

Moyano E, Jiménez M (2005) Los Andaluces y el Medio Ambiente. Ecobarómetro de Andalucía. Resource Document. Sevilla: Junta de Andalucía. Consejería de Medio Ambiente. Retrieved from: http://www.juntadeandalucia.es/medioambiente/consolidado/publicacionesdigitales. Accessed 20 Dec 2016

Navarro-Hoyos J (2010) Declaración de Madrid de la VI Cumbre Unión Europea-América Latina y el Caribe 2010: ¿Una cumbre de resultados? Revista VIA IURIS 9:123–138. Bogotá: Fundación Universitaria Los Libertadores.

Ocaña M, Pérez M, Quijano R (2013) Elaboración y validación de una escala de creencias de los alumnos de educación secundaria obligatoria respecto al medio ambiente. Profesorado. Revista de Currículum y Formación de Profesorado 17(1):431–454

Peza G, González RA (2011) La Gestión Ambiental y la Participación Comunitaria en las Instituciones Gubernamentales. SEMARNAT-SENL, Monterrey

Robinson NA (2012) Beyond sustainability: environmental management for the Anthropocene Epoch. J Public Aff 12(3):181–194

Simons H (2009) Case study research in practice. SAGE Publications, London

Tamayo (1997) El Análisis de las Políticas Públicas. In: Bañón R, Carrillo E (eds) La Nueva Administración Pública, pp. 281–31. Alianza Editorial, Madrid

United Nations (UN) (2015a) Objetivos de Desarrollo del Milenio Informe de 2015. ONU, Nueva York

United Nations (UN) (2015b) Transformar nuestro mundo: La agenda 2030 para el desarrollo sostenible. Resolución aprobada por la Asamblea General. 70. Retrieved from: http://www.un.org/es/comun/docs/?symbol=A/RES/70/1. Accessed 6 Apr 2016

United Nations Educational, Scientific, and Cultural Organization (UNESCO) (2012) Educación para el desarrollo sostenible. Instrumentos de aprendizaje y formación. UNESCO, París. Retrieved from http://unesdoc.unesco.org/images/0021/002167/216756s.pdf. Accessed 4 Apr 2016

United Nations Educational, Scientific, and Cultural Organization (UNESCO) (2015) Repensar la Educación. Hacia un bien común. UNESCO, 2015. UNESCO, París. Retrieved from: http://unesdoc.unesco.org/images/0023/002326/232697s.pdf. Accessed 6 June 2016
Wenger E (1998) Communities of practice: learning, meaning and identity. Cambridge University Press, Cambridge
Yin R (2014) Case study research. Design and methods. SAGE, London

Author Biographies

Gloria Peza-Hernández Ph.D. in Education. Master in Social Sciences and Master in Environmental Education. Primary Education degree. Has published various articles, chapters of books, and books about environmental education, environmental management, teacher training, and others. Gloria has been a research supervisor, collaborated with the SEP in developing curricula and syllabi. Currently, she is the General Coordinator of the Investigation and Innovation programme at the IIIEPE in Monterrey, Mexico.

Ana Laura Barrera-González Ph.D. in Educational Investigation and Innovation. Master in Educational Administration. Primary Education degree. Has been the Director of Continuing Education and Diffusion at the Mexican Secretariat of the Interior, a Coordinator of Academic Affairs at the Vice-Rectory for Social development at Monterrey Institute of Technology and Higher Education. Currently, she is the Coordinator of Strategic Academic Programmes at IIIEPE.

Ana Ilse Benavides-Lahnstein Ph.D. candidate in Education. Master in Education. Industrial Engineering degree. Her work experience includes collaboration with the science education group in CINVESTAV-Monterrey and teaching in higher education.

Green Areas and Environmental Justice: Toward the Urban Sustainability of León, Guanajuato

Jairo Agustín Reyes Plata, María del Carmen Villanueva Vilchis and Arlene Iskra García Vázquez

Abstract Green areas help improve the living conditions for all inhabitants in cities. This study contributes to clarify the relationship between environmental justice and urban sustainability based on the analysis of the rate and distribution of urban green areas particularly in the case of the City of León, Guanajuato, Mexico. A systematic literature review related to the topic of study was conducted. In addition, spatial analysis of the rate and distribution patterns of urban green areas in the city was carried out and some of the basic principles that guide the planning of green areas in the city were discussed. The results showed that there is a rate of green areas in the City of León of 1.74 m^2, which follows a dispersed pattern of urban growth, characterized by a socio-spatial fragmentation of the city. The rate of green areas in the priority polygons is about 1.86 m^2, with unequal concentrations on each zone and an influence range that restricts an equitable access to the benefits of these areas. It has been identified that urban sustainability implies the habitability of the public space, which involves the guarantee of a more equitable, equal, and democratic use of the natural richness or of the one socially generated, where the

J. A. R. Plata (✉)
Licenciatura en Desarrollo Territorial, Escuela Nacional de Estudios Superiores, Universidad Nacional Autónoma de México, Blvd. UNAM 2011, Predio el Saucillo y el Potrero, Comunidad de los Tepetates, León, C.P. 37684 Guanajuato, Mexico
e-mail: jreyes@enes.unam.mx

M. del Carmen Villanueva Vilchis
Departamento de Salud Pública, Escuela Nacional de Estudios Superiores, Unidad León, Universidad Nacional Autónoma de México, Blvd. UNAM 2011, Predio el Saucillo y el Potrero, Comunidad de los Tepetates, León, C.P. 37684 Guanajuato, Mexico
e-mail: vv.carmen@gmail.com

A. I. G. Vázquez
Licenciatura en Desarrollo Territorial, Escuela Nacional de Estudios Superiores, Unidad León, Universidad Nacional Autónoma de México, Blvd. UNAM 2011, Predio el Saucillo y el Potrero, Comunidad de los Tepetates, León, C.P. 37684 Guanajuato, Mexico
e-mail: iskragv@gmail.com

W. Leal Filho et al. (eds.), *Sustainable Development Research and Practice in Mexico and Selected Latin American Countries*, World Sustainability Series,
https://doi.org/10.1007/978-3-319-70560-6_18

green areas constitute a strategic element to reach the environmental, economic and social benefits of the sustainability.

Keywords Green areas · Environmental justice · Urban sustainability

1 Introduction

Cities certainly offer opportunities for employment, education and other basic services for human development; however, pollution, energy expenditure and consumption of natural resources, exclusion, as well as inequality and socio-spatial segregation that characterizes the cities, generate deficient living conditions and environmental, economic and social vulnerability, affecting large segments of the population.

The new approaches for sustainable development seek to understand and manage those factors that influence urban integration and well-being of the population. Strategies and policies are focused on the importance of meeting the needs of the inhabitants, their rights and aspirations, beyond the exclusive administration of the physical, anthropic or natural components of the city. Thus, the presence of sufficient and high quality green areas is one of the main variables that affect the well-being of urban dwellers, as well as their distribution and equitable access.

Different researches have documented the environmental, social and cultural benefits that green spaces provide for the improvement of the living conditions of the population, (Reyes and Figueroa 2010; Vélez 2009; García and Guerrero 2006; Chiesura 2004; Krishnamurthy and Rente 1997) particularly for sectors which are in poverty or social backwardness (Wolch et al. 2014).

Given the benefits provided by the green areas, these spaces gain a strategic relevance to generate sustainable urban environment, especially when considering that according to the United Nations (2016) in 2014, 54.0% of the world population lived in cities, and this number will continue to increase in the coming years. In Mexico, the percentage is even higher, 8 out of every 10 inhabitants were considered urban in 2010 (INEGI 2016).

In 2015 the General Assembly of the United Nations proposed 17 objectives to reach sustainable development, establishing the year 2030 as the deadline to achieve it. The eleventh objective in the Agenda, agrees that the cities and human settings should be inclusive, particularly for women, children, persons with disabilities and older persons (United Nations 2016).

The Quito Declaration (United Nations 2016) on the New Urban Agenda calls to promote the feeling of belonging to the cities, amongst other factors, through the well-connected and well-distributed public spaces within the city. These spaces need to be secure, inclusive, accessible, green and with high quality; contributing to the improvement of the social interaction and intergenerational coexistence, as well as to the free demonstration of the cultural expressions. The intention is to encourage the creation of inclusive, safe, peaceful and plural cities, where the specific needs of the social vulnerable sectors are recognized. The equitable distribution and access to green areas is a matter of environmental justice and therefore important for achieving urban sustainability.

Given this panorama, the study makes a spatial analysis of the distribution of green areas in León, Guanajuato, aiming to clarify the relationship between environmental justice and urban sustainability. The research is based on the premise that sustainability seeks to make the city habitable by generating a suitable environment to fulfill the needs and aspirations of its inhabitants with the intention of improving their quality of life and that the presence of green areas is a decisive element of urban sustainability. For these reasons, this research also discusses some of the basic principles that guide the planning of green areas in the city of León.

2 Equal Distribution of Green Areas: A Matter of Environmental Justice

In the urban field, sustainability is often reduced to the search for balance between the natural and built environments. If the city meets certain environmental criteria, such as water savings, energy savings, waste recycling and efficient transport, it is said to be sustainable. However, the concept goes beyond the conservation of nature and its availability to meet the current and future material needs of the population. Urban sustainability implies to make the city livable, regardless of its size (Lezama and Dominguez 2006), meaning the quality of a place to meet the people's needs and aspirations (Castro 1999 cited by Moreno 2008). Thus, the adaptation of the surrounding environment has a purpose: to improve the quality of life. Pérez Maldonado (1999 cited by Moreno 2008) argues that the quality of urban life involves the feeling of biological and psychosocial comfort of those who inhabit and live in the city, by determining the degree of satisfaction in the use of services and the perception of the urban space as healthy, safe and visually pleasing. The presence of green areas is therefore a conditioning factor of urban livability, resulting in the improvement of the inhabitants quality of life (Alcalá 2007).

An important principle of urban sustainability is undoubtedly social equity, which has its foundations in justice, regarding the distribution of resources and equal access to public services essential for development and enjoyment of life. In urban contexts, equity is related to inclusion in two dimensions: social and environmental. In a society, there will be social equity if there are no discriminatory or exclusionary practices that prevent inhabitants from participating as citizens in the economic, political and social spheres (Pierson 2000 quoted in Dempsey et al. 2011), and there is enjoyment of services and benefits of the ecosystem and a healthy urban environment, which also promotes social and environmental justice.

Urban areas in a poor living environment and with limited or unequal access to public services are areas of social and environmental exclusion and injustice. Key public services to which all urban dwellers must have equitable access, include education, transportation, housing, infrastructure, spaces for culture, exercise and recreation, as well as green areas (Dempsey et al. 2011). However, it is clear that not all inhabitants have equitable access to these services, due to the social and structural conditions generated by the phenomenon of globalization and, on the other hand, to the accelerated process of urbanization that has taken place in many cities. Social and territorial fragmentation has resulted in poorly articulated urban systems, where a large part of the population lives in conditions of marginalization, social exclusion and with no environmental justice.

Environmental justice involves the recognition of the fundamental rights to a healthy environment, as well as the defense of human rights and prerogatives of a social, economic and human development nature, that ensure fair distribution and quality provision among all individuals and population groups to the environmental, social and cultural benefits that an ecosystem provides (Ramirez et al. 2015; Bonne and Fragkias 2013; Steward et al. 2013). A fair distribution means, that the most vulnerable population groups receive the greatest benefits, a goal that requires an equitable system in which public administration establishes compensatory, corrective and guaranteeing strategies and policies for access to environmental services, depending on the nature and context (Cancino 2014; Svara and Brunet 2005).

The provision of quality services, involves the ensuring of the long-term maintenance of ecosystem functions, as well as the transfer and acquisition in optimal conditions, under minimum standards, of the services and benefits for the whole population. In this sense, some sectors may have their own means of achieving a pleasant, dignified and valuable life, but others may not, unless the public administration guarantees it (Svara and Brunet 2005).

In relation to green areas, it has been established that environmental justice is given in the beginning, by the equal assignation of green area per inhabitant or by funds or sources of recreation per capita by neighborhood or socioeconomic status (Bonne et al. 2009). Other authors consider that, in addition, an equitable distribution in the number of green areas throughout the city is necessary to avoid socio-spatial segregation; however, proximity does not ensure itself equitable access or translate into positive effects if people cannot freely experience the place due to social factors in the environment such as crime or heavy vehicular traffic (Jennings et al. 2016). The sociodemographic characteristics of the population, as well as their needs, merits and choices, could also determine the opportunities for equitable access to these spaces (Hay 1995).

To ensure universal access to safe, inclusive and accessible green spaces it is necessary to have indicators that, through quantitative and qualitative variables, evaluate their nature, quality and functionality in the urban fabric (Keng et al. 2015; Haq 2011; Reyes and Figueroa 2010; Vélez 2009; Garcia and Guerrero 2006; Chiesura 2004). For example, for the green area per capita, the World Health Organization (WHO) established a minimum area of 9 m^2 per inhabitant (Saz and Rausell 2008).

The sources available agree that there is a deficit of green areas for the City of León. The ratio of green areas per capita is 1.7 m^2, according to the "Diagnóstico Ambiental de León (In English: "Environmental Diagnosis of León") (IMPLAN 2015). However, the use of this indicator, based on the total concentration of the surface in proportion to the population number, only covers one aspect of the phenomenon, whose reality could vary throughout the city, and, thus, determine the equal access different social groups have to the benefits provided by green areas.

3 Methodology

1. Study area and unit of analysis

The study area is located in León, one of the municipalities of the State of Guanajuato. It has a population of 1,578,626 inhabitants, equivalent to 26.0% of the state population (INEGI 2015); It is part of a metropolitan agglomeration of 1.7 million inhabitants, located to the northwest of the state, within the region known as the Bajío Mexicano, which is characterized by a growing economic dynamism in the automotive sector, which has favored the consolidation of an industrial belt that seems to develop with some autonomy.

Geographically speaking, the municipality is characterized by the presence of steep slopes and hills to the north, large expanses of plains and some hills towards the center, a portion that comprises the urban area and of greater growth of the city and by extensive plains towards the south, where different agricultural activities are performed. The center portion covers about 22 thousand hectares, that is, 17.0% of the total area of the municipality; it concentrates most of the population and presents average densities of over 50 inhabitants per square kilometer. Official data shows that there are 685 green areas located in the urban area of León, between parks and public gardens, covering an area of 228.98 ha. (IMPLAN 2015). In the study area, there are 559 green areas totaling an area of 215.98 ha.

The unit of analysis for this study is the Basic Geostatistics Area (AGEB), which gathers information on the characteristics of the population and housing according to the Census of Population and Housing survey conducted by the Instituto de National Institute of Statistics and Geography (INEGI) in 2010. In the study area, there are 398 analysis units, with a total population of 1,237,418 inhabitants.

2. Analysis of the distribution of urban green areas

The analysis of the distribution of urban green areas in the city of León was made from the classification of the study units into 9 classes, as shown in Fig. 1. The classification obeys three conditions: the presence of green areas, the coefficient of green areas and the degree of urban marginalization. The coefficient expresses the area (square meters) of green area per inhabitant, in two ranges: <9 square meters and ≥ 9 square meters. The computing of the coefficient was obtained from the population data of the Population and Housing Census 2010

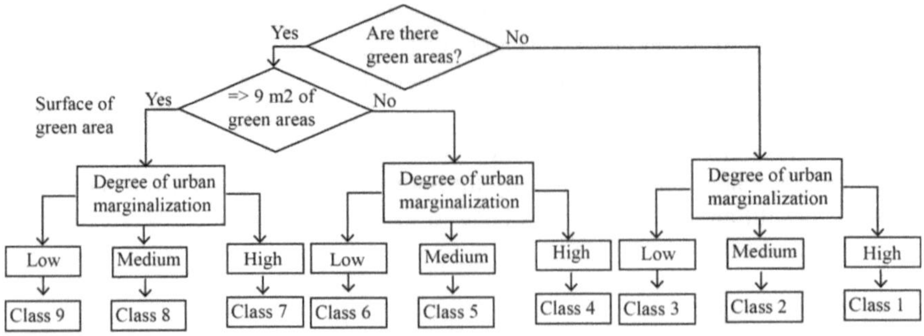

Fig. 1 Classification of the study units by classes

(INEGI 2010) and the total area of green area by AGEB, according to information from the Environmental Map of the Municipality of León (DGGA 2017). The degree of urban marginalization of the units of analysis was obtained from the classification made by the National Population Council (CONAPO 2010), which is a measure that expresses the shortcomings of the population in terms of income, health, education and housing. The degree of marginalization falls into three categories: high, medium and low. Finally, a non-parametric test of chi square was calculated at 95.0% confidence to evaluate the relationship between the degree of urban marginalization and the coefficient of green areas per inhabitant.

3. Analysis of the regulatory framework related to green areas.

The analysis of municipal plans, programs and legislation included the revision of the normative instruments in urban and environmental matters, in force in December 2016, in order to know the conceptual and operational bases that assist the planning of the green areas of León dealing with the endowment, maintenance or conservation of the green spaces located within the urban area of the city.

4 Results

1. Planning and regulation of green areas

Planning and administration of green areas involve the departments of urban, environmental and planning administration in the City of León. It is related to problems in matter of environmental, territorial, landscape and institutional management. Public administration possesses four policy instruments to regulate the management of green areas: "Programa Municipal de Desarrollo Urbano y de Ordenamiento Ecológico y Territorial" (H. Ayuntamiento de León 2015), "Reglamento para la Gestión Ambiental" (H. Ayuntamiento de León 2014), "Programa Municipal de Espacios Verdes Urbanos" (H. Ayuntamiento de León

2016) and "Código Reglamentario de Desarrollo Urbano" (H. Ayuntamiento de León 2010).

The "Programa Municipal de Espacios Verdes Urbanos" (In English: "Municipal Program of Urban Green Spaces) states that green areas are components of the public space, which contains living elements that constitute the physical basis of their environmental value and of the relationship that the community maintains with the natural environment in the urban context. Operationally, the municipal green areas are spaces with plenty of plants in stratification, with layers that go from trees to low vegetation and with urban illumination equipment. Green areas are located within urban land and possess the status of common property in behalf of all citizens because of the environmental, recreational, cultural and educational services they provide and the improvement of the urban image. Its typology includes: urban park (when the surface is equal or greater to ten thousand square meters), public garden (if its surface is less than ten thousand square meters) and landscaped areas.

Strategies of urban policy are based on an environmental approach which addresses two aspects: the increase of green areas surface and the maintenance of the existing ones. Regarding the first aspect, the definition of the term "green area" contains a key concept: "donation area", which reveals the mechanism that is used to promote the global increase of the surface of municipal green areas. According to the "Código Reglamentario de Desarrollo Urbano" (In English: "Regulatory Code on Urban Development"), a "donation area" is defined as: "*the area of land that the developer transmits to the Municipality, destined to the provision of urban equipment and green areas of the subdivisions and condominium developments, in the proportion established in the Territorial Code, with respect to the total area of the authorized project*;" (H. Ayuntamiento de León 2010, p. 8). In this context, the creation of new municipal green areas, urban parks or public gardens, is subject to a process of transmission of goods, from a private to a public one, linking the increase of the green area of common use and public utility with the dynamics of Housing Development market, which is marked by the times and practices between the public administration and the real estate sector. For that situation, there is an unequal access and distribution of the benefits that green areas provide throughout the city and for different population groups.

2. Distribution of green areas along city

A crossing data analysis on the degree of urban marginalization and the coefficient of green areas for each unit of study was held and the results of the spatial analysis show an unequal, inequitable and deficit distribution of the green areas in the study area.

The average coefficient of green areas per inhabitant in the study area was calculated in 1.74 m^2 (215.98 hectares of green area/ 1,237,418 inhabitants). However, for 45.94% of the population there is an absence of green areas in their living surroundings and for only 3.26% of the population, the surface of green areas is higher than 9 m^2 per inhabitant, which is the recommendation issued by World

Health Organization (Saz and Rausell 2008). There is a significant association between the degree of marginalization and the coefficient of green areas ($X2 = 95.27$ $p < 0.001$).

Class 1 and class 9 represent the extremes in terms of the best and worst conditions in the distribution of green areas along the city. 12.52% of the population belongs to class 1, characterized by a high degree of urban marginalization and absence of green areas. It is located in the periphery of the city, in areas of recent urban expansion, as shown in Fig. 2. By contrast, only 2.55% of the population is within class 9, showing a low degree of marginalization and a provision of green areas higher than 9 m^2. This population is grouped in three blocks to the northwest, northeast and south of the city.

The majority of the population (46.3%) belongs to classes 2 and 5; showing medium levels of marginalization and a low or even an absent coefficient of green areas. These classes are located at the center and western parts of the city. Population with high levels of marginalization, classes 1, 4 and 7, constitute 17.41% of the total population. They are characterized by the lowest rates in the provision of green areas, which denote inequity in the distribution of green areas for the population that most need them. However, the absence of green areas affects all population groups as shown in Table 1. 45–94% of the population does not have

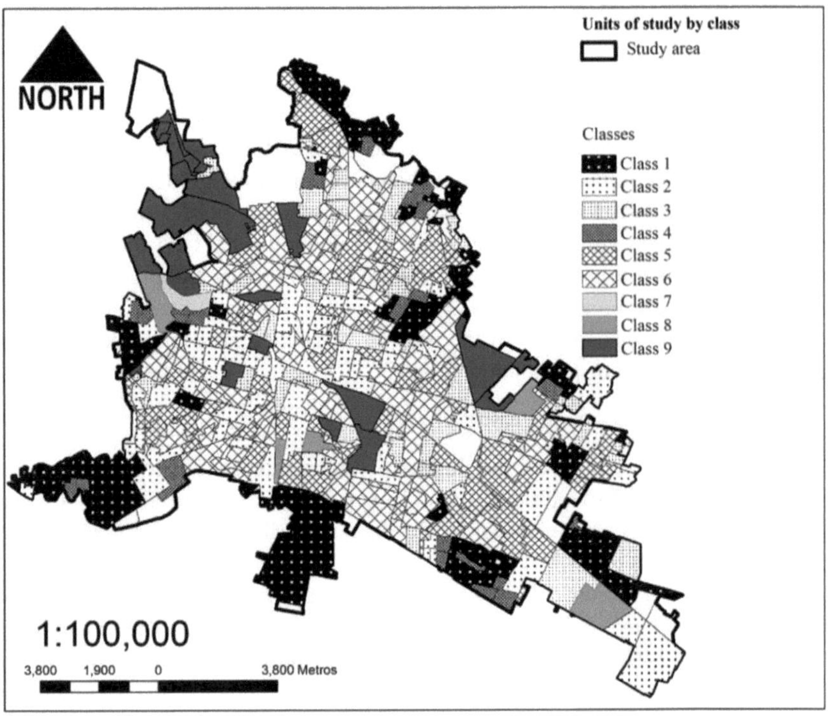

Fig. 2 Distribution of units of study by classes

Table 1 Population classes by the degree of urban marginalization and coefficient of green areas

		Degree of urban marginalization % of population			Total of green areas surface (%)
		High	Medium	Low	
Coefficient of green areas by inhabitant (square meters)	0.00	Class 1 = 12.52 (154,972 inhabitants)	Class 2 = 24.93 (308,546 inhabitants)	Class 3 = 8.48 (104,920 inhabitants)	45.94 (568,438 inhabitants)
	0.01 a < 9.00	Class 4 = 4.75 (58,764 inhabitants)	Class 5 = 21.44 (265,244 inhabitants)	Class 6 = 24.62 (304,665 inhabitants)	50.81 (628,673 inhabitants)
	≥ 9.00	Class 7 = 0.04 (437 inhabitants)	Class 8 = 0.67 (8,349 inhabitants)	Class 9 = 2.55 (31,521 inhabitants)	3.26 (40,307 inhabitants)
Total of population in urban marginalization		17.31 (214,173 inhabitants)	47.04 (582,139 inhabitants)	35.65 (441,106 inhabitants)	100.00 (1,237,418 inhabitants)

Information sources Elaborated with data from the Census of Population and Housing (INEGI 2010), the Environmental Map of the city of León (DGGA 2017), and the classification on degrees of urban marginalization by the National Population Council (CONAPO 2010)

immediate access to green areas. 24.93% and 8.48% in that condition show degrees of marginalization classified in medium and high, respectively. It denotes an uneven distribution in the green areas of the city for all population groups.

Regarding the urban zones with the highest poverty conditions or so-called "priority polygons for human development", the coefficient of green areas per inhabitant was estimated in 1.86 m^2. It is comparatively higher than the city average rate, but still negative. As seen at Table 2, the distribution of green areas among the polygons is unequal. For instance, five zones show low rates below the municipal average and two of them rate at half of the municipal average.

Regarding the maintenance condition of existing green areas, the "Programa Municipal de Espacios Verdes Urbanos (In English: "Municipal Program of Urban Green Spaces") points out that projects, measures and actions need to be developed to ensure maintenance, conservation and improvement of green areas. That requirement is evident, because of that 43.37 ha of green areas shown at Table 3 (20% of the total area of urban green spaces) are neglected, especially in poverty zones as is the case in priority polygons for human development in León, which contrasts with the good maintenance condition of green areas located in the middle and high socioeconomic strata (IMPLAN 2012), reflecting inequalities and exclusion in access to quality public services for the whole population and the need for environmental justice.

Table 2 Green areas within the "priority polygons for human development"

Name of the priority polygons for human development	Population (Inhabitants)	Green areas surface (Hectares)	Quantity of green areas	Surface of green areas per capita. (m^2/ inhabitant)	Negative balance of green areas (Hectares)
Diez de mayo	127,952	23.62	45	1.8	91.53
Jacinto López	47,644	39.77	15	8.3	31
Las Joyas	70,400	8.27	35	1.1	55.08
Los Castillos	73,850	4.21	18	0.5	62.24
Medina	158,113	24.46	38	1.5	117.84
Piletas	43,357	4.85	9	1.1	34.16
San Francisco	74,275	5.88	61	0.8	60.95
	595,591	111.08	183	1.8	42.49

Information source Elaborated with data from Diagnóstico del Municipio de León (IMPLAN 2015)

Table 3 Condition of urban green areas by class

Category	Number of Green urban spaces	Surface (Hectare)	Number of Green urban abandoned spaces	Surface (Hectare)
Urban parks				
Metropolitan parks	6	45.22	0	0
Urban parks	41	67.46	16	26.20
Public gardens				
Neighborhood parks	53	35.48	7	5.13
Neighborhood parks	545	73.48	79	12.04
	645	221.63	102	43.37

Information source "Programa Municipal de Espacios Verdes Urbanos" (In English: "Municipal Program of Urban Green Spaces") (2016)

Planning regulations assess the distribution of green areas along the city through four indicators: surface for the use of green areas, improving surface of green areas, number of linear parks and percentage of urban green spaces in use per year.

Indicators are intended to describe a specific process, condensing information of several attributes, characteristics or important properties of a system, unlike the strictly numerical information (Masera et al. 2000). The changes that occur can be measured by means of qualitative and quantitative variables, which allow the full evaluation of the achievement of an objective, a product or a project. In the case of León, the indicators measure mainly quantitative aspects: the concentration of green area surface in relation to the total urbanized area. This approach leaves aside other variables both quantitative and qualitative that would serve to evaluate quality and functionality of green areas at multiple scales of the territory. Consequently, the instruments that have assisted the urban planning in the city of León have provoked an unequal distribution and access to the green areas, in terms of surface and quality, which is a matter of environmental justice, and also limits the aspiration for achieving urban sustainability in the city.

In recent years, new patterns of urbanization in León have been characterized by the construction of different models of housing development. Some of them, under a design pattern in controlled access clusters and the presence of private green areas, where groups with a medium and high-income level reside. Other housing complexes have a popular character with null or low provision of green areas and are characterized by a population of low income level. In addition, irregular settlements inhabited by a marginalized population have emerged, with less access to essential public services for a better quality of life and better social well-being (Suárez et al. 2015). Thus, the green areas have not been evenly distributed along the city, so that the inhabitants do not have the same possibilities of accessing them and therefore the benefits they offer.

The importance of green areas as a strategic element for achieving urban sustainability is justified if we assume that they fulfill environmental, social and cultural functions that improve the quality of life and social well-being, as illustrated in Fig. 3. Green areas are therefore a strategic element to be considered within planning or urban development programs, contributing to the achievement of the goals set by the United Nations General Assembly and the New Urban Agenda, which, however, until now have been rarely contemplated, as it is the case of the City of León, Guanajuato.

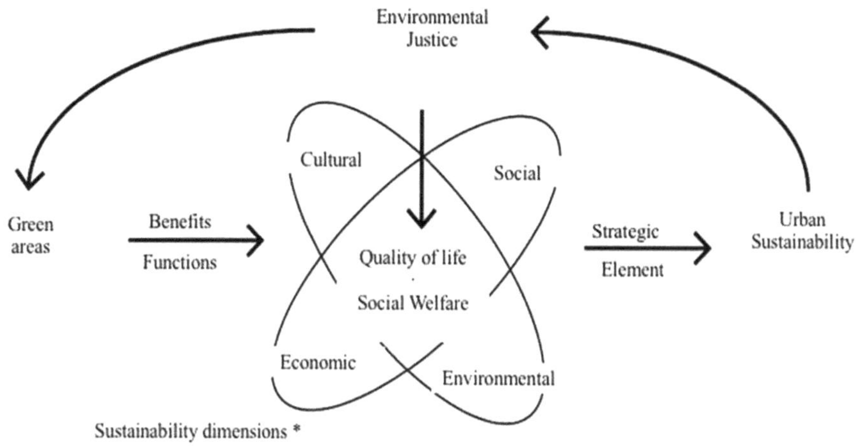

Fig. 3 Importance of green areas for urban sustainability

5 Conclusions

The recent incorporation of the concept of sustainability into the scientific field requires clarifying of its theoretical and methodological implications. This article aims to offer an insight that sustainability involves a multidisciplinary approach, which holistically addresses problems of a complex nature. The reflection on the concept of urban sustainability has led to the conclusion that it is necessary to provide elements for a more equitable, egalitarian and democratic access to a natural or socially generated wealth. This implies ensuring habitability, providing environmental justice, equity and social inclusion. As it has been discussed, green areas are a strategic element for the achievement of these objectives because of their environmental, economic and social benefits that encompass all dimensions of sustainability. Green areas have qualities that lead to improved living conditions, as they contribute to the construction of inclusive urban environments with a healthy environment conducive to coexistence, quality of life and social well-being. The results of the study on spatial analysis based on variables related to the coefficient of green areas per capita and the degree of urban marginalization showed an unequal, inequitable and deficit distribution of the green areas in León. The challenge that lies in this city, which is subject to processes of rapid and dispersed urbanization is to achieve a fair and equitable distribution of green areas, as well as guarantee the accessibility and enjoyment of the benefits they provide, particularly for socially vulnerable groups. Future studies are needed to analyze the accessibility of these groups to green areas and its contributions to the improvement of quality of life and social well-being. The latter would help generate effective strategies and instruments to achieve urban sustainability, with environmental justice.

Acknowledgements This work was supported by UNAM-DGAPA-PAPIIT. Research project "Espacios verdes públicos y sustentabilidad. Estudios culturales, sociales y ambientales".

References

Álcala L (2007) Dimesiones urbanas del problema habitacional. El caso de la ciudad de Resistencia, Argentina. *Revista Invi.* N° 59, Mayo, 22(35): 68

Ayuntamiento de León H (16 de diciembre de 2014). In: Guanjauto PO (ed) Reglamento para la Gestión Ambiental del Municipio de León. Retrieved from http://legismex.mty.itesm.mx/estados/ley-gto/GTO-RM-León-GestAmb2015_10.pdf

Ayuntamiento de León H (16 de octubre de 2015). In: Guanjauto PO (ed) Programa Municipal de Desarrollo Urbano y de Ordenamiento Ecológico y Territorial del Municipio de León. Retrieved from http://implan.gob.mx/downloads/PMDUOET2015.pdf

Ayuntamiento de León H (5 de abril de 2016). In: Guanjauto PO (ed) Programa Municipal de Espacios Verdes Urbanos del Municipio de León. Retrieved from http://periodico.guanajuato.gob.mx/downloadfile?dir=files_migrados&file=PO_55_2da_Parte_20160405_1638_3.pdf

Ayuntamiento de León H (6 de agosto de 2010). In: Guanjauto PO (ed) Código Reglamentario de Desarrollo Urbano para el Municipio de León. Retrieved from http://du.Leon.gob.mx/index/

pdfs/leyes/CODIGO%20REGLAMENTARIO%20DE%20DESARROLLO%20URBANO%2006%20DE%20OCT%202015.pdf

Bonne CG, Fragkias M (2013) Connecting environmental justice, sustainability, and vulnerability. In: Boone CG, Fragkias M (eds), Urbanization and sustainability: linking urban ecology, environmental justice and global environmental change, human-environment interactions 3, Springer Science+Business Media, Dordrecht, USA, 49–59

Boone CG, Geoffrey L, Buckley GL, Morgan G, Sister C (2009) Parks and people: an environmental justice inquiry in Baltimore Maryland. Ann Assoc Am Geogr 99(4):767–787

Cancino M (2014) Fortalecimiento de la justicia ambiental en México, Consideraciones fundamentales. *El futuro de la justicia ambiental en México*. Senado de la Republica, LXII Legislatura, H. Cámara de Diputados, LXII Legislatura, Secretaría del Medio Ambiente y Recursos Naturales, Procuraduría Federal de Protección al Medio Ambiente, Centro de Estudios Jurídicos y Ambientales A.C. y Universidad Nacional Autónoma de México, México

Chiesura A (2004) The role of urban parks for the sustainable city. Landscape and Urban Plann 68:129–138

Consejo Nacional de Población (CONAPO) (2010) Índice de Marginación Urbana 2010. Retrieved from http://www.conapo.gob.mx/en/CONAPO/Indice_de_marginacion_urbana_2010

Dempsey N, Bramley G, Power S, Brown C (2011) The social dimension of sustainable development: defining urban social sustainability. Sustain Dev 19:289–300

Dirección de General de Gestión Ambiental de León (DGGA) (2017) Mapa ambiental de León. Retrieved from http://www.leon.gob.mx/medioambiente/index.php/mapa-verde

García S, Guerrero M (2006) Indicadores de sustentabilidad ambiental en la gestión de espacios verdes. Parque urbano Monte Calvario, Tandil, Argentina. Revista de Geografía Norte Grande, núm. 35, julio, 45–57

Haq SMA (2011) Urban Green Spaces and an Integrative Approach to Sustainable Environment. J Environ Prot 2:601–608

Hay AM (1995) Concepts of equity, fairness and justice in geographical studies. Trans Inst Br Geogr 20:500–508

Instituto Municipal de Planeación (IMPLAN) (2012) Estrategia Integral de Áreas Verdes en la Ciudad de León. Retrieved from https://www.implan. gob.mx/ en/publi caciones/ estudios-planes-proyectos/desarrollo-sustentable/areasverdesypaisajeurbano/ 18-estrategia-integral-areas-verdes-2012/file.html

Instituto Municipal de Planeación (IMPLAN) (2015) Diagnóstico del Municipio de León. Retrieved from https://www.implan.gob.mx/dow nloads/DIAG_MPAL_2015.pdf

Instituto Nacional de Geografía y Estadística (INEGI) (2010) Censo de Población y Vivienda 2010. Retrieved from http://www.beta.inegi.org.mx/proyectos/ccpv/2010/

Instituto Nacional de Geografía y Estadística (INEGI) (2015) Encuesta Intercensal 2015. Criterios de agrupación para generar los tabulados básicos: estatal/municipal. Retrieved from http://internet.contenidos.inegi.org.mx/contenidos/productos//prod_serv/contenidos/espanol/bvinegi/productos/nueva_estruc/702825086657.pdf

Instituto Nacional de Geografía y Estadística (INEGI) (2016) Población Rural y Urbana. Retrieved from http://cuentame.inegi.org.mx/poblacion/rur_urb.aspx?tema=P

Jennings V, Larson L, Yun J (2016) Advancing sustainability through urban green pace: cultural ecosystem services, equity, and social determinants of health. Int J Environ Res Public Health 13(2):196

Keng A, Jordan A, Horsley J (2015) Value of urban green space in promoting healthy living and wellbeing: prospects for planning. Risk Manage Healthc Policy 8:131–137

Krishnamurthy L, Rente N (eds) (1997) Áreas Verdes Urbanas en Latinoamérica y el Caribe. Centro de Agroforestería para el Desarrollo Sostenible. Universidad Autónoma de Chapingo, Texcoco

Lezama JL, Domínguez J (2006) Medio Ambiente y sustentabilidad urbana. Papeles de Población, vol 12, núm. 49, julio-septiembre, 153–176

Masera O, Astier M, López Ridaura S (2000) Sustentabilidad y manejo de recursos naturales. El marco MESMIS. Grupo Interdisciplinario de Tecnología Rural Apropiada A.C, México

Moreno S (2008) La habitabilidad urbana como condición de calidad de vida. Palapa, vol. III, núm. II, julio-diciembre, 2008, 47–54

Ramírez G, Sonia J, Galindo M, Contreras C (2015) Justicia ambiental. Entre la utopía y la realidad social. Culturales, vol. III, núm. 1, enero-junio, 225

Reyes S, Figueroa Aldunce I (2010) Distribución, superficie y accesibilidad de las áreas verdes en Santiago de Chile. Revista EURE - Revista De Estudios Urbano Regionales, 36(109)

Saz SD, Rausell P (2008) A Double-Hurdle model of urban green areas valuation: dealing with zero responses. Landscape and Urban Plann 84:241–251

Steward TA, Boone CG, Mary L, Cadenasso ML (2013) Ecology and environmental justice: understanding disturbance using ecological theory. In: Boone CG, Fragkias M (eds). Urbanization and sustainability: linking urban ecology, environmental justice and global environmental change, human-environment interactions 3, Springer Science+Business Media, Dordrecht, USA, 27–47

Suárez P, Martínez E, García A (2015) La Dinámica económica y cultural de la Zona Metropolitana de León, Guanajuato. Desafíos para el desarrollo humano y territorial. México, UNAM ENES-UNIDAD León y Juan Pablos Editor

Svara J, Brunet J (2005) Social equity is a pillar of public administration. Retrieved from https://www.researchgate.net/publication/249675111_Social_Equity_Is_a_Pillar_of_Public_Administration

United Nations (2016) Goal 11: Make cities inclusive, safe, resilient and sustainable. Retrieved from http://www.un.org/sustainabledevelopment/es/cities/#5daa7124851e142e8

Vélez L (2009) Del parque urbano al parque sostenible. Bases conceptuales y analíticas para la evaluación de la sustentabilidad de parques urbanos. Revista de Geografía Norte Grande, núm. 43, septiembre, 31–49

Wolch JR, Byrne J, Newell JP (2014) Urban green space, public health, and environmental justice: the challenge of making cities 'just green enough, Landscape and Urban Plan, 125, May, pp 234–244

The Need for an Integrated Landscape Management Approach Illustrated by the Analysis of the Participation in the Ecologic Spatial Planning of Benito Juárez, Quintana Roo, Mexico

Anna Lena Di Carlo, Miguel Aguilar Robledo, Carlos Alfonso Muñoz Robles and Pedro Medellín-Milán

Abstract This paper exposes the need to use an integrated landscape management approach for sustainable development through the co-responsible participation of all stakeholders based upon the analysis of the ecologic spatial planning of the municipality of Benito Juárez, Quintana Roo, Mexico. The results of interviews with stakeholders are integrated by a set of indicators and translated into four participation levels and sustainability ranks. In Benito Juárez, most participation is done by government stakeholders and through mechanisms of citizen participation, such as public opinion polls, whereas the lowest result has been found in community participation. Furthermore, a high degree of social and economic development within the municipality, principally due to Cancun tourism, contrasts with a very low score for environmental sustainability. The paper seeks to achieve the understanding of an approach that claims for nothing new, but rather integrates the existing planning instruments and is based on inclusive and equitable decision-making processes.

A. L. Di Carlo (✉)
Universidad Autónoma de San Luis Potosí (UASLP),
Álvaro Obregón #64, Col. Centro, 78000 San Luis Potosí, SLP, Mexico
e-mail: lena.dicarlo@gmail.com

M. Aguilar Robledo
Social Sciences and Humanities, UASLP, Av. Industrias #101-A,
Fracc. Talleres, 78399 San Luis Potosí, Mexico
e-mail: aguilarm@uaslp.mx

C. A. Muñoz Robles
Desert Research Institute, UASLP, Universidad Autónoma de San Luis Potosí (UASLP), Altair num. 200, Col. del Llano, 78377 San Luis Potosí, Mexico
e-mail: carlos.munoz@uaslp.mx

P. Medellín-Milán
Agenda Ambiental de la UASLP, Ave. Manuel Nava 201, 2do piso,
78210 San Luis Potosí, Mexico
e-mail: pmm@uaslp.mx

W. Leal Filho et al. (eds.), *Sustainable Development Research and Practice in Mexico and Selected Latin American Countries*, World Sustainability Series,
https://doi.org/10.1007/978-3-319-70560-6_19

Keywords Sustainable development · Spatial planning · Participation
Integrated landscape management · Cancun

1 The Local Ecologic Spatial Planning Programme: Participation by All Stakeholders

The local ecologic spatial planning programme[1] (hereafter "the Programme") is a public policy instrument that is based on a participation process involving different governmental and societal stakeholders. The idea is that the participants decide together how to use the territory of their municipality towards sustainable development. The purpose is to attribute conservation, protection, preservation, restoration and sustainable use policies to environmental management units, in order to preserve the environment, incentivize the sustainable management of natural resources and protect ecosystems.

The adoption of the Programme and its plan by the executive council, which is made of representatives of the three administration levels—municipal, state and federal—and a citizen representative, is supposed to be based on a consensus, which is elaborated in the technical council. The latter reconciles different interests of communities, civil society organisations, production sectors, and scientific and technical experts. Together, the two councils constitute the committee of the ecologic spatial planning programme (hereafter "the Committee"), which is constituted at the beginning of the planning process. In addition, the use of participatory workshops and a public opinion poll should guarantee the participation of all of these different stakeholders.

The approach that underlies the Programme is currently called integrated landscape management (hereafter "ILM"). The landscape is composed of a mosaic of patches (rivers, lakes, wood land, agriculture, stock farming, protected areas, industry, coast, sea, villages and cities etc.) that are interconnected and interdependent. The approach searches to do justice to this connection by the integration of planning and management decision-making in the landscape through the collaboration between different sectors, the cooperation between stakeholders and the integration between the planning and management instruments. With the common objective of sustainable development, it is thereby possible to ensure the conservation and sustainable use of natural and cultural resources when all parties adopt the ILM approach.

In this context, the participative methodology of the Programme is promising and the importance for sustainable land-use planning indisputable in a country that consists of an exceptional variety of biological diversity in terms of ecosystems, species and genes. However, since the adoption of this instrument of environmental policy in the modification of the federal general law on ecologic balance and the

[1]In Spanish: *Programa de ordenamiento ecológico territorial local.*

protection of the environment[2] in 1996 and its regulation of 2003, of all 2446 existing Mexican municipalities, to date only 71 have published their Programme.

Obviously, no easy answer could explain why more municipalities have not succeeded in the last 20 years in the adoption of Programmes at the local scale.[3] One fact is that the decentralization process of the Mexican administration is not yet complete and funds do not automatically arrive to the municipal level. Also, municipal government periods are of only three years and, as authorities often still lack a culture of responsibility, this does not allow for process continuity from one government to the other (Cabrero Mendoza 2002). Finally, there are many more endogenous and exogenous factors, depending on the territorial features of every entity, which have colluded to the absence of Programmes in the overwhelming majority of municipalities.

This paper seeks to point out one element of ILM that can determine the sustainability of spatial planning: the substantial participation of stakeholders. However, it is important to underline that it presents a selection of variables and indicators, which are entirely based on the focus of the research and do not pretend to be exclusive for evaluating the Programme or sustainability or ILM.

2 Case Study: Caribbean Tourist Paradise Cancun

This research is based on a Ph.D. thesis that is the first of its kind with regard to three aspects: first, the evaluation of the Programme in general; second, in particular its relation to participation; and, third, as based on real case studies. Here we present the case study of Benito Juárez, Quintana Roo (Fig. 1), a municipality loaded with conflicts due to diverging interests between, on the one hand, national and international developers of Cancun´s tourism sector and, on the other hand, environmentalist non-governmental organisations and academic activists who defend the conservation of critical ecosystems and endangered species. The fight for the conservation of ecosystem services struggles against an unsustainable and short-sighted massive entertainment model that was designed in the nineteen-seventies.

Benito Juárez includes two RAMSAR[4] sites -*Parque Nacional Arrecife de Puerto Morelos* and *Manglares de Nichupté*- that comprise of 1,171,114 ha, which represent 14% of total RAMSAR surface of Mexico (Secretaría de Turismo SECTUR 2013). Both are federal Natural Protected Areas and are undergoing problems of contamination and pressures caused by tourism activities. The *Manglares de Nichupté* are made up of four lagoons that are surrounded by the so-called "Cancun Isle", the main hotel

[2]In Spanish: *Ley General de Equilibrio Ecologico y Protección al Ambiente.*

[3]Apart from the local ecologic spatial planning programmes, there is the nationwide general programme, the marine programmes and the regional programmes for one or more states.

[4]"Convention on Wetlands of International Importance especially as Waterfowl Habitat", stipulated in Ramsar, Iran on the 2nd of February 1971 by the United Nations Educational, Scientific and Cultural Organization UNESCO.

Fig. 1 Location of Benito Juárez, Quintana Roo, Mexico in Google Earth

development facing the Caribbean Sea (see Fig. 2). Because of its proximity to the city of Cancun with its exponentially increasing human population and the hotel zone, as well as solid and liquid waste deposits, the wetlands suffer from water contamination and its subsequent decline of biotic populations (Secretaría de Medio Ambiente y Recursos Naturales SEMARNAT 2014).

Cancun is also an important income source for Mexico, representing 14.4% of total tourist arrivals to Mexico in 2008 and 0.4% of world tourist arrivals the same year (Dzul Huchin and Moncada Jiménez 2008). In fact, Cancun´s airport is the second most important by means of arrivals of the country (Secretaría de Turismo SECTUR 2013). Apart from the beaches, Cancun is also a good place of departure for visits to numerous Mayan ruins, Chichén-Itzá and Tulum being the most visited. In addition, Cancun is prepared for more tourism, given that there are 145 hotels—some of which are enormous- that totalled 30,608 hotel rooms in 2014, to which has to be added a vast extension of residential condominiums (Secretaría de Turismo SECTUR 2013).

Due to this pressure on natural resources, the Mexican Federal Attorney for Environmental Protection[5] has repeatedly declared that the capacity of Cancun´s beaches had exceeded its limit and the excessive exploitation was threatening its ecologic balance. The problem is so severe, that the recommendation by the same authority in 2012 was the demolition of hotel rooms (Secretaría de Turismo SECTUR 2013). However, the urban development programme[6] (PDU) for

[5]In Spanish: *Procuraduría Federal de Protección al Ambiente PROFEPA.*

[6]In Spanish: *Programa de Desarrollo Urbano (PDU).*

Fig. 2 Map of Benito Juárez Programme's plan with environmental management units (Beige: sustainable use; Light green: conservation; Medium green: protection; Dark green: preservation; Brown: restoration), natural protected areas (red and italic shaded) and Ramsar sites (straight shaded) (Secretaría de Medio Ambiente y Recursos Naturales SEMARNAT 2017)

2014–2030 updated the land uses in sites of high environmental value and increased the development allowed in such areas (Secretaría de Desarrollo Agrario, Territorial y Urbano SEDATU 2014). This means that Cancun's PDU and Benito Juárez's Programme are not harmonized, although the first should theoretically take the second into account. Nevertheless, it is not surprising that the two planning instruments diverge in substance because their responsibility lies in two different institutions that pertain to two separate sectors -one of development, the other of conservation- and, thus, have distinct legal and operational bases.

3 Sustainability Actors or the Tragedy of the Commons

The tragedy of the commons that Garrett Hardin described in 1968 is precisely about the limits of natural resources and the disastrous consequences of individual decisions based entirely on an independent, rational and free position (Hardin 1968). The relevance of this classical text for our topic is based on two consecutive aspects that the author mentions:

1. In a world were all people only pursue their own personal interests, based on egoistic and short-sighted reasoning, ultimately all suffer the (irrevocable) consequences.
2. Every time that the use of the commons[7] is limited we are also limiting personal freedom; however, decisions and laws taken collectively make the entire humanity freer (Hardin 1968).

[7]Hardin uses the "commons" for common property: goods and land that belong to all (or no one).

Evidently, it is not possible to resolve the tragedy of the commons by leaving the responsibility to the morality of individual public servants, given that the personal eventually prevails over the public interest and makes those officials easily corruptible in return for personal benefits. Therefore, Hardin proposes a government made of laws instead of humans (Hardin 1968). However, based on the experience of the situation in Mexico, having a good legal framework, as well as excellent planning and management instruments, does not necessarily mean that these will be properly used.

In Mexico, the problem of violating or intentionally misinterpreting the institutional processes required by the law is generally linked to the limitations of the democratic practice and attributed to structural problems, such as the lack of equity and the absence of a political will (Blauert et al. 2006). But then, if neither the governors on their own, nor the laws by themselves can resolve the tragedy of the commons: what can be done to avoid it? Nobel prize winner Elinor Ostrom´s theory on the organization of local individuals for a sustainable use of natural resources (Ostrom 1990) is an alternative solution to central-power monopoly on planning and management: the involved population has the best knowledge and the most direct access to the resources they use and therefore are best at managing collectively the commons dilemma.

Theoretically, citizens form a political community and pursue the aim of common wellbeing (Ochoa Arias and Petrizzo 2006). However, the transformation of the free and egoistic individualist into an altruistic and committed citizen is neither a miraculous transformation of the nature of human beings, nor something that occurs automatically the moment people are given their democratic rights. Rather, the construction of citizenship is a slow and non-linear process, since there can be setbacks and discontinuity when the lack of quick results discourages the enthusiasm of participants (Cabrero Mendoza 2002).

Thus, it is important that citizen participation coexists and grows alongside with state responsibility, resulting in the so-called co-responsibility between society and government (Cabrero Mendoza 2002; Oseguera Ponce et al. 2010; García Barrios 2012). The outcome of co-responsibility is a cooperative relation between citizens and their government rather than a fight for power. This implies that citizen participation "claims at the same time the previous acceptance of the rules of the democratic game and the free will of people who decide to participate: the rule of law and the freedom of the people" (Merino 2001).

Some authors propose that environmental problems and territorial conflicts can open up new spaces for participation, because they motivate citizens who are affected directly or indirectly to formulate solutions that correspond to their necessities and demands (Larraguibel Galarce 2002). To get back to the tragedy of the commons, we have arrived at that moment in history when the consequences of individual behaviour are affecting ourselves to the same extent as the rest of humankind (and all other species). Therefore, rational behaviour does not necessarily imply an antithetic position between personal and public benefit. Thus, despite plenty of possibilities for action at the individual level, collective action nevertheless seems most effective for achieving visible changes that are reflected in an improvement of the general conditions.

In order that this new social actor called "sustainability actor" (Larraguibel Galarce 2002) is able to participate substantially, he/she must be part of an informed and organised citizenry that is provided with its democratic rights and duties. Thereby, communities will be capable of taking decisions on the design and implementation of public policies that are relevant for them. Hence, a renewed or new civil society, made of empowered citizens and communities in processes of endogenous development organises new forms of participation and common action spaces.

The State as the unique decision-maker is losing importance for environmental governance in the twenty-first century (Agrawal and Lemos 2007). Rather, the cooperation with new stakeholders and different sectors—as the private sector and communities- substitutes the state monopoly in environmental decision-making. Thereby power is not only exercised top-down as by the State, and in other theories the Market, but other players who take decisions from the bottom up always gain more importance: non-governmental organisations and locally organised civil society—all sustainability actors—play a crucial role in the protection of natural resources and the construction of sustainable development (Arts 2004; Bruch and Czebiniak 2002). They have the capacity of pushing decision-makers by the use of transnational networks (Jelin 2000) becoming important associates, together with civil society, who complement the public administration at all levels (Bruch and Czebiniak 2002).

4 Four Types of Participation

As early as 1969, in the framework of worldwide pacifist and anti-nuclear movements, Sherry Arnstein underlined that participation is of no avail if it does not come along with a real redistribution of power, given that otherwise it continues to be found in the hands of a very small minority (Arnstein 1969). Furthermore, it is important to note that in our societies, which adopted universal franchise, participation is not only an option but, paradoxically, even the lack of participation has a participative impact. Because not participating means "giving a vote of confidence to those who are making the decisions: a blank cheque for others to act" in one′s name (Merino 2001).

In this context, not only is the right to participate in public matters a basic principle of the social and democratic rule of law that has as its premise -citizen participation- in the institutional and political structure of the State (Rio Fernandes et al. 2015). But, furthermore, from the moment that this right exists it also always transfers a responsibility to the citizens. In addition, the exercise of this right and responsibility must be based on the citizens′ trust in the democratic institutions (Serrano Rodríguez 2015). This is how co-responsibility between all stakeholders is based on transparency and decentralization. And this is how it becomes an expression of the recognition and appreciation of all participants, which gives public policies such as the Programme legitimacy and leads to the viability of its implementation.

Thus, only if the Programme is an expression of a consensus between all stakeholders, with rational individual trade-offs and common benefits as a solution to the well-known prisoner′s dilemma,[8] can it contribute to the achievement of sustainable development. What helps to overcome the dilemma in this case, beyond the enforcement of the State and the contract between stakeholders, is the constant participation of all sustainability actors based on co-responsibility.

According to the World Bank, citizen participation is a process through which stakeholders impair and share control over development initiatives and the decisions and resources that affect them (Crespo Flores 1999). In addition, starting with a participation that was limited to the right to vote and other representative instruments, citizen participation then became part of the public policy design process to, finally, be claimed for during implementation and evaluation processes too (Pacheco Vega and Vega López 2001).

For this purpose, citizens have different modes of participation, depending on whether they wait for the State to require them to choose from a menu of options (citizen participation) or they organise with other citizens around a mutual professional or private concern (social participation). In addition, there is the participation of government representatives as part of their functions (public participation) and the participation of community representatives (community participation), who are either elected for that specific purpose or represent their community as members of a distinct decision-making structure.[9] Obviously, these four types of participation are not mutually exclusive and only serve for the purpose of theoretical analysis. They are defined in the table (Table 1).

5 Methodology: Interviews on Participation and Sustainable Development

The process of the Programme is based on the assumption that successful spatial planning relies on the substantial participation of all key stakeholders. The interviews, which were conducted for this research with 21 key stakeholders who have participated in the Programme's process in Benito Juárez, enquire how much each

[8]The prisoner′s dilemma is a classic game of game theory that shows why two isolated rational individuals tend not to cooperate although it would lead them to the most advantageous scenario for both.

[9]The first in Latin-America to distinguish four types of participation (social, community, political and citizen) was Nuria Cunill Grau (Cunill Grau 1991; Cunill Grau 1997). However, the idea of the four types used here for an instrument of environmental management was adopted from a research on participation in the management processes of the river basin of Rio Valles, San Luis Potosí, Mexico (Ferney Leonel et al. 2010).

Table 1 Types of participation as defined in this paper for the ecological spatial planning process

Type of participation	Definition
Social	The broadest category of stakeholders in the Programme is of those who exercise social participation. It includes the private (production) sector, professional councils[a], non-governmental and civil society organizations, academia and the scientific sector. What all these stakeholders have in common is that their members do not participate for the sake of their individual interests. They defend the common interest of all members, being a professional, economic, social or environmental cause that binds them together. Social participation is exercised in the technical council of the Committee and during the public opinion poll.
Community	This type is the most organic because it is based on the assumption that communities are directly -physically and culturally- linked to the territory. Either rural or urban, they are the first in noting changes and the first to suffer alterations. In Mexico, community participation[b] often responds to distinct internal organisation mechanisms that are not legally stipulated. Community representatives often have the most precious knowledge in terms of every-day experience with the environment and specifically defend the vision and interests of their communities. In the Programme, participative workshops allow for this type of participation.
Public	In Mexico, there are three legally stipulated levels of political decision-making, which shall all participate and cooperate in the Programme through government representatives: federal, state and municipal stakeholders. In the case of environmental matters, in this group of participation actors, all those civil servants who work for some environmental government dependency are especially important. Public participation is potentially present in both councils of the Programme's Committee.
Citizen	This is a type of participation that the State convenes, thus where government and society are in contact. Citizens involve themselves directly into public actions within the limits of the legal framework. Elections are a typical example of citizen participation, but also public opinion polls and the representation of civil society by an elected member for example in the two councils of the Programme´s Committee are expressions of citizen participation.

[a] Typically, those who participate are biologists, civil engineers and architects

[b] Usually, and differently to the definition used here, community participation in Mexico is interpreted as a claim for State assistance on the part of the community (Villarreal Martínez 2009)

of the participation types has been used. Quantifiable data was obtained through questionnaires, in order to calculate the participation rate for every participation type.

The qualitative results (possible answers to questionnaire were YES, NO and DON'T KNOW) obtained from the interviews were numerically codified in order to work with the data in Matlab. For every questionnaire, the number of each possible answer was counted per question and interviewee. In addition, we applied Cronbach´s alpha to the results of our pilot study and thereby excluded questions that turned out to be unreliable (Pérez-Tejada 2008).

The participation index (Ip) is generated simply by the weighted sum of the positive responses plus the weighted sum of the negative responses, divided by the total of responses (N). Positive answers are assessed a weight of 1, whereas negative or uncertain answers obtain a weight of 0. This means that a final result of 1 would express a maximum level of participation and 0 would indicate its inexistence. The formula that was used for the calculation of the participation index is the following:

$$Ip = \frac{1}{N}\sum_{n=1}^{N} R_n$$

Finally, the results were translated into participation levels considering the idea of the ladder by Sherry Arnstein (Arnstein 1969) as in Ferney et al. (2010) but with five equal ranges[10] as follows:

0.76–1 high
0.564–0.75 medium-high
0.376–0.563 regular
0.189–0.375 low
0-0.188 very low

It is important to note that the level of participation that results from our research is based on the perception of the key stakeholders and, thus, goes beyond counting register signatures in the Committees′ sessions, workshops and opinion polls. The reason for this methodological choice is that we believe participation is not restricted to being present, and therefore the level of participation of a group cannot be only calculated by counting the number of signatures per event. To measure the participation level, it is not sufficient to know about quantity, or how many members of a group participated, but also about the quality or intensity of their contributions. Furthermore, the perception by participants of ownership is crucial for the success of managing the Programme, given that it determines acceptance or refusal.

In addition to the quantitative results obtained through the application of questionnaires, the interviews with key stakeholders of the ecologic spatial planning process also included a section of semi-structured interviews with open answers. The information that resulted from this part of the interviews helped in the process of interpretation of both participation levels and the levels of municipal sustainable development.

In the same way as for the measuring method of participation, also the indicator set for measuring the level of sustainable development of the municipality has been kept as simple as possible. This is because the idea behind the methodology is that it should be simple to replicate and with easily, freely and constantly available data. Thus, every one of the three classic dimensions of sustainable development (social,

[10]Ferney et al. use three ranges (low, regular and high).

economic, environmental), in addition to the political dimension, which is particularly important in a research on participation, are measured by two indicators each (Table 2). As with participation, also the perception of stakeholders on sustainable development in their municipality was measured and combined with the results from the indicator set. The calculation method for each sustainable development dimension is the same as for the types of participation, as well as the distribution of ranges.

However, given the complexity of all elements of a territorial unit, indicators will always only reflect a small part of reality, and their choice depends on the focus of each analysis. Therefore, here the indicators were exclusively chosen on the basis of our theoretical framework and practical experience, and are not meant to be perfect nor irreplaceable. Three main premises have shaped the methodology of this research from the beginning:

1. Mexico has promising spatial planning and management instruments whose application, however, needs to be evaluated.
2. Sustainable development indicators and indexes are mostly complex and difficult to use. Simple tools that offer a rough idea on the state of a territory within the different dimensions of sustainable development are sometimes sufficient and more viable.[11]
3. When it comes to substantial participation, it is important to include the perception of stakeholders.

6 Results on Spatial Planning Processes

We defined five equally large ranges of participation in the Programme and sustainable development from very low to high. The results from the case study of Benito Juárez, Quintana Roo reported in Table 3, reflect medium-high participation by government stakeholders (public participation) and the citizen representatives (citizen participation) in the executive council of the Committee, as well as medium-high participation by citizens in the public opinion poll (citizen participation). Social participation, mostly exercised in the technical council of the Committee and during the public opinion poll, obtained a medium level and community participation resulted to be very low. The latter implies an insufficient or incorrect use of the participatory workshops, which are supposed to explore the visions and priorities of the communities that form the municipality.

The results of the indicator set on sustainable development reproduced in Table 4, combined with the information retrieved by the open answers of the

[11]This is obviously only true for the planning process. When it comes to the evaluation of the results and impacts of processes that are based on the plan, programme or policy, a more complex and specific set of indicators is needed.

Table 2 Indicators for every dimension of sustainable development

Sustainable development dimension	Social	Environmental	Economic	Political
Indicators	Health index (United Nations Development Programme UNDP 2014) with data by INEGI, 2010 in (United Nations Development Programme UNDP 2014)	Indicator of disturbed areas (Salgado Vega et al. 2011) with data of (Secretaría de Medio Ambiente y Recursos Naturales SEMARNAT 2013) and (Salgado Vega et al. 2011)	Income index (United Nations Development Programme UNDP 2014) with data by INEGI, 2010 and World Bank in (United Nations Development Programme UNDP 2014)	Turnout of voters' index (Flamand 2007) with data by the Municipality of Benito Juárez (Ayuntamiento de Benito Juárez 2016)
	Education index (United Nations Development Programme UNDP 2014) with data by INEGI, 2010 in (United Nations Development Programme UNDP 2014)	Indicator of public investment in environment (Salgado Vega et al. 2011) with data by INEGI, 2010 "urbanización y medio ambiente" from (Instituto Nacional de Estadistica y Geografía INEGI 2015) and (Salgado Vega et al. 2011)	Indicator of employment level (Flamand et al. 2007) with data from (Flamand et al. 2007)	Gender development index (United Nations Development Programme UNDP 2014) with data by INEGI, 2010 in (United Nations Development Programme UNDP 2014)

Table 3 Results for participation in the ecologic spatial planning process of Benito Juárez, Quintana Roo, Mexico

Type of participation	Result	Level
Social	0.543	Medium
Community	0.295	Very low
Public	0.732	Medium high
Citizen	0.619	Medium high

Table 4 Results for the dimensions of sustainable development of Benito Juárez, Quintana Roo, Mexico

Sustainable development dimension	Result	Level
Social	0.796	High
Environmental	0.063	Very low
Economic	0.711	Medium high
Political	0.565	Medium-high

interviews, allows us to assume that Benito Juárez is an economically prosperous municipality with excellent social conditions, due to business and work opportunities offered by Cancun's international tourist expansion. Also, the political dimension shows a good performance, while environmental sustainability dramatically lags behind. The reason for this poor environmental score is rather obvious, considering the consequences in terms of contamination, destruction of ecosystems and fragmentation of habitat that mass tourism implicates. In addition, the perception of stakeholders on sustainable development in Benito Juárez -with a focus on participation and the Programme- also represents a low level.

7 Interpretation of Participation and Sustainable Development in Cancun

Of course, Cancun in the Northeast of the municipality of Benito Juárez is an ideal international tourist destination, favoured by a warm subhumid (tropical) climate with an average annual temperature of 26.6 C. Long white sand strips and turquoise waters make Cancun particularly attractive for beach tourism. This is why, since the seventies, tourism in Benito Juárez has been growing constantly and turned the location into an important generator of income and an attractive destination for work opportunities. Because of its visibility and an economically and intellectually influential settlers' community, political conditions have been improving for citizens.

However, it is striking that this economic, social and political development did not come about with the necessary precaution for what concerns environmental outcomes, especially if considering that local ecosystems are known to be extremely vulnerable. The vegetation of the municipality is predominantly composed of medium and lowland forests, where emblematic mammals such as the jaguar and reptiles like crocodiles are found. Furthermore, large portions of wetlands around the lagoons facing the Caribbean Sea provide important environmental services, such as being a shelter for reproduction of marine species, as well as to serve as a buffer zone for hurricanes. In addition, Benito Juárez´s coral reefs are part of the Mesoamerican Reef System,[12] another ecosystem of unique beauty and vulnerability.

One of the greatest environmental risks in Benito Juárez is that underground water flows directly to the sea. Given the presence of irregular settlements that lack a proper sewage system, wastewater is drained into the groundwater reservoir and moves on to the wetlands and into the sea, facilitated by the subterranean hydrodynamic (Secretaría de Medio Ambiente y Recursos Naturales SEMARNAT 2014). The contamination affects water quality, marine flora and fauna and contributes to local coral bleaching. Furthermore, the connectivity of wetlands is often disrespected, leading to infrastructure built on top of ecosystems of inestimable value.

There are many open questions on the findings presented in this text. And evidently the same amount of answers can be found to explain the complex relations between the infinity of variables that are somehow connected to each other within a landscape. However, one assumption is that there could be a relation between the very low community participation and a very low level of environmental sustainability. Given the relative youth of the municipality and its original design as a mass tourist destination, that gave work opportunities to a huge number of settlers, there could be a relation between the lack of an organic belonging of the population to the territory and their lack of interest in its development patterns through spatial planning processes. This is especially true if we consider that most people depend on jobs from mass tourism and fear that changes to this development model could leave them jobless. In addition, confidence in political institutions and participation processes evidently fall when people see that regulations and agreements are not respected.

However, qualitative information from the interviews, which was triangulated between different stakeholders, offers additional explanations to the much lower community and social participation as compared to citizen and public participation: participatory workshops were apparently not communicated with a sufficient anticipation and frequently realized in the city centre instead of decentralized sites that are more accessible to rural community representatives; and, since the first Programme of Benito Juárez in 2005, social stakeholders noticed that their participation in the form of recommendations and revisions of the proposal presented in the public opinion poll had not been taken into account in the final document. Both circumstances obviously lower the trust and motivation of participants.

Furthermore, from what has been mentioned above, it becomes clear that there is a lack of integration of different management instruments; of natural protected areas and cultural heritage; tourism development; ecologic spatial and urban development

[12]The research was realized in 2015 when Puerto Morelos was still part of Benito Juárez. Today Puerto Morelos is a separate municipality.

planning etc. This demonstrates that there is no common vision on the development and management of the municipality between the stakeholders. Although the participation of government representatives in the Programme was perceived as medium-high, it seems that the Committee of the Programme nevertheless predominantly includes stakeholders from the environmental sector and thus does not achieve a real integration, neither with urban planning nor with tourist development.

8 The Need for Integrated Landscape Management (ILM)

The recent National Vision for ILM and Connectivity by the Mexican environmental sector, together with stakeholders from other sectors of the federal public administration—notably Tourism-, different non-governmental organisations, representatives of academia and experts in the topic, which was presented by the Mexican Secretary for the Environment during the COP13 of the Convention on Biological Diversity in December 2016 in Cancun, is that

> In 2030, on a quarter of Mexico´s surface, spatial management decisions are taken with an articulated and adaptive approach from terrestrial and inland water landscapes –both rural and urban- and seascapes. These are connected and managed integrally through negotiation and collaboration processes that value and link the interests of all stakeholders. Hereby, the conservation and sustainable use of natural and cultural resources are ensured for the benefit of national development. (SEMARNAT, CONABIO, CONAFOR and CONANP, 2017)

The idea of adopting this approach is to overcome the division between conservation and development that has marked the country´s legislation, institutions and public policy until the present day. In the case of Benito Juárez, this would mean a common agenda between tourism, production, environmental and urban sectors for the sake of a sustainable development of the municipality. This would guarantee the conservation of ecosystem services and resources for the satisfaction of the present and future needs of the populations of all species, and, at the same time, allow the long-term economic development of the municipality and its contribution in terms of the spread of wealth via tourism and other activities.

The ILM approach for Benito Juárez implies an example for the need of diversifying tourist activities by the integration of sustainable tourism to the portfolio. The spread of new sustainable work opportunities would quell people´s fears for environmental protection and allow them to value and protect cultural and natural heritage. Also, urban and ecologic planning have to be integrated and take account of the interests and needs of communities, based on the respect for social justice and human rights. The latter includes both the right to participate and the right to a healthy environment, which State and society construct and implement together by exercising co-responsibility.

ILM is not the explicit approach for the Programme, although it shares most of its elements with the theory of spatial planning. The same approach has also different names depending on the sector, for example "sustainable territorial management" in the Mexican productive sector. However, the core of it is the same belief of a need for effective planning, management and evaluation processes that lead to coherence between, on the one hand, the Mexican compromise for sustainable development positioned in policy, laws and planning instruments and, on the other hand, its effective, transparent and inclusive implementation through integrated management. Only when citizens perceive congruency between discourse and facts, between development and conservation, between participation and decision-making, will the conditions be set for common objectives and shared strategies.

9 Conclusions

Based on the assumption that the participation of all stakeholders in the process is meant to be the foundation for its success, the case study illustrates that the methodology for the local ecologic spatial planning programme is still perfectible, considering that in Benito Juárez community participation is much lower than public and citizen participation.

This may be, because in Mexico there is a predominating interpretation of participation convened by the State to ratify decisions previously made and determined by particular and sectorial (economic) interests. Therefore, it is crucial to mainstream a culture of co-responsibility where State and citizens show mutual trust and respect, based on transparency and subsidiarity, as well as processes of empowerment and the internalization of the sustainable development paradigm as a win-win option through mutual trade-offs.

Participation spaces, such as the local ecologic spatial planning programmes, must be used without a sectorial bias in order to foster dialogue between stakeholders with different interests. This is how the existing planning and management instruments are to be integrated in an articulated and adaptive way from the territory. Conservation and development instruments are not in opposition; there is a need for finally unifying them to allow for a path towards sustainable development. Thus, there is a need for integrated landscape management, based on inclusive and equitable decision-making processes among all stakeholders.

The general findings of this paper and the future prospects for its topic can be summarized as follows:

1. Co-responsibility requires that community, public, social and citizen stakeholders have recognized and accessible spaces to participate in decision-making and that their participation is taken into account in a transparent way. If the State "betrays" these terms, participation by all other stakeholders -especially community and social- will fall drastically and can become an opposition force that makes future decision-making much more difficult.

2. Citizen participation is the direct expression of the social and democratic rule of law. However, participation spaces that are convened by the State are not sufficient to guarantee sustainable development because they are too easy to be manipulated. It is therefore crucial for an integrated process to continue engaging interest groups and communities. Even if, because they have sometimes more radical positions -defending a special and direct cause- it can complicate and slower decision-making, the effort is worth it because the difficulties only reflect the complexity of reality.
3. The legitimacy and viability of public policies lies within its acceptation and appropriation by all stakeholders, who only participate in their implementation if they share responsibility. It is necessary that more effort is spent on capacity building and socialization of information and options in a dialogue between scientific and traditional knowledges in order to obtain the expected results for which these policies are formulated.

References

Agrawal A, Lemos M (2007) A greener revolution in the making? Environmental governance in the 21st century. Environment, 36–45

Arnstein S (1969) A ladder of citizen participation. J Am Plan Assoc, 216–224

Arts B (2004) The global-local nexus: NGO's and the articulation of scale. Tijdschrift voor Economische en Sociale Geografie, 498–510

Ayuntamiento de Benito Juárez (2016 November 1) H. Ayuntamiento de Benito Juárez 2016–2018. Retrieved from http://cancun.gob.mx/

Blauert J, Rosas M, Anta S, Graf S (2006) ¿Espacios de deliberación o la toma de decisiones? Lecciones para la participación y las políticas en consejos ambientales en México. In Isunza Vera E, Olvera A, Ramón MA (eds) Democratización, rendición de cuentas y sociedad civil: participación ciudadana y control social, CIESAS, Universidad Veracruzana, Veracruz, pp 597–639

Bruch C, Czebiniak R (2002) Globalizing environmental governance: making the leap from regional initiatives on transparency, participation and accountability in environmental matters. Environmental Law Institute, Washington DC

Cabrero Mendoza E (2002) El premio gobierno y gestión local en México: un mecanismo de institucionalización de la inovación en gobiernos locales. CIDE, Ciudad de México

Crespo Flores C (1999) Políticas públicas, gobierno local y conflictos socioambientales. Oxford Brookes University, Oxford

Cunill Grau N (1991) Participación ciudadana: dilemas y perspectivas para la democratización de los Estados latioamericanos. CLAD, Venezuela

Cunill Grau N (1997) Repensando lo público a través de la sociedad. Nuevas formas de gestión pública y representación social. CLAD y Nueva Sociedad, Venezuela

Dzul Huchin D, Moncada Jiménez P (2008 March) Participación de Cancún en el turismo del mundo y Caribe. Boletín Turístico de Cancún, 1(4)

Ferney Leonel H, Aguilar Robledo M, Reyes Hernández H (2010) Factores y niveles de participación en la gestión de la cuenca hidrográfica del Río Valles, oriente de México. Prospectiva, 263–293

Flamand L, Martínez Pellegrini S, Camacho O (2007 February 23) Metodología de cálculo Índice de Desarrollo Municipal Básico (IDMb). Retrieved from inafed: http://www.inafed.gob.mx/work/siha_2015/4/MetodologiaIDMb-Feb07.pdf
García Barrios R (2012) La disputa por el territorio y su ordenamiento en Cuernavaca (parte 1). Cultura y representaciones sociales 13:136–184
Hardin G (1968) The tragedy of the commons. Science, 1243–1248
Instituto Nacional de Estadistica y Geografía INEGI (2015 November 11) Instituto Nacional de Estadistica y Geografía INEGI. Retrieved from http://www.inegi.org.mx/
Jelin E (2000) Towards a global environmental citizenship. Citizsh Stud 4(1):47–62
Larraguibel Galarce C (2002) De la utopia a la acción: la construcción de la sustentabilidad desde el espacio local (aportes ciudadanos al desarrollo sustentable). Revista Mad, 6
Merino M (2001) La participación ciudadana en la democracia. Istituto Federal Electoral, Mexico
Ochoa Arias A, Petrizzo P (2006) Ciudadanía y Desarrollo Endógeno. Hacia la reconstitución del quehacer social. In: Ochoa Arias A (ed) Aprendiendo en torno al desarrollo endógeno, Venezuela
Oseguera Ponce J, Rosete Vergés F, Sorani Dalbón V (2010) Reflexiones acerca del Ordenamiento Ecológico en México. Investigación ambiental, 32–40
Ostrom E (1990) Governing the commons: the evolution of institutions for collective action. Cambridge Univesity Press, Cambridge
Pacheco Vega R, Vega López O (2001) Dos modalidades de participación ciudadana en política ambiental. Perspectivas teóricas sobre coaliciones ambientales y procesos multiparticipativos. Economía, Sociedad y Territorio, 25–62
Pérez-Tejada HE (2008) Estadística para las ciencias sociales, del comportamiento y de la salud. CENGAGE Learning, México
Rio Fernandes J, Savério Sposito E, Trinca Fighera D (2015) Diccionario de Geografía aplicada y profesional. Terminología de análisis planificación y gestión del territorio. Universidad de León, León
Salgado Vega J, Manguía Ramón J, Alcántera Cruz F (2011) Indicadores de Desarrollo Sostenible. Análisis Comparativo del Distrito Federal y los Municipios de Monterrey, Guadalajara, Toluca y Metepec. Revista de Estudios en Contaduría, Administración e Informática, 48–66
Secretaría de Desarrollo Agrario, Territorial y Urbano SEDATU (2014) Programa de Desarrollo Urbano del Centro de Población Cancún, Municipio de Benito Juárez, Quintana Roo (2014–2030)
Secretaría de Medio Ambiente y Recursos Naturales SEMARNAT (2013) Modificación del Programa de Ordenamiento Ecológico Local del Municipio de Benito Juárez, Quintana Roo. Municipio e Benito Juárez, Quintana Roo, Cancun
Secretaría de Medio Ambiente y Recursos Naturales SEMARNAT (2014) Programa de Manejo Área de Protección de Flora y Fauna Manglares de Nichupté
Secretaría de Medio Ambiente y Recursos Naturales SEMARNAT (2017 September 2) Subsistema de Información para el Ordenamiento Ecológico SIORE. Retrieved from http://gisviewer.semarnat.gob.mx/aplicaciones/
Secretaría de Turismo SECTUR (2013) Estudio de competitividad turística del destino Cancun: Agendas de competitividad de los destinos turísticos de México. Universidad de Quintana Roo, Quintana Roo
SEMARNAT, CONABIO, CONAFOR, CONANP (2017) Visión Nacional de Manejo Integrado del Paisaje y Conectividad. Retrieved from http://www.biodiversidad.gob.mx/corredor/cobiored/images/2016/VNMIPVersionFINAL.pdf
Serrano Rodríguez A (2015 January–April) La participación ciudadana en México. Estudios políticos, 9(34), 93–116
United Nations Development Programme UNDP (2014) Índice de Desarrollo Humano Municipal en México: nueva metodología. United Nations Organisation UNO
Villarreal Martínez M (2009) Participación ciudadana y políticas públicas. Décimo Certamen de Ensayo Político

Mediating Traditional Ecological Knowledge Through Participatory Documentary: Emerging Educational Strategies for Sustainability in Southern Mexico

Juan Carlos A. Sandoval Rivera, Guadalupe Mendoza Zuany and Rob O'Donoghue

Abstract This article reports the development and use of video materials for the mediating of learning in relation to indigenous knowledge practices in southern Veracruz, Mexico. The case study describes the production of a video on traditional livelihood practices with youth and how sharing the video took participants back to cultural experiences and environmental meaning-making. The dialogical work with video enabled in depth insights that exemplified the knowledge within the context of fishing practices in local wetlands, a practice done mainly by women that had receded and was being lost in modern times. The study found that it was the animated representations and dialogic education processes (discussions) amongst learners in Mother Tongue that surfaced much of the rich detail that then articulated into the local environment context and indigenous sustainability practices to enhance human wellbeing.

Keywords Education for sustainability · Indigenous knowledge Participatory documentary

J. C. A. Sandoval Rivera (✉) · G. Mendoza Zuany
Instituto de Investigaciones En Educación, Universidad Veracruzana, Campus Sur, Paseo 112, Lote 2, Sección 2A Edificio B, 2° piso; Colonia Nuevo Xalapa, 91097 Xalapa, Veracruz, Mexico
e-mail: jcsandoval.rivera@gmail.com

G. Mendoza Zuany
e-mail: lupitamendoza.zuany@gmail.com

R. O'Donoghue
Environmental Learning Research Centre, Rhodes University. ELRC, P.O. Box 94, Grahamstown 6140, South Africa
e-mail: r.odonoghue@ru.ac.za

W. Leal Filho et al. (eds.), *Sustainable Development Research and Practice in Mexico and Selected Latin American Countries*, World Sustainability Series,
https://doi.org/10.1007/978-3-319-70560-6_20

1 Introduction

This article is important because show the results of a research project that demonstrates the pedagogical applications of the use of video cameras to document and to disseminate traditional ecological knowledge practices (Grenier 1998; O'Donoghue et al. 2013; Shava 2013). The potential of this method to promote young people's interest in their own culture and local identity was also explored. An analysis was made of the impact and learning (Cundill and Rodela 2012) generated through the process of audio-visual creation carried out amongst indigenous youths in the municipality of Zaragoza in Veracruz, Mexico, a town of around 18,000 inhabitants, over 70% of whom speak Nahuatl language.

The research was composed of two aspects which were carried out in parallel: (1) the documentation and dissemination of traditional ecological knowledge concerning ancestral fishing methods, and (2) an analysis focused on group reflexion (De Zutter 1997) regarding the work that was done during this process.

This collaboration project (Rappaport and Ramos 2005) was aimed at tackling the problem of the loss of knowledge, especially that pertaining to traditional fishing because, as they themselves reported, young people today are not interested in learning about, nor taking part in this activity. Rather, they prefer to study in order to take up a professional career, instead of learning traditional local fishing practices which are sustainable because they are effective, locally tested, inexpensive, culturally appropriate and developed based on the patterns and processes of nature (Grenier 1998; Gasché 2008).

Therefore, the project was developed as a participatory audio-visual production as a means to awaken the young people's interest in their own culture and their community's knowledge of traditional fishing in local wetlands.

2 Participatory Documentary and Ethnographic Video

Since this research was carried out as an educational project with indigenous knowledge (Kincheloe and Steinberg 2008) using audio-visual media, in this section we will define both participatory documentary as an educational process, and ethnographic video as a research tool. By 'participatory documentary' We are following the model proposed by Mosangini (2010) for this kind of collaborative work in which the creation of audiovisual material focusing on participation in the process attempts to transform a specific reality by proposing a horizontal and negotiated methodology. Mosangini (2010) says that a documentary is participatory

> […] when people or social organizations create it collectively. [This] group carries out every step of the production together (script, filming, editing, distribution) […]. The participatory nature blurs the boundaries between the several roles required in documentary making: writing the script, filming and distributing the documentary. Roles overlap as all

> the participants can take part in each stage of production, in contrast to the typical methods of audiovisual production which are built on a strong division of labour (10).

The production process involved in a participatory documentary focusses more on content and less on appearance, and the audiovisual material is created with a specific audience and objectives in mind (Lunch and Lunch 2006). This manner of tackling the creative and participatory creation of audiovisual material began to be developed in the 1960s. Those who have promoted it, and the groups involved, as well as the definitions and meanings that have been assigned to various experiments with this type of documentary production, have varied according to the context and the people or groups behind them (Mosangini 2010). The participatory documentary can be seen as a process and not just a product. In such a process, participants produce audiovisual materials based on negotiated strategies, in which the production process is as important as the final product. A key component in the production of a participatory documentary is its educational aspect. This can be used to capitalise on and strengthen the learning acquired through each stage of the project. In this way, a participatory approach can allow those involved to create their own films, to decide which topics to cover in them according to their own world-view, to choose their political goals, and to take control of how they are represented (Smith 1999).

As for using ethnographic video as a research tool, this can be traced back to two points of origin, one in documentary cinema, and another in anthropology. Uniting these two fields gives rise to creative possibilities that transform the way in which ethnographic audiovisual productions have traditionally been carried out:

> […] in ethnographic film-making and visual anthropology there is a shift in the focus of interest to the analysis of the components of the communicative process. We could say that attention shifts on the one hand from the process of production to that of reception of a film, and on the other from the film-maker to the subject of the film. In […] previous models the process came to an end with the product, and the subject was merely the object represented but did not take part as a producer, neither was their active role in the processes of production and reception of the film usually recognized. They lent their image, collaborated with the ethnographic film-maker, but had no control over the product. This is one of the demands of the participatory current: the role of subjects as active components in the production of a film about themselves, and their participation as a co-producer at every stage of the decision-making regarding the film (Ardèvol 1994, p. 107).

In this way, the participation of the filmed subject in the film-making process implies that the final product is directed towards the subject participants themselves, as they are in the end the first consumers of the material produced.

So, research using ethnographic video has two distinct levels: firstly, self-description by local people comes to the fore because it is the very subjects who express themselves in the film: "we want films to be made by the subjects themselves, autobiography and self-reference, we want to film the members of the cultural group talking in their own terms about themselves and what they think they are doing (Ardèvol 1994, p. 108)". Secondly, ethnographic video also implies looking towards:

> [...] an examination of the form taken by the research and film-making process, the form of ethnographic presentation, the change that occurs in the subject-researcher or ethnic film-maker in the field. The researcher uses fieldwork methods to create an ethnography of the research process. The ethnic film-maker films her or himself filming. The product demonstrates the process by which it has been created (Ardèvol 1994, p. 108).

This self-observation process permits the generation of new levels of reflexion concerning audiovisual production and ethnographic research, with regard to debates concerning the objectivity or subjectivity of the work, since the researchers are also observed during the production process that they themselves carry out.

3 Towards Relevant Research with a Social Impact

As mentioned earlier, this research focused on an analysis of the process of passing on ancient knowledge (from older ladies to the youth) concerning nature in an indigenous community. During the first phase of fieldwork it was noted that whilst the elderly in the community possess this kind of knowledge, they no longer take part in these activities due to their advanced years. At the same time, it was also noted that most of the community's youth was not interested in learning the traditional ecological knowledge of their grandparents and that it was becoming less and less common to see youngsters approach their grandparents to ask about their ancient cultural practices.

During the second fieldwork phase, we carried out thirty interviews with high school students through a random selection process. We wanted to speak to them because we had hoped that the youth would be interested in what their grandparents knew and what they might learn from them, even if they gave more importance to that which 'modernity' offers them. What we discovered as a result of these interviews was something that we already suspected: these youths stated that they were more interested in technology than in talking to their elders, and they would rather go to university than work in the countryside in the manner of their grandparents.

In interviews with ten elder women of the community it was found to be important to them that their ancestral wisdom should be included in local educational systems in order to highlight at school the importance of this kind of knowledge for their local culture. However, at school, at least at the official school, this wisdom was seen to be completely marginalized and erased. The school seemed to offer knowledge to give the children better opportunities in the future, and knowledge from outside the school, such as that which we were trying to document, was not seen as being relevant nor legitimate enough to make up part of the curriculum. During these initial considerations, we realised that part of the problem in the transmission from generation to generation of ancestral wisdom was in fact due to the relationship between the community and the school. The problem that this stage of the research identified was that the elders wanted to pass on their knowledge, that this knowledge still exists, but what was missing was a means by which the youth could reconnect with that knowledge, by which it could be given the value and importance that it has for their identity, for their culture, for an understanding and conservation of their land and for the relevance of the traditional ecological knowledge for local sustainability.

4 Using Video as a Means to Collaborate with Indigenous Youth

In a meeting with indigenous fisherwomen from Zaragoza it was mentioned that videos are an excellent way to connect with youngsters, as they love to watch films, documentaries and television. It was suggested that they could use video to make young people aware of, and teach them about, their local culture, what their elders know, and what is happening in their community. In this situation, my specific suggestion was that we could carry out a collaborative audiovisual production in order to reach out to the youth. However, they were not best pleased with this idea. The fisherwomen said that they did not feel capable of operating a video camera and making a film. Whilst they said they could help by giving testimony and showing us what they knew about their community's culture, they were doubtful that they would be able to take part directly in the design and production of the documentary. They suggested making a video together with the youngsters to allow them to see and experience directly what the fisherwomen do and know: "We can support them, they can come to our houses or see us fish. In that way, the youth will be involved, it could be more useful for them." Making the most of the support from the fisherwomen, we decided to ask for volunteers among the young people in the community to take part in a documentary-making project. The way in which we intended to create dialogue about knowledge and between generations in this project was through the use of participatory documentary.

As a result, a workshop was held with the title 'Traditional knowledge and participatory video' which took place over a total of 40 h, combining practical and theory sessions, and 10 undergraduate youths from the community took part. The topics to be covered in the workshop were decided collectively with the young participants. They worked on aspects of planning and production, technical aspects of video-camera operation, and cinematographic language. Themes of social and political relevance of the use participatory video and the importance of documenting local knowledge regarding land use in accord with the local world-view were also incorporated.

In this framework, the youths taking part interviewed several members of the community before collectively analysing the interviews. Three possible themes to study were identified in this analysis: (1) the views of the youth regarding traditional fishing today, that is, to show how young people are no longer interested in this traditional activity because their personal development goals involve studying to take up a profession; (2) the transformation of traditional fishing methods over time, that is, to show how fishing was carried out before and how it is done today; and (3) the relationship between fishing and crocodiles and local oral tradition. Once the topics to be covered in the documentary had been established, the script was planned and the storyboard structured to plan in detail each of the sequences and how to shoot them during the filming stage.

During filming, material was shot to correspond to each planned sequence. Interviews with fisherwomen and students from the community were filmed. Locations around the town of Zaragoza were also filmed, as well as preparations for fishing, and on the last day of filming a group fishing activity by women from the town was recorded. Once filming was over, the material obtained was reviewed and the contents analysed with regard to the topics addressed and the technical quality of the video and sound recordings. Following this, in the post-production stage a documentary called 'The heirs of *Achaneh*'[1] was created.

When the documentary was completed, the dissemination stage started. In this stage, the documentary was screened in universities and in the high school of Zaragoza. The methodology adopted in the dissemination of the documentary consisted of presenting the production team to the audience, and explaining the objectives of disseminating the documentary and the reasons that motivated us to create it. After this presentation, the documentary would be shown. Following the screening, the audience had the chance to ask questions about the film. Most questions sought to go more deeply into, or to clear up, elements of the local oral tradition, the production of the video, or the traditional fishing methods of the women from Zaragoza.

At the end of the question and answer session we would thank those who gave us a platform to screen the film and those who gave their time and attention to come and see it and to ask questions. At every showing a copy of the film was donated to the educational institution where it was shown, in order to distribute the material and leave a copy for their video libraries, to be used by staff and students.

5 Results of the Experience

In these sections, we will present the ground of the analysis developed, the results of the production and dissemination of the documentary and the reflections on the educational potential of participatory documentary-making as an educational tool.

The analysis carried out is based on the proposal of Smith (1999) for educational research projects with indigenous peoples. She proposes several guiding principles which allow us to imagine indigenous aspirations and the importance of the indigenous knowledge in the field of education. We took three of those principles, which help us to analyse the results of the educational process for sustainable development that unfolded during the experience narrated in this article. We would

[1]https://vimeo.com/126888373

particularly like to underline with these three principles how the perceptions of the young film-makers concerning their community, local environment, identity, and ability to organise themselves to face community challenges were transformed.

The first of the principles proposed by Smith that we take up for the analysis is 'intervening'. This implies that research is understood as a proactive process concerning change. In this sense, the community itself invites the project in and sets out its parameters. The researchers involved in local projects must be ready to adapt, redirect strategies, design appropriate actions, and be able to work with diverse groups in different ways. Intervening intends to change institutions that manage projects with indigenous communities and not to change the indigenous communities so that they fit the institutional structures (Smith 1999, p. 147). We also observed an impact on the way the youngsters work and *intervene* in other contexts, for example, with the group of youths who give extra classes to children at primary school in the community. In this context, it was observed that the sensibility developed by the young people who participated in the project does not just influence them, rather they begin to pass it on to the children they work with as they are teaching them to take care of the environment even if this topic is not part of the objectives of these extra classes challenging the school as institution.

The second considered principle is 'networking'. This principle is a process of building relationships and disseminate knowledge and information (Smith 1999, p. 157). This is a way to establish contact between marginalised communities. By definition, their marginalisation excludes them from participating in the activities of the hegemonic, non-indigenous culture that controls the ways and means of communication. Networking has also become an effective way to stimulate the co-creation of socioecological knowledge and to recognize its problems within communities. The strengthening of local networks helped to build knowledge based on relationships and connections. The documentary and their participation in its creation allowed the young film-makers to widen their perspective regarding their community and its problems, and to make them sensitive to what they see in it. This is because, whilst they recognise activities that they do with their families, they realise that this is not enough and that improving the living conditions of their community requires action at the community level. Furthermore, this type of comment about their role in the community allowed the youths to see that it is not only important to propagate the knowledge of their community, but that it is important for them to get involved in solving the problems that face their community such as the loss of local traditions and local language, as well as the pollution of the local freshwater bodies.

The third principle is 'creating'. This refers to transcending the basic means of survival through the use of resources or abilities that the whole indigenous community has maintained since before colonisation: the ability to create and to be creative. Indigenous communities have a lot to offer the non-indigenous world, which is not precisely ancient or from pre-colonial times. 'Creating' is about how collective indigenous creativity produces solutions to indigenous and non-indigenous problems. Smith mentions that this creativity has been a characteristic of indigenous peoples before, during and after colonialism: "Throughout the period of colonization indigenous peoples survived because of their imaginative spirit, their ability to adapt and to think around a problem" (Smith 1999, p. 158). The documentary production process that we analyse it is creative, and there is a creative potential in intergenerational collaboration (fisherwomen and youths). The youths' statements demonstrated that they valued the opportunity to learn directly from the fisherwomen, and that they also value the knowledge they discovered with them. Similarly, it was seen that participation in fishing develops sensitivity and connection with this activity, especially amongst young women since there is a strong gender element involved, which is passed on from mothers to daughters. New knowledge and practices of the indigenous peoples can be applied not only within indigenous contexts, but also offer clues that may help to build sustainable societies elsewhere.

6 Learning Through the Documentary Production: Voices from Film-Makers and Fisherwomen

Based on the analysis we carried out we were able to identify, from the youngsters' own declarations, learning outcomes that were developed through the experience of creating the documentary. It is worth mentioning that much of the traditional lore of indigenous communities is passed on by word of mouth, nevertheless, it is also passed on during communal work such as, in this case, fishing, which the youths took part in during filming:

> One important thing I learned during the process was the values: living together, working together, the human feeling that is employed. What I liked about the project was the collaboration, the feeling of participation that everybody has, the way we coordinated to make the project happen (Conrado, film-maker).

Other declarations help us identify a re-evaluation of the traditional teachings and practices of the community. In the same way, the teachings that local legends have for the community and especially their messages as community regulators of local values, were recognised:

> For example, I had heard that people went fishing, but my parents never took me. My dad used to go fishing, but years ago, but, well I never went, and so I didn't know how to do it. [...]. Since I saw the video my opinion has changed because it's a very beautiful practice that makes up, or made up, part of our parents' and grandparents' diets. Zaragoza used to be a very poor town, it's still very poor, but they had to find a way to feed themselves. And so in this way I can value my grandparents more because they made an effort to bring up their children and, well I really identify with that because it's part of my town's past and I identify with Zaragoza. And in the video, to see how the women organise themselves, how they enjoy doing it, well that makes me identify with Zaragoza even more. [...] And well, also what the legends tell us, that you have to be faithful to your wife, and you think that you do have to be faithful because if not you won't be able to catch any fish, and that is why it's something that regulates; and you can believe it or not but just in case you don't do it, and if I did it, well I'd better not go now [laughter] (Irene, film-maker).

The importance of the video as a point of reference for the transmission of oral history is also reflected. The video brings a further representative element that allows the transmission of knowledge not to depend solely on the oral tradition, rather it can also be based on audiovisual material to pass on ancestral wisdom:

> What sticks in my mind is not just what I have been told but also what you see in the video, which is what I think I will tell my son; those are the versions that I have been told and that I have seen myself. To be honest I don't believe in the story but I do believe that it exists as a legend (Mariela, film-maker).

An outside view, also from a lecturer at the university where the youths study, who has worked with them for longer, allowed me to observe other aspects related to the lessons that the young participants learned by collaborating on this project:

> I can see they are more sure of themselves. They have become more critical when faced with social problems. And I like that, I see them. Because now they organise themselves, they say to me, "Hey, we have to arrange some activities so that the children aren't left with nothing to do in the holidays," "How can we teach them but without charging them?" and, "Let's see how we can deliver the course." They are innovating, based on the idea that in a group, communally it's better. I think that […] they are acquiring a kind of mentality and sensitivity, and I think they are involved because they have the idea that they want to do more things, they are not indifferent. I have told them that every group, at the end of the day, is there to do something for, in this case for the environment, for education, even for party politics (Antonio, university lecturer).

Another area of interest in the analysis of this work was a discussion with the participants, both the documentary makers and the fisherwomen, about the relevance of the use of video as a medium and as an educational process. We would like to make it clear that this tool is analysed based on criteria that the participants themselves came up with following different stages of the process. As points of reference they took the planning of the documentary, the distribution process and the film itself as an educational resource.

Through this analysis we picked up several points of view regarding this discussion. On the one hand, the use of video as a tool was related to the propagation of community values and knowledge:

> I think the video is useful because it teaches us about values. Because, for example, now the young people can understand how things were organised in the past, know what the town was like in the past. People say us young people should know our community. There are a lot of people who don't know, people who have no idea what Zaragoza was like in the past. I think it is important to know what it was like in the past, because it is important to preserve its essence, because that essence transmits values such as help and respect. Because in the past people respected each other, but not anymore. Organisation was important in the past, but not now, back then people organised themselves and I think that's something worth recovering. That's what the video's for (Lucia, film-maker).

The video was seen as a complementary tool to support the oral tradition, a tool in which cultural processes can be observed which otherwise would only be imagined:

> The way I see it it's an excellent tool that goes beyond story-telling, because you see the people working, organising themselves, what their ideas are, or what they do while they are fishing, as in this case (Irene, film-maker).

It is also seen as an educational resource which can be used to connect with a public that prefers audiovisual media to the written word:

> A video is more fun for a lot of people, we don't read books, rather we go to the cinema or watch television. I think young people are more interested in a video than in reading, and trying to record an idea in video was very fruitful, it gave results because many people have seen it and been interested, but if we were to sell books about our topics, it would be more difficult because the public reads less (Gustavo, film-maker).

In the declarations we also discovered an interesting debate concerning the relationship between technology and culture. Here we analysed the way in which technology has typically displaced traditional wisdom, and in this instance, we took advantage of technological progress to re-evaluate culture and strengthen identity. Thus we forged a new view of the role of technology in supporting cultures that have historically been erased:

> I think we created a useful link because we tried to unite technology with cultural expressions. In many cases, technology pushes cultural considerations to one side, because of stereotypes surrounding culture. But in this instance we are using it in order to preserve culture. Technology shouldn't come and push out the culture from a place, rather we should use it to conserve our culture (Conrado, film-maker).

The youngsters who took part realised that the audiovisual tool can be used as an educational resource:

> From the educational perspective, I think there is a lot to be done with video; […] video helps, for example, as an educational resource, because knowledge is being lost, it isn't being transformed, it is being lost (Alberto, film-maker).

Both the film-makers and the fisherwomen picked up on the idea that documentary making is a tool that can help strengthen identity, since it can be used to show aspects that it would otherwise be difficult to show:

> I think the video helps a lot to show our identity, especially to young people, because they ought to know our traditional ways, our traditions, how things were in the past and how they are now, that would help them a lot because that gives us our identity (Felícita, fisherwoman).

Another interesting aspect that the youths identified regarding the importance of documentary making and their experience in the project relates to using video as a means to raise awareness in outsiders, such as teachers, with regard to aspects of the community's culture. They see that with a resource that they can show to teachers who come to the community, these teachers could be more sensitive to the knowledge and community practices that are important to local people and that are often looked down on by teachers who come from outside the indigenous community:

> A video can also raise awareness amongst teachers who are in an indigenous community and show them that in these places, there is also knowledge, and that this knowledge, whilst it is empirical, in the end agrees with science. It doesn't have a scientific base, but it has a base with a moral lesson. A video can help us greatly in training people who see the need to care for the environment and preserve their traditions and customs, I think that in this way it is very useful (Pedro, film-maker).

7 Conclusions

From an academic standpoint it is interesting to analyse the participatory process in every aspect in order to identify the successes and failings in this kind of approach to a collective building and generation of knowledge and audiovisual resources. From the political and group collaboration standpoint, it is important to analyse participation, because the way in which this is configured can show us the moments and the ways in which decisions are taken. Similarly, it is important to analyse the way in which a unilateral or hierarchical decision can affect both the success of the process, and the final product.

During the dissemination stage it was interesting to discuss the film with those involved in its screening. On the one hand, discussions with university students allowed us to debate contrasting realities in neighbouring but little-known communities. During screenings for students in Zaragoza, discussions between the audience and film-makers were not as relevant as they were with university students, rather on these occasions the screenings were marked by surprise and admiration. It was a special moment for the documentary presenters to see the curious looks on the faces of the spectators during the screening and the validation of the documentary during the question and answer sessions between the audience and the documentary makers.

As for the individual and group effects of taking part in the project, the positive effects on the young film-makers of their participation in this project should be highlighted. Of all the effects one of the most important was the political profile that the youngsters managed to develop as a result of making the film. Similarly, the knowledge generated during the collaborative production of the documentary enabled some debate concerning traditional oral transmission within the community, its relevance regarding local sustainability and how this contrasts with new ways of showing traditional knowledge and recording local memory through video. This research process allows us to visualize how this audiovisual production constituted a methodological strategy for the strengthening of indigenous peoples' own cultural identities. Besides, the involvement in this process also helped to develop and strengthen organisational and decision-making capabilities amongst the youngsters. In future educational research projects, it would be relevant to explore the use of audiovisual production such as visual and transmedia storytelling, photovoice, shared ethnographic filmmaking, etc., as methodological strategies for the achievement of the Sustainable Development Goals in the post-2015 agenda.

References

Ardèvol E (1994) La mirada antropológica o la antropología de la mirada; el uso de la cámara de video en la investigación etnográfica. Dissertation, Universidad Autónoma de Barcelona

Cundill G, Rodela R (2012) A search for coherence: social learning in natural resource management. In: Lotz-Sisitka H (ed), (Re) views on social learning literature. Grahamstown/ Howick: E.L.R.C., Rhodes University, pp 12–26

De Zutter P (1997) Historias, saberes y gentes: de la experiencia al conocimiento. Escuela para el desarrollo-Horizonte, Lima

Gasché J (2008) La motivación política de la educación intercultural indígena y sus exigencias pedagógicas. ¿Hasta dónde abarca la interculturalidad? In: Educando en la diversidad. Investigaciones y experiencias educativas interculturales y bilingües. Abya-Yala, Quito

Grenier L (1998) Working with indigenous knowledge: a guide for researchers. IDRC, Ottawa

Kincheloe JL, Steinberg SR (2008) Indigenous knowledges in education: Complexities, dangers, and profound benefits. In: Denzin NK, Lincoln YS, Smith LT (eds) The handbook of critical and indigenous methodologies. Thousand Oaks, SAGE, pp 135–156

Lunch N, Lunch C (2006) Una mirada al video participativo. Manual para actividades de campo. InsightShare, Oxford

Mosangini G (2010) Documentales para la transformación. Guía para la elaboración de documentales sociales participativos. ACSUR, Las Segovias

O'Donoghue R et al (2013) African heritage knowledge in the context of social innovation. Learning contributions of the regional centres of expertise on education for sustainable development. United Nations University—Institute of Advanced Studies, Yokohama

Rappaport J, Ramos A (2005) Una historia colaborativa: reto para el diálogo indígena-académico. Historia Crítica 29:39–62

Shava S (2013) The representation of indigenous knowledge. In: Stevenson RB, Brody M, Dillon J, Wals A (eds) International handbook of research on environmental education. Routledge, New York. pp 384–393

Smith LT (1999) Decolonizing methodologies. Research and indigenous peoples. Zed Books, New York

Author Biographies

Juan Carlos A. Sandoval Rivera is lecturer of the Institute of Educational Research at Universidad Veracruzana (Mexico) in the research unit on Environmental Education. In his research on intercultural and environmental education he has been working participatory video with indigenous youth, visual anthropology and social learning in indigenous knowledge systems.

Guadalupe Mendoza Zuany is lecturer of the Institute of Educational Research at Universidad Veracruzana (Mexico) in the research unit on Intercultural Studies. In her research on intercultural education she has been working with basic education in indigenous contexts, educational indigenous policies, intercultural curriculum and indigenous students in higher education.

Rob O'Donoghue is an Emeritus Professor of the Environmental Learning Research Centre at Rhodes University (South Africa). In his research on environmental education processes of learning-led change, he has given close attention to indigenous knowledge practices, social theory and environmental learning in post-colonial curriculum and community contexts.

Evaluation of Climate Change Impact as a Tool for Sustainable Development: A Case Study of a Mexican Basin

Rodrigo Dávila-Ortiz and Juan Alberto Velázquez-Zapata

Abstract Nowadays, the conservation of water resources is one of the main challenges of sustainable development in Mexico. In the future, the expected climate change would increase the water supply issues in the country. This work shows the evaluation of the climate change impact on water resources as a tool for sustainable development at local scale. This research topic is important as the assessment of the expected changes in temperature, precipitation and streamflow provides valuable information for society and decision makers in order to be prepared for this challenge, especially in such regions with water scarcity conditions. The case study is the Valles River Basin, which is located in the Mexican Sierra Madre Oriental mountain chain. The Valles River Basin is a region that is already affected by water shortage and by the water stress caused by the inadequate management of its water resources. The methodology of this study includes the use of climate simulations to feed a hydrological model in order to estimate future streamflow. In this research, climate simulations were obtained from the Canadian Regional Model under two representative concentration pathways (RCP4.5 and RCP8.5) over a reference (1971–2000) and future (2041–2070) periods. The results indicate an important decrease on future streamflow, especially during the wet season. All in all, the estimation of the future conditions of the basin's water resources would help local water managers in order to prepare adaptation strategies.

Keywords Climate change impacts · Hydrology · Sustainable development

R. Dávila-Ortiz (✉)
El Colegio de San Luis, Parque de Macul no. 155, C.P. 78294 San Luis Potosí, Mexico
e-mail: rodrigo.davort@hotmail.com

J. A. Velázquez-Zapata
Consejo Nacional de Ciencia y Tecnología - El Colegio de San Luis,
Parque de Macul no. 155, C.P. 78294 San Luis Potosí, Mexico
e-mail: javelazquezza@conacyt.mx

W. Leal Filho et al. (eds.), *Sustainable Development Research and Practice in Mexico and Selected Latin American Countries*, World Sustainability Series,
https://doi.org/10.1007/978-3-319-70560-6_21

1 Introduction

Climate change is one of the most studied topics in recent decades; hence, international organizations, governments and research groups have joined efforts to mitigate its effects. Climate, which is the average weather condition over long periods, has been modified by the high concentrations of greenhouse gases (GHG) in the atmosphere as a result of industrial activities. In consequence, the global temperature has increased and the precipitation patterns have been modified; in fact, such modifications are larger than the climate's natural variability. For this reason, the United Nations' 2030 Agenda for Sustainable Development considers climate change as one of the main threats for global sustainability in the 21st century (United Nations 2015).

The Intergovernmental Panel on Climate Change (IPCC) brought together experts on climatology, geosciences, economy and social sciences in order to study the potential impacts of climate change. According to the IPCC Fifth Assessment Report (IPCC 2014) it is expected with a high level of confidence, that the hydrologic regime of rivers will be modified as a result of changes in the atmospheric water content and precipitation, the increase of the temperature and evapotranspiration and the cryosphere reduction (Bates et al. 2008). Similarly, it is estimated that such modifications in the rainfall runoff process will have negative impacts in the future water availability; in addition to that, there will be possible changes on the frequency of hydro-meteorological extremes in the future.

Arreguín et al. (2015) claim that sustainability should be the basic principle of the Mexican Government Water Policy for ensuring water supply for future generations. In that aspect, the sustainability of Mexican water resources in the country is threatened by climate change (RMCC 2015).

Vulnerability has a key role in the expected climate change impact on future water supply. In particular, the IPCC (2014) considers vulnerability as the political, economic and social conditions that predispose people to be adversely affected by climate change. Vulnerability is a complex concept which is difficult to estimate, for it requires the assessment of interactions between the climate change impacts on water resources and the social actors (Conde and Saldaña 2007).

Several studies have evaluated the vulnerability in a climate change context. For instance, Rivas (2015) performed a spatial analysis of the hydrological vulnerability to climate change and proposed a vulnerability index that takes into account economical and meteorological variables. Results show that those Mexican regions which are currently under hydric stress conditions, such as the northern dry regions and the largest cities, would be the most affected regions by climate change. Therefore, water supply in a climate change context would be a challenge since the Mexican population is expected to grow from 112 million in 2012 to 179 million in 2100 (INEGI 2017; CEPAL 2015).

Some studies have focused on climate change impact on the hydrological regime of large Mexican rivers; for example, Rivas et al. (2010) evaluated the expected changes on three large basins (Lerma-Chapala, Conchos and Grijalva); moreover, Velázquez et al. (2015) and Molina et al. (2016) assessed climate change impact on the water resources of the Tampaón River basin and the Guadalupe River basin respectively.

The main scope of this study is the assessment of the climate change impact on the hydrology of a small catchment located in central Mexico by using one climate model under two representative concentration pathways which represent different assumptions of the future society. The evaluation of climate change impact at regional level is necessary, for it provides valuable information to local people and decision makers about the possible future changes that would affect their sources of water. Therefore, the information generated in this study can contribute to sustainable management of water resources in the Valles River Basin.

The manuscript is organized as follows: first, the study area is presented, including the context of the water use in the study basin; second, a description of the methods and data is given; third, relevant results of the evaluation of the climate change impacts over the hydrological regime are presented, and, finally, concluding remarks close the manuscript.

2 Study Area

The Valles River Basin is located in the Huasteca region, mainly in the Mexican state of San Luis Potosí (Fig. 1). The study basin covers an area of 3521 km^2 (IMTA 2016), and it is a sub-catchment of the Pánuco River Basin (hydrologic zone 26; SIATL 2016) which discharges into the Gulf of Mexico.

The Valles River Basin is situated in the Mexican Sierra Madre Oriental mountain chain (INEGI 2001), thus presenting a rugged relief with an elevation that ranges from 69 to 1918 m. The complex topography is a natural barrier to the wet sea winds from the Gulf of Mexico; for this reason, the region has an important precipitation. The study basin has three main climatic zones (Fig. 2) according to the climate Köppen–Giger classification (Peel et al. 2007): Cwa (Temperate, dry in winter and hot summer), Am (Tropical Monsoon) and Aw (Tropical Savannah).

The study basin is formed mainly by sedimentary rocks of marine origin (SEDUE 1986). In addition, there are geological faults in the Huasteca region, so the basin presents underground water exchange (Santacruz and Ramos 2010).

Figure 3 shows the observed mean monthly precipitation, temperature and streamflow in the study catchment for the 1971–2000 period. The mean temperature ranges from 15 °C in January to 25 °C in June.

The rainy season presents itself between May and October, so July is the month with the most important mean precipitation (280 mm per month); in the dry season (from November to April), the monthly mean precipitation is below 50 mm per month. The streamflow ranges between 6 and 92 m^3/s, being July and September the months with the highest discharges. Also, the midsummer drought, period in which the temperature increases and precipitation decreases, occurs in August.

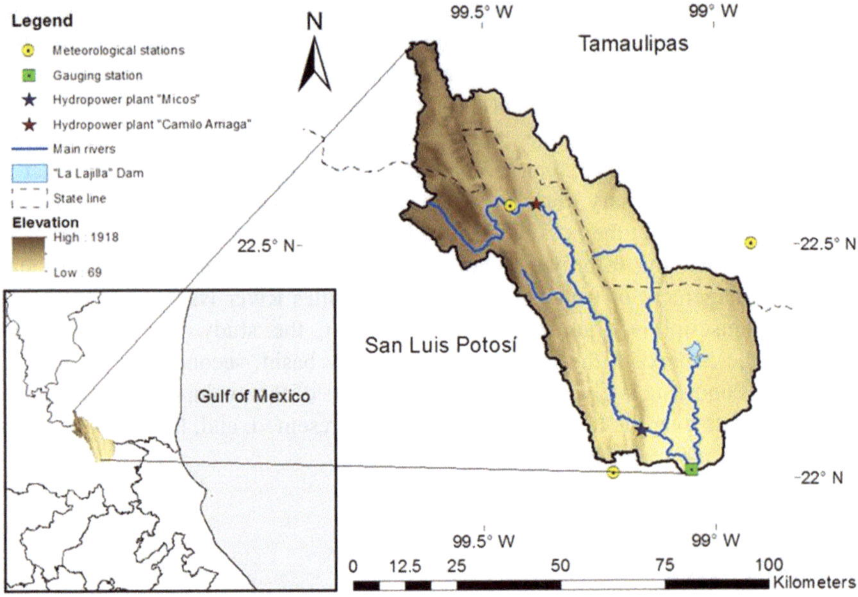

Fig. 1 Location of study catchment

Fig. 2 Köppen–Geiger climate classification in the Valles River Basin. Based on the work of Peel et al. (2007)

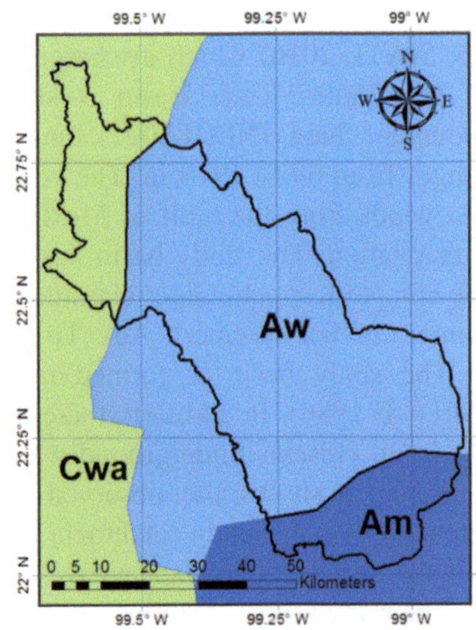

Regarding water use in the basin, Santacruz (2012) claims that water consumption is distributed as follows: agriculture uses 89.2%, the industrial activities, mainly the cane sugar refining, use 7.5%, and the rest is reserved for water supply.

The current context of the water use has caused shortage in the Valles River Basin, affecting the most vulnerable people. In that aspect, Jabardo and Padilla

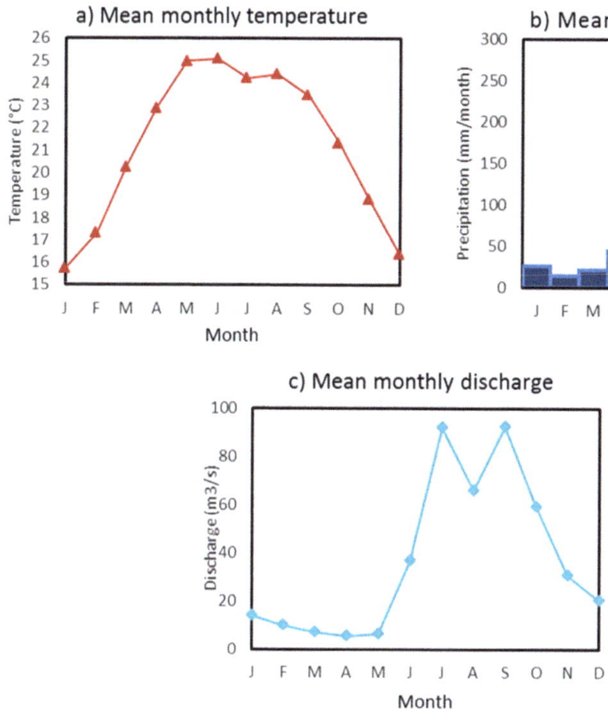

Fig. 3 Mean monthly observed temperature (**a**), precipitation (**b**) and discharge (**c**) for the Valles River Basin over the 1971–2000 period

(2016) argue that an inadequate management of water resources in the Huasteca region has caused severe scarcity conditions. Water stress is also conditioned by climate variability; for example, Algara (2009) identified drought conditions in recent years in the region.

3 Methods

In this research, the evaluation of the climate change impact on the hydrological regime followed several steps: firstly, climate simulations (daily precipitation and temperature time series) were obtained from a climate model; then, they were statistically bias corrected. Secondly, a hydrological model was calibrated and validated with observed data for a historical period. Thirdly, the bias-corrected climate simulations in reference (1971–2000) and future period (2041–2070) were used to feed the hydrological model. Finally, monthly streamflow averages were computed and compared in order to evaluate the climate change impact on the water resources of the study basin (e.g. Velázquez et al. 2015).

The daily time series of precipitation (*P*) and minimum and maximum temperature (*Tmin* and *Tmax*) were taken from the three meteorological stations depicted in Fig. 1. The daily discharge data come from the gauging station Santa Rosa. The databases used in this research are CLICOM (CLICOM 2016) and BANDAS (IMTA 2016), for meteorological and discharge data respectively, over the 1971–2000 period.

The climate simulations were taken from the Canadian Regional Model (CanRCM4, Scinocca et al. 2016). Moreover, two representative concentration pathways (RCPs) were considered for this study, namely RCP 4.5 and RCP 8.5, which are two possible radiative forcing values in future (2041–2070) period. Therefore, there are two future scenarios which take into account different economic, demographic and climatic conditions (Moss et al. 2010; Van Vuuren et al. 2011). Indeed, RCP 8.5 considers that future climate will face a higher concentration of GHG than RCP 4.5, thus, leading to a greater climate change impact.

The climate simulations are affected by systematic biases (i.e., differences between the simulations and observations) that preclude their direct use on hydrological models (Troin et al. 2015). In this study, the statistical bias correction method proposed by Lenderink et al. (2007) was applied to correct climate simulations in both periods. This bias correction method has been largely used in climate change impact studies (e.g. Levison et al. 2014; Troin et al. 2015).

The conceptual lumped hydrological model GR4J (Perrin et al. 2003) was selected to conduct this investigation. The model requires mean daily precipitation and potential evapotranspiration (ETP) as input data. GR4J was chosen for its ability to reproduce observed streamflow. For instance, the GR4J performs well in calibration and validation periods for the Valles River Basin in terms of the Nash-Sutcliffe (NS, Nash and Sutcliffe 1970) efficiency coefficient (Table 1). Most importantly, the model has been used in several climate change impact studies on water resources (e.g., Seiller and Anctil 2014; Velázquez et al. 2015). The hydrological model shows a good performance in the simulation of daily streamflow; nevertheless, Velázquez et al. (2015) report, for a Mexican catchment, that GR4J slightly underestimates the mean monthly streamflow in the rainy season.

Table 1 The Nash-Sutcliffe (NS) coefficient obtained in calibration and validation periods for the Valles River Basin. A Value of NS = 1 corresponds to a perfect match between observations and simulations

Calibration		Validation	
Period	NS	Period	NS
1989–1995	0.750	1973–1979	0.779

4 Results and Discussion

The climate change signal (CCS) is the change in the value of a given variable between the future period (2041–2070) and the reference period (1971–2000). Figure 4 shows the CCS on monthly temperature and precipitation (computed with the bias-corrected climate simulations) for the two RCPs. From this figure, it can be seen that, for RCP4.5, the change in temperature ranges between 1.6 and 1.8 °C. Moreover, the estimated change in precipitation for the rainy months (June to September) is mixed with values that range from −10 to +10%; however, the dry months show an increment in precipitation (e.g. +50% in March). Regarding RCP8.5, Fig. 4 shows an increment of temperature that ranges from 1.9° to 2.4°. It also could be noted that the warmest months in the historical period (April and May) estimate the largest increment in temperature. The CCS in precipitation is mainly negative, with values that range between −12 and −30% in the rainy season.

RCP8.5 considers a higher GHG concentration than RCP4.5; therefore, the former RCP estimates a larger increment of temperature in the future period for the study basin. Also, RCP8.5 evaluates a large negative CCS in precipitation, whereas RCP4.5 estimates a smaller CCS with positive and negative values. The estimated increment in temperature and evapotranspiration, combined with changes in precipitation, would lead to changes in the hydrological regime of the River Valles Basin.

The bias-corrected climate simulations were used as input data for the hydrological model GR4J in reference and future periods. Figure 5 shows the mean monthly streamflow as simulated with GR4J. The figure depicts that, when considering RCP4.5, the hydrological model estimates similar values of discharge in

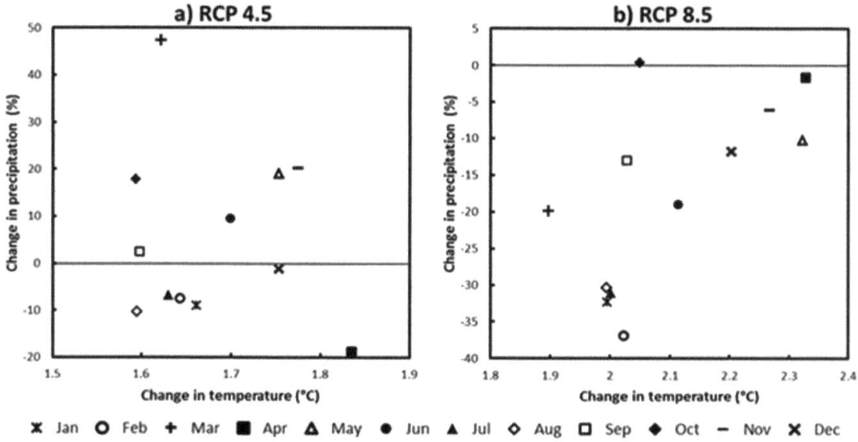

Fig. 4 Climate change signal for temperature and precipitation for the two considered RCPs

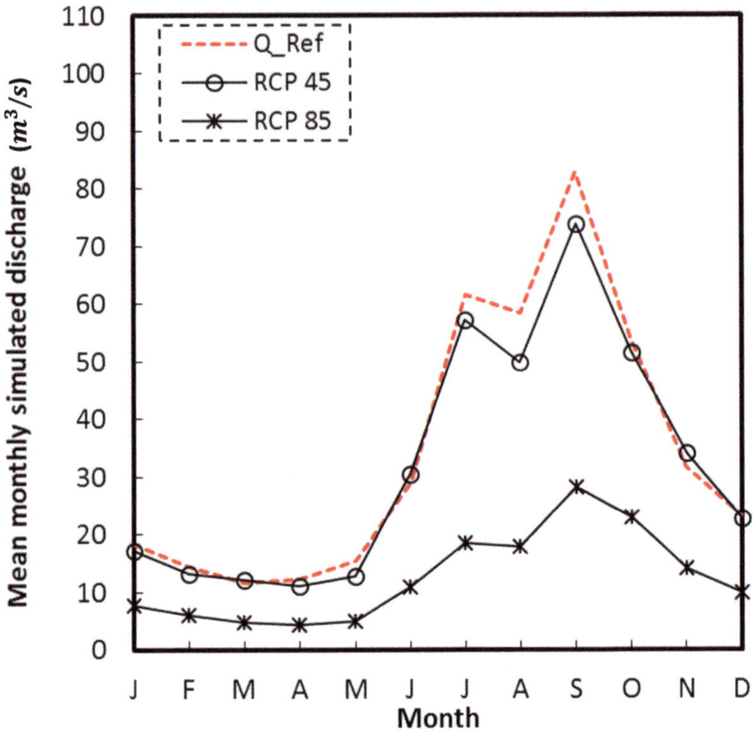

Fig. 5 Mean monthly discharges for the Valles River Basin obtained with climate simulations in the reference (Qref) and future periods (RCP45 and RCP85)

reference and future periods from October to June; whereas a decrement of streamflow is evaluated in the rainy season (from June to September) in future period; for instance, the mean monthly discharge in July decreases from 82 to 67 m^3/s. The results show that the decrease of streamflow in the rainy season is related to the decrease in precipitation estimated with this RCP.

Regarding the results obtained with RCP8.5, Fig. 5 shows an important decrement of the mean monthly discharge as a result of the general negative change in precipitation and the increment of temperature. For example, the discharge in September decreases from 82 to 28 m^3/s. The decrease is also important in the dry season. The results show that the changes in temperature and precipitation could modify the hydrological regime. It is important to mention that the evaluation of climate change impact is subject to several sources of uncertainty that arises from the selection of the climate model, the RCPs, the hydrological model, and the bias correction method. For this reason, the assessment of the climate change impact on water resources as a tool for sustainable development should be taken with caution.

5 Conclusions

The potential impact of climate change on water resources needs more attention from policy-makers, as the changes in precipitation and temperature could lead to future water scarcity issues. In that aspect, the evaluation of the climate change impact at local level is a tool that provides valuable information about the future local conditions to generate water policies, and therefore, to achieve sustainability in water resources.

This study aims to generate knowledge about the potential impacts of climate change on water resources at local scale; in particular, the Valles River Basin, a Mexican basin that is already affected by water shortage and by the water stress caused by the inadequate management of its water resources (Jabardo and Padilla 2016). Our study shows that, for RCP 8.5, the increase in temperature is estimated between 1.9 and 2.4 °C, and the change in precipitation is estimated between −12 and −30% for the period 2041–2070 compared with the period 1971–2000. As a consequence, the decrease of streamflow in the basin is estimated from 82 to 28 m^3/s in the rainy season. Therefore, the climate change impact would strongly increase the water issues in the basin. In that context, the adaptation to climate change can significantly reduce the adverse impacts, by planning water management and, thus, having the potential to reduce vulnerability (Smit and Pilifosova 2003).

This work is focused on the climate change impact on the hydrological regime of a small basin in central Mexico based on climate simulations from one climate model. Future work should include simulations from several models in order to take into account the variability in the estimation of future precipitation and temperature. Therefore, the use of several models and climate scenarios will give us a more complete perspective of the challenges that local society will face due to climate change. In addition, this work evaluates only the changes on the hydrological regime, so interdisciplinary work with social sciences specialists, decision makers and water users is needed to reach the sustainability of water resources in a climate change context.

Acknowledgements The authors thank the research groups that made the hydro-meteorological data and the climate simulations available. Rodrigo Dávila wishes to thank 'El Colegio de San Luis A.C.' for the financial support of his research internship.

References

Algara M (2009) Propuesta metodológica para medir el impacto del fenómeno de la sequía en la Huasteca Potosina y propuesta general de Manejo (Tesis para obtener el grado de Doctor en Ciencias Ambientales). Universidad Autónoma de San Luis Potosí, México

Arreguín FI, López M, Rodríguez O, Montero MJ (2015) Atlas de vulnerabilidad hídrica en México ante el cambio climático. Instituto Mexicano de Tecnología del Agua, Jiutepec, México

Bates BC, Kundzewicz ZW, Wu S, Palutikof JP (eds) (2008) Climate change and water. Technical Paper of the Intergovernmental Panel on Climate Change, IPCC Secretariat, Geneva, 210p
CEPAL (2015) Comisión Económica para América Latina y el Caribe. http://www.cepal.org/celade/proyecciones/basedatos_bd.htm?wptouch_preview_them=enabled. Accessed 12 de Enero de 2017
Conde C, Saldaña S (2007) Cambio climático en América Latina y el Caribe: impactos, vulnerabilidad y adaptación. Ambiente y desarrollo 23(2):23–30
Datos climáticos diarios del CLICOM (2016) Servicio Meteorológico Nacional a través del CICESE. http://clicom-mex.cicese.mx. Accessed 28 de Marzo de 2016
IMTA (2016) Banco Nacional de Datos de Aguas Superficiales (BANDAS). Instituto Mexicano de Tecnología del Agua. http://www.conagua.gob.mx/conagua07/contenido/documentos/portada%20bandas.htm. Accessed 28 de Marzo de 2016
INEGI (2001) Información cartográfica en formato vectorial escala 1:1, 000,000, México
INEGI (2017) Volumen y crecimiento. Población total por entidad federativa, 1895 a 2010 http://www3.inegi.org.mx/sistemas/sisept/Default.aspx?t=mdemo148&s=est&c=29192. Accessed 12 de Enero de 2017
IPCC (2014) Impacts, adaptation, and vulnerability. Part B: Regional aspects. Contribution of working group II to the fifth assessment report of the intergovernmental panel on climate change [Barros VR, Field CB, Dokken DJ, Mastrandrea MD, Mach KJ, Bilir TE, Chatterjee M, Ebi KL, Estrada YO, Genova RC, Girma B, Kissel ES, Levy AN, MacCracken S, Mastrandrea PR, White LL (eds)]. Cambridge University Press, Cambridge, United Kingdom and New York, NY, USA, 2014, 688p
Jabardo V, Padilla LS (2016) La escasez de agua en la Huasteca Potosina (México): conflictos locales. Revista Universitaria de Geografía 25(2):133–165
Lenderink G, Buishand A, Van Deursen W (2007) Estimates of future discharges of the river Rhine using two scenario methodologies: direct versus delta approach. Hydrol Earth Syst Sci 11(3):1145–1159
Levison J, Larocque M, Fournier V, Gagné S, Pellerin S, Ouellet MA (2014) Dynamics of a headwater system and peatland under current conditions and with climate change. Hydrol Process 28(17):4808–4822
Molina E, Hallack M, Martínez S, Ramírez J, Mungaray A, Sastre A (2016) Hydrological modeling and climate change impacts in an agricultural semiarid region. Case study: Guadalupe River basin, Mexico. Agric Water Manag 175:29–42
Moss RH, Edmonds JA, Hibbard KA, Manning MR, Rose SK, Van Vuuren DP, Meehl GA et al (2010) The next generation of scenarios for climate change research and assessment. Nature 463(7282):747–756
Nash JE, Sutcliffe JV (1970) River flow forecasting through conceptual models. Part I—A discussion of principles. J Hydrol 27(3):282–290
Peel MC, Finalayson BL, McMahon TA (2007) Updated world map of Köppen-Gieger climate classification. Hydrol Earth Syst Sci 11:1633–1644
Perrin C, Michel C, Andréassian V (2003) Improvement of a parsimonious model for streamflow simulation. J Hydrol 279(1):275–289
RMCC (2015) Reporte Mexicano de Cambio Climático, Grupo I Bases Científicas. Modelos y Modelación. México: Universidad Nacional Autónoma de México. Programa de Investigación en Cambio Climático
Rivas I, Güitrón A, Ballinas HA (2010) Vulnerabilidad hídrica global: aguas superficiales. In: Martínez PF, Patiño C (eds) Atlas de vulnerabilidad hídrica: efectos del cambio climático en los recursos hídricos de México. Jiutepec, México: Instituto Mexicano de Tecnología del Agua. pp 81–113
Rivas I (2015) Efectos del cambio climático en el recurso hídrico de México (agua superficial) In: Arreguín FI, López M, Rodríguez O, Montero MJ (eds) Atlas de vulnerabilidad hídrica en México ante el cambio climático. Jiutepec, México: Instituto Mexicano de Tecnología del Agua. pp 89–98

Santacruz G, Ramos JA (2010) Estudio del comportamiento hidrogeológico de la Reserva Sierra Abra Tanchipa y su zona de Influencia. Reporte Técnico GTZ-CONANP-COLSAN-IPICYT, San Luis Potosí, México

Santacruz G (2012) El espejismo de la gestión de los recursos hídricos superficiales por cuenca hidrográfica. El caso de río Valles. La Huasteca, México. San Luis Potosí: El Colegio de San Luis, A.C

Scinocca JF, Kharin VV, Jiao Y, Qian MW, Lazare M, Solheim L, Dugas B (2016) Coordinated global and regional climate modeling. J Clim 29(1):17–35

SEDUE (1986) Estudio de calidad y clasificación del Río Tampaón, San Luis Potosí y Querétaro, México. Secretaría de Desarrollo Urbano y Ecología, México

Seiller G, Anctil F (2014) Climate change impacts on the hydrologic regime of a Canadian river: comparing uncertainties arising from climate natural variability and lumped hydrological model structures. Hydrol Earth Syst Sci 18(6):2033

SIATL (2016) Simulador de Flujos de Agua de Cuencas Hidrográficas. http://antares.inegi.org.mx/analisis/red_hidro/SIATL/#. Accessed 04 Feb 2016

Smit B, Pilifosova O (2003) Adaptation to climate change in the context of sustainable development and equity. Sustain Dev 8(9):9

Troin M, Velázquez JA, Caya D, Brissette F (2015) Comparing statistical post-processing of regional and global climate scenarios for hydrological impacts assessment: a case study of two Canadian catchments. J Hydrol 520:268–288

United Nations (2015) Transforming Our World: The 2030 Agenda for Sustainable Development. Resolution adopted by the General Assembly on 25 Sept 2015. United Nations, New York

Van Vuuren DP, Edmonds J, Kainuma M, Riahi K, Thomson A, Hibbard K, Masui T et al (2011) The representative concentration pathways: an overview. Clim Change 109(1–2):5

Velázquez JA, Troin M, Caya D (2015) Hydrological modeling of the Tampaon River in the context of climate change. Tecnología Ciencia del Agua 6(5):17–30

Author Biographies

Rodrigo Dávila-Ortiz graduated in Environmental Engineering from the Autonomous University of San Luis Potosí. Currently, he is a research assistant at El Colegio de San Luis A. C. His research interests are environmental and climate change impact assessment.

Juan Alberto Velázquez-Zapata received a Civil Engineering degree and an Environmental Engineering Master's degree from the Autonomous University of San Luis Potosí, Mexico. He received his Ph.D. in Civil Engineering from Laval University at Québec City, Canada. Currently, he is a CONACYT (National Council of Science and Technology) researcher at El Colegio de San Luis. His research interests are related to hydro-meteorological risk and climate change impact on water resources.

From Asset to Liability: The Sustainability of Waterscape Transformations in the Santiago River

Lourdes Marcela López Mares, Filiberto Adrián Moreno Mata, Benjamín Fidel Alva Fuentes and Joel Hernández Martínez

Abstract In Mexico and many other parts of the world, it is common practice since the late 19th century to pave riverbeds for public health and urban growth purposes. However, evidence suggests that costs associated with this practice outweigh the expected benefits, hindering the long-term environmental sustainability of growing cities. This case study undertaken in San Luis Potosí, Mexico, presents some urban, social and environmental problems resulting from paving one of the city's river, the Rio Santiago, in the 1980s. We approached the Rio as a waterscape: a socially produced landscape where the presence and management of water is central. From this perspective, the paper's main purpose is to analyze the Rio's waterscape transformations within the context of the Metropolitan Region's accelerated urban growth, including its watershed exploitation. We argue that the transformation of the Rio reflects a historical change insofar as waterways are presently perceived as a source of illness and pollution as opposed to a source of life and wellbeing. This paper concludes that the River's transformations not only spatialized the conflictive relation society-nature but also the social inequities that yield from uneven urban growth.

Keywords Waterscape · Urban rivers · Urban planning · Urban sustainability

L. M. López Mares (✉) · F. A. Moreno Mata · B. F. Alva Fuentes · J. Hernández Martínez
Facultad del Habitat, Universidad Autonoma de San Luis Potosi,
Niño Artillero 150, CP 78290 San Luis Potosí, SLP, Mexico
e-mail: marcela.lopez.mares@gmail.com

F. A. Moreno Mata
e-mail: adrian.moreno@uaslp.mx

B. F. Alva Fuentes
e-mail: balvaf@gmail.com

J. Hernández Martínez
e-mail: joelh062@gmail.com

W. Leal Filho et al. (eds.), *Sustainable Development Research and Practice in Mexico and Selected Latin American Countries*, World Sustainability Series,
https://doi.org/10.1007/978-3-319-70560-6_22

1 Introduction

Paving riverbeds in the name of public health and sanitation became a common practice under the hygienist paradigm in many parts of the world as of the late 19th century. From this paradigm, the paving of riverbeds is considered an urban management strategy that allows municipalities to hide rivers' sanitary problems while improving urban mobility (González et al. 2010). Examples of this strategy abound in Mexico. For example, watershed draining throughout Mexico City began in 1608, yet it was not until the 1930s that the social perception of rivers changed, from being sources of fresh water to sources of risk and pollution. This social change was spatially reflected on the waterscape: in 40 years 83 km of rivers were "erased" (i.e. piped and paved) in the city (Musset 1996; González et al. 2010: 34). Similar, close to 88% of Guadalajara's rivers and water streams have been piped (El Informador 2010), not to mention comparable interventions in Aguascalientes (Arellano and El Cedazo water streams, SEMADESU 2011), San Luis Potosí (Santiago and Españita rivers), Morelia (Boulevard Garcia León) and Puebla (San Francisco river), among other Mexican cities.

Such river interventions as the ones mentioned before are done according to specific discourses (e.g. hygienist and modernization) and agendas benefiting some and affecting others unevenly. This research addresses these issues by analyzing some of the urban, social and environmental consequences of the Santiago River's urbanization in the 1980s. We approach the Rio as a waterscape, a socially produced landscape where the presence and management of water is central. From this perspective, the paper's main purpose is to analyze the Rio's waterscape transformations in the context of the Metropolitan Zone San Luis-Soledad's (ZMSLP) watershed exploitation and its accelerated urban growth. In sum, the paper presents some of the complex historical elements that tame society-nature relationships in order to question the sustainability of our current urban growth patterns.

We argue that the transformation of the Rio reflects the way in which our society currently deals with waterscapes by subordinating nature to urban and economic growth. Covering riverbeds with concrete transforms natural landscapes into domesticated urban spaces. As a corollary of this strategy, natural absorption along rivers is sealed and the way for private vehicle users paved. This reflects an urban investment trend, currently reproduced by planners and policy makers alike that addresses the needs of the better off and leaves behind vulnerable communities and low-income city dwellers.

To address some of the social, environmental and urban consequences of river transformations, the paper first explores the notion of waterscape and its relationship to broader urban phenomena. Under this framework, we then briefly analyze the water history in the ZMSLP in order to understand the Santiago River's transformations as a socio-economic process. Subsequently, we analyze some of the current social and urban consequences of those transformations. Finally, we conclude by highlighting the main lessons learned from the research and outlining future research prospects.

2 Waterscape Transformations

Paving and piping rivers is a common strategy in urban settings because it allows for the management of water flows, particularly in the case of intermittent rivers whose runoff is not constant along the year: "as our cities have grown, the risk to life and property from flooding has increased. Engineering strategies for managing runoff and preventing floods have centered on directing the water elsewhere as quickly as possible, while largely ignoring the potential value of that water" (Dallman et al. 1999: 3). These engineering strategies have nonetheless perverse consequences on the natural ecosystem, on the water cycle and on the urban context. Paving rivers seals the permeable soil which naturally allows the infiltration of water to both the shallow and deep aquifer. Paving a waterway also accelerates the flow of water into flood-prone areas, pollutes runoff water, destroys the riparian ecosystem and reduces biodiversity (Dallman et al. 1999). Finally, this type of strategy disregards the potential ecological and urban benefits of rivers as sources of clean water, landscape, identity and history artifacts, urban connectors and linear axis of public spaces and facilities (González et al. 2010; Perlo 2010).

Several cases around the world have shown the possibility of rehabilitating urban rivers in order to recuperate their social and natural functions. In Korea, the Cheonggyecheon River was reconverted into a waterway after being a major highway. The Tiete river in Brazil, Los Angeles River in the U.S, the Medellin River in Colombia and Santa Lucia River in Monterrey, México are examples of how a natural element can be integrated into the urban fabric for the public good (Hernández Vera 2017; Perlo 2010). In these cases, land use changes, public space, facility provision and mobility strategies were central to connect these rivers with the urban encroachment around them and to promote the production of humanized (walkable, legible, usable), democratic spaces. These interventions, however, are costly and require a strong political will to take place, along with powerful discourses to back interventions and the economic capacity to implement them.

What these cases show us is that urban rivers' transformations are tightly linked to specific social contexts, to political power and economic resources seeking to transform and manage nature. The complex relationship between these factors has been studied from different perspectives. From an urban fabric perspective, water urbanism approaches the physical structure of the city in relationship with hydraulic elements, in the words of Shannon (2009: 65): "Water urbanism is possible when urban planning and design, civil and sanitary engineering and landscape architecture are folded into one another as are concerns for mobility, health, recreation and scenery". This perspective, however, fails to integrate the role of power and social relations into the equation, approaching the production of water landscapes from a purely technical standpoint. Conversely, the concept of hydro-social territory looks precisely at the way in which power plays a central role in understanding who has access to water and how. This concept dates back to the 50s, and was used to understand irrigation systems as means of production and social control (Valencia and Héctor 2015). The concept, however, has been used to understand in more general terms the historical relation between societies, the use of water-related resources and the transformation of natural landscapes. Valencia and Héctor (2015) refer to this relation as the hydro-social cycle, defining it as "the complex and diachronical way in which human action (and its inherent power) reconfigures water's natural cycle and vice versa" (p. 102, translated by the authors). The hydro-social cycle has been used to study landscapes (visible characteristics of an urban or natural space) and territories (socially and politically defined areas).

In this research, we adhere to the notion of waterscape, neologism used in the Anglo-Saxon literature to approach both landscapes and hydro-social cycles. This notion sheds light on the complex socio-political processes that shape nature and, in an urban setting, the urban fabric. The intricate relationship of nature production and society allows us to unveil the power relations at play that facilitate not only specific spatial transformations but also the production of discourses, images and symbols (Swyngedouw 1999).

According to Harvey (1989), in urban studies, we need to recognize that "…the production of images and of discourses is an important facet of activity that has to be analyzed as part and parcel of the reproduction and transformation of any symbolic order. Aesthetic and cultural practices matter, and the conditions of their production deserve the closest attention" (Harvey 1989: 355). Along these lines, we ought to understand the intersection of discourse and space production, by paying special attention to "structural relations of power" and the way in which these produce discourses and the knowledge needed to control waterscape transformations (Mele 2000: 629; Foucault and Gordon 1980).

Through discourse, spatial transformations are framed as "normal and beneficial", the social externalities of transformations are justified and new place identities are constructed (Mele 2000: 632). In the case of the Santiago River we argue that practices and discourses around water exploitation, management and hygiene, fostered the transformation of the city's waterscape and benefited the population selectively: "The material conditions that comprise urban environments are controlled, manipulated and serve the interests of the elite at the expense of marginalized populations". (Mehta and Karpouzoglou 2015: 160). Socially produced natural environments do not scape these manipulations, conversely, they are class-struggle arenas that show how power distribution shapes environmental injustices, uneven access to resources and weak decision making power (Mehta and Karpouzoglou 2015; Swyngedouw 1999).

This research analyzes the way in which the relationship society-nature has changed over time. This relationship has been spatialized through waterscape transformations, some of them more visible (e.g. Santiago River) than others (e.g. wells) but all produce uneven geographies reflective of problems at a larger urban scale.

3 Methods

This research stems from an academic project called "A Motorway Called Rio": Liquid Urbanism in San Luis Potosí. The project started in 2015 with an international workshop in which students and professors from the Oslo University and the UASLP tested possible futures for the Santiago River and finally produced a wide range of landscape and public facility projects to tackle some of the problems found. Within this project a master's thesis, a public exhibition, newspaper articles, a discussion forum and a book in print were done. This paper develops part of the project's findings, focusing on how the relationships between society, environment and urban growth change across time.

The project is qualitatively oriented and data was obtained from secondary sources, public databases (i.e. census bureau INEGI), interviews with key informants from the municipal water company (INTERAPAS), neighbors and river users, a photographic survey of the river and neighboring communities, on-site observation and surveys applied to Boulevard users (n = 96; a detailed description of the survey can be seen in Hernández Vera 2017). Descriptive analysis was used to summarize quantitative results, content analyses to interpret documents and interview data and spatial GIS analyses to cross-reference maps. Observation and interview data were codified to understand how social problems are spatially distributed, how water flows through the city and how the natural landscape transforms as the city grows. The Santiago River, as a case study, serves as a springboard to understand city-wide urban phenomena.

Research limitations include the interview's reduced sample size (n = 15) and limited historical information on the river and its transformations. Further interview research with a representative sample of neighbors, users and public officials is needed to understand the River's history, how different actors perceive it and its role within the city.

4 Rio Santiago: From Asset to Liability

In this section, we will briefly review the history of the hydraulic systems in the city, in order to understand how the socio-political context played a role in the transformation of the River's waterscape. The section highlights the changing discourses that supported the material transformations of the waterscape, from the Spaniards to current days. Paraphrasing Dallman et al. (1999) this journey shows how the Santiago River changed from asset to liability. In other words, from source of life and reason for the city's foundation to a health and material risk that needed to be paved and managed.

5 Waterscape Transformations in San Luis Potosi

The history of the city of San Luis Potosi clearly reflects the tensions existing between natural resources and urban growth, as well as entrenched urban problems such as spatial inequality and segregation characteristic of many Latin American countries. Accelerated urban growth, expansion and marginalization challenge resource management including water as well as service provision: "…the use of water in urban areas of countries from the region [referring to Latin America], is more and more intensive and conflictive. Many cities exceed the capacity of their 'economic' water provision sources and have increasing economic, environmental and social costs to attain new sources" (Dourojeanni and Jouravlev 1999: 11).

A case in point is the ZMSLP, the 11th biggest urban conglomeration in Mexico with close to 1200 inhabitants in two municipalities: San Luis Potosi-Soledad de Graciano Sánchez (Moreno and López 2016). As many other Latin American cities, the ZMSLP has rapidly grown since 1960. San Luis Potosí alone multiplied its population by 3.5 between 1960 and 2000 (from 193,670 to 670,532 inhabitants) and Soledad by 14.3 (from 12,591 to 180,296 inhabitants) (Peña 2013) (Fig. 1).

In this context of accelerated growth, water access is at risk. In 1960, 60% of the ZMSLP's water supply was obtained from superficial sources (rivers and dams) and 40% from underground sources (shallow and deep aquifer). Nowadays, the percentages are reverted and more than 90% of the water for the Zone's consumption is extracted from underground sources (Peña 2013). According with the State's Water Authority, El Salado, the 2000 km^2 endorheic watershed that feeds the ZMSLP is over-exploited, with a deficit between recharge and extraction of 75.3 million cubic meters per year (CONAGUA 2015). The main collector of this watershed is the Santiago River, whose runoff floods into two lagoons and the "Tinaja" plains, 40 km east of the ZMSLP (Contreras and Galindo 2008).

In addition to overexploiting the watershed and as a result of an increasing demand derived from demographic explosion and industrial uses, the city's development patterns are promoting changes to the hydrological cycle, depleting the precious resource and polluting the shallow aquifer. Urban expansion for example, promotes land use changes that accelerate deforestation and consumption of the vegetal cover as well as pavement of infiltration areas. In fact, while the population multiplied by 4.12 in average, between 1960 and 2000, the ZMSLP's area multiplied by 8, from 1760 ha to 14,000 (INEGI 2000; Contreras and Galindo 2008). In addition to these problems, failures in the management and water distribution system contribute to the over exploitation of the deep aquifer. Studies have shown that the water consumed in the city is ancient which means it has been stored for more than a thousand years and is not being refilled (Contreras and Galindo 2008). However, according with a study conducted by COTAS and the UASLP, there are traces of modern water in proximity with the Santiago River which means that before it was paved, the river infiltrated the runoff naturally into the deep aquifer (Hernández Vera 2017).

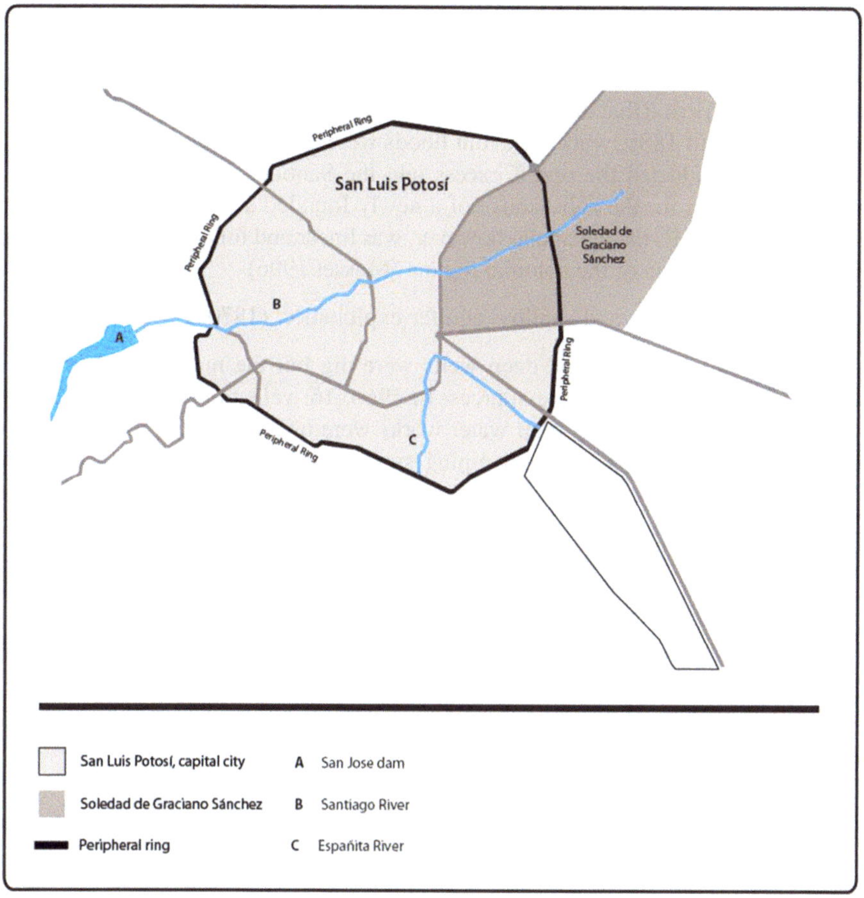

Fig. 1 Location of the Santiago River

In sum, access to water in the ZMSLP is tightly linked to the political economy of urban growth. In the same vein, the transformation of the Santiago River from source of water and natural drainage into a source of pollution and risk can only be understood in relation to economic, social and political changes, traced back to the city's early stages. These can be roughly divided in four key moments (Peña 2013; Padron Moncada 2014):

1. Foundation of the city, mining industry and city growth: superficial water consumption (1592–1870)

The Spaniards founded the city of San Luis Potosí in 1592, due to its superficial water sources, needed to secure their subsistence and to facilitate economic growth (Salazar 2010). Surface water was used to grind stone, wash the metals and, in some cases, as a source of energy (Urquiola 2004). It was nonetheless until the 19th century when several efforts to collect and manage runoff and surface sources of water were undertaken. In 1828, for example, an aqueduct, fountains and a water tank (*caja del agua*) were built to bring water into the city's downtown from the south east. It was also during this period that the San José dam was projected and started its construction (Padron 2014). In 1888, works to avoid floods were done including the "Corriente" channel that conducted the runoff excess into the Santiago River. In sum, surface water was central for the subsistence of a newly founded settlement and a nascent economic system, its management, however, was linear and followed an exploitation mindset characteristic of the colonial regime (Musset 1996).

2. Water management and shallow aquifer exploitation (1870–1960)

During this period the first deep wells were dig but the main source of water were shallow wells and surface sources. In 1960, 16 wells not deeper than 200 m were dig. Additionally, important water works were undertaken to provide the city with clean water such as the first piping system (1883) and a purification plant (1938). In parallel, the San José dam, in the Santiago River's basin was finished (1905) and El Peaje dam was built (1949–1950). Finally, important management changes also took place in this period: in 1894 the Water Anonymous Company was founded and 44 years later, in the midst of a political turmoil due the city's 1933 great flooding, this private company was acquired by the state government and then municipalized in 1959. Currently, water supply is managed by INTERAPAS, an inter-municipal organism that serves San Luis, Soledad and Cerro de San Pedro (Padron 2014). In sum, during this period, the water administration system was consolidated and water exploitation relied mainly on surface and shallow aquifer sources.

3. Deep aquifer exploitation, paving and piping superficial sources (1960–2010)

Since 1960 the city expanded towards the north of the Santiago River. This growth was facilitated by water access through well drilling. In fact, the number of wells expanded from 16 to 370 in this period (CONAGUA 2015).

In parallel, in 1962, a presidential decree vetoed the exploitation of the San Luis Potosí Valley's water. This decree limited water extraction to domestic and cattle-related uses and in 1985 an additional decree expanded the veto to cover infiltration areas (CONAGUA 2015). As a result, records show that the larger percentage (67.7%) of water consumption from wells in the ZMSLP is domestic but in fact, part of this water is employed for industrial uses (Contreras and Galindo 2008).

In addition to deep aquifer exploitation and control, in this period, both the municipal and state water companies built two additional dams, "El Potosino" (1985–1988) in the Santiago basin and "La Cañada del Lobo" (1986–1987) in the Españita River basin (CONAGUA 2015), paved two important waterways (Santiago River 1985 and Españita River 1995) and launched efforts to treat wastewater, building seven plants since 1998. Currently, the construction of El Morro plant is under way on the riverside of the Santiago River, in Soledad (Diaz de León and Courjaret 2006) (Fig. 2).

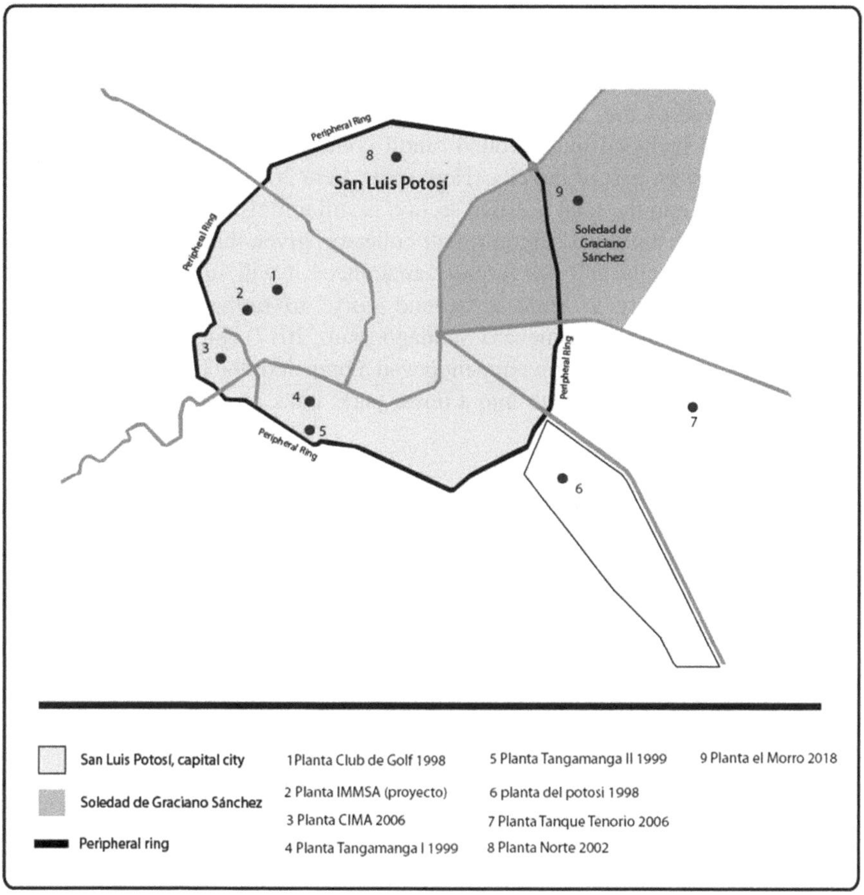

Fig. 2 Water treatment plants in the ZMSLP. Adapted from Diaz de León and Courjaret (2006)

After the construction of the three dams on the Santiago river (San José, Potosinos and el Peaje), the runoff significantly decreased, drying the Santiago River's bed most of the year. Urban growth, poor public services and lack of maintenance further detonated the river's deterioration since it became repository of sewerage and trash. As a result, since the 50s, the state government intended to transform the River under both hygienic and urban expansion discourses: "the Communications and Public Works Ministry intends to take advantage of the Santiago river by building on it a boulevard that will facilitate access to the Central Road…By taking advantage of the river, the city avoids another garbage dump…" (Sol de San Luis 1954). However, it was not until 1985, that the State, under Governor Jonguitud Barrios, could afford the cost of such an expensive project. In fact, parallel to the project that was implemented, the head of the Urban Development Ministry at the time, presented an alternative landscape project that included a linear park with public facilities, a central channel to carry runoff water and vehicular lanes. This project however, was rejected in favor of a less expensive, short-term engineering proposal.

The first stage of the "Santiago river regeneration project" covered 3.3 km of paved road and included underground runoff collectors and sewage infrastructure serving the northern part of the city (Hernandez Vera 2017). A second section was paved in the 90s and the final Soledad section is still unfinished. Currently, the state government is reinstalling a central runoff collector, given that the original quickly overpassed its capacity and has recently announced it will subcontract a study to evaluate the feasibility to build a "second story" to further "facilitate" private vehicle circulation in the Boulevard Santiago (Ruiz 2017) (Fig. 3).

Five years later, the state government paved a branch of the Rio Españita in 1995 and transformed another branch into a linear park. Lack of maintenance, however,

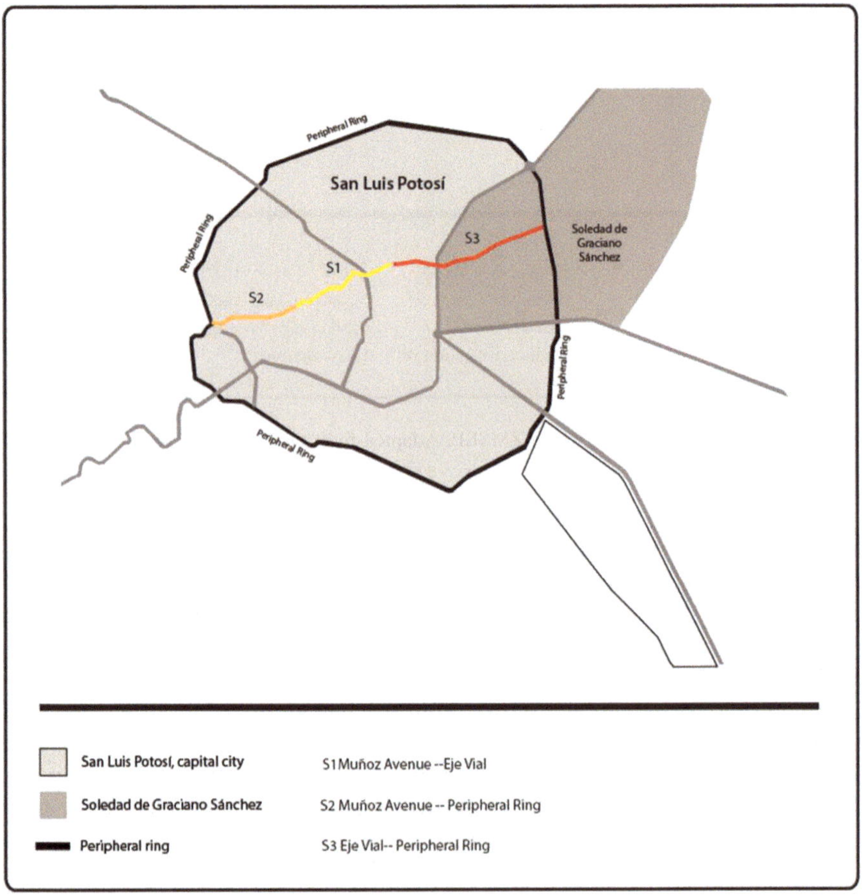

Fig. 3 Santiago River transformation stages

promoted the deterioration of this public space. To date, only the Paisanos River has escaped urbanization and kept wild but its decline emulates that of the Santiago River before its intervention.

4. New water frontiers: water import (2010)

Under current conditions, the watershed in the ZMSLP is over-exploited and the resource is non-renewable. Therefore, the water frontier was expanded in 2012 when the state of San Luis started importing water from *El Realito* dam in Guanajuato, 123 km west of the urban zone. Studies foresee that the dam will supply 1 m^3 per second, serving approximately 400,000 inhabitants, a third of the ZMSLP's current population (Peña 2013). According to Contreras and Galindo (2008), repairing 40% of the system's leeks could secure the same amount of water.

Importing water entails large costs and increases energy consumption. One liter of El Realito water costs the INTERAPAS $15, $5 of which are paid by consumers, and the remaining $10 are subsidized by the municipal government (interview with INTERAPAS official, March, 2016).

In addition to *El Realito* and foreseeing future water scarcity, the CEA announced last year, plans to import water from the Panuco river basin, 375 km east of the ZMSLP (La Jornada 2016). This indicates that the city will continue conquering new territories by searching for natural resources in the name of urbanization and development. This trend is evidently a strategic response to industrial and urban growth that relies on water availability to secure profits.

In sum, the transformation of the Santiago River is merged in a context of water exploitation that seemingly has no territorial limits. In this context, underground water sources have become essential suppliers and superficial waterways the city's backyards and recipients of waste. This water-use pattern however, is promoting an unsustainable exploitation and river transformations that result in different urban, social and environmental problems.

6 The Santiago River: From Asset to Liability and Source of Urban Problems

Currently, the Santiago River is the main driveway that connects the ZMSLP from east to west with an estimated traffic count of 45,000 vehicles per day (Hernandez Vera 2017). Its 12.7 km of pavement link the municipalities of San Luis and Soledad in approximately 20 min (Hernández Vera 2017). Access to the boulevard, however, is restricted to pedestrians, non-motorized vehicles and public transportation systems. As a result, around 30% of the privileged few that cover their trips on a private vehicle are the only users of this key piece of infrastructure (ITDP 2017).

Restricted access to alternative modes of transportation is not the only urban problem in the boulevard. Floods caused by runoff, as well as the San Jose dam's overflow water released and channeled through the boulevard can shut the Santiago River off during the rainy season for more than 15 consecutive days (interview with INTERAPAS official, march, 2016). In addition to this, the poor design of the boulevard (slopes, bridges, accesses and exit ramps) and lack of public lighting and pavement maintenance, combined with high speed, promotes traffic accidents (around 5% of the city's total) (Hernández Vera 2017). Boulevard users take this road because it is fast, except when it regains its natural waterway function. When it floods, users take alternative routes that double or treble their traveling time (Hernández Vera 2017) (Fig. 4).

While the boulevard was an urban border until the 70s, when the city grew towards the north it became an urban divider. The surrounding built environment reinforces this idea by turning its back to the River and by turning inwards through gates. However, the River holds great potential to become a connector that stitches together very dissimilar parts of the city through shared public spaces.

Fig. 4 Flooded river. *Source* Motorway Called Rio Workshop, taken by Femke Peters

Along its way, from west to east, the river puts on stage an urban patchwork of social disparities. In general, the west is richer, better equipped and greener. This contrasts with a dense, poorly equipped low-income middle as well as with a semi-rural and also poor end. This social depiction of an urban setting is also reflected on the water's cycle: clean upstream and dirty downstream (Fig. 5).

Environmentally, and in line with the literature (Dallman et al. 1999), paving and piping the boulevard reduced water infiltration into the deep aquifer, where modern water was found (Hernández Vera 2017) and accelerated the flow of water, instead of retaining it. This increased flooding in *La Tinaja*, one of the most affected communities down east of the ZMSLP, where the works to avoid floods have not been sufficient (Garcia 2016). This, of course, cannot be only attributed to the River; the city's accelerated growth largely contributed to increasing the amount of

Fig. 5 Where the paved river becomes a natural river. *Source* Motorway Called Rio Workshop, taken by Kong Chin Wai

impervious areas, reducing retention and infiltration in places such as the foothill of the San Miguelito range (Moreno and Lopez 2016).

Paving and piping the Santiago River, as a strategy to deal with the public health problems it caused, disregarded the potential urban and environmental benefits of its preservation or transformation according with green infrastructure principles. The media has criticized the Santiago River's transformation since the 90s, and in fact, climate changes, current urban growth and increasing number of automobiles have further accrued its problems. These problems highlight an enduring conflictive relation society-nature in the ZMSLP.

7 Conclusions

This research approached the reconversion of a river into a boulevard from a hydro-social perspective, arguing that hydrological resources and social processes are inseparable. The relationship between these two allows one to understand how space transforms in particular ways and how discourses facilitate these transformations and justify their externalities.

This paper's historical narrative shows how water-related natural resources gradually absorbed urban growth's externalities. Water exploitation has a long history that first tackled superficial water sources and then the deep aquifer, eventually drying both.

Different actors undertook efforts to control the resource from early water exploitation stages. Nonetheless, these remain limited in the face of accelerated urban growth. Strategies to bring fresh water into the city and get rid of the waste fast and far are also old. However, strategies to treat and reuse water are recent and still fall short. This shows how the colonial mentality of resource exploitation has not changed radically, on the contrary, our urbanization and unsustainable water exploitation patterns point to a short-term natural disaster currently pushing us to search for clear water farther and farther.

Currently, the supply offered by the deep aquifer and the dam system is not enough to cover the city's demand. The industrial sector continues consuming drinkable water for its processes and the shallow aquifer is full but polluted and hence not suitable for consumption. This unsustainable exploitation of water is further reinforced by system failures such as thousands of irregular connections and leaks (Contreras and Galindo 2008).

Watershed exploitation paralleled waterscape transformations. However, while surface source transformations such as dams, aqueducts, water reservoirs and river urbanizations affect greatly the image and social imaginary of the city, shallow and deep watershed exploitation is barely visible but devastating.

The history of the Santiago River reflects the contradictory relationship society-nature. While society depends on water, it exploits, pollutes, hides underground and disposes of the resource. From early stages, the Santiago River facilitated urban growth; however, from asset and source of water, it became a liability, more useful as a vehicular boulevard than as natural element. Water, however, claims its presence in the city in unearthly ways, flowing quickly on paved surfaces, conquering and flooding vulnerable areas and re-appropriating the spaces that once hosted it.

In addition to flooding, the River reproduces different urban, social and environmental problems found across the city such as exclusion of soft and massive modes of transportation, uneven distribution of services and facilities, unequal access to public space, segregation and lack of connection. Further, paving the river reduced water infiltration into the deep aquifer and accelerated the flow of water into flood-prone areas, not to mention the destruction of a riparian ecosystem that citizens do not seem to remember. Currently, the River is mainly perceived as a flooding road and not as a paved river. In sum, the River's transformations not only spatialized the conflictive relation society-nature but also the social inequities that yield from uneven urban growth.

The future of the river and the watershed can only be sustainable if the Metropolitan Zone points its efforts towards the New Urban Agenda that prioritizes sustainable growth, preservation of water and ecosystems, alternative modes of transportation, inclusion and urban resilience, among many other objectives (ONU Habitat 2016). Adopting these objectives entails the sustainable management of the watershed, instead of conquering new territories to supply the resource and promoting sustainable mobility options instead of investing on private vehicle infrastructure such as a second floor in the River. Future research will be necessary to better understand how urban planning, water urbanism and social participation and justice can be better aligned with urban growth and resource consumption.

References

CONAGUA (2015) Acuerdo por el que se dan a conocer los estudios técnicos del acuífero 2411 San Luis Potosí, en el Estado de San Luis potosí. Diario Oficial de la Federación, San Luis Potosí. Retrieved on 01 April 2017, from: http://dof.gob.mx/nota_detalle.php?codigo=5150943&fecha=07/07/2010

Contreras Servin C, Galindo Mendoza MG (2008) Abasto futuro de agua potable, análisis espacial y vulnerabilidad de la ciudad de San Luis Potosí, México. Cuadernos de Geografía 17:127

Dallman S, Piechota TC, Green D (1999) Storm water: asset not liability Los Angeles. The Los Angeles & San Gabriel Rivers Watershed Council, California

Diaz de Leon Barroso U, Courjaret C (2006) Saneamiento Integral y Reúso Industrial y en la agricultura en San Luis Potosí Urbano. Paper presented at the 4th World Water Forum, presentation Retrieved on 01 April 2017, from: http://www.bvsde.paho.org/bvsacg/e/foro4/17marzo/business/sanea.pdf

Dourojeanni A, Jouravlev A (1999) Gestión de cuencas y ríos vinculados con centros urbanos. Comisión Económica para América Latina y el Caribe (CEPAL). Retrieved on 01 April 2017, from: http://repositorio.cepal.org/handle/11362/31384

El Informador (2010) Entubados, 88% de ríos y arroyos en la ciudad. El Informador.mx. Retrieved on 24/04/2017, from: http://www.informador.com.mx/jalisco/2010/257928/6/entubados-88-de-rios-y-arroyos-en-la-ciudad.htm, 20 Dec 2010

Foucault M, Gordon C (1980) Power/knowledge: selected interviews and other writings, 1972–1977. Pantheon, New York

Garcia C (2016) Río Santiago inunda La Tinaja: Es la zona más afectada en temporada de lluvia en SLP. Plano Informativo. Retrieved on 03 April 2017 from http://planoinformativo.com/nota/id/475200/noticia/rio-santiago-inunda-, 14 Aug 2016

Gonzalez A, Hernandez L, Perlo M, Zamora I (2010) Rescate de ríos urbanos: propuestas conceptuales y metodológicas para la restauración y rehabilitación de ríos. Universidad Nacional Autónoma de México, Coordinación de Humanidades, Programa Universitario de Estudios Sobre la Ciudad. Retrieved on 01 April 2017, from http://www.puec.unam.mx/pdf/publicaciones_digitales/rescate_rios_digital.pdf

Harvey D (1989) The condition of postmodernity: an enquiry into the origins of cultural change. Blackwell, New York

Hernandez Vera R (2017) Rehabilitación de ríos urbanos a través de la infraestructura verde (undergraduate thesis). Universidad Autónoma de San Luis Potosí, Facultad del Hábitat, San Luis Potosí

Instituto Nacional de Estadistica, Geografía e Informatica (INEGI) (2000) XII Censo General de Población y Vivienda, ITER, Estado de San Luis Potosí. INEGI, México. Retrieved on 03 April 2017, from: http://www.beta.inegi.org.mx/proyectos/ccpv/2010/

ITDP (2017) Invertir para movernos, interactive map. Retrieved on 03 April 2017, from http://itdp.mx/invertirparamovernos/#!/mapa

La Jornada (2016) Agua del río pánuco para varios municipios de SLP: CEA. La Jornada. Retrieved on 01 Mar 2017, from: http://lajornadasanluis.com.mx/politica-y-sociedad/agua-del-rio-panuco-varios-municipios-slp-cea/, 21 Aug 2016

Mehta L, Karpouzoglou T (2015) Limits of policy and planning in peri-urban waterscapes: the case of Ghaziabad. Habitat Int 159–168

Mele C (2000) The materiality of urban discourse: rational planning in the restructuring of the early twentieth-century ghetto. Urban Aff Rev 35(5):628–648

Moreno AM, López LMM (2016) Vulnerabilidad, Dispersión y Fragmentación de la Zona Metropolitana de San Luis potosí. El Caso de Una Vialidad Llamada Río. In: Moreno Mata A (ed) Sistemas Metropolitanos en Crisis. Editorial Académica Española, S.L.P, pp 209–235

Musset A (1996) Mudarse o desaparecer. Traslado de ciudades hispanoamericanas y desastres (siglos XVI–XVIII). In: García Acosta V (ed) El estudio histórico de los desastres. Tercer Mundo Editores, Colombia, pp 41–69

ONU HABITAT (2016) New Urban Agenda, Draft outcome document for adoption in Quito, October 2016. Retrieved on 03 April 2017 from https://www2.habitat3.org/bitcache/97ced11dcecef85d41f74043195e5472836f6291?vid=588897&disposition=inline&op=view

Padron Moncada J (2014) Los Hombres del Agua en San Luis Potosí. INTERAPAS, San Luis Potosí. Retrieved on 29 Mar 2017, from: http://www.interapas.mx/files/cultura_del_agua/publicaciones/Los_Hombres_del_Agua.pdf

Peña F (2013) La sed urbana la ciudad como construcción hidráulica. El Colegio de San Luis, San Luis Potosí, México

Perlo C (2010) La apuesta para el desarrollo inmobiliario: Recuperación de Ríos Urbanos. In: González A, Hernández L, Perló M, Zamora I (eds) Rescate de ríos urbanos: propuestas conceptuales y metodológicas para la restauración y rehabilitación de ríos. Universidad Nacional Autónoma de México, Coordinación de Humanidades: Programa Universitario de Estudios Sobre la Ciudad, pp 50–54

Ruiz P (2017) Analizan construir segundo piso en Blvd. Rio Santiago. Pulso, San Luis Potosí. Retrieved on 03 April 2017 from http://pulsoslp.com.mx/2017/03/11/analizan-construir-segundo-piso-en-el-blvd-rio-santiago/, 11 Mar 2017

Salazar Gonzalez G (2010) Colonización y Estructura del territorio. In: Guadalupe Salazar (coord) Estudios del espacio arquitectónico y del territorio en San Luis Potosí. Universidad Autónoma de San Luis Potosí, San Luis Potosí, pp 20–65

SEMADESU (2011) Programa de rehabilitación y rescate de microcuencas "revive tu arroyo". Secretaria del Medio Ambiente Y Desarrollo Sustentable, Aguascalientes. Retrieved on 03

April 2017 from http://www.agscumple.gob.mx/phpSIME/evidencias/PROGRAMA%20DE%20REHABILITACION%20Y%20RESCATE%20DE%20MICROCUENCAS%20ajustado.pdf
Shannon K (2009) Water urbanism: hydrological infrastructure as an urban frame in Vietnam. In: Feyen J, Shannon K, Neville M (eds) Water and urban development paradigms. Towards an integration of engineering, design and management approaches. CRC Press, London, pp 55–66
Swyngedouw E (1999) Modernity and hybridity: nature, regeneracionismo, and the production of the Spanish waterscape, 1890–1930. Ann Assoc Am Geogr 89(3):443–465
Urquiola Permasan JI (2004) Agua para los ingenios: San Luis Potosí y el valle de San Francisco a inicios de la época colonial. Estudio introductorio y documentos sobre ingenios de beneficio de metales. El Colegio de San Luis, San Luis potosí, San Luis
Valencia D, Héctor G (2015) Redefiniendo territorios hidrosociales: control hídrico en el valle de Ica, Perú. Cuadernos de Desarrollo Rural 99–134

Children's Perception to Environmental Risks to Health, Key Element in the Design of Environmental Health Intervention Programs

Ana Cristina Cubillas-Tejeda, Alejandra León-Gómez, Juan Carlos Torrico-Albino and Luz María Nieto-Caraveo

Abstract In San Luis Potosí, México, there is environmental contamination in different places, produced as much by human activities, as by natural pollution. The characteristics of susceptibility and vulnerability in children, as well as their importance in Sustainable Development Goals, give them a priority in the considerations for measures to protect the population. An intervention strategy to improve children's environmental health could be a Risk Communication (RC). In order for RC to be successful, it is essential to know children's opinions, perceptions and concerns; give them voice to express themselves and be heard by researchers. Based on the above, the objective was to conduct an analysis of perception of exposure to environmental risks to health, in children living in sites with different risk scenarios, as a preliminary step for the design of RC programs according to each site, in order to improve the environmental health of the population. This study analyses risks perceptions using drawings as a data collection tool; research was conducted with 170 children from 4 schools located in the study sites. The activity was carried out during the "*II Children's Congress: Cuidando nuestra gran canica azul*", held in April 2012. Results showed that there is

A. C. Cubillas-Tejeda (✉)
Facultad de Ciencias Químicas, Universidad Autónoma de San Luis Potosí,
Av. Dr. Manuel Nava no. 6, Zona Universitaria, 78210 San Luis Potosí, SLP, Mexico
e-mail: acris@uaslp.mx

A. León-Gómez
Programas Multidisciplinarios de Posgrado En Ciencias Ambientales,
Universidad Autónoma de San Luis Potosí, Av. Dr. Manuel Nava no. 6,
Zona Universitaria, 78210 San Luis Potosí, SLP, Mexico
e-mail: biol.alejandralg@gmail.com

J. C. Torrico-Albino
Bolivian Institute for Economics and Agrarian Policy, Calle Presbítero
Medina no. 2597, La Paz, Bolivia
e-mail: torrico@web.de

L. M. Nieto-Caraveo
Agenda Ambiental, Universidad Autónoma de San Luis Potosí, Av. Dr. Manuel Nava
No. 201 Segundo Piso, Zona Universitaria, 78210 San Luis Potosí, SLP, Mexico
e-mail: lmnieto@uaslp.mx

W. Leal Filho et al. (eds.), *Sustainable Development Research and Practice in Mexico and Selected Latin American Countries*, World Sustainability Series,
https://doi.org/10.1007/978-3-319-70560-6_23

awareness in children about the dangers to which they are exposed in their environment; however, there was a lack of knowledge about some previously identified risks. Children's risk perception was different according to their place of residence.

Keywords Risk perception · Risk communication · Environmental health Children · Health promotion

1 Introduction

The Sustainable Development Goals (SDGs) recognize that *ending poverty must go hand-in-hand with strategies that build economic growth and addresses a range of social needs including health, education, while tackling climate change and environmental protection* (United Nations 2015). In the same way, the relationship between the environment and health has been recognized by the World Health Organization (WHO); it has been estimated that 23% of the total mortality burden and 26% of deaths in children under 5 years of age, are attributable to environmental factors (Prüss-Ustün et al. 2016). Environmental risks are all those physical, chemical, biological and social factors that can affect health; these environmental factors can be modified, so they can be approached in different areas, and a fundamental one is prevention through changes in behaviours and lifestyles that reduce the risks to health.

The susceptibility and vulnerability of the population are two important elements related to human health. Susceptibility is understood as the intrinsic individual predisposition to environmental risks; the vulnerability of a person or population is determined by extrinsic factors such as inequity, poverty, cultural practices, education level, and effects by climate change, among others (Bolte et al. 2010). There are different types of vulnerabilities (Wilches-Chaux 1993), among them, the environmental vulnerability that is understood as the particular disadvantage that some population groups have in face of chemical, physical or biological threats. In environmental vulnerability scenarios, it is common that contamination, be due to mixtures of toxic chemical and biological agents, affect several groups of a community (Díaz-Barriga et al. 2010). Due to their characteristic of being extrinsic elements to individuals or population, vulnerability levels can be modified if all key actors are involved, and the necessary resources as well as corrective measures are applied (Prüss-Ustün et al. 2016).

Children, who are considered to be under 18 years of age (UNICEF 2006), are susceptible to environmental risks, mainly due to their biological and physiological characteristics; together with their mental development process, their limited responsiveness to exposure to environmental hazards and their daily habits, such as playing with dirt, poor hygiene or sucking toys (WHO 2006). Furthermore, socioeconomic level has a direct influence on their vulnerability and, due to their susceptibility children are one of the most affected strata of the population. The characteristics of susceptibility and vulnerability in children, as well as their importance in SDGs, give them a priority in the considerations for measures to protect the population (United Nations 2015; Prüss-Ustün et al. 2016).

Enjoying a healthy environment during childhood is a human right that is essential for the full physical, mental and emotional development of people in their adult years (UNICEF 2006). Ensuring healthy lives for all at all ages is essential to sustainable development (United Nations 2015). It is imperative, therefore, that all actors involved provide adequate conditions to maintain the environmental health mainly of the child population. To achieve this, it is important to identify environmental determinants of health and to combat environmental risks through intervention strategies.

2 Antecedents

In México there are numerous sites with pollution problems caused by different sources. In areas with high mining or metallurgical activity there is contamination by metals and metalloids, such as lead (Pb), mercury (Hg), cadmium (Cd) and arsenic (As), among others. There are also sites in which there are brickyards that represent a serious problem of environmental contamination; other areas that deserve attention are those located near municipal garbage dumps or sites where hazardous waste is disposed. The burning of garbage is a common practice in peri-urban areas, which causes great air pollution by a toxic mixture.

In the state of San Luis Potosí, México, there are localities with a long history of mining, agricultural and industrial activities that promoted growth and economic improvement; however, these activities had negative repercussions in the environmental area and in human health. Within the state there are problems of natural contamination of water with fluoride or arsenic, causing diseases in the population such as fluorosis and osteoporosis (Ortiz-Pérez 2011). In all the above cases there is a health risk in the exposed population, mostly in areas where there are several of these problems at the same time.

Researchers from environmental health area of Programas Multidisciplinarios de Posgrado en Ciencias Ambientales (PMPCA) from Universidad Autónoma de San Luis Potosí (UASLP), have identified sites where the severity of environmental risks is a priority for the application of intervention measures. Based on the prioritization of sites in the state of San Luis Potosí, four sites were selected for this research, three in the municipality of San Luis Potosí and one in the municipality of Mexquitic de Carmona (Fig. 1, Table 1). At these sites, health risk assessment studies have been carried out and it has been shown that the problems of environmental contamination have a negative effect on the health of the population (Carrizales et al. 2006; Ortiz-Pérez 2011; Bocanegra-Salazar 2011; Flores-Ramírez et al. 2012, 2017; Zuki-Orozco 2012; Domínguez-Cortinas et al. 2013).

These scenarios represent a social problem that deserves to be addressed immediately since generally the most affected people are children. In according to goal 3 of DSGs, measures should therefore be established to reduce health risk by

Fig. 1 Localization of the selected sites, San Luis Potosí, Mexico, 2012. Map modified by the authors

Table 1 Selected sites and schools, San Luis Potosí, Mexico, 2012

Sites	Environmental risks	Pollutant probable	Participating children and schools
Rincón de San José, municipality of Mexquitic de Carmona	Use of tap water for drinking and cooking Use of firewood for cooking Confinement of hazardous waste Mercury recycling company Asphalt manufacturer	Fluorine PAHs Benzene Lead POPs	Children: 35 Primary school EEZ
Colonia Morales, municipality of San Luis Potosí	Use of tap water for drinking and cooking Metallurgical mining company High vehicular traffic	Fluorine PAHs Benzene Lead Arsenic	Children: 35 Primary school IMM
Colonia Real de Peñasco-Las Terceras, municipality of San Luis Potosí	Use of tap water for drinking and cooking Presence of 148 brickyards that use various fuels for the cooking of the bricks	Fluorine PAHs Benzene Lead POPs	Children: 70 Primary school PDRV
Colonia Bellas Lomas, municipality of San Luis Potosí	Use of tap water for drinking and cooking High vehicular traffic Car battery recycling companies Paint and tinplate companies	Fluorine PAHs Benzene Lead	Children: 60 Primary school PFMR

Abbreviations: polycyclic aromatic hydrocarbons (*PAHs*), persistent organic compounds (*POPs*)

decreasing exposure to contaminants (United Nations 2015). Therefore, alternatives for intervention should be sought and one of them is the Risk Communication (RC), aimed mainly at the most exposed and vulnerable population groups. The RC is a process of collective and plural awareness to persuade, inform and to influence the target population about those factors and threats that endanger the health (Moreno-Sánchez et al. 2016). Our group has implemented RC programs (RCP) (Cubillas-Tejeda et al. 2011; Coronado-Salas et al. 2012; Meza-Lozano et al. 2016), whose results have allowed us to propose the elements that can be articulated in a model aimed at improving environmental health (Cubillas-Tejeda and González-Mares 2015).

In order for RCP to be successful, it is essential to know the target population, and if the target group are children, it is fundamental to know their opinions, perceptions and concerns; give them voice to express themselves and be heard by researchers (Payne 2012; Börner et al. 2017). Besides that, children are critical agents of change and with great capacities for activism into a promotion of a better environmental health. Likewise, people who are vulnerable must be empowered, those whose needs are reflected in the SDGs, include children (United Nations 2015). Contextualization allows a better understanding of people's reasons and responses to environmental risks and helps to understand their risk perception. Risk perception refers to the subjective appreciation of the characteristics and severity of a hazard within a given socio-environmental context, based on the emotions, attitudes and knowledge of a situation (Slovic et al. 2004; Moreno-Sánchez et al. 2016).

There are several strategies to analyse risk perception, its selection depends on the target population. For example, working with children, short and simple strategies are preferable (Piko and Bak 2006; Börner et al. 2015; Terán-Hernández et al. 2016; Moreno-Sánchez et al. 2016). In this study, the target group was children, so the technique selected was drawing, which has already been tested by our group (Torres-Nerio et al. 2010; Coronado-Salas et al. 2012; Meza-Lozano et al. 2016).

Based on the above, the objective of the present investigation was to conduct an analysis of perception of exposure to environmental risks to health, in children living in sites with different risk scenarios, as a preliminary step for the design of RCP according to each site, in order to improve the environmental health of the population.

3 Study Strategy

On April 19, 20 and 21, 2012, the "*II Congreso infantil: Cuidando Nuestra Gran Canica Azul*", was held at Laberinto Museum of Science and Arts in San Luis Potosí, SLP, México. All the practical work of this research was done during this event. The organization of the event was through the *Consorcio Académico: Niño, Casa, Ambiente y Salud* (CANICAS), integrated by researchers and students of the

Faculty of Chemical Sciences and the Faculty of Medicine of the UASLP. The objective of the congress was to inform children about the risks of exposure to environmental pollutants, risks associated with climate change, the relationship between the environment and health, and about different forms of expression and communication; so that through the knowledge the participants develop protective behaviours, and learn to express what has been learned. The congress was a RC strategy, which was designed based on the results of the health risk assessment carried out at each selected site. The infantile congress had a participation of 200 children from 4 primary schools located at the selected sites (Fig. 1, Table 1). The number of children participating was different in each school, due to the limited capacity of resources to carry out the congress (Cubillas-Tejeda 2012).

4 Methods

4.1 Study Population

Of the 200 children who attended the congress, only 170 of the children made all the drawings (age 6–12 years). The proportion of boys and girls could not be controlled given the invitation procedure and inclusion criteria used. In Fig. 2 shows the percentage of children that were considered in the analysis of the drawings, by age, school and sex.

4.2 Ethical Considerations

The study was approved by the Bioethics Committee of the Medicine Faculty at the UASLP. Written consent was provided by the involved educational authorities and by the parents of each participant. When presenting the project, we stressed that participation was voluntary. Apart from the parents' written consent, students had to give their oral consent in order to take part in the study. We also communicated clearly that they were allowed to withdraw from the study at any time.

4.3 Risk Perception Analysis

The selected tool was the drawing technique, which was carried out on the first and last day of the children's congress. This technique has been used by our group in different scenarios of risk (Torres-Nerio et al. 2010; Cubillas-Tejeda et al. 2011; Coronado-Salas et al. 2012; Meza-Lozano et al. 2016; Terán-Hernández et al. 2016; Börner et al. 2017). The purpose of this activity prior to the beginning of the

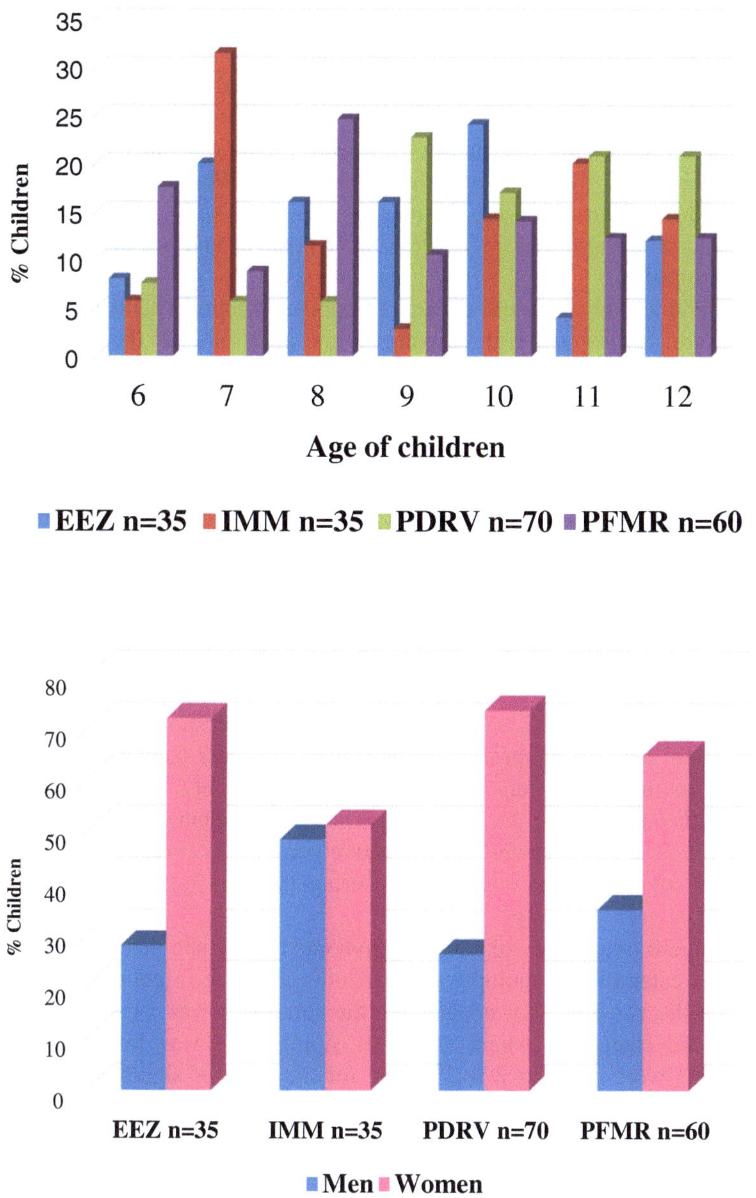

Fig. 2 Distribution of the children by age, gender and school, San Luis Potosí, México, 19 April 2012

congress was to know the risk perception and concerns of children before providing any kind of information. At the end of the congress, the aimed was to find out what were the main topics of interest related to the environmental problems of their localities, and to identify which were the themes of the congress that had the most impact on the participants.

Based on the previous risk assessment of the children's health in the study sites, two exposure scenarios were defined: *At home* and *Outside home*. Each child was asked to answer the following questions, by developing drawings: (1) *What is good for you at home?* (2) *What is bad for you at home*? (3) *What is good for you outside your home*? (4) *What is bad for you outside your home*? The words *Good* and *Bad* were included in the questions to know their positive and negative perception of their environment.

Each child was given sheets, pencils, colours and rubber. The instructions for the activity were: (1) respond with drawings to the questions, (2) the drawing is free, (3) it is not a test, (4) there are no correct or incorrect answers, (5) this is not a drawing competition, (6) the activity is individual and (7) the time to perform it is one hour. At the end of the activity, was checked with each of the children, what meant the elements they drew (especially the abstract and symbolic drawings). Their responses were recorded on the back of the sheets, for further analysis.

Data analysis was conducted using a mix-method approach combining quantitative and qualitative content analysis (Álvarez-Gayou 2003; Curry et al. 2009; Torres-Nerio et al. 2010; Börner et al. 2017), to this purpose, elements in each drawing were observed, and on the basis of them, inductive and excluding categories were established. Drawings generally contained several elements; therefore, the individual elements were coded and assigned to the different categories. The categorization was carried out separately for each question. We conducted the coding as a peer-based process which included peer-feedback by a second researcher on the coding framework and the assignation of the different elements to the respective categories. After the classification of elements, frequencies were counted for each category and the respective community. Frequency analysis was carried out per individual without considering the total number of drawings per participant in each category.

To compare frequencies of children from each study site that drew elements in established categories, a multiple comparison was made in contingency tables of $2 \times k$ and the value of χ^2 was obtained, the value of k was 4 (study sites). In the cases of the categories where statistical significance was found, independent comparisons were made in 2×2 tables, for the level of significance Bonferroni correction was used, whereby the level of significance was $p \leq 0.0125$. To compare the frequencies before and after the infantile congress, the obtained frequencies were submitted to χ^2 test with a level of statistical significance of p 0.05. The software JMP® 10 was used for this analysis.

4.4 Limitations of the Study

Regarding the technique of drawing, a limitation was the inability of some children, both motor and cognitive, to answer with drawings to the questions asked. In this sense, it is recommended that the drawing technique be complemented with other methodologies for obtaining information, such as questionnaires or interviews that help to deepen and understand the children's risk perception, knowledge and concerns. Another limitation was the lack of equal numbers of children for the four study sites, as well as an equal number between male and female.

5 Results

For issues of relevance in terms of environmental risks to health, the present document only presents the results of the scenarios *At home, Bad* and *Outside home, Bad.*

5.1 Results from Scenario At Home, Bad

For this scenario, twelve categories were stablished (Fig. 3), and quantitatively no statistical differences were found between the children's perception according to the place of residence. Qualitatively, in this scenario most of the children of all schools drew in the *Activities* category, the drawings referred mainly to sedentary activities, such as watching television, playing video games and using the internet (Fig. 4). In the category of *Risk of fire or burn*, the drawings showed a negative perception towards the kitchen environment and highlighted the stove ignited as the main element of risk. They also perceived several flammable items as negative for their well-being, for example boilers, gas tanks, gas lighters, matches, firecrackers, and

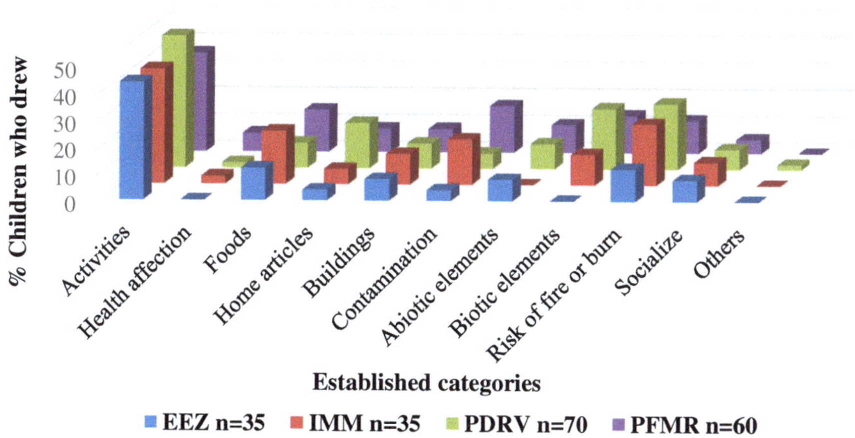

Fig. 3 Percentage of children per school that drew in the categories established for the scenario *At home, Bad*; San Luis Potosí, México, 19 April 2012

Fig. 4 Drawing of a 10 year old girl. EEZ school, scenario *At home, Bad*, 19 April 2012

Fig. 5 Drawing of a 12 year old girl. PDRV school, scenario *At home, Bad*, 19 April 2012

candles (Fig. 5). In the *Food* category many children drew; the drawings referred to "junk" food as fried foods, candies and soft drinks.

5.2 *Results from Scenario* Outside Home, Bad

In this scenario most of the children drew in the *Contamination* category (Fig. 6). Qualitatively most of the drawings indicated contamination of water, air and soil, but the contaminant was not specified. However, in some cases the contaminant, source, route of exposure and even health effects were indicated (Fig. 7). Although the children from distinct sites perceived contamination as negative, there is a higher percentage of children who drew in this category in the community of

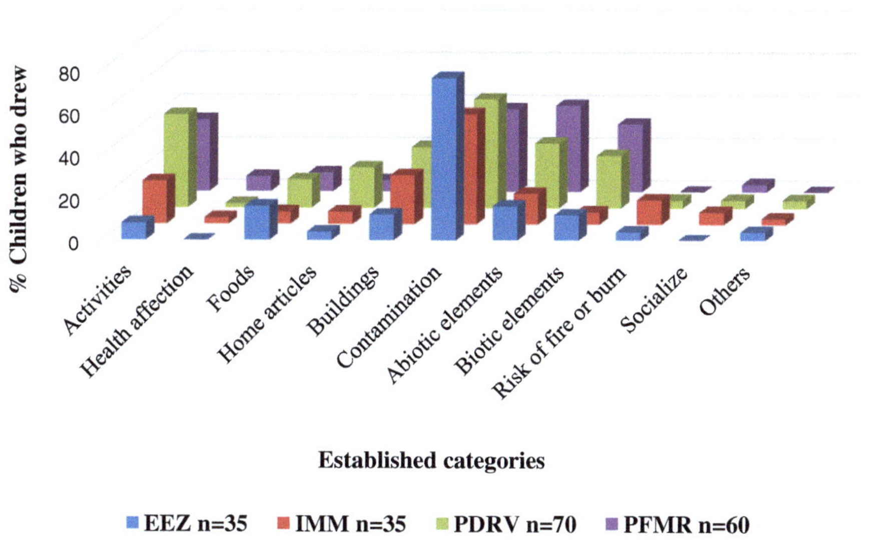

Fig. 6 Percentage of children per school who drew in the categories set for scenario *Outside, Bad*; San Luis Potosí, México, 19 April 2012. Activities $\chi^2 = 13.93$, $p = 0.0031$; Contamination $\chi^2 = 9.75$, $p = 0.0208$; Abiotic Elements $\chi^2 = 9.43$, $p = 0.024$; Biotic elements $\chi^2 = 10.32$, $p = 0.016$

Fig. 7 Drawing of a 10 year old girl. EEZ school, scenario *Outside home, Bad,* 19 April 2012

Rincón de San José in Mexquitic (EEZ) compared to the children of the colony Bellas Lomas (PFMR) ($p = 0.0035$).

Qualitatively, in the *Abiotic elements* category, children drew items such as rain, clouds, storms, and the sun. Some children perceived as negative to get wet in the rain, because it can cause illnesses such as flu, cough and fever (Fig. 8). The negative perception of the sun can be related to the high temperatures that are registered in dry climates and that can be a reason for sufferings like insolation, burns of first degree, headaches, heatstroke, among others. In this category, we found significant differences in the frequency of children who drew depending on the place of residence (Fig. 6), mainly among the children of the colony Bellas Lomas (PFMR) and the Morales children (IMM) ($p = 0.0102$).

Fig. 8 Drawing of a 9 year old girl. EEZ school, scenario *Outside home, Bad*, 19 April 2012

Regarding the *Food* category, qualitatively the drawings referred to "junk" food and food that can be purchased on the street; some children also referred to alcoholic beverages and soft drinks. In the *Constructions* category, children's drawings showed a tendency to relate urban spaces with negative places for their well-being; the elements drawn were mainly industries, houses and roads. In the *Activities* category, a difference in the frequency of children who drew was found, in relation to the area of residence (Fig. 6). For this category 4 subcategories were established (Fig. 9). In the subcategory *Risky activities* the drawn elements were related to the negative perception to be alone in the street, to cross streets or roads, and to interact with strangers. The schools that had the highest percentage were PDRV and PFMR, both located in peri-urban zones. In the subcategory *Addictions and crimes*, children from IMM, PDRV and PFMR schools drew elements that express violence and insecurity in the streets, for example gangs, assaults, guns, fights, use and sale of drugs, among others (Figs. 10 and 11).

5.3 Comparison of Risks Perceived Before and After the Children's Congress

The objective of the perception analysis after the congress was to assess whether the information provided was captured and understood by the children, as well as to know the main environmental health issues that interested them. It was not sought to evaluate the change of knowledge and habits of the children from the information that was given in the congress, since for this a longer period of time must pass.

In the four study scenarios no statistically significant differences were found before and after the congress in any of the established categories. However, qualitative changes were found as new elements appeared in the drawings. On the *At home, Bad* scenario, after the congress elements appeared that refer to the risk that children see on contact with toxic substances and cleaning products such as chlorine (Fig. 12). In the scenario *Outside home, Bad*, elements appeared related to the

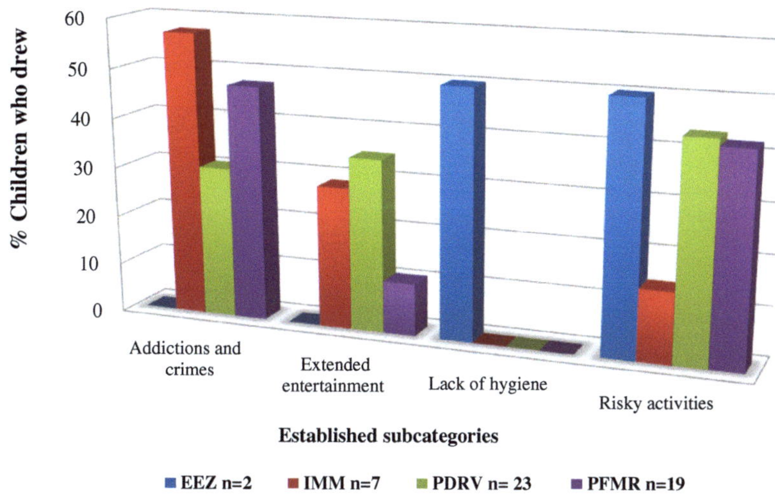

Fig. 9 Percentage of children who drew in the subcategories established for the category Activities, scenario *Outside, Bad*; April 19, 2102, San Luis Potosí, México

Fig. 10 Drawing of a 12 year old boy. PDRV school, scenario *Outside home, Bad*, 19 April, 2012

contamination of rivers, to take water of rivers, to drink tap water (Fig. 13) and the air pollution by burning trash and companies.

5.4 *Comparison of Environmental Risks Identified by Researchers and Perceived by Children*

In order to design an effective RCP, it must be adapted to the audience to be addressed. Therefore, it is necessary to know if the children perceive the risks

Fig. 11 Drawing of an 11 year old boy. PFMR school, scenario *Outside home, Bad*, 19 April, 2012

Fig. 12 Drawing of a 9 year old girl. PFMR school, scenario *At home, Bad,* 21 April 2012

Fig. 13 Drawing of a 9 year old girl. PFMR school, scenario *Outside, Bad,* 21 April 2012

identified by researchers, if they have knowledge about what measures to take in order to reduce risks; or if they perceive other risks that were not identified by researchers (Cubillas-Tejeda and González Mares 2015; Moreno-Sánchez et al. 2016). This is important because the way of approaching the RC will be very different.

Table 2 summarizes the risks previously identified by researchers, and the risks that were perceived by children. It was found that there is awareness in children about the dangers to which they are exposed in their environment. In particular, children perceived the risks of contamination by bricklayers, the burning of garbage, pollution with garbage, exposure to animals and pests, as well as social problems, such as alcoholism and addictions. However, there was a lack of knowledge about some previously identified risks, including water contamination with fluoride. On the other hand, the children perceived some risks that had not been analyzed by the researchers, such as those related to the use of new technologies and sedentarism, among others. These risks identified by the children are relevant so in the RCP for each site these will be considered.

Table 2 Comparison of risks identified by researchers with risks perceived by children; San Luis Potosí, México, April 2012

	Schools											
Risks	EEZ			IMM			PDRV			PFMR		
	I	P	%	I	P	%	I	P	%	I	P	%
	Risks at home											
Indoor air pollution												
(a) Smoke generated by the use of firewood for cooking	Yes	No	–	ND	No	–	Yes	No	–	Yes	Yes	1.8
(b) By smoking	Yes	No	–	Yes	Yes	2.9	Yes	No	–	Yes	No	–
Use of glazed earthenware for cooking	Yes	No	–	Yes	No	–	Yes	No	–	Yes	No	–
Consumption of contaminated water	Yes	No	–	Yes	No	–	Yes	No	–	Yes	Yes	1.8
Noxious wildlife	Yes	No	–	Yes	No	–	Yes	Yes	17.0	Yes	Yes	3.5
Malnutrition	Yes	Yes	4.0	Yes	No	–	Yes	Yes	1.9	Yes	No	–
Poor hygiene measures	Yes	Yes	8.0	Yes	No	–	Yes	No	–	Yes	Yes	1.8
Sedentary lifestyle	NA	Yes	44.0	NA	Yes	54.3	NA	Yes	52.8	NA	Yes	36.8
Soil contaminated by waste	Yes	No	–	Yes	Yes	8.6	Yes	No	–	Yes	Yes	8.8
Alcoholism	Yes	Yes	4.0	Yes	Yes	5.7	Yes	No	–	Yes	Yes	1.8

(continued)

Table 2 (continued)

	Risks outside home											
Outdoor air pollution												
(a) By burning trash	Yes	Yes	4.0	ND	No	–	Yes	Yes	1.9	Yes	No	–
(b) By industrial activities or brickworks	Yes	Yes	12.0	Yes	Yes	5.7	Yes	Yes	30.2	Yes	Yes	8.8
Water contamination	Yes	Yes	16.0	Yes	No	–	Yes	Yes	3.8	Yes	Yes	5.3
Street animals	Yes	Yes	8.0	Yes	No	–	Yes	Yes	5.7	Yes	Yes	8.8
Meteorological hazards												
(a) Rain or storm	NA	Yes	16.0	NA	Yes	5.7	NA	Yes	28.3	NA	Yes	35.1
(b) High temperatures/ effects of the sun	NA	Yes	12.0	NA	Yes	5.7	NA	Yes	17.0	NA	Yes	19.3
Soil pollution												
(a) By domestic waste	Yes	Yes	24.0	Yes	Yes	22.9	Yes	Yes	11.3	Yes	Yes	12.3
(b) From companies residues	Yes	No	–	Yes	No	–	Yes	No	–	Yes	Yes	1.8
Social risks												
(a) Addictions	Yes	No	4.0	Yes	Yes	2.9	Yes	Yes	5.7	Yes	Yes	7.0
(b) Delinquency	Yes	No	–	Yes	Yes	8.6	Yes	Yes	11.3	Yes	Yes	10.5

Abbreviations: identified by researchers (I), perceived (P), not analyzed by researchers (NA), not detected by researchers (ND), Percentage of perception by the children of each school (%)

6 Conclusions

Due to their high vulnerability, children are strongly affected by the interplay of social inequality and environmental risks. Therefore, it is important to recognize children as key actors in environmental health interventions based on their everyday interaction with their environment.

In the present study it was confirmed that the drawing technique was useful to know the risk perception, knowledge and concerns of the children about the environmental health risks of the places where they live. It is also a tool that allowed them to express themselves in a free and fun way, and be heard so that their contributions are taken into account by the researchers for the design of the RCP.

The results showed that there is awareness in children about the dangers to which they are exposed in their environment; however, there was a lack of knowledge about some previously identified risks. Children's risk perception was different according to their place of residence. It was also found that the information provided in the children's congress did have an impact on them, and although the differences were not quantitative, qualitative changes were found, since in the drawings new elements appeared related to the information that was provided.

The information provided by the children, the contextualization of each site and the risk assessment by the researchers, are the basis for the CPR design that will be implemented in each site, according to each problem and scenario.

Acknowledgements We thank the children, parents, teachers and directors of the schools who supported this research.

Funding This work was supported by the Fondo Sectorial de Investigación en Salud y Seguridad Social from the Consejo Nacional de Ciencia y Tecnología (CONACyT) under grant SALUD-142064.

References

Álvarez-Gayou JL (2003) Cómo hacer investigación cualitativa. Editorial Paidós Mexicana, México, D.F

Bocanegra-Salazar M (2011) Muestreo pasivo de hidrocarburos aromáticos Policíclicos en las hojas de la especie vegetal Eriobotrya Japonica (Tesis de doctorado). Programas Multidisciplinarios de Posgrado de Ciencias Ambientales de la Universidad Autónoma de San Luis Potosí. San Luis Potosí, México

Bolte G, Tamburlini G, Kohlhuber M (2010) Environmental inequalities among children in Europe evaluation of scientific evidence and policy implications. Eur J Public Health 20(1):14–20

Börner S, Torrico-Albino JC, Nieto-Caraveo LM, Cubillas-Tejeda AC (2015) Exploring Mexican adolescents' perceptions of environmental health risks: a photographic approach to risk analysis. Ciência & Saúde Coletiva 20(5):1617–1627

Börner S, Torrico-Albino JC, Nieto-Caraveo LM, Cubillas-Tejeda AC (2017) Living with everyday environmental risks: giving a voice to young people in the design of community-based risk communication programs in the city of San Luis Potosí, México. Children's Geographies. doi:10.1080/14733285.2017.1310369. Accessed 9 June 2017

Carrizales L, Razo I, Téllez-Hernández JI, Torres-Nerio R, Torres A, Batres LE, Cubillas-Tejeda AC, Díaz-Barriga F (2006) Exposure to arsenic and lead of children living near a copper-smelter in San Luis Potosi, Mexico: importance of soil contamination for exposure of children. Environ Res 110(1):1–10

Coronado-Salas C, Díaz-Barriga F, Moreno-Sánchez AR, Carrizales-Yáñez L, Torres-Nerio R, Cubillas-Tejeda AC (2012) La comunicación de riesgos como una herramienta para disminuir la exposición infantil a plomo y arsénico en la zona contaminada de Villa de la Paz-Matehuala, San Luis Potosí, México. Revista Internacional de Contaminación Ambiental 28(2):167–181

Cubillas-Tejeda AC (2012) Cuidemos nuestra gran canica azul. Resource document. Revista Universitarios Potosinos. http://ninive.uaslp.mx/jspui/bitstream/i/2712/2/08_201211.pdf. Accessed 19 July 2017

Cubillas-Tejeda AC, González-Mares MO (2015) La comunicación de riesgos como estrategia de intervención para mejorar la salud ambiental en poblaciones vulnerables. In: Mendieta A, Testas P (eds) ¿Legitimidad o reconocimiento? Las investigadoras del SIN: retos y propuestas. Benemérita Universidad Autónoma de Puebla, Ediciones La Biblioteca, México, DF, pp 505–512

Cubillas-Tejeda AC, Torres-Nerio R, Díaz-Barriga F, Carrizales-Yáñez L, Nieto Caraveo LM, Coronado-Salas C et al (2011) Diseño y Aplicación de un Programa de Comunicación de Riesgos para la Salud Ambiental en un Sitio Contaminado con Plomo y Arsénico. Ciencia & Saúde Coletiva 16(10):4115–4126

Curry LA, Nembhard IM, Bradley EH (2009) Qualitative and mixed methods provide unique contributions to outcomes research. Circulation 119: 1442–1452. http://circ.ahajournals.org/content/119/10/1442.full.pdf+html. Accessed 19 July 2017

Díaz-Barriga F, García S, Corra L (2010) Prevención de riesgos en poblaciones vulnerables. In: Galvao LA, Finkelman LJ, Henao S (eds) Determinantes ambientales y sociales de la salud. Organización Panamericana de la Salud, McGraw–Hill Interamericana, Distrito Federal, México, pp 139–153

Domínguez-Cortinas G, Díaz-Barriga F, Martínez-Salinas R, Cossío P, Pérez-Maldonado IN (2013) Exposure to chemical mixtures in Mexican children: high-risk scenarios. Environ Sci Pollut Res 20(1):351–357. doi:10.1007/s11356-012-0933-x

Flores-Ramírez R, Rico-Escobar E, Núñez-Monreal JE, García-Nieto E, Carrizales-Yáñez L, Ilizaliturri-Hernández C, Díaz-Barriga F (2012) Exposición infantil al plomo en sitios contaminados. Salud pública de México 54(4):383–392

Flores-Ramírez R, Pérez-Vázquez FJ, Rodríguez-Aguilar M, Medellín-Garibay SE, Van Brussel E, Cubillas-Tejeda AC, Carrizales-Yáñez L, Díaz-Barriga F (2017) Biomonitoring of persistent organic pollutants (POPs) in child populations living near contaminated sites in Mexico. Sci Total Environ 579(2017):1120–1126

Meza-Lozano B, Ortiz-Pérez MD, Ponce-Palomares M, Castillo-Gutiérrez SG, Flores-Ramírez R, Cubillas-Tejeda AC (2016) Implementación y evaluación de un Programa de Comunicación de Riesgos por exposición a flúor en la comunidad de El Fuerte, Santa María del Río, San Luis Potosí. México. Revista Internacional de Contaminación Ambiental 32(1):87–100

Moreno-Sánchez AR, Cubillas-Tejeda AC, Guerra-García A, Peres F (2016) Risk communication in Latin America. In: Galvao LAC, Finkelman J, Henao S (eds) Environmental and social determinants of health. Pan American Health Organization (PAHO), Washington, DC, pp 335–360

Ortiz-Pérez MD (2011) El agua de consumo humano en México. http://www.colsan.edu.mx/investigacion/pays/archivo/El_agua_para_consumo_humano_Mexico-2015-07.pdf. Accessed 19 July 2017

Payne R (2012) 'Extraordinary survivors' or 'ordinary lives'? Embracing 'everyday agency' in social interventions with child-headed households in Zambia. Child Geographies 10(4):399–411

Piko BF, Bak J (2006) Children's perceptions of health and illness: images and lay concepts in preadolescence. Health Educ Res 21(5):643–653

Prüss-Ustün A, Wolf J, Corvalán C, Bos R, Neria M (2016) Preventing disease through healthy environments: a global assessment of the burden of disease from environmental risks. WHO, Geneva, Switzerland. http://apps.who.int/iris/bitstream/10665/204585/1/9789241565196_eng.pdf?ua=1. Accessed 15 July 2017

Slovic P, Finucane ML, Peters E, MacGregor DG (2004) Risk as analysis and risk as feelings: some thoughts about affect, reason, risk and rationality. Risk Anal 24(2):311–322

Terán-Hernández M, Díaz-Barriga F, Cubillas-Tejeda AC (2016) Diagnóstico de salud y percepción de riesgos, elementos clave para una propuesta de intervención en comunidades indígenas. Revista Panamericana de Salud Pública 39(2):106–114

Torres-Nerio R, Domínguez-Cortinas G, van't Hooft A, Díaz-Barriga F, Cubillas-Tejeda AC (2010) Análisis de la percepción de la exposición a riesgos ambientales para la salud, en dos poblaciones infantiles, mediante la elaboración de dibujos. Salud Colectiva 6(1):65–81

UNICEF Comité Español (2006) Convención sobre los derechos de los niños. http://www.un.org/es/events/childrenday/pdf/derechos.pdf. Accessed 27 June 2017

United Nations General Assembly (2015) Transforming our world: the 2030 Agenda for Sustainable Development (resolution 70/1). http://www.un.org/ga/search/view_doc.asp?symbol=A/RES/70/1&Lang=E. Accessed 14 Aug 2017

WHO (2006) Principles for evaluating health risks in children associated with exposure to chemicals (Environmental health criteria 237). First drafts prepared by Louis G, Damstra T, Díaz-Barriga F, Faustman E, Hass U, Kavlock R, Kimmel C, Kimmel G, Krishnan K, Luderer U, Sheldon L. WHO, Geneva, Switzerland. http://www.who.int/ipcs/publications/ehc/ehc237.pdf. Accessed 19 July 2017

Wilches-Chaux G (1993) La vulnerabilidad global. In: Los desastres no son naturales. Red de Estudios Sociales en Prevención de Desastres en América Latina. http://www.desenredando.org/public/libros/1993/ldnsn/. Accessed 23 June 2017

Zuki-Orozco AB (2012) Evaluación de la exposición a benceno e hidrocarburos policíclicos aromáticos en población infantil de San Luis Potosí (Tesis Maestría). Programas Multidisciplinarios de Posgrado de Ciencias Ambientales de la Universidad Autónoma de San Luis Potosí. San Luis Potosí, México

Author Biographies

Ana Cristina Cubillas-Tejeda M.Sc. in Immunology at Universidad Autónoma de Nuevo León, México (1993). Ph.D. in Basic Biomedical Sciences at UASLP (2000). She is currently a Full Time Research Professor at the Faculty of Chemical Sciences of the UASLP since 1992 to date. Her research area and teaching is the Risks Communication and Environmental Health, and Immunology.

Alejandra León-Gómez Master in Environmental Sciences, at PMPCA, UASLP, Mexico/ Master's in Technology and Resource Management in the Tropics and Subtropics, Institut Technologie-und Ressourcenmanagement in den Trope und Subtropen (ITT), Cologne University, Germany (2012). She is currently teaching in Biology, Ecology and Environment at the Lázaro Cárdenas Federal High School in Tijuana, Baja California, México.

Juan Carlos Torrico-Albino M.Sc. in Agricultural resources management (2003). Ph.D. Agricultural science (2007). Coordinator for GIZ-Bolivia Continuing Education. Senior researcher of the Bolivian Institute of Economics and Agrarian Policy. His research area is food security and rural development and curriculum development for postgraduate courses in the water and agriculture sector.

Luz María Nieto-Caraveo Master in Education Sciences (1985) at Centro Interdisciplinario de Investigación y Docencia en Educación Técnica (CIIDET), SEP-SEIT, Querétaro, México. She was Full Time Research Professor at the Faculty of Engineering, UASLP since 1984 to 2017. Her research area and teaching is Environmental education and sustainable development in technical and higher education levels.

Risk Communication as an Alternative Intervention to Improve the Environmental Health in Children in an Area with Various Environmental Problems

Claudia Davinia Monsiváis-Nava, Ismael García-Cedillo, Leonardo Ernesto Márquez-Mireles, Rogelio Flores-Ramírez and Ana Cristina Cubillas-Tejeda

Abstract Children are the most vulnerable group to certain environmental risks, which is why this project focused on those children living in a marginalized area located north of the city of San Luis Potosí. It was found that in that area, children were exposed to multiple environmental health risks. The objective of this project was to design, implement and evaluate a Risk Communication Program (RCP) to improve children's environmental health. Based on the health risk assessment and risk perception analysis, an RCP was designed. The topics of healthy eating and

C. D. Monsiváis-Nava
Universidad Autónoma de San Luis Potosí, Programas Multidisciplinarios de Posgrado en Ciencias Ambientales (PMPCA), Agenda Ambiental, Avenida Dr. Manuel Nava 201, segundo piso, Zona Universitaria, 28210 San Luis Potosí, SLP, Mexico
e-mail: claudia.dmn@gmail.com

I. García-Cedillo
Facultad de Psicología, Universidad Autónoma de San Luis Potosí, Carretera Central Km.424.5, Zona Universitaria Oriente, 78494 San Luis Potosí, SLP, Mexico
e-mail: ismaelgace@yahoo.com.mx

L. E. Márquez-Mireles
Facultad de Ciencias Sociales y Humanidades, Universidad Autónoma de San Luis Potosí, Av. Industrias 101-A Fraccionamiento Talleres, Zona Universitaria Oriente, 78399 San Luis Potosí, Mexico
e-mail: leonardoemm@uaslp.mx

R. Flores-Ramírez
Cátedra CONACYT-Universidad Autónoma de San Luis Potosí, Centro de Investigación Aplicada en Ambiente y Salud/CIACYT, Avenida Sierra Leona No.550, Colonia Lomas Segunda Sección, 78210 San Luis Potosí, SLP, Mexico
e-mail: rfloresra@conacyt.mx

A. C. Cubillas-Tejeda (✉)
Facultad de Ciencias Químicas, Universidad Autónoma de San Luis Potosí, Av. Dr. Manuel Nava No. 6, Zona Universitaria, 78210 San Luis Potosí, SLP, Mexico
e-mail: acris@uaslp.mx

W. Leal Filho et al. (eds.), *Sustainable Development Research and Practice in Mexico and Selected Latin American Countries*, World Sustainability Series, https://doi.org/10.1007/978-3-319-70560-6_24

hygiene were addressed, as well as the risks of exposure to fluoride and lead. In the first stage, the target audience was made up of 62 parents and 10 teachers from two schools. The second stage, which had support and participation of parents and teachers, involved working with 338 students (3–15 years of age). Upon concluding the implementation of the program, knowledge integration and decrease in exposure to fluoride and lead were found in children in the area; as well as changes in habits in the participating families.

Keywords Environmental risks · Environmental health · Child health
Health communication · Vulnerable groups

1 Introduction

The conditions and lifestyles in which people grow, live, work and grow old, determine their quality of life and health; those who live in rural areas or marginalized peri-urban communities face scenarios of greater inequality and social injustice (Díaz-Barriga et al. 2010). Because of a variety of factors, boys and girls are in a greater state of vulnerability. Such is the case of those who live in developing countries, where they lose, on average, eight times the amount of healthy life years per inhabitant because of diseases caused by the environment (Chelala 1999; Prüss-Üstün and Corvalán 2006). This is why programs which focus on prevention and improving children's health must be improved, contributing to decreased morbidity due to modifiable environmental factors.

As such, this project was aimed at boys and girls from "Las Terceras" neighbourhood, a marginalized area located at the north of the city of San Luis Potosí, in turn located in the state by the same name, in México. According to previous studies conducted by researchers in the area of Salud Ambiental Integrada (SAI), Programas Miltidisciplinarios en Ciencias Ambientales (PMPCA) from the Universidad Autónoma de San Luis Potosí (UASLP), children were exposed to multiple risks, including exposure to fluoride in the water used for drinking and cooking, exposure to pollutants generated by burning waste and different material such as wood, sawdust and plastic, which are used to produce brick. They are also exposed to lead found in pots or glazed clay containers which are used for cooking or storing food. Because of the absence of some hygiene related behaviours, as well as lacking paved roads and drainage, there is an increase in the probability that children in the area will develop parasitism and inadequate dietary habits in the families which are associated with the heredofamilial factors of Type 2 Diabetes Mellitus.

2 Antecedents

Risk Communication (RC) is a strategy which can be used as a way to approach vulnerable groups, with the goal of contributing to decreasing population morbidity and mortality, communicating messages to community members in a planned, sensible manner throughout the risk management process (Moreno-Sánchez et al. 2010; Cubillas-Tejeda et al. 2011; Coronado-Salas et al. 2012; Meza-Lozano et al. 2016). Because of the previously exposed information, designing an intervention strategy to decrease risk exposure in the area was considered necessary.

For this reason, a RC program (RCP) was designed to tackle different environmental risks to which children are exposed, focused at improving children's environmental health of the population. Parents, teachers and school officials from two learning centres in the area of focus were involved in this project. They were selected because of the role they play in the health of their children and students, and in order to increase work in community education. An agreement was reached with the population in the area of study about the topics which would be addressed in this project; it would be focused at decreasing exposure to lead and fluoride because of the effects it has in boys' and girls' development and the irreversible consequences it brings. Hygiene and diet would also be addressed transversally throughout the implementation of the RCP. The goal of this work was to design, implement, and evaluate a RCP aimed at improving children's environmental health in a locality facing multiple health risks in the city of San Luis Potosí, México.

"Las Terceras" is a marginalized zone located in the north periphery of the city, there are small unpaved streets and passages, and in some areas, it resembles more rural areas than urban ones. Also, there are clandestine dumps and brickwork that work without regulation, houses built of sheet and wood, some others with bricks or concrete block. Besides, some parts of the land are used for cultivation and breeding of livestock (Erbe 2011).

In 2009, in the metropolitan area of San Luis Potosí, a study was carried out by researchers from the SAI of the PMPCA, whose main objective was to compare and identify different health problems associated with environmental injustice, for which a comprehensive health assessment was carried out through the application of 36 social, environmental and health indicators. "Las Terceras" showed evidence of the complex health problem faced, due to the comprehensive diagnosis of health and vulnerability, expressed in a "Community Health Index". Among the indicators that were found, the prevalence of dental fluorosis, malnutrition, anemia, exposure to lead, exposure to fluoride and polycyclic aromatic hydrocarbons; social and environmental risks such as low schooling, lack of access to health services, burning of garbage and smoking, high rates of violence, alcoholism, drug addiction, child labor, among others (Domínguez-Cortinas 2009).

In the study area, work was done in two schools, the municipal preschool "Salvador Nava Martínez" and the federal primary "Profra. Dolores Reyes Velázquez", in both schools, the population was exposed to different environmental risks that affected their health, being children the most affected.

3 Methodology

Mixed methods were used; the qualitative methodology in this case was useful, as it allowed for a better approach and understanding of the problems, as well as the perceptions, knowledge, and habits that the people involved in the project had (Flick 2002; Curry et al. 2009). On the other hand, the quantitative research gave us the opportunity to generalize the results from a numerical perspective and assess the magnitude of these numbers, analyzing and studying the data through statistical procedures. The study complied with the ethical principles of the 2008 Declaration of Helsinki, and was approved by the Bioethics Committee of the Faculty of Medicine of the UASLP. Written consent was provided by the involved educational authorities and by the parents of each participant. When presenting the project, we stressed that participation was voluntary. Apart from the parents' written consent, students had to give their oral consent in order to take part in the study. We also communicated clearly that they were allowed to withdraw from the study at any time.

3.1 Design and Implementation of the RCP

Based on the health risk assessment and the risk perception analysis previously conducted (Domínguez-Cortinas 2009; Domíngez-Cortinas et al. 2013; Börner et al. 2015; Flores-Ramírez et al. 2017), the RCP was designed; the approach followed was that of self-managed, participatory, adaptable and collective education; and the principles of the Universal Design for Learning (Gronneberg and Johnston 2015) which propose the elaboration of a design which seeks to achieve meaningful learning.

Risk Communication as an intervention strategy seeks to involve the people in the community throughout the process (Cubillas-Tejeda and González Mares 2015), so the first step involved working with parents and teachers, due to the role they play in habit, knowledge and attitude acquisition in their children and students, and to give continuity to the program so it can be extended to include their families and future generations. As such, the target audience in this first stage was made up of 62 parents and 10 teachers who were part of two learning centres. The risks to which the children are exposed and which would be the focus of the RCP were selected by parents and teachers through the Health Risk Evaluation tool, as well as their own opinions and needs. Lead and fluoride exposure and child parasitism were chosen as the targets. In order to approach the risks chosen by the population, the following topics were selected: personal hygiene, hygiene in the household, hygiene in public spaces, hygiene in schools and healthy diet. The implementation of the RCP was through the *Consorcio Académico: Niño, Casa, Ambiente y Salud* (CANICAS), integrated by researchers and students of the Faculty of Chemical Sciences and the Faculty of Medicine of the UASLP.

To better communicate the messages to parents and teachers, different tools and strategies were utilized. The channels used to transmit messages were: Power Point presentations, presentations using flipcharts and markers to write with, debate and reflective groups, videos, games, images, didactic material, role-playing games, drawing pictures, and reflective writing. In the meetings, we had the support and participation of the parents, since the workshops built upon what teachers and families knew about the topics, and their knowledge, perceptions, beliefs, habits and desires were later studied, reaching a favourable exchange of information. Once the first stage of key message transmission to adults was achieved, we worked in the construction of communications of the PCR aimed at the children.

The second stage, which had the support and participation of parents and teachers, involved working with 338 students (3–15 years old). To design and implement the PCR directed at children, work was conducted with teachers, authorities in both schools and parents. Experts in the reviewed topics were also consulted, considering work in community education and the characteristics and needs of the neighbourhood and school. The messages designed for the RCP aimed at students were built with the help of parents and teachers, based on the premise that each student learns differently and that, in the same group, there are different learning paces. As such, the first challenge the team had was to design activities that the teachers could tailor to a students' needs and grade, i.e. to achieve a contextualized and meaningful learning. The result was a RCP manual composed of over 35 activities to work through with the children at both schools. The means of communication used were tailored to the students'characteristics and care was taken to make them appealing so they could reach proper transmission of the key messages. The channels of communication used were theatre, movies, videos, images, pictures, stories, experiments, specific material, games, posters, leaflets, drawings, thermoses so they could drink water, tablecloths and refrigerator magnets. The application of the RCP began in October 2013 and concluded in May 2014. Each teacher, considering the characteristics of their group and students, chose and adapted the activities to be conducted based on the RCP manual. Each activity was video recorded and later, with the group of teachers, was revised and made changes that were included in the manual, also result non-participant observation notes were taken.

3.2 RCP Assessment

The assessment was conducted throughout the process systematically and continuously; formative, process and outcome assessments were used. With regards to the formative assessment, the topics and subtopics were evaluated so adjustments could be made if necessary, the designed activities, materials and teaching resources were verified. For this kind of evaluation, the particular characteristics of each teaching centre, grade and group in which the strategy was implemented were considered. In terms of the process evaluation, each teacher's compliance with programed activities was evaluated, as well as the participation of everyone involved in the process.

In outcomes assessment, the population of children were evaluated through three different strategies: drawings before and after RCP to compare changes in knowledge and perception surrounding the topics which were worked on; focus groups before and after the RCP to compare knowledge, attitudes, perceptions and habits; and lastly, observations conducted at different times in the school day (arrival, during class, before lunch, during recess, after physical education, end of school day). The observation was conducted throughout the entire process, with the goal of identifying any change in students' habits. Through these evaluations, a triangulation of the obtained data was made through the design and implementation processes. Another way to evaluate program effectiveness was through obtaining children's levels of fluoride in urine and lead in blood.

3.3 Drawings Before and After the RCP

The procedure followed consisted of asking a question, which children had to answer with a drawing. Once the drawings were finished, each child was asked about the elements in the drawings, and notes were taken when necessary. Afterwards, the drawings were subjected to content analysis (Álvarez-Gayou Jurgenson 2003).

Content analysis is a useful technique to analyse several communication processes in different contexts. First, all the drawings were reviewed and, based on the elements uncovered, thematic categories were established. Afterwards, the elements in the drawings were classified in the established categories and they were tallied to get the frequency with which elements were drawn in the established categories (Sánchez-Aranda 2005; Meza-Lozano et al. 2016). The categorization process was conducted in a Microsoft Office Excel spread sheet. Figure 1 is an example of a drawing which can help to explain how the categorization of the elements was done.

To determine if there was a difference in the proportion of children who drew in the different categories, before and after RCP, Chi squared test or Fisher's exact test

Fig. 1 Drawing made by a boy in preschool about what he thinks hygiene means

were used. Significance level was set at $p \leq 0.05$. The software JMP® 10 was used for this analysis.

3.4 Focus Groups Before and After RCP

Focus groups were conducted which were directed at the topics worked through by the RCP, before implementation of the program and after, so results could be compared. In elementary school and preschool, each focus group was made up of 10 students, trying to keep the same children engaged in both sessions. The estimated time spent was 20–30 min. To analyse the information, each meeting was recorded so that the most relevant information could later be transcribed (Onwuegbuzie et al. 2011) Notes were taken in the session as well, which helped to complete the information in the recordings. The questions were the same during the two times that the focus groups were applied, and looked for all participants to share their opinions and points of view.

3.5 Observations

Observations conducted at different points in the program were an element which provided valuable information regarding how the RCP was working. After each visit or activity, notes were taken regarding events considered important, obtaining an overall picture of the participants, children and adults, involved in the RCP.

Observations before, during and after the CRP were made during recess, taking notes of the products they sold at the school store, the food the students consumed most frequently, and also what the teachers consumed during their recess. Also, hygiene habits related to hand washing before consumer foods or after going to the bathroom were noted. In addition to the notes on these moments already described, photographs were taken. Another important moment was after the physical education class, since it is a time of physical activity, the children sought to hydrate themselves, so it was noted if the students had a bottle to drink water or went to the bathroom to take water from the tap.

Observation was also carried out within the classroom, mainly in the execution of the didactic activities of the RCP, to take note of the effectiveness or failure of the activities. During observation within the classroom, consideration was given to whether the materials were ready, if the teacher knew the development of the activity and knowledge on the subject, and the students' response. Observations of the activities were accompanied by photographs and videos or audios.

The annotations were written in a notebook, taking special care in identifying data, such as date, place of observation, observed moment and participants, so that data could be systematically analyzed throughout the process.

4 Results

4.1 Knowledge Integration

In preschool, the instruction "*Draw everything you think hygiene means*" (Fig. 2), assessment prior to the RCP revealed that only 31% of children drew elements related to hygiene; after the RCP, 95% of the children drew more than one element which corresponded to one or more categories, with a statistically significant difference ($p < 0.0001$) (Figs. 3 and 4). After the program, subcategories like tooth brushing, hand washing using soap and not eating toothpaste (Fig. 4).

With regard to the instruction "*Draw everything you think a healthy diet means*", (Fig. 5) new categories and elements were found after the RCP. "*Drinking/cooking with bottled water", "Eatwell plate", "Not eating junk food*" "*Super protein (Legume + cereal)", and "Handwashing before eating*". After the RCP, no children drew in the "*Sugar and Fats*" category, associating them with healthy food, in comparison with 12% who did before the RCP ($p \leq 0.05$). The percentage of children who drew "bottled water" in association with healthy drinks, went from 24 to 83% before and after the intervention (Fig. 6), with a statistically significant difference of ($p < 0.0001$).

In the elementary school, when instructed to "*Draw everything you think hygiene means*", the children drew elements associated with the concept 67–98% before and after the program respectively, with a statistically significant difference of ($p < 0.0001$) (Fig. 7). Other categories that had a statistically significant difference ($p \leq 0.05$) before and after the program were "*Personal hygiene*", "*Hygiene inside and outside the house" "Bottled water consumption", "Washing and disinfecting food"* (Figs. 7 and 8). When asked to "*Draw everything you think a healthy diet*

Fig. 2 "Draw everything you think hygiene means". Description: Drawing made by a girl in the third grade of preschool before and after RCP

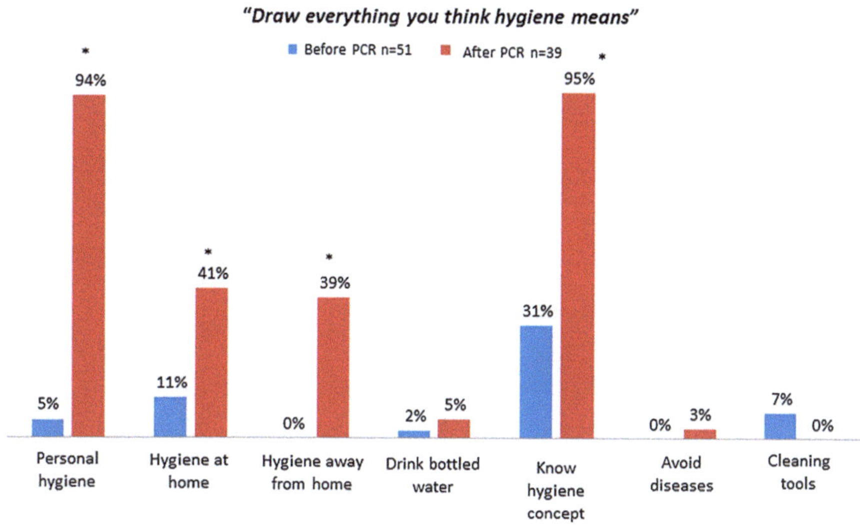

Fig. 3 Percentage of children who drew in the established categories regarding hygiene. "Salvador Nava Martínez" preschool ($^{*}p \leq 0.05$)

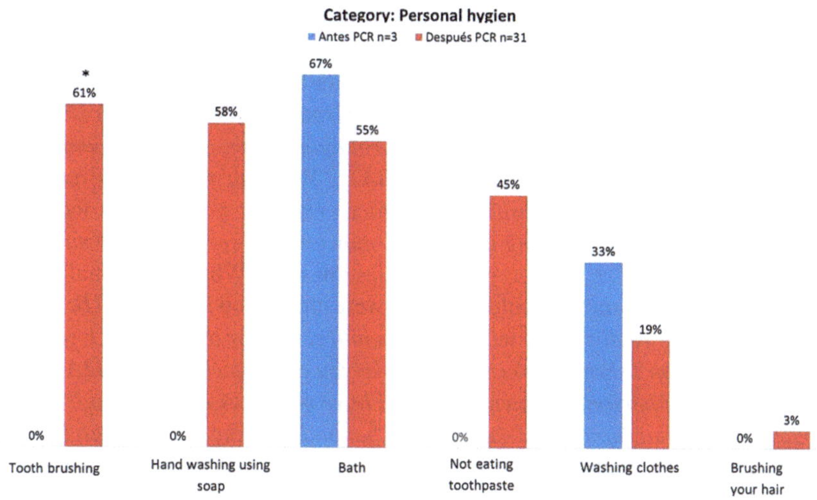

Fig. 4 Percentage of children who draw in the established subcategories regarding personal hygiene. "Salvador Nava Martínez" preschool ($^{*}p \leq 0.05$)

means", new categories after the RCP appeared, "*Exercising*" and "*Avoiding cooking in glazed clay pots*". Elements like "super protein", "hand washing before eating" and "Drinking and cooking with bottled water" also appeared.

Fig. 5 "Draw what healthy eating means". Description: Drawings of a 3-year-old preschool girl before and after RCP

4.2 Focus Groups

To analyse the information categories were formed for each of the questions, in which the answer was compared before and after the RCP. Here, we note the answers a child in preschool gave before and after the RCP to the question: What does Hygiene mean to you? ***Hygiene Focus Group Before (HFGB)****: "Hygiene means to sweep, mop, clean, being clean, changing your clothes, cleaning your shoes".* ***Hygiene Focus Group After (HFGA)****: "Washing your hands and showering, brushing our teeth, washing your face, washing your hands with soap, showering using soap, scrubbing our heads, showering every day" "not sucking on things because they are dirty".* Another question was: Why is personal hygiene important? The answers the children gave were the following: ***HFG1B****: "So we aren't dirty, if we are dirty they will not let us into school or they can kick you out of school... if you don't shower, you will look like you are homeless"* ***HFG2A****: "If you don't take care of personal hygiene, you will have microbes in your hands because you don't wash your hands and you eat like that and then the microbes go into your stomach and that is why it hurts, that is why you wash with soap, so the microbes fall off", "if we shower we will look very pretty and smell nice and you will not be itchy", "we have to wash our hands and shower so our stomachs won't get sick because we ate microbes".*

In elementary school, results obtained before and after about "Healthy diet" were as follows. The groups were asked the same questions before and after; what do you think a healthy diet is? Their answers were: ***HFG1B:*** *"Eating a little bit of everything", "eating fruits and vegetables", "if you eat a lot of junk food, you can get obesity and that will kill you"* ***HFG2A****: "Eating fruits and vegetables, and less fats" "eating a variety of foods, a bit of everything and not eat meats in excess",*

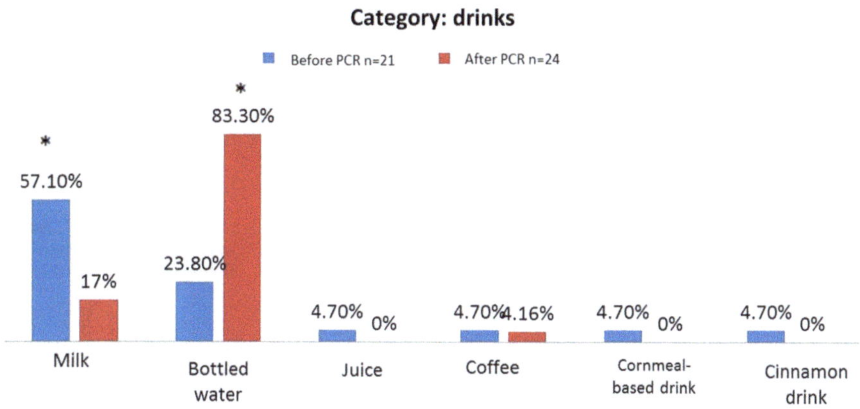

Fig. 6 Percentage of children drawing in the subcategories established in the Beverages category: Preschool "Salvador Nava Martínez" ($^{*}P \leq 0.05$)

"eating a lot of beans, bean 'taco' because when there is no meat, beans and tortilla have vegetable protein, the super protein", "not putting so much salt or sugar in food" "watching what you eat", "not eating so much 'churro', soda, potatoes, candy, 'valentina' sauce, 'salchipapas'", "drinking less soda", "eating well to avoid some diseases like diabetes, because then you have to get injections and that person has to take care of themselves too much, we have to take advantage of our health or you have to take a lot of medication", "not using glazed clay pots". The children in preschool and elementary school could understand the relationship that exists between hygiene and health.

4.3 *Observation*

A change in habits was observed in some children during and after the program and a fundamental factor in acquisition and maintenance of new behaviours was different in those children who had the support of an adult who was important to them, teacher or parent, being much more positive and lasting (Fig. 9).

It was also noted that at the conclusion of the PCR application during recess, the students included in their lunches beverages such as natural water or flavoured water, there were still pupils who drank juice or soda, but they were less students once the intervention was finished. Also, teachers during the implementation of the program began to carry other types of food, more varied and healthy since some of them said that by example was also modeled, and that they could not talk about healthy eating to their students if them owns habits were not healthy. Although not

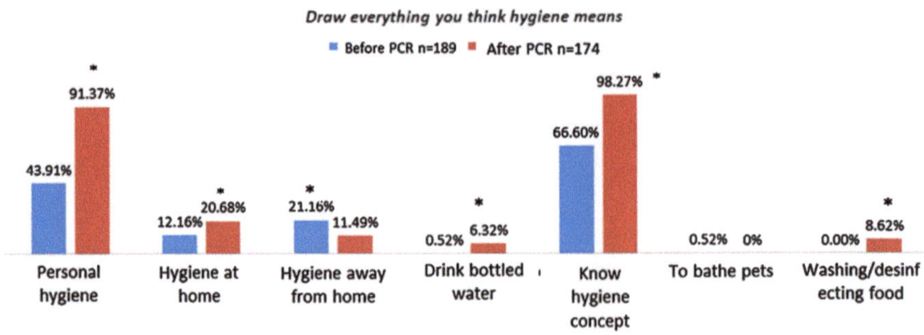

Fig. 7 Percentage of children who draw in the established categories for Hygiene "Profra Dolores Reyes Velázquez" Elementary school ($^{*}p \leq 0.05$)

Fig. 8 Draw everything you think hygiene means. Eight year old student

everyone and not always the teachers were carrying healthy food, there was an important difference. Another point in favor was that the school store began to offer other foods at the request of students, such as fruits or chopped vegetables and natural water.

As for hand washing, it was also observed in most of the participating families that children were washing their hands correctly and with soap. The most important observation is that those families who participated actively during the whole intervention had better results, observing more personal hygiene in their children and greater care in the food they took to eat during the recess (Fig. 9).

4.4 External Evaluation of the Project

In the preschool, the assessment prior to the RCP was conducted in 2011 with 22 children. After the RCP, 23 children were assessed in 2014 (Table 1). Results showed that a statistically significant decrease in mean level of fluoride after RCP ($p < 0.001$). In terms of levels of lead, the mean and concentration range decreased as well. The percentage of children above the limit also decreased, however, this difference was not statistically significant. In the elementary school, 13 children participated (Table 2) and the results showed a statistically significant decrease in mean level of fluoride after the RCP ($p = 0.003$). In terms of lead levels, a statistically significant decrease in mean after the RCP was also found ($p < 0.001$). A parasitical stool test was conducted on the children who participated in biological monitoring before and after the RCP. Forty five percent of children in the preschool tested positive before the RCP and 43% tested positive after. In the elementary school, percentage of infected children went from 40 to 22% after the program.

Fig. 9 Child and his mother practicing the washing of teeth; Girl in front of the mirror brushing her teeth (*Photo* Claudia Monsiváis)

Table 1 Descriptive statistics of neurotoxic levels before (2011) and after the RCP (2014). Salvador Nava Martínez Preschool

	Parameter	N	Mean and SD	Range	Reference value (RV)	% > RV
Before	Fluoride in Urine (mg fluoride/g creatinine)	22	1.72 ± 1.11	0.60–4.40	1.5	12 (55%)
After		23	0.85 ± 0.30	0.32–1.45		1 (4.3)
Before	Lead in blood (μg Pb/dl Blood)	23	5.41 ± 3.32	<3.0–17.79	50.0	12 (52%)
After		23	3.98 ± 2.29	<3.0–9.8		7 (30.43%)

Table 2 Descriptive statistics of neurotoxic levels before (2011) and after the RCP (2014). Profesora Dolores Reyes Elementary School

	Parameter	n	Age range	Mean and SD	Range	Reference value (RV)	% > RV
Before	Fluoride in Urine (mg fluoride/g creatinine)	13	6–9	1.64 ± 0.83	0.34–3.53	1.5	6 (46%)
After		13	8–11	0.75 ± 0.25	0.4–1.4		1 (7.6%)
Before	Lead in blood (μg Pb/dl Blood)	13	6–9	5.13 ± 3.27	<3.0–13.41	5.0	7 (53.8%)
After		13	8–11	2.71 ± 2.21	<3.0–7.8		2 (15.38%)

4.5 *Risk Exposure Questionnaire for Parents*

When parents were asked about the use of glazed clay pots, before the RCP, 60% didn't use them, after the RCP, 89% of parents mentioned not using them. When asked about the use of purified water for drinking, there was a difference of 67–91% before and after RCP. The use of purified water for cooking also had a difference of 18–66% before and after RCP.

4.6 *Limitations of the Study*

One of the limitations of the present study was that the PCR didn't integrate aspects that are also important to address, for example, the violence that exist in this area urban-marginalized, or child labour, among others. However, this work was a great help for further works in the same study area, helping the prevention of different environmental and social risks and thus, contribute to the best quality of life of children in the "Las Terceras" neighbourhood.

5 Conclusions

Based on the results obtained an education model based on risk communication was designed to improve environmental health. It can be easily adapted to different learning centres and health risks. This model, which is directed at children, can be contextualized at different learning centres and levels, from preschool to high school, aiming to improve environmental health in populations exposed to similar risks as those in the studied population in this project.

The model used in this RCP can be adapted at different contexts, keeping in mind economic, political, social and environmental factors in each learning centre. Within the necessary components of this model is a previous diagnosis of the school and locality in which it is found, it is also advised that the characteristics of the families and the teachers which integrate this learning centre are assessed. In the target audience evaluation, it is important that the habits and risky and protective behaviours which the members of the school undertake are known, as well as their beliefs and risk perceptions. The evaluation must be continuous and systematic throughout the whole process. In this particular project, formative, process and results assessment were used. To establish goals for the RCP, the opinion of all those involved were considered; this was also the case for the content elaboration and communication construction, relying on different inclusive, constructivist teaching strategies in order for participants to achieve meaningful learning. Additionally, because it is self-managed, there is a bigger chance that the RCP can continue through the years, being implemented in the different education centres and in the participating families.

One of the biggest contributions the work conducted has is to have achieved work in community education through schools, where parents can bring forth their proposals and solutions to other problems. In terms of teachers who actively participated throughout the process, a significant achievement was that they could take initiative in approaching topics regarding the immediate context, make their students engage in reflective thinking, propose new contextualized activities and notice differences in their students, know that, aside from working in the assigned contents, they can approach topics with a more relevant impact in their students' lives. And in terms of parents, they had the satisfaction of having worked in a team, and living the differences in terms of habit change in their children and families.

The situation in the neighbourhood is complex, and despite working with some of the risks identified by the population, there are still significant problems in the area that need to be addressed. However, by getting parents, children and teachers to work in the community and by living the positive changes that this project had, a first step was taken towards working for the place that most children, families and teachers want.

Acknowledgements We thank the children, parents, teachers and directors of the schools who supported this research.

Funding

This work was supported by the Fondo Sectorial de Investigación en Salud y Seguridad Social from the Consejo Nacional de Ciencia y Tecnología (CONACyT) under grant SALUD-142064.

References

Álvarez-Gayou Jurgenson JL (2003) Cómo hacer investigación cualitativa. Editorial Paidós Mexicana, México, D.F

Börner S, Torrico-Albino JC, Nieto-Caraveo LM, Cubillas-Tejeda AC (2015) Exploring Mexican adolescents' perceptions of environmental health risks: a photographic approach to risk analysis. Ciência & Saúde Coletiva 20(5):1617–1627

Chelala C (1999) Impacto del ambiente sobre la salud infantil. En: Salud infantil ambiental. Organización Panamericana de la Salud, Washington

Coronado-Salas C, Díaz-Barriga F, Moreno-Sánchez AR, Carrizales-Yáñez L, Torres-Nerio R, Cubillas-Tejeda AC (2012) La comunicación de riesgos como una herramienta para disminuir la exposición infantil a plomo y arsénico en la zona contaminada de Villa de la Paz-Matehuala, San Luis Potosí, México. Revista Internacional de Contaminación Ambiental 28(2):167–181

Cubillas-Tejeda AC, Torres-Nerio R, Díaz-Barriga F, Carrizales-Yáñez L, Nieto Caraveo LM, Coronado-Salas C et al (2011) Diseño y Aplicación de un Programa de Comunicación de Riesgos para la Salud Ambiental en un Sitio Contaminado con Plomo y Arsénico. Ciencia & Saúde Coletiva 16(10):4115–4126

Cubillas-Tejeda AC, González-Mares MO (2015) La comunicación de riesgos como estrategia de intervención para mejorar la salud ambiental en poblaciones vulnerables. In: A. Mendieta P. Testas (ed), ¿Legitimidad o reconocimiento? Las investigadoras del SIN: retos y propuestas (p.p. 505–512). Benemérita Universidad Autónoma de Puebla, Ediciones La Biblioteca, México, DF

Curry LA, Nembhard IM, Bradley EH (2009) Qualitative and mixed methods provide unique contributions to outcomes research. Circulation 119: 1442–1452. http://circ.ahajournals.org/content/119/10/1442.full.pdf+html. Accessed 27 Feb 2017

Díaz-Barriga F, García S, Corra L (2010) Prevención de riesgos en poblaciones vulnerables en: Determinantes ambientales y sociales de la salud, Organización Panamericana de la Salud. LA Galvao L, J Finkelman, S Henao (ed). McGraw–Hill Interamericana, Organización Panamericana de la Salud, Distrito Federal, México, pp 139–153

Domínguez-Cortinas G (2009) Evaluación del impacto del fenómeno de iniquidad ambiental en la salud de poblaciones infantiles en San Luis Potosí, SLP, México. PMPCA. Tesis de Doctorado, Universidad Autónoma de San Luis Potosí

Domínguez-Cortinas G, Díaz-Barriga F, Martínez-Salinas R, Cossío P, Pérez-Maldonado IN (2013) Exposure to chemical mixtures in Mexican children: high-risk scenarios. Environ Sci Pollut Res 20(1):351–357. doi:10.1007/s11356-012-0933-x

Erbe SO (2011) Technical, economical and organizational analysis of informal brick production in Tercera Chica, SLP, México. PMPCA. Tesis de Maestría, Universidad Autónoma de San Luis Potosí

Flick U (2002) Introducción a la investigación cualitativa. Morata S. L, Madrid, España

Flores-Ramírez R, Pérez-Vázquez FJ, Rodríguez-Aguilar M, Medellín-Garibay SE, Van Brussel E, Cubillas-Tejeda AC, Carrizales-Yáñez L, Díaz-Barriga F (2017) Biomonitoring of persistent organic pollutants (POPs) in child populations living near contaminated sites in Mexico. Sci Total Environ 579(2017):1120–1126

Gronneberg J, Johnston S (2015) 7 things you should know about universal design for learning [Brief]. Educause Learning Initiative. Retrieved from http://www.educause.edu/library/resources/7-things-you-should-know-about-universal-design-learning

Meza-Lozano B, Ortiz-Pérez MD, Ponce-Palomares M, Castillo-Gutiérrez SG, Flores-Ramírez R, Cubillas-Tejeda AC (2016) Implementación y evaluación de un Programa de Comunicación de Riesgos por exposición a flúor en la comunidad de El Fuerte, Santa María del Río, San Luis Potosí, México. Revista Internacional de Contaminación Ambiental 32(1):87–100

Moreno-Sánchez AR, Cubillas-Tejeda AC, Guerra-García A, Peres F (2010) Comunicación de Riesgos en América Latina. In: Galvao LAC, Finkelman J, Henao S (ed) Determinantes ambientales y sociales de la Salud. Organización Panamericana de la Salud, McGraw–Hill Interamericana, México, D.F, pp 279–300

Onwuegbuzie AJ, Dickinson WB, Leech NL, Zoran AG (2011) Un marco cualitativo para la recolección y análisis de datos en la investigación basada en grupos focales. Paradigmas 3: 127–157

Prüss-Üstün A, Corvalán C (2006) Ambientes saludables y prevención de enfermedades: hacia una estimación de la carga de morbilidad atribuible al medio ambiente. Organización Mundial de la Salud. Recuperado de: http://whqlibdoc.who.int/publications/2006/9243594206_spa.pdf
Sánchez-Aranda J (2005) Análisis de contenido cuantitativo de medios. En: María Berganza y José Ruiz (ed) Investigar en comunicación. Guía práctica de métodos y técnicas de investigación social en comunicación. McGraw-Hill, Madrid, pp 207–228

Author Biographies

Claudia Davinia Monsiváis-Nava, studied the bachelor's degree of psychology at the Universidad Autónoma de San Luis Potosí (UASLP), later, she studied a Masters at the Programas Multidisciplinarios en Ciencias Ambientales in the UASLP. Actually, she is currently studying for a doctorate in environmental science.

Ismael García-Cedillo, has a Master and Doctorate in psychology at the Universidad Nacional Autónoma de México (UNAM), México D.F. He is currently a researcher and full time professor of psychology at the Universidad Autónoma de San Luis Potosí.

Leonardo Márques-Mireles, studied the bachelor's degree in the faculty of anthropology UAEMex. Later, he holds his master's and doctorate in the area of social anthropology in the Universidad Iberoamericana. He is currently a full-time research professor at the UASLP at the faculty of social sciences and humanities.

Rogelio Flores-Ramírez, studied the bachelor's degree of Pharmacochemical Chemistry at the UASLP (2007). He holds a Master of Science degree in Environmental Science from the UASLP, México (2010) and obtained a Ph.D. in Environmental Sciences from UASLP (2014). He is research fellow at the Catedras-CONACYT program. His research area is Analytical Chemestry and risk assessment to human health for exposure to toxic chemicals.

Ana Cristina Cubillas-Tejeda, studied the bachelor's degree of Pharmacochemical Chemistry at the UASLP, (1988). She holds a Master of Science degree in Immunology from the Universidad Autónoma de Nuevo León, México (1993) and obtained a Ph.D. in Basic Biomedical Sciences from UASLP (2000). She is currently a Full Time Research Professor at the Faculty of Chemical Sciences of the UASLP. Her research area and teaching is Risk Communication and Environmental Health.

Integrative Project of Converging Knowledge of Sustainability, with Focus on Migratory Processes, Pest Management and Practices of Traditional Medicine in an Educational Model of Interculturality

Edilma De Jesus Desidério, Marja Liza Fajardo Franco and Laurentino Lucas Campo

Abstract This paper contributes to the current debate on sustainability and interculturality, supported on practices carried out in community linkage, practices that were based on the model of intercultural education in Mexico, in which students and research teachers take part, building a dialogue of scientific knowledge and traditional insight, which converge towards a project that integrates communities and identities. We talk about a theoretical-methodological strategy that converges themes, approaches and problems related to return migratory processes, sustainable management of pests and diseases in coffee crops, and traditional medicine practices of native peoples (Totonac). The results include learning, strengthening of cognitive capacities and positions committed to provide answers that would benefit both the rural community environments and the generation and application of knowledge conducive to possibilities of social, scientific and humanistic transformation.

E. De Jesus Desidério (✉)
Division of Social Processes, Universidad Intercultural del Estado de Puebla, Street Calle Principal a Lipuntahuaca s/n, location Lipuntahuaca, Huehuetla, Puebla, CP 73475, Mexico
e-mail: edilma.dejesus@uiep.edu.mx; edilmabr@gmail.com

M. L. Fajardo Franco (✉)
Academy of Sustainable Development, Universidad Intercultural del Estado de Puebla, Street Calle Principal a Lipuntahuaca s/n, location Lipuntahuaca, Huehuetla, Puebla CP 73475, Mexico
e-mail: marjaliza.fajardo@uiep.edu.mx; azilmar@gmail.com

L. L. Campo (✉)
Academy of Language and Culture, Universidad Intercultural del Estado de Puebla, Street Calle Principal a Lipuntahuaca s/n, location Lipuntahuaca, Huehuetla, Puebla CP 73475, Mexico
e-mail: laurentino.lucas@uiep.edu.mx; laurenlucas53@yahoo.com

W. Leal Filho et al. (eds.), *Sustainable Development Research and Practice in Mexico and Selected Latin American Countries*, World Sustainability Series,
https://doi.org/10.1007/978-3-319-70560-6_25

Keywords Integrative project · Sustainability and interculturality
Return migration · Pest management · Traditional medicine

1 Introduction

In the processes of learning and mastery of capacities from different education levels, new information and communication technologies (ICT) have been introduced so that groups of students acquire skills and improve their skills. However, alongside the inclusion of new technologies, current models of education and development of science and technology are incorporating the revaluation of traditional insights, whose most significant component is the rescue of local capacities, the value and use of traditional knowledge existing in different communities.

With the development of educational policies presently, and the advance of knowledge in Mexico and Latin America, institutions of elementary, middle and higher education often have to expand the spaces that may yield in fertile ground for scientific and technological innovation. This requires that the scientific communities have a dialogue of both technological competencies and sciences towards a much more integrative and inclusive proposal.

As a possibility of advance in scientific knowledge, it is taken into account that what is constructed from the social and political must contemplate the commitment to the cultural and identity contexts, and of course to involve the being (or subject in the process) and its meaning in the world in which s/he participates as a producer and as a product of the new knowledge, derived from teaching and learning.

This paper contributes to the current debate on how to build an integrative project of converging knowledge that includes social sciences, natural sciences and traditional insight, which brings together themes and methodological designs, which is based on a specific intercultural educational model.

The work is structured in sections. The first, "*Convergent knowledge and the dialogue of knowledge with a perspective of sustainability and interculturality*", addresses the theoretical and methodological framework that is built directing the reflection towards the so-called "integrative project of communities of practice".

The second section, "*Participation in the construction of communities of convergent practices: the academic field, linkage and the community*" advances in the contextualization on the integrative project from the different knowledges, with pedagogical practices oriented to exploratory and mainly participative activities that are based on the Intercultural Educational Model.

In "*social learning dialogues and nuances of sustainability and interculturality: about migratory processes, pest management and practices of traditional medicine*", from three research projects that were developed in municipalities that established an agreement with the University, which are also related to lines of generation and application of knowledge, cultivated by researchers, the process of constructing meanings and identities is explored and explained, taking as a model the intercultural community linkage in which teachers, students, social actors of communities, governmental and non-governmental institutional agents and community in general take part.

The three research projects mentioned as examples are based on the intercultural educational model, which with the presence of students of native languages and cultures in the communities of practice incite their commitment in these places, stimulating the participation of residents, fomenting social commitment and providing a fruitful exchange of knowledge.

The sub-section "*Research-action in the integrative project: practices about migratory processes, pest management and traditional medicine practices*", describes the types of learning that are intended to be stimulated with the university linkage; in addition, a configuration is structured, which relates the scope on *what is learned* and *what results* in/from a community linkage.

The presented results contain an approximation of the set of practices that is gathered in those three cases and subjects of study, according to their specific objectives and methodological designs, conforming the universe of intervention experiences produced by the stays of linkage in communities of the region of Sierra Norte of the state of Puebla, Mexico.

2 Converging Knowledges and the Dialogue of Insights, with a Perspective of Sustainability and Interculturality

The scientific community on which knowledge is built from social practice and the dialogue of science and insights at present, has reached a greater potential to generate answers which are closer to the everyday reality, although the implementation of certain techniques or technologies may result in a low or no incidence when results take a course far away from everyday life, and said results do not return to social issues.

The theoretical construction of the integrative project that seeks the convergence between knowledge and dialogues of insight requires to be approached from different perspectives, derived from theories of social practice and that are constructed from the intentions of producing and reproducing ways of participating in the world, in the conformation of daily life scenarios.

Alongside the organization and coordination of groups based on their knowledge, we find identity, cultural interpretation, creation of belonging and the use of affiliation to a thematic line or area of scientific and technological knowledge that guide them in the development of practices.

The integrative project as an ontological strategy is, therefore, the space for dialogue between scientific knowledge and traditional insight that, precisely, supports the collaborative and participatory work.

Another theoretical perspective that is taken into account are the concepts of sustainability and interculturality. It is about a strategy that contains in its genesis a progressive philosophy based on values and principles on the grounds of the constructing a fair, equitable global society, open to dialogue amongst cultures. Contemplating these concepts and bringing them into the practice of community linkage becomes crucial, since both concepts are related to ways of life and the scope of "good living"; and it is essential to teach and learn how to lead to overcome the human needs by means of objectives sustained generationally.

The integrative project also promotes a methodological and epistemological change, since in social relations, which are simultaneously material (between things) and semiotic (between concepts), there are personal and idea interactions, skills, techniques and technologies; all of this, associated, manifest the way of acting and put societies and communities closer, resulting in synergies of knowledge.

Methodologically, the universe of an integrative project is designed from the construction of actor-network and participatory research (Latour 2008); in addition, the acting parties must be involved in a network of associates, and be open to identify, during the progress of the process, what is relevant to continue being taken into consideration and/or to produce decolonizing ruptures of concepts and insights (Santos 2014).

To a large extent, Wenger (2001) indicates that commitment processes in the construction of knowledge gained through participatory research activities result in a "*mutual commitment in a shared practice*", although the scope of results is subject to "*an intricate process of constant adjustment between experience and competence*." Of course, this is not an easy task when the different communities enter into a practice of "*social participation as a process of learning and knowing*"; however, the possibilities for converging dialogues become greater.

The integrative project between the community of practice and the social learning dialogue contains a theoretical framework that conceives "*learning as social participation*", whose action and interaction is given "*in an active way in the practices of social communities*", which leads to construct "*identities in relation to these communities*." Such participation, as the author points out, not only shapes what we do but also shapes who we are and how we interpret what we do (Wenger 2001).

To the theoretical framework is added the epistemology of inter-scientific dialogue that "*starts from the principle that all systems of knowledge in the world are sciences*" that are developed within their own dynamics of interaction and learning, with exchange of methods and research results (Haverkort et al. 2013).

In this search for answers to adapt their own paradigms and jointly create what would be a "*plurality of sciences*", the author points out, it is generated a meaning of complementarity which is, at the same time, "*a previous step to an inter-scientific dialogue*"; that is, the construction of intercultural dialogue is based on the revaluation of local insights and the ancestral wisdom of the native indigenous nations and it is then a pillar that gives meaning to what would be another scale: intraculturality.

On the other hand, opening up to the dialogue of social learning also means to enter into a relational process that includes practices associated with concepts such as interculturalism, which are linked to the understanding of the other from beliefs, insights, norms, aims and values (Olivé 2004; Arguetta et al. 2012). This conception finds place in other notions such as plurality, transcendence, criticality, consequence and temporality, as Paulo Freire (1996) points out. In this way, paths to dialogue are open and ways of facing old and new paradigms strengthen.

In the development of the project to construct community practices, academic and everyday life are linked, which makes that the different actors and the realities of their places interact, activating important elements such as tongue, language and culture. This connection occupies a very significant dimension in the process of approaching the identities of native peoples.

What is the purpose and who are we? The configuration of the integrative project is related to a multi-community clustering system (Fig. 1).

- *Interdisciplinary and multidisciplinary scientific community*: in this context, it is formed out of the development of dialogues of inter and multidisciplinary knowledge, which has to do with belonging to a identity or specific subject area, but also enables an approach towards the generation of scientific alliances of different disciplines and epistemologies allowing, in a more precise way, the realization of what we are calling an integrative project of converging knowledges.
- *Student community of cultural and linguistic identity linked to native peoples*: it contains a double identity (specific cultural and linguistic identity), and belonging (university and disciplinary); it incorporates into the system by

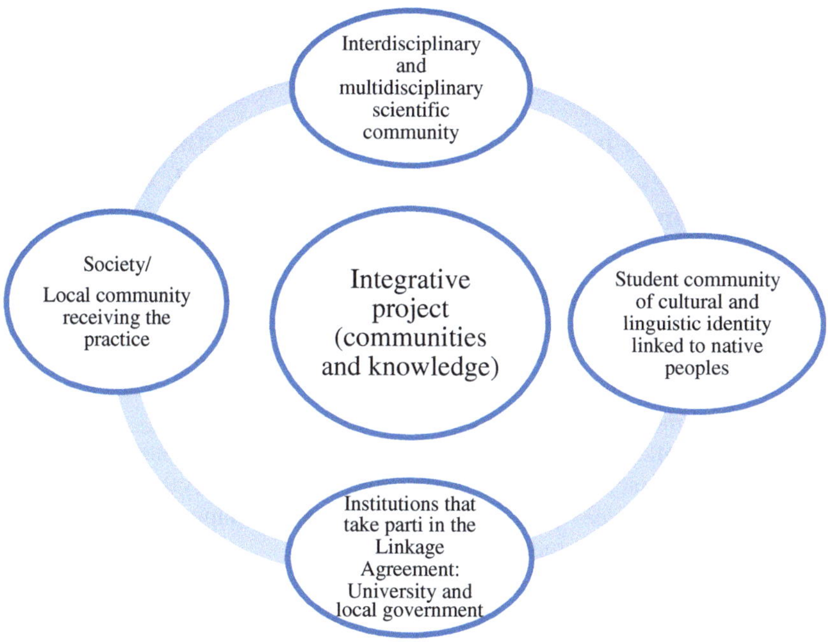

Fig. 1 Configuration of the integrative project of practice communities

integrating on the one hand, with what it brings as a baggage of traditional knowledge, culture and native tongue that becomes, of course, a significant tool when linking; and on the other, the learning acquired in a classroom that becomes social when carrying the knowledges (theoretical and practical) to the receiving society, the rural areas, the agricultural fields, and the daily life.

- *Institutions that take part in the Collaboration Agreement between the University and local government*: they join the project by opening roads or paths that make it possible to materialize knowledge, relationships and exchange of learning from the political, operational, strategic management and collaboration with local governments (Municipalities and localities).
- *The local community and/or society receiving the practice*: rural or product of the changes derived from the new rurality, local societies and communities take an extraordinary importance in favoring particular elements of its historical, social, economic and cultural construction, its modes of production, crop management, worldview and demographic singularities; it is the place of the simultaneity for the production of relations of social learning, exchange and possibility of hegemonic ruptures of insights and knowledges.

The development of an educational model based on these new paradigms of inter-scientific dialogue and traditional intercultural insights is what would promote the creation of other ways of avant-garde learning, by making possible the inclusion of contents, concepts and ideas like bridges amongst practice communities, scientific and social as well, organizational, institutional-educational, pedagogic. All the above, grounded on traditional insights with a sustainability perspective and an intercultural approach.

3 Participation in the Construction of Communities of Converging Practices: The Academic World, Linkage and Community

The project of building knowledge and dialogues focus the interest in participation because it manifests deep repercussions for what it means to understand, support and live the moment of university-community linkage; that is to say, for the different actors, to participate and contribute to the practices of their communities acquires a real meaning.

For the scientific communities, on the other hand, to valuate and take advantage of the space-temporary exchange of knowledges and insights is the most significant thing since it guarantees participation, interest and commitment from the new generations of members (students and teachers).

For organizations (institutions, local governments and civil society of the receiving places), to produce ways of interconnecting with communities of practice, through which an organization knows what it knows and, consequently, makes the experience effective and valuable within the integrative project.

Mexico, as one of the pluricultural countries in Latin America, has required actions by the government educational sectors in favor of the development and strengthening of intercultural education. This approach, of course, constitutes a mechanism for opening up and expanding spaces for linguistic inclusion of a young population of indigenous origin, allowing among them, the promotion of a taste for scientific research related to intercultural education, as pointed out by the Programa Especial de Educación Intercultural (the '*Special Intercultural Education Program*') (2014–2018) in its diagnosis.

The Intercultural Universities are institutions attached to the state governments and are born linked to autonomous educational institutions; those are product of various social processes, as well as of specific educational policies, both national and international (Jablonska 2010; Bastida Muños 2012; Dietz and Mateos 2013).

They constitute a substantive part of the Intercultural Education Model; its pedagogical proposal, with an exploratory and participative orientation, whose principles are based on the establishment of relationships and formative activities of teaching and learning, involve internal and external social actors in actions of knowledge exchange and practices to promote community development, in all its areas.

Based on the Model of Intercultural Education to enrich the integrative project, it is sought to establish a relational channel between university education and learning, enabling capacities to interact with community social actors. Practices provide the possibility to participate in a moment where the student and teaching communities carry out didactic-pedagogical actions and exchanges of knowledge such as: approaches to problems that can stimulate studies, exploration and gathering of information to generate community diagnoses and training to become a manager or consultant in the community.

The practice contains specific objectives such as: developing areas of synergy with the community, supporting organizations through the implementation of training activities, thematic workshops, etcetera; that is to say, the academic work is addressed to the receiving communities, with the aim of giving them attention in certain problems that may require knowledge that comes from the scientific academic environment; this, in fact, enriches the trajectory of the student community.

As a methodological tool in the teaching-learning process, the project offers the opportunity for an advance and/or maturity in the development of critical thinking of students, and a thoughtful look at everyday practice.

In addition to the theoretical framework based on the dialogue of knowledge and converging knowledge, the integrative project has required a methodological design based on the Research-Action-Participation (Teppa 2012), where the IAP (for its acronym in Spanish) is applied as a strategy committed to the social context in which the practice is carried out.

In this context, the development of activities are directed to knowledge and exchange of capacities and insights, including, among other things: contributing to the search for possible solutions to problems related to daily life and according to the competencies of students, resulting in the strengthening of the intercultural and multidisciplinary identity within the community of practices in the group and collaborative environment.

4 Dialogues of Social Learning and Nuances of Sustainability and Interculturality: The Origin of the Integrative Project

The process of development and maturity of the community integrative project and practices of knowledge, which has been theoretically and methodologically grounded, keeps a direct relationship with the creation of knowledge generation and application lines (LGAC, for its acronym in Spanish), that consists of grouping thematic fields in which the trajectories of a community of researchers that integrate a certain basic academic core of an educational program converge, as well as work done by students, from a systemic perspective.

From a perspective of sustainability and interculturality, projects are elaborated, focused on different themes such as those taken as examples in this reflection that deal with: "migratory processes", "management of pests and diseases in coffee crops" and "practices of traditional medicine". The form of convergences of knowledge and learning is reflected precisely in the results of an institutional project called Community Linkage, developed by all intercultural universities.

In general, the view of an integrative project is constructed by relating it to the project of linkage given the similarity in some stages that can be defined as: (a) exploratory, (b) falling in love, (c) commitment and participation, and (d) construction of identities; all above strengthened by two significant frameworks that are sustainability and interculturality.

What is sought? To be a dialectical process of knowing-acting; that can generate enough synergy to achieve the commitment with the community to implement transformative actions in a short time; to promote sustainable processes, that is, that contacts that are made become satisfactory, and the follow-up so the present generations and those that will become incorporated in a future to the university, remain in love with the practices; it is also expected to sensitize the different social actors on the importance of establishing productive dialogues and co-responsible between university and social environment.

To learn by means of participating in an interactive process involves teachers, students and social actors who receive in their places and *homes*, the brigades that perform the linkage. At this moment, the practice becomes the means by which is explored and falls in love with a learning that each community contributes to the project. It is also the transforming moment, in which affiliation or rejection are manifested, the incorporation of an identity, commitment, action and reaction since the experience is built day by day over two weeks per semester.

In the execution of the integrative project, the methodological design that is prepared in advance, albeit strategic as a guide, takes second place when faced significant questions of relevance, time, space and power; because, as Wenger (2001) points out: "*learning cannot be designed: it can only be facilitated or frustrated.*"

To linkage processes join different actors: government, who sign the agreements and provide the availability of its infrastructure to carry out the work, civil society, other actors (of security, of health, etc.) that allow the coming together with the resident population. But, it is the towns with their customs, knowledge of its bio-cultural, gastronomic riches, and the warmth with which it receives the grouping of communities what contributes to the dialogue of intercultural learning.

In the search for dialogue, much more than elements to build an institutional response, the origin of the integrative project is, therefore, in the construction of identities and learning that are strengthened with interdisciplinary commitments and benefits communities in which intercultural practices are performed, with new and revitalized fragments of knowledge, in which the student groups have participated.

Together and through different disciplinary paths, the groups are shaping scientific identities in the integrative project of social learning, whose main commitment is to return knowledge to the social milieu and to communities of traditional knowledge.

4.1 Configuring the Social University Linkage with Practices on Migratory Processes, Pest Management and Traditional Medicine Practices

For the purposes of this work, we have called the *Integrative Project* a didactic-pedagogical strategy that, by conforming and putting into practice themes and dialogues developed by social and natural scientists under a model of higher education in an academic environment, whose mission and vision are based on the axes of sustainability and interculturality, are constructing socially, culturally and linguistically committed interventions in rural, semi-rural and indigenous populations and communities.

Material and symbolically, the project contemplates the promotion of educational practice and intercultural dialogue between the communities that emit and receive knowledge. It is spatially-temporary configured in an environment of exchange of insights, which is called community linkage, in which, periodically, groups of students put into practice their proficiencies based on aptitudes and values, knowledge and skills acquired in the classrooms that should be applied in specific subjects and problems of the social and cultural reality of the places.

The intercultural educational model is deeply related to joint efforts (government, local and community) to promote a quality higher education, with cultural and linguistic relevance, as it was from 2004 the creation of a Subsystem of Intercultural Universities (SUI by its acronym in Spanish).

The purpose of the UIs has been to constitute themselves as an advance in the decolonization of a university system that does not reach to include a part of the population, in its majority with diverse linguistic origin; in that context, bringing higher education institutions closer to or relatively close to their places of residence, to regions and communities with high levels of poverty, marginalization and vulnerability, would represent an historic effort to include indigenous language speakers.

The project of community linkage that is carried out within the Intercultural Education Model (MEI, by its acronym in Spanish) is, as Casillas and Santini (2006) points out, "both a formative space, and a permanent interaction tool with surrounding communities"; it contains different forms of both elaboration and understanding as of execution of the project, as required by the Subsystem of Intercultural Universities, when putting into practice the dialogue of insights and cultures.

The *Model of Linkage with the Community* constitutes a substantive part of the Intercultural Universities and in the different local governments where they are located. Its pedagogical proposal is based on principles and the establishment of relationships and formative activities of teaching-learning that involve the diverse actors in actions of investigation, systematization, formulation of programs and projects; all with the aim of promoting community development in many areas.

Within the curriculum map, the linkage is considered as the methodological axis of the MEI to be carried out with theoretical classes and specific practical workshops since all IUs have to be taken into account, although each institution develops it under its own pedagogy and the purpose to carry out the extra college stays.

Community linkage is important because of the research approach that serves as a framework to be carried out under the methodology of Research—Participatory action; its purpose is to improve the capacities of the actors involved, especially through a continuous process of analysis, reflection and action.

Another significant aspect of the linkage is that it promotes horizontal relations among the project members and the community, promoting and strengthening attitudes of solidarity and collaboration among the participants, within a framework of inclusion and respect; thus, the students function as facilitators of the processes of change, feed backing the experience, which is product of the dialogue of insights with the actors and their disciplinary tasks.

Based on three topics related to generation and application of knowledge lines, the experience of community linkage is made known, taking into account that what strengthens the integrative project of practice communities and converging knowledge, is also reflected in the educational programs in which they are inserted, such as Community Forestry Engineering, Sustainable Development, Alternative Tourism, Language and Culture, Law with an Intercultural and Nursing approach, as an educational offer of the UIEP.

4.1.1 Research-Action in the UIEP Integrative Project: Practices on Migratory Processes, Pest Management and Traditional Medicine Practices

The research-action, as the methodological axis that underlies intercultural education, is strongly worked both in the training field of students and, above all, in the practice of linkage, which, in the case of the Intercultural University of the State of Puebla (UIEP, for its acronym in Spanish), is carried out in the municipalities and rural, semirural and indigenous communities of the Sierra Norte region of the state.

The pedagogical model contemplates the development of practices through two types of intervention:

I. Linkage Projects;
II. Dissemination and University Linkage.

These types of intervention are subdivided into two categories:

1. Formative Projects of Research-action;
2. Collaboration Projects

The Research-action Training Project is the first stage of an approaching between the community and the student linkage brigade whose objective is to develop interest in research as a form of learning and, at the same time, to promote actions or tasks implemented in the same community of students, which shall be a product of the skills obtained in the classroom.

The Collaboration Project basically comprises the coverage of students in two aspects: (1) the training needs of students in different levels or academic cycles and (2) the needs of the community concerning some demand in specific, either by the institutional agent who signs the collaboration agreement with the university, or the sectors of the communities in the municipality.

In addition to these categories, with the promotion to the generation and application of knowledge that the university's teaching community has developed, taking advantage of the moment of community linkage becomes important for the production of studies and information gathering for research projects. This is specifically the case of studies carried out on "return migration", "sustainable management of pests and diseases in coffee crops" and "traditional medicine practices in Tutunakú indigenous peoples."

Regardless of the issues that are under study in the realization of community linkages, the specific objectives of each proposal in all their dimensions and projections unify actions such as:

- The development of activities related to the issues
- The generation of specific attention strategies related to demands of problems corresponding to each community, in both sustainable development and intercultural issues
- The delivery of possible solutions built along the linkage practices.

The studies are grounded on aptitudes to be developed, which are:

- Apply acquired knowledge to solve problems in new or little known environments within broader (or multidisciplinary) contexts related to the object of study.
- Perform a critical analysis, evaluation and synthesis of new and complex ideas.

In the research projects carried out in the municipalities of the Sierra region of the state of Puebla that have signed agreements with the Intercultural University, whose approaches are: (a) *Return migration processes and social-work reintegration*, (b) *Dialogue of knowledge from the Tutunakú medicine*, and (c) *Sustainable management of pests and diseases in coffee crops*, the scope in the development of aptitudes by the participants are addressed to:

1.To know: Understand, analyze and apply in the practices, the basic concepts approached in the thematic units of the classes	**2.Knowing** how to do: A critical and synthetic analysis of the development of the issues and problems that are faced in the daily practices of the communities receiving the linkage	**3.Being:** Capable to propose actions and interventions in the environments experienced in the practices of community linkage

In this context, is presented the matrix about what is learned and what results from an integrative project in the community linkage, for the three experiences, that within the results contemplate, as a whole:

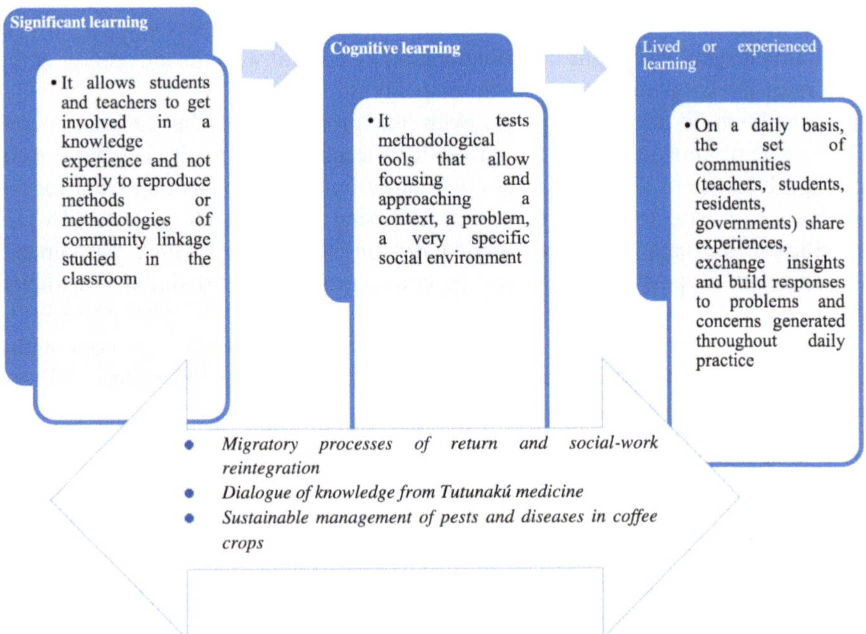

- **Main results summarized in the three experiences**

In the three researches, the methodological design contemplates actions and strategies such as: (a) realization of a diagnosis as a first approach to the practice of daily life of the peoples, (b) later, the application of pedagogical strategies with approaches of scientific or technological intervention as required by the problem or situation to be solved.

A point of intersection in all three cases is that practices are based on the epistemological pluralism that recognizes the knowledge system of said peoples (Olivé 2009; Zambrana 2014) and contributes to the collection and exchange of traditional and scientific knowledge.

Another common aspect that is expected is that the capacities of the groups of students be strengthened so that they recognize the accumulation of insights that have generated historically the native peoples, and that in the future they, by themselves, be those who would carry out research and interventions based on their professional formation, valuing their belonging to rural and native communities, thus allowing them to become native researchers (Bertely 2011).

In the learning process interaction-experience, identities, cultures and native tongues are articulated and these elements become significant to detect the strengths of some practices that coexist in the community linkage concerning agriculture, food preparation, health care, education, conservation of the environment, among others.

In all three cases, it is possible to promote learning through the provision of service in community contexts or learning based on a specific service (Díaz Barriga 2006), or sector that is related to the profile of the graduate of its educational program.

The projects include training in certain tools such as the application of interviews conducted by the student community that make up brigades, especially from the mother tongue (Tutunakú or Nahuatl) and the methodological designs previously studied, that will be applied in the practices to social actors, receptors of the places where the linkage is carried out, who share their experience.

The way in which the practices are carried out modifies the understanding of many students in terms of knowledge and insights about what they perceive as meaning by establishing intercultural and interdisciplinary dialogues that positively affect their social, professional and human performance in the future.

In all three cases, it is perceived the concern for the fact of no longer reproducing in the young generations the custom of keeping their traditional practices in operation. This represents an important pedagogical intervention gap for the scientific community.

Derived from the tools applied during the stays of university linkage in the social aspect, it is evident that, at each period, there is a greater interest from the resident population to know the techniques and technologies that are being taught in the courses, reason why a didactical strategy that is validated in this process is giving workshops, where the academic community, from its different areas of scientific knowledge, together with the university learning community, can share with the resident population, in different spaces, the knowledge that they possess on the different topics approached.

The main motivating aspect is the constant search for dialogue with community actors that contribute to the construction of knowledge, to the promotion of actions that result in possible solutions to local problems and needs, making possible for native and non-native youth a spirit of interculturality and open to the dialogue of insights.

The linkage stays foster the integrative project and allow establishing important and significant ties to carry out collaboration works directly with the community, and to give continuity to the findings derived from the many activities, with the purpose of generating a proposal for the improving of their problems and their empirical responses within the production of converging knowledge, according to the socioeconomic, linguistic and cultural conditions of those who are relevant elements in the configuration of an integrative project.

5 Conclusions

Based on the main elements that give form and content to the integrative project, we recover the main dimensions that go through the theoretical construction of what would be said project and the methodology that takes advantage of problematic issues and is enriched with a model of intercultural education.

Since it is an epistemological proposal that requires dialogues of insights, it effectively feeds on both pedagogical and empirical strategies, which is why it involves joint actions and greater participation of social and institutional actors and, above all, the contribution of the receiving peoples of the social learning community and their research projects.

The experience that is developed in the localities of the northern range of the state of Puebla, for all the groups, has the purpose of reaching an integrative project; however, it has resulted in important events such as the rescue of traditional techniques and technologies, and the construction of university identities committed to provide answers to rural communities, which are their places of origin, of course.

In fact, what is developed as a commitment to issues and problems of the rural communities visited during linkage becomes truly integrative in its execution, above all by what it contributes to the beneficiaries and residents that participate.

As a precision, and illustrating the particular case of the three experiences, we have that from the different knowledge that are integrated in a dialogue of social learning, there have been identified important levels of development of the identities and strengthening of the research groups that, because of the same pedagogical dynamics of the project, are moving the insights, the doing and the task of educators and learners.

In the interaction between the young students of the university institution, the population as well as the scientific community, mediated by the relations with the municipal authorities, are carrying out possibilities of changes in the local governmental agendas that, as decision makers, may open ways to envision the academic community as an actor with a certain degree of social influence.

Through this exercise of integration, students approach to reality of situations, to the phenomena and problems that are contained in a specific social community. That is why each participation especially enhances the dialogue of social learning.

It is considered that to the extent that the participation of the students within the university brigade becomes constant, assuming their belonging as a product and producer of knowledge committed to the native peoples, the possibilities of continuing to generate research processes addressed towards the problems of their own rural communities get increased.

As it is a set of strategies, the integrative project seeks, basically, that during the teaching and learning process, the field of study focus on the reality where students are inserted from their community contexts.

Last, but not less important, we highlight the importance of the relationships that occur within this process of convergence of tasks that lead to a multiplicity and manifestation of efforts (teaching, researching, collaborative and reciprocity), on the one hand, and on the other, reinforces the recognition of plural epistemic thought in which the insights of the original peoples, that historically have generated the reflection and understanding of the world from its own cultural references, especially in places and regions where interculturality acquires special importance from their linguistic, normative, world vision, etc. systems, awarding daily senses and meanings the conception of the world and the richness of its natural social environment.

References

Argueta Villamar A, Gómez Salazar M, Antezana JN (coords.) (2012) Traditional knowledge, innovation and social reappropriation. Siglo XXI, México

Bastida Muños MC (2012) Public policies of institutions and intercultural higher education programs in Mexico in Mato D. (coordinator) (2012) Higher education and indigenous and Afro-descendant peoples in Latin America. Standards, policies and practices. UNESCO, IESALC

Bertely M (coord.) (2011) Inter-learning among indigenous people. How educators learn community knowledge and meaning in intercultural contexts. Intercultural Inductive Education Network (REDIIN), CIESAS, UPN, Mexico

Casillas Muñoz ML, Santini Villar L (2006) Educational model. Intercultural University, SEP, México

Díaz Barriga F (2006) Teaching located. Link between school and life. McGraw Hill Interamericana, Mexico

Dietz G, Mateos LS (2013) Intercultural and intercultural education in Mexico. An analysis of national and international discourses on their impact on national educational models. SEP-CGEIB, México

Freire P (1996) Pedagogy of autonomy: knowledge necessary to educational practice. Peace and earth editor, Sao Paulo, Brasil

General Coordination of Intercultural and Bilingual Education (CGEIB) Special Program of Bilingual Education 2014–2018. http://dof.gob.mx/nota_detalle.php?codigo=5342484&fecha=28/04/2014

Haverkort B, Delgado BF, Shankar D, Millar D (2013) Towards inter-scientific dialogue. Building from the plurality of world views, values and methods in different knowledge communities. Agruco, Plural editors, Bolivia

Jablonska A (2010) The intercultural educational policy of the Mexican government in the framework of the recommendations of the international organizations. In: Construction of intercultural educational policies in Mexico: debates, tendencies, problems, challenges. UPN, México

Latour B (2008) Reassembling the social: an introduction to actor-network social theory. Manantial Press, Buenos Aires

Olivé L (2004) Interculturalismo y justicia social. UNAM, México

Olivé L (2009) For an authentic interculturality based on the recognition of epistemological plurality. In: Epistemological pluralism. CLACSO, Muela del Diablo, Bolivia

Teppa S (2012) Research-participatory action in the daily pedagogical praxis: to intervene the pedagogical practice to transform society and achieve the evolution of the teacher-researcher. Spain Academic Editor, Spain

Santos BS (2014) Epistemologies of the South. Justice against Epistemicide. Paradigm Publishers, Boulder, Londres

Wenger E (2001) Communities of practices: learning, meaning, and identity. Paidós Ibérica Press, Barcelona

Zambrana A (2014) Epistemological pluralism. Reflections on higher education in the plurinational state of Bolivia. FUNPROEIB-Andes, Bolivia

Water-Worlds: How to Research Under the Umbrella of Sustainable Development Being Aware of Its Multiple Ambiguities?

Javier Taks

Abstract The United Nations definition of Agenda 2030 re-launched sustainable development as a planetary horizon for eradicating world poverty while at the same time preserving Earth life-support processes. Since the 90s many scholars, activists and politicians have critically assessed sustainable development and considered it an oxymoron in the context of current global capital accumulation. This paper takes the matter seriously and explores the limits and possibilities of researching water management towards Sustainable Development Goal 6: "Ensure availability and sustainable management of water and sanitation for all". The analysis will touch upon three main fields of enquiry, namely the creation of a world water crisis regime, the encounter of diverse water ontologies while dealing with water management, and the raising of the human right to water and sanitation as a counter-point to the privatization of water resources. Sustainable development requires a stronger inclusion of human rights principles to become a more inspiring narrative for theoretical analysis and transformative interventions. It is argued that embedding sustainable development together with the political and cultural struggles of the human rights idiom, as exemplified in the case of the human right to water and sanitation, could provide a better framework to make sustainable development a useful tensional concept to reflect upon for building more equalitarian societies, and thus to care for life and the environment, within and outside universities.

Keywords Human rights · Sustainability · Culture · Water · Global regime

---

J. Taks (✉)
Universidad de la República, Av. Uruguay 1695, 11200 Montevideo, Uruguay
e-mail: javier.taks@gmail.com

W. Leal Filho et al. (eds.), *Sustainable Development Research and Practice in Mexico and Selected Latin American Countries*, World Sustainability Series,
https://doi.org/10.1007/978-3-319-70560-6_26

1 Re-Launching Global Sustainable Development

In 2015, countries adopted the 2030 Agenda for Sustainable Development and its 17 Sustainable Development Goals (SDGs). Paragraph 7 of the UN Declaration reads as follows:

> In these Goals and targets, we are setting out a supremely ambitious and transformational vision. We envisage a world free of poverty, hunger, disease and want, where all life can thrive. We envisage a world free of fear and violence. A world with universal literacy. A world with equitable and universal access to quality education at all levels, to health care and social protection, where physical, mental and social well-being are assured. *A world where we reaffirm our commitments regarding the human right to safe drinking water and sanitation and where there is improved hygiene*; and where food is sufficient, safe, affordable and nutritious. A world where human habitats are safe, resilient and sustainable and where there is universal access to affordable, reliable and sustainable energy. (United Nations Organization 2015; the italics are mine)

Many believe that this declaration is just a list of good intentions, with the pasteurized idiom of sustainable development as coined and used in mainstream developmental discourse since the Bruntland Report in 1987 (Pierri 2005). Back in the early 90s, Sachs (1993) stated that the Bruntland Report of 1987 "incorporated concern for the environment into the concept of development by erecting 'sustainable development' as the conceptual roof for both violating and healing the environment". He pointed out that sustainable development was another example of a conceptual stretching strategy from those who promote unlimited economic growth as the only viable path to wellbeing: when the destructive effects of economic development were recognized, "the concept was stretched in such a way as to include both injury and therapy" as in the case of growth and poverty, or growth and gender equity, and so on. In sustainable development the aim is, he added, to continue boosting the GNP but at the same monitor and manage water, soils, air and energy utilization to warrant their availability to increase production and consumption and "contain the environmental disaster for the generations to come". This line of thought believes that sustainable development, and sustainable economic growth as one of its main dimensions, is actually an oxymoron (Guimaraes 2003).

More recently, anthropologist Jason Hickel from the London School of Economics, critically assessed Agenda 2030 in similar manner, indicating that it does not recognize the need for the substitution of the current global economic model, with its trend to unlimited growth and deep concentration of wealth, as part of the causes behind unsustainability, and thus the main constraint for achieving the SDGs (Hickel 2015). Moreover, in Latin America and elsewhere there are increasing numbers of scholars and activists who contest the Sustainable Development paradigm with other conceptualizations centred on *Buenvivir* (the Good Life), Degrowth Theory or Ecosocialism.

Nevertheless, the current UN special rapporteur on the human rights to safe drinking water and sanitation, hydrologist Leo Heller (Brown et al. 2016), and many other international water activists (see Joint Statement 2015) celebrated, though with caution, the inclusion of the human right to water and the right to sanitation (HRtWS) as integral parts of the SDGs, as expressed in SDG 6 "Ensure availability and sustainable management of water and sanitation for all". Despite the many ambiguities they acknowledged, they also showed hopeful advances in Agenda 2030 as compared with the Millennium Development Goals (see also Unceta 2015).

In this article, I wish to explore the limits and potentialities of the relation between sustainable development and the HRtWS for a real betterment of life for all human and non-human becomings (Ingold 2012). In order to do it, I highlight three aspects of water management and governance that became an arena for epistemological and conceptual debate: namely, the water crisis, the different place allocated to water in diverse ontologies, and the increasing consideration of water as a fundamental human right. I suggest that these debates might shed light on the manner how we may still understand and use the concept of sustainable development, not only in our environmental research and teaching projects at University, but also in our more general awareness of modern science as both cause and solution of (un)sustainability. In other words, I want to demonstrate theoretically—using the example of water—that sustainable development requires a stronger inclusion of human rights principles to become a more inspiring narrative. Yet, both, the definitions of sustainability and humanity need to be open-ended to be inclusive of all cases of current human-nature configurations preventing the risk of a one sided imposed hegemony.

My methodological approach is based on a critical review of recent UN documents on the SDGs and the HRtWS in the light of public debates mostly, although not exclusively, in Latin America among scholars and water justice activists concerning the management and governance of water mainly for human consumption. In other words, how the perspectives of sustainability and human rights might establish a dialogue regarding better educational and research practices aiming at a transformational vision and action in our common planetary home. It must be said, that this article does not contrast sustainable development against other already mentioned alternative currents in the region. Furthermore, to take into account only the Latin American debate might occlude other realities.

2 World Water Crisis Regime

The Agenda 2030 for SD is clear about the manifestation of a water crisis of planetary scale: "fresh water scarcity [exacerbated by climate change] is part of the most relevant challenges which humanity faces" (United Nations Organization 2015). In the new agenda, governments are therefore "determined to conserve and sustainably use freshwater resources" and "tackle water scarcity and water pollution" (United Nations Organization 2015). It is thus not just by chance that one of the targets to meet SDG 6 is "[b]y 2030, substantially increase water use efficiency across all sectors and ensure sustainable withdrawals and supply of freshwater to address water scarcity and substantially reduce the number of people suffering from water scarcity" (United Nations Organization 2015).

Freshwater sources are unevenly distributed in Earth, and for millennia humans have tried to counter this fact through social organization, technological ingenuity, resource investments and the use of power. Nevertheless, the idea of scarcity of water in a modern sense, according to Vandana Shiva, gradually developed together with a two-stage process of commoditization of natural resources. The first during colonial capitalism, when, faced with an abundance of resources, colonizers exploited natural resources in a predatory manner, appropriating as much as they could. This resulted in a physical limitation for production and consumption that led to the second stage: natural resource management to respond to the scarcity of fodder, water, minerals, and so on. In Shiva's words: "It was this violation of nature's limits that then brought forth the most recent phase in the ever-changing development recipe—the notions of 'sustainable development' and 'sustainable growth'. New limits are now to be imposed on nature's processes in order to sustain development and growth. The crisis of scarcity is now being formulated in the language of sustainability" (Shiva 1993). In the same year that sustainable development became hegemonic in the Rio Summit, the Dublin Statement on Water and Sustainable Development declared that "Fresh water is a finite and vulnerable resource" (Principle 1) and that "Water has an economic value in all its competing uses and should be recognized as an economic good" (Principle 4) (The Dublin Statement 1992).[1] Following Polanyi (2001), there we can recognize a significant step in the establishment of water as a "fictitious commodity", a resource to fragment, manage and economically speculate with. Water scarcity might therefore be resolved with more technology, market mechanisms and better governance.

Against the idea of a global water scarcity crisis, researchers and activists involved in the Water Justice social movement have declared "that the principal form of the water crisis is not a shortage of water, nor failures of government, but the many injustices in access to, the allocation of, and the quality of water. The global water crisis is not likely to be resolved by the provision of more water. Redressing injustice is a more promising approach. That requires a critical rethinking and transformation in how water, water rights and authority are distributed" (Santa Cruz Declaration 2014).

These ideas according to which (in)equality and allocation of water rights are the basis of the current water crisis should be more seriously approach by research on sustainable development, and this means politicizing the debate on sustainability.

[1]The other principles are, on the one hand, the need for a participatory approach to water management and, on the other, the need to recognize the central role of women in the provision, management and safeguarding of water.

3 The Encounter of Diverse Water Ontologies When Dealing with Water Management

The Agenda 2030 for SD envisages "a world of universal respect for human rights and human dignity, the rule of law, justice, equality and non-discrimination; of respect for race, ethnicity and cultural diversity; and of equal opportunity permitting the full realization of human potential and contributing to shared prosperity" (United Nations Organization 2015). Moreover, signatories acknowledged "the natural and cultural diversity of the world and recognize that all cultures and civilizations can contribute to, and are crucial enablers of, sustainable development" (United Nations Organizations 2015). Though the Agenda 2030 affirms the more traditional three dimensions of sustainable development, namely the economic, the social and the environmental, cultural diversity certainly appears in the literature as a fourth dimension of sustainable development (Guimaraes 2003; The Hangzhou Declaration 2013). Cultural diversity is a challenge for a paradigm with clear normative observations.

Anthropologists have shown that there are numerous collective attitudes to and representations of "nature" and "humanity". It is therefore problematic to define "people", "planet" and their relations in singular forms. On the other hand, how to walk together towards global sustainable development if different human groups might comprise diverse natures and multiverses (as opposed to a singular universe)? (Escobar 2016).

The water question cannot be excluded from this dilemma. There are different ontologies that place water in many diverse domains, sometimes contradictory. For many people, water is a live entity not just a natural physical and chemical resource that sustains human life and more. Even in modern contexts, a river can be considered as a relative, and is thus included as an extension of the social rather than the natural realm. Moreover, the idea of water as separated entity already denotes an ontologically fragmented realm of natural resources against a more holistic or integrated vision of life and the environment. Many water conflicts arose due to differing manners of appreciating water, particularly in its sacred or non-sacred substantiation (Hassan 2004), and the possibility of conceiving water as separate from the land, as expressed in the more neoliberal legal frameworks (Ávila-García 2016).

Research on sustainable development needs to consider the diverse ontologies based on human-nature co-evolution and promote space for negotiation between them, parting from the principle of general recognition and respect, but also facing the challenge of conflict transformation, which in turn acknowledges cultural and ethnic politics. The human right to water has, for example, been contested by Aymara and other South American indigenous scholars and activists, who see it as an anthropocentrically imposed perspective, whilst they consider water as a need, and sometimes a right, for all creatures and landscapes, according to a trans-human epistemology (Diego Quispe 2016, personal communication). While it is important to collect, register and compare different water ontologies, it could be argued that it is equally important to reflect upon how to transcend the idea of non-commensurable world-views without imposing any particular local vision of what water is and how we should deal with its many challenges. The following debate about the need to overcome the Western liberal definition of human rights might shed light on how to find cosmopolitan answers to our current water dilemmas in the face of the various and diverse existing water-worlds.

4 The Raising of the Human Right to Water and Sanitation

In 2010 the General Assembly of the United Nations Organization recognized "the right to safe and clean drinking water and sanitation as a human right that is essential for the full enjoyment of life and all human rights" (UN General Assembly 2010). The adoption of this resolution was the output of a long process that lasted at least three decades when governments, social organizations and social movements, legal experts and scholars found the theoretical and legal bases to advance towards a more socially and environmentally just water management model, where each human person might realize his/her right to access clean and safe water and to live with water to be able to realize his/her other rights as individuals and communities, in agreement with the human rights paradigm.

The HRtWS is a bridge between the second generation of economic, social and cultural rights (*égalité*) aimed to satisfy the so called basic needs, and the third generation of solidarity or group rights (*fraternité*), more concerned with the right to peace, the right to a clean and healthy environment, and the right to humanitarian disaster relief (Weston 2014).

As in all other human rights' definitions, the HRtWS includes the guiding principles of non-discrimination and equality; free, active and relevant participation in social life; transparency (i.e. access to information), and the required progressive advancement in its realization. But the HRtWS also promotes normative contents, namely: (i) the universal right to water for personal and domestic uses; and (ii) the fact that water must be physically accessible, safe, culturally acceptable and affordable in all domains of a person's daily life (home, work and study spaces, public spaces and elsewhere).

According to Conka (2005) the HRtWS is the outcome of a global partnership between norm entrepreneurs and social movements. Undoubtedly, the HRtWS is part of a symbolic framework promoted by anti-commoditization forces and anti-privatization of water and sanitation utility networks since the 90s to date. In this regard, I suggest that while the HRtWS functions as a critique of neoliberal principles and policies, the SDGs generally remain neutral, when not in favour of private sector empowerment. This tension seems to be more present in SDC6 than in any other Goal.

Many Latin American governments, mainly Bolivia, Ecuador and Uruguay, were quite influential in relation to the General Assembly resolution regarding the HRtWS. Water conflicts and the transition to formal democracy at the beginning of the twenty-first century showed an articulation between the experience of civil society's defence of human rights during and after the military regimes, and environmental justice theories and insights regarding the importance of water to meet basic needs and for the continuation of human life. The above mentioned countries pioneered the inclusion of the HRtWS in their Constitutional Laws.

The region nevertheless shows hugely unequal progress towards the realization of the HRtWS. A recent study highlights that only half of the Latin American countries count with a legal framework recognizing these fundamental rights, and the right to sanitation is the most neglected. Furthermore, in most countries, the HRtWS dimensions, have actually been reduced solely to physical access to potable water, leaving aside the other intrinsic components (affordability, cultural acceptance, social participation, non-discrimination, and so on) (Portuguez and Dubois 2015).

Despite an increasingly consensual narrative of the obligations and benefits to states and citizens when pursuing the HRtWS, the Latin American experience introduces a relevant debate over the possibility of the HRtWS to contribute to address sustainable development. On the one hand, the HRtWS has been seen by many critical authors as part of a process that disembeds water from its immediate geographical, social and cultural relationships, as shown for instance in the current privatizing water law in Chile (Ávila-García 2016), in line with the current trend to create "modern water" (Linton 2010), an abstract entity known as H_2O, or a natural resource to be managed, as the only ways of conceiving this world wide fluid; this in turn brings up the discussion about the Western expansionist claim for a human rights paradigm and the relativist-universalist debate.

A possible derivation of this disembedding trend is the previously mentioned focus on the warrant for accessing potable water as equivalent to the whole human right to water which has allowed water corporations, mainly transnational, to present themselves as HRtWS promoters while stressing that states and local communities, mainly in poorer countries, are unable to comply with the need to expand utilities to meet the goal of water for all. Experts and activists contest this view, stating that the HRtWS is part of the definition of water as a common good, and that its defence is part and parcel of the protection of all other common goods (such as knowledge, land and labour, among others) from private property and profit-seeking organizations. The latter, they argue, would not guarantee a comprehensive vision of the HRtWS involving access to relevant information, accountability, participation, and so on (Portuguez and Dubois 2015). The question regarding the role of different agents (state, community, private corporations) in the realization of the HRtWS is part of the work towards the achievement of SDG 6. On the other hand, several states in Latin America have limited and even forbidden non-state water management systems such as customary indigenous systems (Dwinell and Olivera 2014) in the name of the HRtWS. According to state representatives, these non-statutory systems cannot meet quantity and quality standards as stated by international organizations, for instance the World Health Organization; they therefore believe that a more centralized state intervention is needed, leaving aside the search for alternatives based on local knowledge, cultural heritage and the peoples' expertise developed when the state was oblivious of those territories. Once again, this contested arena will also arise with the advances in the implementation of SDG 6.

5 Binding Sustainability and Human Rights in Our University Practices

How can we at University help to continue with the necessary task of changing paradigms and meanings for SD while including the human rights perspective?

First, by avoiding the defence of cultural relativism in its extreme expression. In the case of the HRtWS we must recognize the values of cultural diversity, but at the same time take into account a universal intercultural definition of who is entitled to claim these rights: every human becoming, men and women, communities, future generations. We don't need an essentialist definition of humans, and must create an increasingly inclusive open-ended definition of human becomings, with their huge variations. Human rights should be defined through an exercise of what Santos calls diatopic hermeneutics towards an intercultural reconstruction of human rights (Santos 2010).

Second, by recognizing the contradictory nature of human engagement with other living and non-living entities, meaning that "sustainability" requires massive efforts to continue, and these may never end. Yet, this ambiguity, this consciousness that SD would not be achieved for once and for all, cannot veil the fact that some social groups work towards unsustainability while others do not accept that we will live forever in a degraded, unhealthy, ruined Nature.

Finally, we must consider the HRtWS as a means to impact on the SDGs with less expertocracy. The process of emergence of the human right to water in Latin America and elsewhere was democratic, from bottom up, linking indigenous people, urban inhabitants, workers unions, politicians and researchers. In this dialogue of knowledges, we should not view science as evil. We need to analyse scientific knowledge as situated as any other form of knowledge. As anthropologist Tim Ingold puts it:

> Far from abandoning science (…) or opposing the knowledge of inhabitants to scientific knowledge, we need to find ways in which they can work together. This calls for both a revaluation of the environmental experience and creative interventions of lay practitioners and an acknowledgment that science and technology, too, are grounded in practices of habitation. (Ingold 2013)

In our University activity we should keep our feet on the ground without refusing the enjoyment of enskilled abstraction and universal thinking to contribute to sustainable futures.

6 Conclusion

Sustainable development has been re-launched into the international arena. It could have served to contest the more economicist Green Economy vision that disputed narrative centrality to deal with the environmental and social challenges of our time. Carbon bonus, the commoditization of water and financing nature, seem opposed to the necessary political and cultural motivation to create alternative models of production and consumption, a post Anthropocene era. By now it seems that there is no other "big enough story" (Haraway 2015) than Sustainable Development to lead diverse actors like "scientists, entrepreneurs, politicians, labourers for humanity… governments, institutions, the private sector, workers and societies" (United Nations Organization 2015) to work together towards a different earth for people "to live and die well" (Haraway 2015).

The SDGs must be understood as a moral imperative rather than, or at least not only, a rational guideline. In this way, the oxymoron can become productive in a transitional stage. People all over the world cannot stop production and consumption-as-usual from one day to another, but need to realize that they must work in this direction, while inventing new production and consumption relationships to generate different motivations to act in society and in the environment. I would suggest that this is the main reason why many of us, who have been teaching long years a critical approach to the hegemonic definition of sustainable development, are still hopeful that it is worth giving battle for the appropriation of meaning. A new lexicon should be used to talk about sustainability: sufficiency, responsibility, and care, instead of infinite growth, natural capital and competitiveness.

In this sense, the idiom of human rights, as exemplified in this chapter with the case of the HRtWS, provides "a normative basis and constitutes a source of authority and legitimacy for realizing universal and fair access to [water]" (The Pontificial Academy of Sciences 2017); a binding legal obligation for all, ranging from the state to private transnational corporations, in order to prevent water supply —and other socio environmental services—from falling under the influence of powerful and pro-profit minority groups.

Further theoretical and empirical research would be needed to see how the HRtWS are realized within SDG6. Moreover, an intercultural human rights perspective should be use to measure the advance of all other SDGs.

References

Ávila-García P (2016) Hacia una ecología política del agua en Latinoamérica. Revista de Estudios Sociales (Colombia) 55(18–31):2017. doi:10.7440/res55.2016.01 Accessed 27 March

Brown C, Neves-Silva P, Heller L (2016) The human right to water and sanitation: a new perspective for public policies. Ciência & Saúde Coletiva 21(3):661–670

Conka K (2005) Governing water contentious transnational politics and global institution building. MIT Press, Cambridge
Dwinell A, Olivera M (2014) ¡El agua es nuestra, carajo! Agua bien común en Bolivia. In Santos C et al (ed.) Las canillas abiertas de América Latina III. El agua como bien común y derecho humano. Luchas y desafíos a 10 años del Plebiscito del Agua en Uruguay. Casa Bertolt Brecht, Montevideo, pp 49–56
Escobar A (2016) Desde abajo, por la izquierda y con la Tierra. http://blogs.elpais.com/contrapuntos/2016/01/desde-abajo-por-la-izquierda-y-con-la-tierra.html. Accessed 27 Mar 2017
General Assembly of the United Nations Organization (2010) The human right to water and sanitation. Resolution 64/292. http://www.un.org/ga/search/view_doc.asp?symbol=A/RES/64/292. Accessed 27 Mar 2017
Guimaraes R (2003) Tierra de sombras: desafíos de la sustentabilidad y del desarrollo territorial y local ante la globalización corporativa. CEPAL, Santiago de Chile, Serie Medio Ambiente y Desarrollo Nº 67
Haraway D (2015) Staying with the trouble: anthropocene, capitalocene, chthulucene. In: Moore JW (ed) Anthropocene or capitalocene? Nature, history, and the crisis of capitalism. PM Press, Oakland, pp 34–76
Hassan FA (2004) Water and ethics: a historical perspective. UNESCO, Paris
Hickel J (2015) Five reasons to think twice about the UN's sustainable development goals. http://blogs.lse.ac.uk/southasia/2015/09/23/five-reasons-to-think-twice-about-the-uns-sustainable-development-goals. Accessed 19 Feb 2017
Ingold T (2012) Ambientes para la vida. Trilce, Montevideo
Ingold T (2013) Designing environments for life. In: Hastrup K (ed) Anthropology and nature. Routledge, London
Joint Statement (2015) Joint statement from amnesty international, human rights watch, and wash united on UN general assembly resolution 70/169 on the human rights to water and to sanitation. https://www.hrw.org/news/2015/12/17/joint-statement-amnesty-international-human-rights-watch-and-wash-united-un-general. Accessed 11 Mar 2016
Linton J (2010) What is water? The history of a modern abstraction. UBC Press, Vancouver
Pierri N (2005) Historia del concepto de desarrollo sustentable. In: Pierri N, Foladori G (eds) ¿Sustentabilidad? Desacuerdos sobre el desarrollo sustentable. Universidad de Zacatecas/Porrúa, México, pp 60–81
Polanyi K (2001) The great transformation. Beacon Press, Boston
Portuguez JM, Dubois V (2015) Implementación del derecho humano al agua en América Latina. Buenos Aires: CAF. https://www.caf.com/media/2630071/implementacion-derecho-humano-agua-america-sur-caf.pdf. Accessed 22 Mar 2016
Sachs W (1993) Environment. In: Sachs W (ed) The development dictionary: a guide to knowledge as power. Zed Books, London, pp 26–37
Santa Cruz Declaration on the Global Water Crisis (2014) Water Int 39 (2):246–261
Santos B (2010) Para descolonizar Occidente. Más allá del pensamiento abismal. CLACSO/Prometeo, Buenos Aires
Shiva V (1993) Resources. In: Sachs W (ed) The development dictionary: a guide to knowledge as power. Zed Books, London, pp 206–218
The Dublin Statement on Water and Sustainable Development (1992) http://www.un-documents.net/h2o-dub.htm. Accessed 27 Mar 2017
The Hangzhou Declaration (2013) Placing culture at the heart of sustainable development policies. http://www.unesco.org/new/en/culture/themes/culture-and-development/hangzhou-congress. Accessed 27 Mar 2017
The Pontificial Academy of Sciences (2017) Final statement of the workshop on the human right to water. http://www.pas.va/content/accademia/en/events/2017/water/final_statement.html. Accessed 27 Mar 2017
Unceta K (2015) De los ODM a los ODS. http://www.otromundoestaenmarcha.org/lanbroan/?p=33. Accessed 25 Mar 2017

United Nations Organization (2015) Transforming our world: the 2030 agenda for sustainable development. https://sustainabledevelopment.un.org/post2015/transformingourworld. Accessed 27 Mar 2017

Weston BH (2014) Human rights. Encyclopaedia Britannica. https://www.britannica.com/topic/human-rights#ref742089. Accessed 27 Mar 2017

Public Policy to Promote Sustainable Tourism in the State of Veracruz, Mexico

Ingrid Patricia López-Delfín, María Ángeles Piñar-Álvarez, Jorge Alejandro Negrete-Ramírez, Astrid Wojtarowski-Leal and El Colegio de Veracruz

Abstract This research focuses on the search for sustainable touristic services through the design of a participatory public policy proposal that prioritize business requirements and their social and environmental responsibility in the territory. The proposal goes for the municipalities of Xalapa, Xico and Coatepec, in order to enhance their tourism competitiveness under a sustainable framework, by using management systems based on the ISO 9001: 2008 and 14001: 2004 standards, to certify the processes under which tourism services are provided. The purpose of this is to encourage the tourism entrepreneurs to take on their involvement, in international quality standards, and commitment to adopt environmental oriented care protocols. Methodologically, the establishment of quality and environmental management systems based on ISO standards underwent testing by using three measurement tools, quantitative and qualitative, applied to the tourism industry in the three priority tourist destinations in central Veracruz. Findings show the inputs needed to create a touristic and sustainable public policy proposal that affects beneficially in the competitiveness and sustainability of the target region in the Ecological Corridor of the Sierra Madre Oriental of Veracruz, Mexico.

Keywords Sustainable tourism · Management systems of touristic quality
Tourism competitiveness · Touristic public policies

1 Introduction

The promotion of tourism is key to the development of Veracruz, it brings with it numerous benefits such as job creation and economic growth (Datatur 2015). With that in mind, public policies should ensure that activities related to tourism do not

I. P. López-Delfín · M. Á. Piñar-Álvarez (✉) · J. A. Negrete-Ramírez ·
A. Wojtarowski-Leal · E. C. de Veracruz
Carrillo Puerto 26, Zona Centro, C.P. 91000 Xalapa, Veracruz, Mexico
e-mail: apinara@colver.edu.mx; angelespinaralvarez@gmail.com
URL: https://www.colver.edu.mx

W. Leal Filho et al. (eds.), *Sustainable Development Research and Practice in Mexico and Selected Latin American Countries*, World Sustainability Series,
https://doi.org/10.1007/978-3-319-70560-6_27

affect the environment; furthermore, it is also vital that this industry has to set the conditions and quality services for national and international visitors.

The feasibility of developing a public policy under a focus of sustainability, entailing management systems based on the regulation ISO 9001:2008 and 14001:200 to certify tourism services and thus promote the provision of competitive high-quality services with environmental management protocols, requires a priority study area already consolidated as a tourist destination in the federal entity. This concerning Xico and Coatepec (Pueblos Mágicos "Magical Villages Program"), as well as Xalapa (state capital) See Fig. 1.

In the state of Veracruz, 5173 companies provide tourism services. For the study region—Xalapa, Xico and Coatepec—145 hotels, 578 restaurants and 84 tourism companies are located in areas of "other tourist establishments" (travel agencies, spas, car rental companies, golf courses and convention centers). However, according to the information provided by the state of Veracruz (Secretaría de Turismo y Cultura or Secturc), in these three municipalities are only 64 tourist companies, 7.8% of the total, with quality certifications aimed at quality services provision, customer service and protocols follow up that minimize environmental impact due to the provided services (GOEV 2011; GOEV 2012; Secturc 2016). These companies will be our subject of study.

In this context, the collaboration between El Colegio de Veracruz and the Secturc approached those few companies who provide services, under a high-quality scheme and management in order to investigate the elements that can increase business, structured through public policies aimed at meeting the most pressing needs of the productive sector in the municipalities that are the subject of

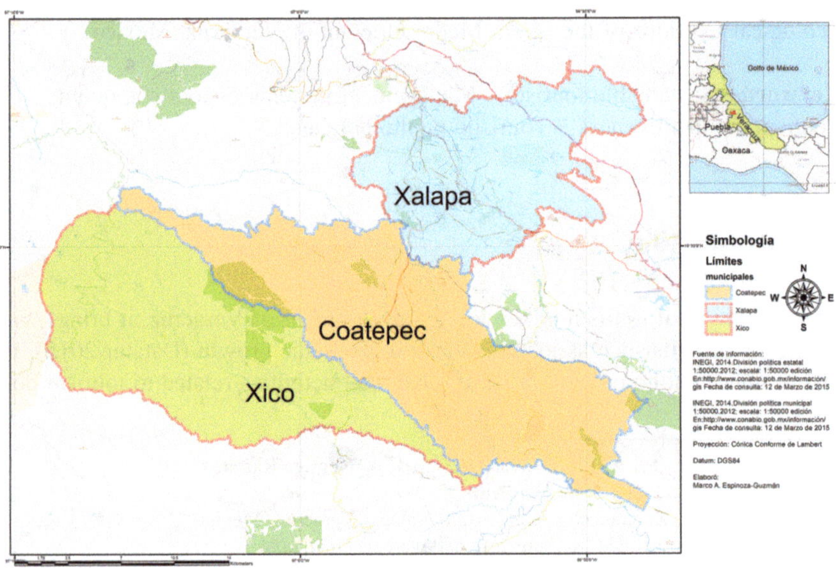

Fig. 1 Study area (Xalapa, Coatepec, Xico)

study. These are nature destinations established in Veracruz (Piñar-Álvarez 2012), with a vast variety of natural, cultural and historical-monumental attractions, which counteracts with the weak environmental leadership of their companies and the strong weaknesses in terms of competitiveness measured by nine factors, especially supply, infrastructure and destination management (Piñar-Álvarez and Arredondo 2014). This encourages to analyze how the quality standards of tourism services should be enhanced, underneath a scheme that promotes sustainability in its three dimensions: social, economic and environmental. This allows us to shed light on the potential of urban-rural tourism as the key element for the biocultural preservation, in the Ecological Corridor of the Sierra Madre Oriental of Veracruz (Fuentes 2013; Negrete-Ramirez and Piñar-Álvarez 2015).

1.1 Research Goals

The main goal of sustainable development research was to design a proposal for the implementation of environmental and quality management systems based on the ISO 14001:2004 and ISO 9001:2008 standards. The specific goals were three:

1. Analyzing the quality and environmental management systems in the public and private market, so that the aim is boosting sustainable tourism policies.
2. Diagnosing the status of those companies who are certified with the quality and environmental management systems in Xalapa, Xico and Coatepec.
3. Designing a public policy based on the implementation of the quality management system under the ISO 9001:2008 standards, and the environmental management system under the ISO 14000:2008 standards, in the tourism sector of the studied municipalities.

1.2 Methodological Approach

The methodological approach of the research was a crossover of quantitative and qualitative instruments with descriptive and explorative nature.

For the first goal, the quality and environmental management systems in the public and private markets, focused on promoting sustainable public policies for the tourist sector, were described in an analytical way. The used technique was an content analysis and the applied instrument was a guide of content analysis through which a bibliographic review of the literature. It was related to the following certified systems, regarding environmental management and quality of the public institutions, was carried out: Certificado Calidad Ambiental Turística (SEMARNAT), the certification Acreditación de Guías de Turistas, Sello de Calidad Punto Limpio, Programa de Calidad Tesoros de México, Programa de Calidad Distintivo S, Programa de Calidad Moderniza, Programa de Manejo Higiénico de los Alimento Distintivo H, Distintivo Moderniza, all of them belonging to the Ministry of Tourism (Sectur 2015) as well as to the private quality systems of the company EarthCheck (2015) and the NGO Rainforest Alliance (Rainforest Alliance 2014).

Concerning the second goal, the conditions of the certified companies with quality and environmental management systems in Xalapa, Xico and Coatepec were diagnosed through an analysis of their strengths and weaknesses. The approach of this goal is mixed (qualitative-quantitative) and was carried out through three instruments:

a. *Semi-structured interview guide*: applied to 23 key informants of certified companies that have environmental certificates or quality management systems in Xalapa, Xico and Coatepec, in order to know the opinion of the managers of these companies regarding the operation of the quality management system enforced in their company;
b. *Likert Scale*: applied to 60 clients, 30 of them of companies with quality certifications, and the other half were companies without quality certifications, in order to know their awareness regarding the services that are offered to them in companies with quality management systems;
c. *Non-participant observation guide*: applied in 40 certified companies of the study area, in order to inquire into the circumstances in which these companies are and the terms in which they render their services.

With regard to the third goal, an analysis of documentary information, opinions of entrepreneurs of the tourism sector was carried out, contrasting the application of the management systems with the practice of the companies. With this information, a public policy design model was applied to promote a sustainable and touristic public policy in the municipalities of study, based on the implementation of the Quality Management System (QMS) in the study area.

2 Results or Findings

2.1 *Quality and Environmental Management Systems in the Public and Private Market*

Tourism is an activity of great significance for Veracruz, and various elements are needed to develop its potential. These factors have much in common with the competitiveness of tourism enterprises and the provision of services under quality standards and protocols for environmental protection, because otherwise the operation of these organizations wouldn't be sustainable.

The supply of quality and environmental management systems in the public (SEMARNAT-SECTUR) and private market (Earthscheck, Rainforest) is sufficient. They also are an ideal way to achieve this combination since its implementation requires the adoption of official quality standards, although rigorous exercise from the company is implied, that will be worth the effort when evolving into the continuous improvement of the provided services. In addition to this, all of the quality distinctives were analyzed in this research demand respect to multiple environmental laws. The quality distinctives are focused on the visitor's attention, to the improvement of a company's processes and they include provisions to reduce the environmental impact of certified tourism companies. With the Moderniza Program, the most driven one in Veracruz, it has been able to encourage the participation of micro, small and medium-sized enterprises in the tourism sector to thereby implement the **Moderniza Quality Management System** in their companies. With a contribution of the federation and the state (65–80%) and another contribution on the part of the entrepreneur (20–55%) who invests 6500 pesos (micro), 12,075 pesos (small) or 16,425 pesos (medium company), with a 3-month program of training, a simplified Quality Management System starts to be implemented based on the requirements of ISO 9001 and 9004 standards. This affects organizational and economic performance of enterprises. Being directed to the senior management of companies (owners, directors and/or managers), the successful implementation is guaranteed. At the country level, it suffers from measuring the impact of programs in companies. The impact of the Moderniza Quality Program has been measured that much, which recognizes the improvement of the management of the companies, that there are 10 measure indicators:

1. greater customer satisfaction,
2. higher productivity,
3. further staff training,
4. greater innovation in product and services design,
5. increase in the number of customers,
6. sales growth,
7. increased profits,
8. less waste,
9. lower costs,
10. less staff absenteeism.

All this is happening in companies with Distinctive "M" at a national level. In other words, it is not about creating new distinctives but to push decisively (by the federation and the state) that what is already known: Moderniza is an excellent tool that meets the requirements of ISO 9001 and ISO 9004 standards, which are the ones that affect organizational and economic performance of enterprises the most.

3 Diagnosis of Certified Companies with Quality and Environmental Management Systems

3.1 Interviewing Entrepreneurs in Municipalities

Xalapa. From the application of the interview in the municipality, the following results are obtained:

- *Motivation* and commitment of the members of the organization are appreciated, as well as from those responsible for maintaining the GSC.

- No company in Xalapa gave a figure of what is *spent*, but they said it was expensive.
- For interviewees, distinctive "H" is a strict SGC, in terms of its processes, and it translates into *better services for the customer.*
- Interviewees are considered included within the organization, and their work contributes to the *achievement of quality policies* of the certified company.

All the interviewed companies considered that their SGC *contributes great value to the organization*, so they are interested in preserving it. A 94% of respondents expressed their interest in renewing their certificate, 6% no, because of the cost.

The interviewees showed abilities to be *environmental leaders*, just as being aware of the influence exerted by their company on the territory, but only two companies in this municipality support sustainable local production projects. All the interviewees indicated that the *main challenge* they have faced in the implementation of their QMS is the human factor or internal customers. As for the *expectations* they had in implementing their QMS, all respondents were satisfied.

Xico. With regard to the interviews results, the following elements are highlighted:

- Regarding companies *motivation* to implement and maintain their certificate, 75% said that competition drives them, and 25% of them need to improve their processes.
- All respondents believe their customers appreciate their efforts to maintain their QMS. A great environmental care, as well as commitment of the certified companies towards the community in which they are immersed, is appreciated.
- Those in charge of implementation are members of the organization with support from external consultants.
- The cost of the QMS in all 40 certified companies is $1000.00 pesos per month.
- As to *benefits* of the GSC, 75% consider it very positive and 25% think that it avoids rigorous inspections for them.
- As for the *challenges* that companies have faced, they also expressed the human factor or internal clients; likewise, the amount of time that has to be invested in the maintenance of the system.
- Expectations of all respondents are satisfied, so they are interested in renewing their certificate.
- As to the corporate social responsibility and the company's influence in the territory, 75% consider themselves environmental leaders and socially responsible companies and 25% of them consider that they do make efforts for the environment, but insufficient.

Coatepec. The findings of the municipality, according to the interviews, show the following:

- The respondents' *motivation* is: facing the great competition that exists in the municipality.
- All interviewees stated that their QMS is very *expensive* and not one company gave a figure of what it spends on implementing or maintaining its SGC.

- There were complaints about the process by which they implemented the distinctive M, that is, through the invitation from the city council.
- The interviewees pointed out that their *SGC is useful* for the organization, but they consider that the client does not perceive a difference because of the quality services they offer.
- As for the *challenges* they face to implement their QMS, 75% said that it is the human factor and 25% training new staff, who eventually will go work for another company in the same municipality.
- All the certified companies of Coatepec have the distinctive M, however, it is appreciated as partially implemented and without the strictness that a SGC must have.
- The implementation *managers*, as members of the organization, complain about the lack of support from external consultants.
- Concerning the benefits of SGC, all respondents said they are positive. Although their expectations have been met in the case of 75% of the companies, 25% are not satisfied.
- A 50% wants to renew the certificate and a 50% doesn't.
- As for the companies influence and social responsibility in the territory, 75% are not considered environmental leaders, but they are socially responsible because they provide local employment. A 25% of these companies have environmental care protocols.
- None of the interviewed companies of the municipality reported their efforts to protect the environment and the visitors.

3.2 Xalapa, Xico and Coatepec: Common Elements After Interviewing the Companies

- The "H" and "M" distinctives, as well as the Sello de Calidad Turística Veracruzana (Touristic Quality Seal of Veracruz), are directed towards customer service, to the improvement of the business processes, containing provisions to reduce the environmental impact of certified tourist companies.
- 91% of the interviewed companies considered that its QMS brings value to the organization and are interested in keeping it. The rest of the companies (9%) have no interest in keeping the distinctive.
- Certified companies, by regulations, follow guidelines in their activities towards environmental sustainability, however, only 23% or nearly one in four, shares it with the visitors.
- 74% of the interviewed companies acknowledged that their day-to-day operations did not include protocols to reduce the environmental impact of their activities, until they began implementing their QMS.
- In Xalapa a greater commitment to process quality and customer satisfaction was observed, while in Xico greater commitment to the community and the

environment was appreciated. In Coatepec, the implementation of the distinctive "M" is not consolidated for appreciation nor for the rigor that QMS requires.

- Interviewees in Xalapa and Xico are considered to be included in the organization and are aware of their work contributing to the quality policy of the certified company. In Coatepec companies on recognized that the SGC does improve their processes, but they don't believe that the customer perceives it.
- In Xalapa and Coatepec no figures of the cost of the QMS were provided, but stated that it was expensive. In Xico, the interviewed companies said they spend $1000.00 pesos per month in maintaining its QMS.
- In Xalapa and Xico all of the interviewed companies were satisfied with their certification and consider that the SGC meets their expectations.

3.3 *Observing the Companies and Surveying Their Customers in Xalapa, Xico and Coatepec*

The application of the observation guide shed light on the situation that the observed companies face, before applying interviews with their managers and/or responsible for management which have quality certifications (Distinctive "H", Distinctive "M", Distinctive "Clean Point" and Sello de Calidad Turística Veracruzana), they had agreed to being interviewed. There are five aspects that stand out in the study region:

1. 97.5% of the companies *exhibit* their governing documents and/or quality certificates to the visitors.
2. In the 100% of the companies, *compliance with the Civil Protection Guidelines* (signage, extinguishers and evacuation routes) is observed.
3. 77.5% of the companies have *customer satisfaction surveys* (via print or electronic). Other companies verbally inquire about the quality of the service (22.5%).
4. In 100% of the companies there is *order, cleanliness and generally adequate conditions* of all facilities.
5. Only 5% of the certified companies in the three municipalities *communicates its efforts towards the environment.*

In general terms, it stands out that customers are satisfied but not aware with the social and environmental responsibility within the sale of services because companies do not report them (Table 1).

3.4 *Representative Features of the Quality Management Systems (CGCMA) of the Region*

Hereunder, a table with the representative features of the SGCMA implemented in the study area is presented. It should be noted that the level of implementation of the SGCMA was measured according to the following Assessment Matrix, with three levels of implementation, shown below

- 3 or High implementation level
- 2 or Medium implementation level
- 1 o Low implementation level.

To measure the level of implementation of the SGCMA in the company, the Table 2 shows the differences.

Under this assessment, according to the aforementioned assessment, the level of implementation of the SGCMA in the municipalities of Xalapa and Xico is at a High level, while in Coatepec it is seen at a Low level.

Piñar-Álvarez and Del Castillo (2014) demonstrated how leading towards sustainability by the design of public policies starts off with a diagnosis that weighs the productive and environmental needs, prioritizes the problems and the effects of each of them to solve others issues, establishes the aims and the strategies, and allows to program concrete measures and specific management and performance indicators.

4 Design of Public Policies that Leads Towards Touristic Sustainability

A public policy project that stems from a participatory and inclusive process of the productive and social actors muss be involved in the territory (García-Falcón 2010). The failure of public policies is due to using the wrong method and the frivolity of raising government actions that aren't planned based on the most pressing needs of the community, without analyzing the impacts a problem can have for the resolution of others.

The preparation of the diagnosis was based on the diagnosis of the Plan Veracruzano de Desarrollo (GOEV 2011), the Programa Sectorial de Turismo (GOEV 2012) and the municipal plans of the municipalities of Xalapa (City Council of Xalapa 2014), Xico (City Council of Xico 2014) And Coatepec (City Council of Coatepec 2014). With this information, alongside the results of the prepared measurement instruments, a SWOT analysis was carried out. It yielded the necessary inputs to conduct the participatory public policy proposal.

Only in the case of Xalapas' City Council a municipal development plan, elaborated with the participation of the productive sector (hotels, restaurants, guides, civil society), can be cherished. This is thanks to workshops of strategic planning and forecast of the possible scenarios, performed within the framework of the Agenda de Competitividad Turística of Xalapa (Piñar-Álvarez and Arredondo 2014; SECTUR-GEMES 2013).

Already in this case it is possible to assess the design (Corzo 2013), the planning for competitiveness (Deming 1999) and the public policy management carried out

Table 1 Features of the quality and/or environmental management system

Town/City	Features of the quality and/or environmental management system (SGCMA)	Implementation level
Xalapa	– Motivation and commitment of the members of the organization and responsible for maintaining the SGCM is *appreciated* – The implementation of the SGCMA is expensive but it rewards in customer satisfaction and benefits – The SGCMA is rigorous in its requirements, which translates into better services – The interviewees are considered included within the organization and their work contributes to the *achievement of quality policy* of the certified company – The clients value the effort of the organization to maintain the SGCMA – The interviewees consider themselves environmental leaders and are aware of their company's commitment to the community	High
Xico	– There is motivation of the members of the organization to maintain the SGCMA – The implementation of the SGCMA is accessible – The SGCMA is rigorous in its requirements, but it is based on the quality of the services offered to the public – The certified companies show commitment with the community in which they are immersed and with the care towards the environment – The clients value the effort of the organization to maintain the SGCMA – The interviewees consider themselves environmental leaders and are aware of the company's commitment to their community, although they consider that they could do much more	High
Coatepec	– Those responsible for implementing the SGCMA are motivated solely by the competition – The interviewees consider that the SGCMA is rigorous in terms of its requirements but those in charge of implementing it did not fulfill their commitments – Respondents stated that their SGC is very expensive – The interviewees are not considered included within the organization – The interviewees pointed out that their SGCMA is *useful* for the organization but consider that the client does not perceive the difference – The clients do not perceive the effort of the organization to maintain the SGCMA – The interviewees are not considered environmental leaders and are not aware of the company's commitment to its community	Low

Table 2 Assessment matrix to measure the level of implementation of the SGCMA in the company

Assessment	Motivation for the maintenance or implementation of their SGCMA	Investment for the maintenance or implementation of your SGCMA	Requirements of the SGCMA	Inclusion of members and quality policy	Customer Perception	Environmental leadership/community engagement
3	The members of the organization are motivated	The members of the organization consider that their SGCMA requires an investment that is rewarded with customer satisfaction	The members of the organization consider that the requirements to implement or maintain their SGCMA are strict	Interviewees are considered included within the organization and their work contributes to the achievement of the quality policy of the certified company	Customers value the effort of the organization to maintain the SGCMA	Respondents are considered environ-mental leaders and are aware of their commitment to the community
2	The members of the organization are partially motivated	The members of the organization consider that their SGCMA is expensive	The members of the organization consider that the requirements to implement or maintain their SGCMA are not rigorous	Interviewees are considered partially included within the organization, but consider that they contribute to the achievement of the quality policy of the certified company	Some clients value the effort of the organization to maintain the SGCMA	Respondents know that they have a commitment with the community but feel they still have many things to do to become environmental leaders
1	The members of the organization are not motivated	The members of the organization consider that their SGCMA is very expensive and that the investment that they carry out is not profitable	Members of the organization are not aware of the requirements of the SGCMA	Interviewees are not considered included within the organization and are not aware of the quality policy	Customers do not value the effort of the organization to maintain the SGCMA	Respondents are not considered environmental leaders and are not aware of their commitment to the community

for over 4 years, as Mejía (2004) points out, given that there are indicators of management and performance.

In the case of Xico and Coatepec we find desktop documents where social and business participation is the great absentee.

An overwhelming majority of those who have theorized about public policies indicate the importance of the diagnostic and evaluation phase of a public policy (Villanueva 1993; Tamayo 1997; Parson 2007). In terms of public tourism policy, the participation of the tourism sector in the diagnosis and planning phases enables the corporate responsibility with the quality of the services it offers and also within the community where the company is inserted (Buades 2010; Wojtarowski-Leal et al. 2016).

If there are no indicators to measure planned actions, a public policy cannot be evaluated. At this point accountability is essential, and part of the established goals of the public policy, on a participatory basis since the diagnosis and the planning stages. Meaning, the management based on the results of a public policy must start from a point of social dissatisfaction or weaknesses in economic, social, environmental and institutional matters. Quality management within a company also requires a decisive drive from those responsible for a tourism public policy, a drive that leads to sustainability. In our view, the quality label Moderniza is an excellent tool oriented to quality planning inside the company (Juran 1990). The results of its implementation since 2004, for more than a decade (Canaco 2016), and the results of the present research show the need for a more determined impulse by the SECTUR—federal and state. Moderniza gives the benefits shown in the satisfaction of internal and external customers, improving productivity and increasing sales and profits, reducing staff costs and staff absenteeism, decreasing time and processes waste as well as the innovation of products and services in the state of Veracruz.

5 Conclusions

Tourism is an activity of great importance for Veracruz State, generating jobs and large scale income, however, it still needs various elements to develop its potential. These factors have much to do with the competitiveness of tourism companies and the provision of services under quality standards and environmental protection protocols, otherwise the operation of these organizations would not be sustainable.

Therefore, quality certificates seem to be an ideal way to achieve this combination since its implementation requires the adoption of official quality standards, although rigorous exercise from the company is implied, that will be worth the effort when evolving into the continuous improvement of the provided services to internal customers (personal) and external customers (visitors). In addition to this, all of the quality distinctives that were analyzed in this research demand respect to multiple environmental laws. It is encouraging to know the findings of the application of one of the measurement instruments where information was collected, which shows that the quality distinctives are oriented towards the public eye and the enhancement of the processes of a company, as well as having provisions to reduce the environmental impact of certified tourism companies in the involved municipalities.

Likewise, nine out of ten interviewed companies (91%) considered their SGC to be of great value to the organization and are interested in keeping it. On the other hand, seven out of 10 interviewed companies (74%) acknowledged that their day-to-day operations did not include protocols in their activities to minimize the impact on the environment, until they began implementing their Quality Management System. The fact that certified companies admit that they began to follow environmental care protocols as a result of starting with the implementation of their quality certificate is a big plus in favor of these quality distinctives. However, in each analyzed destination (close to each other) a different commitment with socio-environmental quality and sustainability was observed: in Xalapa a greater commitment to process quality and customer satisfaction, while in Xico a stronger involvement with the community and the environment was appreciated, and in Coatepec the tourism quality certifications were partially implemented.

This may be linked to the certification companies and their lack of resources. In the case of the Distinctive Moderniza, the certifying companies complain about the resources lack of continuity provided by both the federation and the state.

There is also a lack of control by the Secturc in these companies workability, companies that are dedicated to the implementation of their quality distinctives. Any organization, large or small, is susceptible to adopt the quality philosophy as a management method, all that is needed is effective leadership on the part of public institutions by public policy decision-makers and the companies' owners. To take on a quality management system represents hard work for the company that pays off organizational and productivity results, to which is added the benefits that come along with the achievements in terms of environmental management. Leading towards the sectors sustainability in the territory is possible. With this aim in mind, when the use of quality and environmental certification based on ISO standards is considered, it emanates on a credibility ethic of the companies accredited by the International Organization for Standardization, same which verifies the compliance with the requirements of this standard within certified organizations. As for the fulfillment of the processes, this verification is carried out by an external agent to the organization that will test if the company complies with the international standards of quality. A collegiate body belonging to the International Organization for Standardization shall thereafter recommend the certification of a company.

The common thread among the interviewed companies' owners was to point out that their distinctive was expensive because of the investment made in training and physical structure of the organization. Naturally, the company has to invest in its Quality Management System (QMS): quality costs, but it costs more not to have it.

The proposal made here is that the members of the organization—through public policy and a proper training—shall be responsible for deploying their QMS based on the ISO standards, which will allow them to reduce costs in order to obtain their certificate.

It should be noted that not all tourism companies will be candidates to obtain this ISO certification. Smaller organizations can start with the processes reengineering to improve their day-to-day operations and gradually obtain ongoing progress, which will lead them on the way of certified quality. The companies that are candidates to use SGC, based on ISO standards, are those that have 30 members or more. With these it is necessary to work with the processes reengineering that will lead them to systematize the provision of their goods and services, towards ISO based certification. In general, it is a question of all companies in the study area being able to boost their tourist activity, move towards improving their processes and offer quality services to the customer, in order to be competitive.

Meanwhile, competitiveness is substantive but it is not the only aspect that needs to be considered: sustainability in its three dimensions (social, cultural and environmental), associated with corporate social responsibility, is fundamental. During the interviews it was observed that few organizations consider themselves to be environmental leaders, even if they are making laudable efforts to take care of the environment. Furthermore, all companies certified by regulations have to follow environmental sustainability guidelines in their activities and yet one in four companies (23%) communicates their efforts in environmental improvement to their customers: there is an opportunity for companies providing tourism services in the region and the entity.

It is essential to make tourism companies aware of the role they play in their community and of the important management they have over the territory in which they operate: these provide employment for the benefit of the community and take care of the environment. Clients prefer a tourist company immersed in an environment free of poverty and social and cultural marginalization, as well as of employees displeased with their work. During the interviews, the behavior of dissatisfied employees became evident in certain tourist companies: the external customer or visitor notes the discontent shown by a worker who does not feel included in an organization.

In that case, in pursuit of sustainability, the role of the client and its transforming power is essential: each of us must be conscious and responsible when choosing a company to stay, to spend a recreational day or to consume food. It is the external clients who choose or not socially responsible companies that are making efforts to offer quality services and whose operation is carried out under processes that do not contaminate the environment, immersed in improving their community in socio-cultural terms within a model that leads to a sustainable territory. The customer has the final say and the power to consume both goods and services that take care of their own health and of the environmental health.

As we have seen within the social, cultural and environmental dimension of sustainability, the institutional dimension must be added. While there is a legal structure that requires care and access to a healthy and balanced environment, compliance with environmental laws is often a fallacy at municipal, state and federal levels. Two clear examples: weaknesses in the treatment of all municipal wastewater and in the integrated management of municipal solid waste of the three municipalities. At this point public policies, such as the translation of legal obligations and the decisions of the authority to address the satisfaction of social and environmental needs, are included.

Indeed, Mexico has a long and sad history of failed public policies. This requires searching for the causes and finding the creators of the issues in order to prioritize and decide which ones the economic, social and environmental needs are. The failure of designing public policies is due to using the wrong method and the frivolity to elaborate government plans and programs, which see social participation in diagnostics, and planning as a threat rather than an opportunity to glimpse at the primary needs in economic, social, environmental and institutional matters in the territory to transit towards sustainability.

References

Buades J (2010) Turismo y bien común: De la Irresponsabilidad Corporativa a la Responsabilidad Comunitaria. Albasud, Investigación y Comunicación para el Desarrollo 7(9/2010):1–31

Cámara Nacional de Comercio (2016) Beneficios del Sistema Moderniza para las empresas turísticas. http://www.canaco.net/servicios/mexicoemprende-programa-moderniza.html. Accessed 11 Mar 2017

City Council of Coatepec (2014) Plan Municipal de Coatepec (2014-2017). Ayuntamiento de Coatepec, Xalapa, México

City Council of Xalapa (2014) Plan Municipal de Xalapa (2014-2017). Ayuntamiento de Xalapa y El Colegio de Veracruz, Xalapa, México

City Council of Xico (2014) Plan Municipal de Xico (2014-2017). Ayuntamiento de Xico, Xalapa, México

Corzo JF (2013) Diseño de Políticas Públicas. IEXE Editorial, México

Datatur (2015) Análisis integral del Turismo. Secretaria de Turismo, México. http://www.datatur.sectur.gob.mx/SitePages/CompendioEstadistico.aspx. Accessed 11 Jun 2016

Deming JE (1999) Calidad, productividad y competitividad, la salida de la, crisis. Díaz de Santos, Madrid, España

EarthCheck (2015) Productos y servicios. http://earthcheck.org/products-services/. Accessed 28 May 2016

Fuentes PT (2013) Programa Conservación de la Biodiversidad en la Sierra Madre Oriental de México Corredor Ecológico de la Sierra Madre Oriental (CESMO). Diagnóstico de Gobernanza y de Instrumentos Económicos en las Zonas Prioritarias del CESMO. Consultor. Informe técnico del estado de Veracruz. Llave. Veracruz, México

García Falcón R (2010) Diseño de Políticas Públicas. Escuela de Ciencias Políticas y Administración Pública de Oriente, México

GOEV. Gaceta Oficial del Estado de Veracruz-Llave (2011) Plan Veracruzano de Desarrollo 2011-2016. Gobierno del Estado de Veracruz de Ignacio de la Llave. Tomo CLXXXIII Xalapa-Enríquez, Veracruz.10 de mayo 2011 Núm. Ext. 138. 305p. http://www.veracruz.gob.mx/programadegobierno/files/2013/07/plan_veracruzano_web.pdf. Accessed 06 Dec 2016

GOEV. Gaceta Oficial del Estado de Veracruz-Llave (2012) Programa Veracruzano de Turismo, Cultura y Cinematografía 2011-2016. Gobierno del Estado de Veracruz de Ignacio de la Llave. Secretaría de Finanzas y Planeación. Xalapa-Enríquez, Ver., 18 de abril de 2012. 282-418. Núm. Ext. 129. 143p. http://www.veracruz.gob.mx/finanzas/files/2012/08/tf07-ps-pvd-11-16-pvtcc-13.pdf. Accessed 12 Dec 2016

ISO International Organization for Standardization (2004) Norma ISO 14001:2004. Secretaría Central de la International Organization for Standardization, Ginebra, Suiza

ISO International Organization for Standardization (2008) Norma ISO 9001:2008. Secretaría Central de la International Organization for Standardization, Ginebra, Suiza

Juran JM (1990) Juran y la planificación para la, calidad edn. Díaz de Santos, España

Mejía LJ (2004) La evaluación de la gestión y las políticas, públicas edn. Porrúa, México

Negrete-Ramírez JA, Piñar-Álvarez MA (2015) El papel de la educación ambiental en proyectos de desarrollo sustentable con enfoque de cuenca. Subcuencas de los ríos Pixquiac y Texolo en la Cuenca Alta del Río la Antigua, Veracruz, como casos de estudio. In: Agüero-Rodríguez J C y Torres-Beristain B (eds) Educación Ambiental desde la Innovación, la Transdisciplinariedad e Interculturalidad. U.Veracruzana y Ecorfan, Xalapa, México, pp 322–333

Parsons W (2007) Políticas Públicas, Una introducción a la teoría y la práctica del análisis de políticas, públicas edn. FLACSO, México

Piñar-Álvarez MA (2012) Turismo y medio ambiente: interacción, análisis y perspectivas en el Estado de Veracruz. En: Vela Martínez y González-Rebolledo (Coords.) Propuesta para el desarrollo estratégico de Veracruz 2010-2020. Secretaría de Educación de Veracruz, Xalapa/ México, pp 247–304

Piñar-Álvarez MA, Arredondo-Tapia R (2014) Agenda de Competitividad del destino turístico Xalapa. Secretaría de Turismo, Secretaría de Turismo y Cultura de Veracruz y El Colegio de Veracruz. Xalapa, México. http://www.sectur.gob.mx/programas/gestion-de-destinos/productos-turisticos/virreinal/xalapa/. Accessed 06 Jun 2016

Piñar-Álvarez MA, Del Castillo JE (2014) Agenda de Competitividad del destino turístico Tlacotalpan, Patrimonio de la Humanidad. Secretaría de Turismo, Secretaría de Turismo y Cultura de Veracruz y El Colegio de Veracruz. Xalapa, México. http://www.sectur.gob.mx/programas/gestion-de-destinos/productos-turisticos/virreinal/tlacotalpan/. Accessed 18 Jun 2016

Rainforest Alliance (2005) Buenas prácticas para turismo sustentable. In: Woo CV et al. (comp.) Programa de Turismo Sostenible. Rainforest Alliance

Rainforest Alliance (2014) Buenas prácticas para el turismo sostenible. Ministerio de turismo del Ecuador. Agencia de los Estados Unidos para el Desarrollo Internacional (USAID). Department of Environment, Food and Rural Affairs (DEFRA), Guatemala

Sectur. Secretaría de Turismo (2015) Distintivos turísticos en México. http://www.sectur.gob.mx/tramites-y-servicios/certificacion-turistica/distintivo-h/. Accessed 16 May 2016

Sectur. Secretaría de Turismo (2016) Distintivos de calidad en la industria del turismo). http://www.nl.gob.mx/servicios/certificacion-de-calidad-y-procesos-distintivo-m-moderniza. Accessed 18 Feb 2016

Secturc. Secretaría de Turismo y Cultura (2016) Empresas de turismo en el estado de Veracruz. Datos proporcionados por la Dirección General de Turismo de Aventura y Naturaleza, Secretaría de Turismo y Cultura, Xalapa, México

Sectur-Gemes. Secretaría de Turismo-Grupo Empresarial Estrategia (2013) Modelo de Implantación y establecimiento de las Agendas de Competitividad por Destinos Turísticos. GEMES – Secretaría de Turismo, México

Sectur and Cestur. Secretaría de Turismo y Centro de Estudios Superiores de Turismo (2007) Elementos para evaluar el impacto económico, social y ambiental del turismo de naturaleza en México. México: Sectur y Cestur. http://www.sectur.gob.mx/index.jsp. Accessed 14 Oct 2016

Tamayo SM (1997) El Análisis de las Políticas Públicas en La Nueva Administración Pública. In: Bañón IR, y Carrollo E (comps). Ed. Alianza Universidad; Madrid, 249 p

Villanueva A (1993) Antologías de Política, Pública edn. Porrúa, México

Wojtarowski-Leal A, Piñar-Álvarez MA, Silva-Rivera E, Negrete-Ramírez JA (2016) Responsabilidad Social Empresarial como pieza clave en la transición hacia el desarrollo sustentable en el sector turístico. Revista de turismo y patrimonio cultural 14(1):127–139

Development of Sustainable Agriculture Through Protocols of Organic Fertilization and Its Comparison with Traditional Method Applied to the Cultivation of Radish (*Raphanus sativus*)

Baltazar Vera Juan Carlos, Méndez Valencia Dellanira, Ramos García Alondra, Martínez Rodríguez Olga Karina, Morales López Gladys, Salas Galván Ma Eugenia and Torres Arteaga Iovanna Consuelo

Abstract The development of farming technologies that allow the obtaining of agricultural products free of chemical compounds that are harmful to the agricultural environment as well as the consumers is a challenge that several research groups have addressed. In several case studies the effect of various treatments of organic fertilization on quality variables of the agricultural product to be produced has been analyzed. It is for this reason that the development of fertilization techniques that allow the obtaining of harmless agricultural products and of friendly production for the agricultural environment is very important for the sustainable development of a community. In this work a comparative study was carried out between two fertilization protocols (traditional/organic) in order to observe the effect of these on parameters of radish quality and in this way generate organic fertilization proposals for the development of sustainable agriculture in Mexico. The analysis of the effect on variables related to the quality of production of radish (*Raphanus sativus* L) which occurs in the northern region of the Guanajuato state was carried out. Treatments of traditional fertilization, organic fertilization and mixing of these were analyzed in ratios 1/0, 0.9/0.1, 0.8/0.2, 0.7/0.3, 0.6/0.4, 0.5/0.5, 0.4/0.6, 0.3/0.7, 0.2/0.8, 0.1/0.9, 0/1 (Traditional/Organic), the results show that in terms of yield, root length, equatorial growth, of the product of interest (bulb) there are significant differences in the treatments with greater fraction of

B. V. J. Carlos (✉) · M. V. Dellanira · R. G. Alondra · M. R. O. Karina · S. G. M. Eugenia · T. A. Iovanna Consuelo
Universidad Politécnica del Bicentenario, Carretera Estatal Silao-Romita Km. 2, Silao Gto, Mexico
e-mail: jbaltazarv@upbicentenario.edu.mx

B. V. J. Carlos · M. L. Gladys
Departamento de Ingeniería en Minas, Metalurgia y Geología, Universidad de Guanajuato, Ex hacienda San Matías s/n, 36000 Guanajuato, Gto, Mexico

W. Leal Filho et al. (eds.), *Sustainable Development Research and Practice in Mexico and Selected Latin American Countries*, World Sustainability Series,
https://doi.org/10.1007/978-3-319-70560-6_28

organic compound. In terms of the difference between root length and equatorial growth, the control was the experiment that showed the lowest value. Regarding the number of true leaves and stem height of the radish plant the results show that there is no tendency towards traditional or organic treatment. The study revealed that when fertilization tends toward the organic protocol, the most adequate results are obtained in terms of quality parameters.

Keywords Fertilization · Organic · Sustainable · Radish

1 Introduction

The use of synthetic chemical fertilizers as soil improvers has generated harmful side effects in the agricultural environment, the above is due to the ionic character of the species that make up these products (Laines and Hernández 2017). In addition to the above, the effects of risks to human health produced by the protocols of nutrition and pest control in crops have increased considerably in recent years (Souza 2009). It is for this reason that the development of protocols for fertilization and pest control that is environmentally friendly and without negative effects on human health is of vital importance.

The vermiculture was developed in the 19th century with Charles Darwin who dedicated 40 years of his life to the study of the worm. In 1881 published the book called "The formation of the vegetal earth by the action of the earthworms", in this sense it is considered that from this period begins the study of vermiculture and Charles Darwin is considered as the pioneer of the same (Quintero 2004).

Vermiculture as an agro-technological alternative was born in the United States in the late 1940–1950, but moved to Europe and established in Italy where it was well developed and subsequently spread to other European countries (Ramírez 2002). At the beginning of the decade of the eighties it returns to America, establishing itself in Chile and for 1986 they had near 900 producers distributed throughout the country, which allowed to establish a market for the commercialization of earthworm fertilizer to European countries such as Holland, Belgium and Germany, among others (Delfín 2007).

Vermiculture begins its development in Mexico, with the introduction of a medium-scale project in the State of Mexico for rural areas, almost at the same time, projects were developed in Veracruz at the Institute of Ecology of Jalapa, but it does not reach the expected magnitude, even though our country has all the conditions to achieve a good development. Due to the above the vermiculture in Mexico has had the support of the private enterprise and the government (Delfín 2007).

The Hummus of Worm is one of the fertilizers that began to be used in the last years, both in the crops to improve its production and also for the regeneration of the soils that have been affected by the use of chemical fertilizers that have deteriorated it. It is important to note that the profitability of the crops has been increased with the use of hummus compared to the action of other organic fertilizers, this because hummus has a very rich composition and nitrogen is considered as the most essential component to improve the quality and profitability of crops (Suquilanda 1996). At present the *Eisenia foetida* is the most cultivated species in the whole world, given its rusticity, tolerance to environmental factors (pH, temperature, humidity), reproductive potential and grouping capacity (Rodríguez 2016).

The Californian red worm (*Eisenia foetida*) has the ability to recycle organic matter to produce hummus (vermicompost) used as fertilizer for agricultural use and earthworms added to the gardens or farms improve soil conditions as well as being used as a protein supplement for animals (Díaz 2002).

The vermicompost is an organic fertilizer bioregular and corrector of the ground, is biostable, which means that it does not give rise to fermentation and therefore of fast assimilation, is of blackish color, homogenous and with smell of forest mulch. Worm hummus has a high beneficial microbial load and a mineral rich source containing a high percentage of humic and fulvic acids, produces hormones that stimulate the development of plants, improves the physical-chemical characteristics of the soil, protects it from erosion and regulates the activity of nitrites, Finally, hummus neutralizes the presence of chemical contaminants (Ruiz 2011).

Radish (*Raphanus sativus* L), are from Asia (Casseres 1980), generally require a medium climate, although the high temperatures can give rise to spicy flavors in its roots, its productive cycle is short and can vary between 20 and 70 days, according to the variety.

The radish is an annual or biannual herbaceous plant with a flowering stem that reaches 80 cm in height; Its leaves are rough, large, trimmed in the shape of a tooth and green; Its flowers are white, yellow or violet and are grouped in clusters and its fruit is a silica in whose interior has several round seeds of reddish color. The root is about 3.5 cm in diameter, fleshy sometimes round or somewhat elongated, white or red according to variety and spicy taste (Gómez 2001). The radish plant has between 2 and 3 pairs of true leaves and a stem size that can reach 100 cm (Gómez 2001).

The appropriate temperature ranges from 18 to 22 °C, requires soil rich in organic matter with abundant moisture (Casseres 1980), can be adapted to any type of deep, clayey and neutral reaction soil (Montero et al. 2006). Radishes become hollow and spicy depending on the nutrients available in the substrate and usually crack if there is little moisture. The characteristic red color is given by anthocyanin's in pericyclic cells.

In Mexico, about 6 million tons of radish are produced per year, of which 50,000 tons are produced in the state of Guanajuato, is for this reason and due to the importance that this crop has in Mexico and in the region, the development of sustainable alternatives in terms of fertilization to obtain crops free of harmful agents for man and the environment is attractive.

2 Methodology

The development of this project was carried out in the experimental field belonging to the Polytechnic University of the Bicentennial. In the first instance the field was prepared in a traditional way using a hand tool to carry out the conditioning of the land for the sowing (see Picture 1). Said treatment consists of removing the substrate from the field to decompress it as well as the manual removal of stones that prevent the homogeneous development of the crop. For the removal and leveling of the terrain, agricultural tools of the Truper brand were used.

Seeding was carried out in the field considering 350 cm rows and 25 cm spacing between grooves, taking a total of 12, in each groove were planted 15 radish seeds of the Geneseeds brand, in the sowing a heavy irrigation of 10 L per groove was applied to start with the sprouting process (see Picture 2). The distance between seeds seeded was 23 cm and as a measuring instrument was used a Truper $\pm$ 0.5 mm flexometer of 5 m maximum length. The volumetric flow of the water supply hose (20 L/min) was used to calculate the amount of water per groove and the irrigation time per groove was calculated, which was 30 s.

Once the field was conditioned by means of a heavy irrigation to favor the germination, we proceeded to design the protocol of fertilization by means of the proper nutritional calculation for the cultivation of radish, This nutritional requirement was carried out by scaling the standard demand of Nitrogen, Phosphorus and Potassium (N, P, K). The standard N, P, K standard of the radish crop is 80, 120, 80 kg/ha, respectively (Gómez 2001). The amount of nutritive compounds was divided as follows: 1/0, 0.9/0.1, 0.8/0.2, 0.7/0.3, 0.6/0.4, 0.5/0.5, 0.4/0.6, 0.3/0.7, 0.2/0.8, 0.1/0.9, 0/1 (Traditional Fertilization/Organic Fertilization). As traditional fertilization was used fertilizer N, P, K brand fertiberia and as organic fertilizer was used worm hummus brand Multiagro. The mixtures corresponding to the aforementioned fractionation were prepared using a Uline brand digital scale H-1116 with a capacity of 14 kg $\pm$ 0.0009 kg and applied 2 days after heavy irrigation in cultivation of radish (see Picture 3).

The crop progression was followed by light irrigation of 5 L per grooves every third day, additional fertilization was applied to the 22 days of planting (half of the

Picture 1 Preparation of the experimental field of the Polytechnic University of the Bicentennial

Picture 2 Application of heavy irrigation in the field planted

Picture 3 Application of fertilization mixtures in the field planted

production cycle) And the growth of the plant was constantly monitored in order to prevent diseases and/or pests. The desired product was harvested after completing the sowing, growing and maturation cycle, which was 46 days and proceeded to take the measurements of product quality reported in this study (yield, root length and equatorial size). For measurements a Vernier brand Truper model 14,388 with a capacity of 160 mm $\pm$ 0.06 mm was used. Statistical analysis was performed using the Statgraphics Centurion 2014 software using multi-range analysis to determine whether there is a significant difference between treatment, this analysis was performed by taking 10 replicates of each treatment with a total sample of 120 elements.

In order to determine if there is a significant difference between the two fertilization protocols and their respective mixtures, a discussion of results was carried out based on a statistical analysis of the case studies. In this sense the open-field environmental conditions in the city of Silao were 24 °C with a relative humidity of 55%.

3 Results

3.1 Yield

The product of interest was cultivated on day 46, the yield obtained for each fertilization treatment is shown in Table 1.

Table 1 shows that for the cases 0/0, 0.9/0.1, 0.8/0.2, 0.7/0.3, 0.6/0.4 (traditional/organic) yield is 70%, for cases 1/0, 0.5/0.5, 0.4/0.6, 0.3/0.7 is 80%, for relation 0.2/0.8 is 90% and for 0.1/0.9 y 0/1 yield is 100%. The above is indicative that the greater quantity of product obtained tends towards the treatments where the nutritional contribution to the crop is contributed mainly by the organic fertilizer.

3.2 Root Length

Table 2 shows the average results of root length concerning the various fertilization protocols.

The results corresponding to root length of the radish root (product of interest) show that the best results are found in the relationships where the organic fertilizer fraction is higher, having that for the relations where it has predominance of the traditional fertilizer 1/0, 0.9/0.1, 0.8/0.2, 0.7/0.3, 0.6/0.4 (Traditional/Organic) has a root length of the product of 4.97, 5.46, 5.72, 5.83 y 6.01 cm respectively, which indicates that as the traditional nutrient fraction decreases the root length of the radish increases, for the relation 0.5/0.5 (Traditional/Organic) root length of radish is 5.96 which is superior to the cases previously exposed except for the relation 0.6/0.4. For the cases with greater fraction of organic fertilizer 0.4/0.6, 0.3/0.7, 0.2/0.8, 0.1/09, 0/1 (Traditional/Organic) values of root length of 6.36, 6.38, 6.87, 7.13 y 6.93 cm respectively, which shows that the organic nutrition element fraction

Table 1 Yield (%) depending on the different fertilization protocols

Fraction of fertilization (traditional/organic)	Yield (%)
0/0	70
1/0	80
0.9/0.1	70
0.8/0.2	70
0.7/0.3	70
0.6/0.4	70
0.5/0.5	80
0.4/0.6	80
0.3/0.7	80
0.2/0.8	90
0.1/0.9	100
0/1	100

Table 2 Root length according to the various fertilization protocols

Fraction of fertilization (traditional/organic)	Root length (cm)	Standard deviation	Variance
0/0	3.47	0.094	0.009
1/0	4.97	0.096	0.008
0.9/0.1	5.46	0.217	0.047
0.8/0.2	5.72	0.154	0.024
0.7/0.3	5.83	0.131	0.017
0.6/0.4	6.01	0.145	0.021
0.5/0.5	5.96	0.177	0.031
0.4/0.6	6.36	0.189	0.036
0.3/0.7	6.38	0.091	0.008
0.2/0.8	6.87	0.182	0.033
0.1/0.9	7.13	0.175	0.056
0/1	6.93	0.141	0.02

increases the root length increases, experimentation showed that the best relation is 0.1 of traditional fertilization with 0.9 of organic fertilization. It can be observed that all cases of fertilization obtained results above that reported by the control, which was 3.47 cm. In this sense, all the products presented values of root length superior to the value reported in the literature which is of 3.25 cm (Gómez 2001).

The statistical analysis performed for the various treatments indicates the following: The ANOVA table when decomposing the variance of repetitions into two components, an inter-group component and an in-group component. The F-ratio, which in this case is equal to 145,247, is the quotient between the estimated between-groups and the estimated within-groups, which is higher than the theoretical value (Montgomery 2014) therefore it can be elucidated that there is a significant difference between treatments. In this sense, because the P-value of the F-test yielded a value of 0.01321, which is less than 0.05, indicates that there is a statistically significant difference between the mean of repetitions for one level of treatment and another, with a level of 95.0% confidence. To determine which means are significantly different from others, we performed the multiple-rank test, the result of which is shown in Table 3.

The results of the multiple range test indicate that the treatments: 0/0, 1/0, 0.9/0.1 and 0.6/0.4 (Traditional/Organic) present a significant difference with respect to the other treatments. The treatments that have statistical homogeneity according to the study can be located in three groups, the first one corresponds to the treatments: 0.8/0.2, 0.7/0.3 and 0.5/0.5 (Traditional/organic); the second group corresponds to 0.4/0.6 and 0.3/0.7 (Traditional/organic) and the third group corresponds to cases: 0.2/0.8, 0.1/0.9 and 0/1 (Traditional/organic). The study shows that there is a homogeneous group with a trend towards traditional fertilization and two with a trend towards organic nutrition, in this sense it is worth mentioning that the best values of root size correspond to the group with greater tendency to organic treatment, which

Table 3 Statistical summary of homogeneous groups and significant difference between treatments for root length

Fraction of fertilization (traditional/organic)	Repetitions	Average (cm)	Homogeneous groups
0/0	10	3.47	X
1/0	10	4.97	X
0.9/0.1	10	5.46	X
0.8/0.2	10	5.72	X
0.7/0.3	10	5.82	XX
0.5/0.5	10	5.96	X
0.6/0.4	10	6.01	X
0.4/0.6	10	6.36	X
0.3/0.7	10	6.38	X
0.2/0.8	10	6.87	X
0/1	10	6.93	XX
0.1/0.9	10	7.13	X

is indicative that this parameter tends to its best value when applying this treatment and according to the statistical analysis there is a clear significant difference with respect to the cases where the fertilization tends to be mostly traditional.

3.3 *Equatorial Growth*

Table 4 shows the average results of the equatorial growth concerning the various fertilization protocols.

The results corresponding to the equatorial growth of radish root show that the best results are found in the relations where the organic fertilizer fraction is higher having that for the relations where it has predominance of the traditional fertilizer 1/0, 0.9/0.1, 0.8/0.2, 0.7/0.3, 0.6/0.4 (Traditional/Organic) there is an equatorial growth of the product of 4.02, 4.46, 4.94, 5.01 y 5.18 cm respectively, which indicates that as the traditional nutrient fraction decreases the equatorial growth of radish root increases, for relation of 0.5/0.5 (Traditional/Organic) Equatorial growth of radish root bulb is 5.5 cm which is superior to cases previously exposed. For the cases with greater fraction of organic fertilizer 0.4/0.6, 0.3/0.7, 0.2/0.8, 0.1/09, 0/1 (Traditional/Organic) values of equatorial growth of 5.54, 5.67, 6.00, 6.2 and 5.75 cm respectively, which shows that as the organic nutrition element fraction increases the equatorial growth of radish, taking that for all experiments the best relation is 0.1 of traditional fertilization with 0.9 of organic fertilization. In these schemes of experiments, it can be observed that all cases of fertilization obtained results above that reported by the control, which was 3 cm. In this sense all the products presented values of equatorial growth superior to the value reported in the literature which is of 3.25 cm (Gómez 2001).

Table 4 Equatorial growth according to the various fertilization protocols

Fraction of fertilization (Traditional/ Organic)	Equatorial growth (cm)	Standard deviation	Variance
0/0	3.00	0.194	0.037
1/0	4.02	0.131	0.017
0.9/0.1	4.46	0.189	0.036
0.8/0.2	4.94	0.236	0.056
0.7/0.3	5.01	0.255	0.065
0.6/0.4	5.18	0.257	0.066
0.5/0.5	5.50	0.176	0.031
0.4/0.6	5.54	0.126	0.016
0.3/0.7	5.67	0.141	0.02
0.2/0.8	6.00	0.151	0.022
0.1/0.9	6.20	0.187	0.035
0/1	5.75	0.184	0.033

The statistical analysis performed for the various treatments suggests the following: The ANOVA table when decomposing the variance of repetitions into two components: an inter-group component and an in-group component. The F-ratio, which in this case is equal to 217.31, is the quotient between the estimated between-groups and the estimated within-groups, which is higher than the theoretical value (Montgomery 2014). Therefore it is possible to infer the existence of a significant difference between treatments. In this sense, because the P-value of the F-test yielded a value of 0.008303, which is less than 0.05, indicates that there is a statistically significant difference between the mean of repetitions for one level of treatment and another, with one 95.0% confidence level. To determine which averages are significantly different from others, we performed the multiple range test, the result of which is shown in Table 5.

The results shown by the multiple range test indicate that the treatments: 0/0, 1/0, 0.9/0.1 and 0.6/0.4 (Traditional/Organic) present a significant difference with respect to the other treatments. The treatments that have statistical homogeneity according to the study can be located in three groups, the first one corresponds to the treatments: 0.8/0.2 and 0.7/0.3 (Traditional/organic), the second group corresponds to the cases: 0.5/0.5, 0.4/0.6, 0.3/0.7 and 0/1 and the third group corresponds to the treatments: 0.2/0.8 and 0.1/0.9 (Traditional/organic). As in the previous case, the study shows that there is a homogeneous group with a trend towards traditional fertilization and two with a tendency towards organic nutrition, in this sense it is worth mentioning that the best values of equatorial size correspond to the group with the highest tendency to Organic treatment, which is indicative, as in the previous case that this parameter tends to its best value when applying this treatment and according to the statistical analysis there is a clear significant difference with respect to the cases where the fertilization tends to be mostly traditional.

Table 5 Statistical summary of homogeneous groups and significant difference between treatments for equatorial length

Fraction of fertilization (traditional/ organic)	Repetitions	Average (cm)	Homogeneous groups
0/0	10	3.00	X
1/0	10	4.02	X
0.9/0.1	10	4.46	X
0.8/0.2	10	4.94	X
0.7/0.3	10	5.01	X
0.6/0.4	10	5.18	X
0.5/0.5	10	5.50	X
0.4/0.6	10	5.54	XX
0.3/0.7	10	5.67	XX
0/1	10	5.75	X
0.1/0.9	10	5.98	X
0.2/0.8	10	6.00	X

3.4 *Number of True Leaves*

Table 6 shows the results corresponding to the number of true leaves of the harvested product.

The results corresponding to the number of true leaves of the radish plant show that there is some homogeneity in the number of leaves average, due to that the average of true leaves in all cases oscillates between values close to 4 and 7, The above agrees with the values reported in the literature (Gómez 2001). In this sense, the average values of 3.8, 6.7, 4.3, 5.7, 4.6 and 5 are obtained for the relationships 0/0, 1/0, 0.9/0.1, 0.8/0.2, 0.7/0.3, 0.6/0.4 (Traditional/Organic) respectively. In relation to 0.5/0.5, 0.4/0.6, 0.3/0.7, 0.2/0.8, 0.1/09, 0/1 the values obtained were: 3.5, 6.2, 5.9, 3.8, 3.8 and 6.1 respectively.

The statistical analysis performed for the various treatments indicates the following: The ANOVA table by decomposing the variance of repetitions into two components: an inter-group component and an in-group component. The F-ratio, which in this case is equal to 6.97817, is the quotient between the estimated between-groups and the estimated within-groups, which is higher than the theoretical value (Montgomery 2014), therefore it is possible to infer the existence of a significant difference between treatments. In this sense, because the P-value of the F-test yielded a value of 0.00593, which is less than 0.05, indicates that there is a statistically significant difference between the mean of repetitions for one level of treatment and another, with one 95.0% confidence level. To determine which means are significantly different from others, we performed the multiple-rank test, the result of which is shown in Table 7.

The results shown by the multiple range test indicate that the treatments: 0.5/0.5, 0/0, 0.2/0.8, 0.1/0.9, 0.9/0.1 and 0.7/0.3 (Traditional/Organic) form a homogeneous

Table 6 Number of true leaves for various fertilization protocols

Fraction of fertilization (traditional/organic)	True leaves	Standard deviation	Variance
0/0	3.8	1.751	3.066
1/0	6.7	1.888	3.564
0.9/0.1	4.3	1.159	1.343
0.8/0.2	5.7	2.359	5.564
0.7/0.3	4.6	0.699	0.488
0.6/0.4	5.0	0.942	0.887
0.5/0.5	3.5	1.269	1.611
0.4/0.6	6.2	1.032	1.065
0.3/0.7	5.9	0.737	0.543
0.2/0.8	3.8	1.229	1.518
0.1/0.9	3.8	1.229	1.518
0/1	6.1	0.875	0.765

Table 7 Statistical summary of homogeneous groups and significant difference between treatments for the number of true leaves

Fraction of fertilization (traditional/organic)	Repetitions	Average	Homogeneous groups
0.5/0.5	10	3.5	X
0/0	10	3.8	X
0.2/0.8	10	3.8	X
0.1/0.9	10	3.8	X
0.9/0.1	10	4.3	XX
0.7/0.3	10	4.6	XXX
0.6/0.4	10	5.0	XXX
0.8/0.2	10	5.7	XXX
0.3/0.7	10	5.9	XX
0/1	10	6.1	XX
0.4/0.6	10	6.2	X
1/0	10	6.7	X

group where there is no difference Significant (A), the cases corresponding to: 0.6/0.4 and 0.8/0.2 (Traditional/Organic) make up a homogeneous group (B). The treatments 0.3/0.7, 0/1, 0.4/0.6 and 1/0 (Traditional/Organic) form another group that does not present significant difference (C). The table shows that there is a relationship between the boundaries that separate the groups A, B, C, which suggests that the significant difference between groups is small and that for this parameter the total sample tends to homogeneity.

3.5 Stem Height

Table 8 shows the results corresponding to the stem height of the harvested product.

The results corresponding to the stem height of the radish plant do not present a well defined trend of this parameter with respect to the treatments. In this sense it is observed that the average height of the stem of the product oscillates between values close to 11 and 16 cm which is below the reported (Gómez 2001). This can be a consequence of the time of the harvest cycle, which was lower than reported, but yielded adequate values of yield, root size, equatorial size and number of leaves. In this sense we have for the relations 0/0, 1/0, 0.9/0.1, 0.8/0.2, 0.7/0.3, 0.6/0.4 (Traditional/Organic) we have average values of: 10.8, 14, 14.5, 15.75, 12.5 and 12.2 cm respectively. In relation to the values 0.5/0.5, 0.4/0.6, 0.3/0.7, 0.2/0.8, 0.1/09, 0/1 the values obtained were: 10.05, 10.25, 13, 11.25, 13.6, 12.53 cm respectively.

The statistical analysis performed for the various treatments indicates the following: The ANOVA table by decomposing the variance of repetitions into two components: an inter-group component and an in-group component. The F-ratio, which in this case is equal to 12.13, is the quotient between the estimated between-groups and the estimated within-groups, which is higher than the theoretical value (Montgomery 2014). Therefore it is possible to infer the existence of a significant difference between treatments. In this sense, because the P-value of the F-test yielded a value of 0.00430, which is less than 0.05, indicates that there is a statistically significant difference between the mean of repetitions for one level of treatment and another, with one 95.0% confidence level. To determine which averages are significantly different from others, we performed the multiple-rank test, the result of which is shown in Table 9.

Table 8 Stem height for various fertilization protocols

Fraction of fertilization (traditional/organic)	Stem height (cm)	Standard deviation	Variance
0/0	10.8	1.7029	2.8998
1/0	14.00	1.0540	1.1109
0.9/0.1	14.50	0.9718	0.9443
0.8/0.2	15.75	1.2304	1.5139
0.7/0.3	12.50	1.5092	2.2777
0.6/0.4	12.20	1.7738	3.1464
0.5/0.5	10.05	1.0124	1.0250
0.4/0.6	10.25	1.3794	1.9027
0.3/0.7	13.00	0.7071	0.5000
0.2/0.8	11.25	2.8698	8.2358
0.1/0.9	13.60	2.6749	7.1551
0/1	12.53	2.3754	5.6425

Table 9 Statistical summary of homogeneous groups and significant difference between treatments for stem height

Fraction of fertilization (traditional/organic)	Repetitions	Average	Homogeneous groups
0.7/0.3	10	10.05	X
0.8/0.2	10	10.25	X
0.1/0.9	10	10.80	XX
0/1	10	11.25	XXX
0.6/0.4	10	12.20	XXX
0.5/0.5	10	12.50	XX
0.9/0.1	10	13.00	XX
1/0	10	13.60	XXX
0.2/0.8	10	14.00	XX
0.3/0.7	10	14.50	XX
0.4/0.6	10	15.75	X

The results shown by the multiple range test report that there is some degree of homogeneity among the different treatments among which 3 groups can be elucidated: (A) conformed by the treatments 0.7/0.3, 0.8/0.2, 0.1/0.9, 0/1 (Traditional/Organic), group B consists of cases: 0.6/0.4, 0.5/0.5, 0.9/0.1/Organic) and group C conformed by treatments: 1/0, 0.2/0.8, 0.3/0.7, 0.4/0.6 (Traditional/Organic). The table shows that as in the previous case there is a relation between the boundaries that separate the groups A, B, C, which suggests that the significant difference between groups is small.

3.6 *Comparison Root Length and Equatorial Growth of Radish for the Different Fertilization Protocols*

Figure 1 shows a comparison between root length and equatorial growth of radish for the different fertilization protocols.

Figure 1 shows that the root length is greater than the equatorial growth for all products taking that for cases where there is a larger fraction of the traditional fertilizer 1/0, 0.9/0.1, 0.8/0.2, 0.7/0.3, 0.6/0.4 (Traditional/Organic) the difference between equatorial growth and root length is 0.95, 1.00, 0.78, 0.82, 0.83 cm respectively, for the relation 0.5/0.5 the difference is 0.46 cm. In this sense for the cases in which the fraction of organic fertilizer is higher 0.4/0.6, 0.3/0.7, 0.2/0.8, 0.1/09, 0/1 (Traditional/Organic) differences are 0.82, 0.71, 0.87, 0.93 y 1.18 cm respectively taking the case with the greatest difference was the ratio 0/1 (Traditional/Organic) with 1.18 cm and the case that presented less difference was the control with a value of 0.47 cm. According to the reported, the equatorial growth and the root length should be very similar, so that the difference between both parameters should be 0. In this sense the results show that the greater difference was obtained between the products that presented bigger dimensions which can be explained that to have greater mass and area receive greater compression by the substrate which may lead to an elongation in the product.

The products presented acceptable qualitative characteristics (color, without signs of stress and healthy aspects in stem and leaves) which allows to elucidate that the obtained product has quality conditions, together with the parameters measured in this research indicated that the best results are found in the treatments with the greatest tendency towards organic fertilization, which can be seen in Pictures 4 and 5 which show the products with only traditional fertilization and only organic fertilization respectively. Regarding the number of true leaves, the study did not show a tendency towards a type of treatment, showing that for all the cases there are

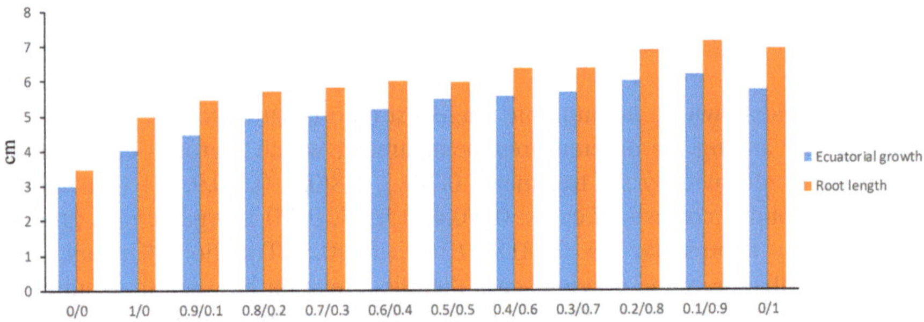

Fig. 1 Comparison between root length and equatorial growth of radish for the different fertilization protocols

Picture 4 Fertilized radishes with traditional compound

Picture 5 Fertilized radishes with organic compound

between two and three pairs of these. In what corresponds to the height of stem we have that as in the number of true leaves cannot be elucidated some tendency towards some treatment having that the average of height oscillates between 11 and 16 cm.

4 Conclusions

Radish was successfully cultivated by applying traditional and organic fertilization protocols and various mixtures of these. The results indicate that for all the fertilization protocols yield percentages were obtained between 70 and 100%, since the experiments with the highest amount of organic fertilizer had the best results in this parameter Being the relation 0/1 and 0.1/0.9 (Traditional/organic) those who reported the highest value which was 100. In this sense with regard to root length of bulb also for organic fertilization the highest results were obtained by observing that the best value is for the relation 0.1/0.9 (Traditional/organic) whose value is 7.13 cm. For the equatorial growth of bulbs, it can be observed that, as in the previous cases, organic fertilization showed the best results, again having the relation 0.1/0.9 (Traditional/organic) reported the best result being 6.2 cm. With regard to stem height, the best ratio was 0.8/0.2 (traditional/organic), reporting a value of 15.75 cm, For the number of true leaves the best treatment was 0.8/0.2 (Traditional/Organic) with a total of 7 leaves. The analysis of results of number of true leaves and height of stem indicates that the various treatments do not generate a trend of improvement towards one case or another. When analyzing the difference between root length and equatorial growth, it was found that the control showed the smallest difference and the organic treatments showed the greatest difference. This can be attributed to the compression exerted by the substrate on the product which is higher in the cases where there is greater weight and area. In this work it can be observed that the best quality parameters were obtained in the products that were fertilized mainly with organic compound (worm hummus), which may be due to the fact that said protocol generates a greater mechanical affinity between the fertilizer and the substrate as well as greater retention of water which favors the absorption of nutrient by the crop. In this sense this work generates an alternative proposal to develop crops with high competitiveness replacing chemical compounds that have harmful side effects for man and his environment.

This work was carried out in open field conditions, which allowed the testing of fertilization protocols proposed in conditions close to large scale agricultural production methods, in this sense the project presents future work options, among which we can highlight the carrying out of bromatological and microbiological studies for the various case studies and make a comparative analysis to determine which protocols provide the best product.

Acknowledgements The authors thank SEP and Prodep for the support provided through the project 19743-UPBIC-CA-1-2015-1-14570.

References

Casseres E (1980) Producción de hortalizas. Instituto interamericano de ciencias agrícolas. San José, Costa Rica 3(1):272–275

Delfín V (2007) La lombricultura como alternativa de diversificación para elevar la rentabilidad y el desarrollo sustentable de los agronegocios en el municipio de El Marqués, Querétaro. Querétaro, Qro. Tesis de maestría. Universidad Autónoma de Querétaro pp 3–4

Díaz E (2002) Guía de lombricultura/Lombricultura una alternativa de producción para emprendedores y productores del agro. Agencia de Desarrollo Económico y Comercio Exterior Municipio Capital de La Rioja. La Rioja, España. Cap pp I 5–12

Gómez L (2001) Evaluación del cultivo de Rábano (*Raphanus sativus* L.) bajo diferentes condiciones de fertilización orgánica e inorgánica, Saltillo Coah. Tesis de licenciatura. Universidad Autonoma Agraria Antonio Narro pp 30–65

Laines J, Hernández L (2017) Evaluation of the effect of bio-fertilizers and chemical fertilizer on the growth of Uspi (*Couepia polyandra*) and Barnacles (*Ormosia macrocalix*). Tabasco México, JEEOS 1(1):35–50

Montero S, Sing B, Taylor R (2006) Evaluación de seis estructuras de producción hidropónica diversificada en el trópico húmedo de costa rica. San Jose de Costa rica, Costa rica. Tierra Tropical 2(1):27–37

Montgomery D (2014) Diseño y análisis de experimentos. U.S.A. Limusa, Wiley, ISBN: 9687270608 1(1):45–48

Quintero R (2004) La lombricultura como una alternativa para el aprovechamiento de los residuos sólidos orgánicos. Caso de estudio "Barrio de Santiaguito", municipio de Texcoco, Estado de México. Tesis de maestría. Instituto Politécnico Nacional pp 4–5

Ramírez R (2002) Efecto de la micorriza arbuscular y vermicomposta sobre el desarrollo en vivero y campo de *Carica papaya* L. Tecomán, Colima. Tesis de Maestría. Universidad de Colima pp 12–13

Rodríguez L (2016) Proyecto comunitario ejidal de producción y aplicación de lombricomposta, como modelo de trabajo profesional del agrotecnólogo. tesis de licenciatura. Silao. Gto. Universidad Politecnica del Bicentenario pp 30–45

Ruiz M (2011) Taller de elaboración de lombricomposta: porque tener lombrices nos beneficia a todos. México. México D.F. Universidad Iberoamericana, A.C. Available in www.uia.mx/web/files/publicaciones/

Souza J (2009) Sobre la utilización de plaguicidas y le registro de las intoxicaciones. Available in www.rap-al.org

Suquilanda V (1996) Agricultura orgánica, alternativa tecnológica del futuro. Fundagro. Quito, Ecuador 1(1):650–660

The Communal Assets of Santo Domingo Huehuetlán as an Example of Sustainable Community Development

María Concepción López-Téllez, Antonio Fernández-Crispín, Hugo Rodolfo Molina-Arroyo, Gonzalo Yanes-Gómez, Valeria García-Loza and Valentina Campos-Cabral

Abstract In contrast with its biological and cultural wealth, the community of *Bienes Comunes de Santo Domingo Huehuetlán* is characterized by high levels of poverty and marginality. The aim of this study was to analyze the history, organization and strategies which have enabled this community to defend and make sustainable use of its common pool resources. Participatory methodologies were used to create meeting and socialization spaces for the exchange of experiences and the documentation of the decision-making processes applied for resource management and social, cultural and economic activities. The processes of governmental intervention were observed, giving insight into the communal management of common pool resources, as well as traditional knowledge-based community organization, in which common pool resources are regulated by local rules and agreements. The community's spatial planning takes into account areas of production, conservation, and use, and the defense of its land against multinational mining projects. While community development has its limitations, it is a viable

M. C. López-Téllez · A. Fernández-Crispín (✉) · H. R. Molina-Arroyo · G. Yanes-Gómez · V. García-Loza
Facultad de Ciencias Biológicas, Benemérita Universidad Autónoma de Puebla, Blvd. Valsequillo y Av. San Claudio, Edificio 112-A, Ciudad, Universitaria, Col. Jardines de San Manuel, CP. 72570 Puebla, Pue, Mexico
e-mail: anfern64@yahoo.com.mx

M. C. López-Téllez
e-mail: concepcionlopeztellez@gmail.com

H. R. Molina-Arroyo
e-mail: hugo_molina_arroyo@hotmail.com

G. Yanes-Gómez
e-mail: gonzalo_y2@hotmail.com

V. García-Loza
e-mail: pq_20@hotmail.com

V. Campos-Cabral
Facultad de Derecho y Ciencias Sociales, Av San Claudio y 22 Sur, Cd Universitaria, Col. Jardines de San Manuel, CP. 72570 Puebla, Pue, Mexico
e-mail: valiacc@gmail.com

W. Leal Filho et al. (eds.), *Sustainable Development Research and Practice in Mexico and Selected Latin American Countries*, World Sustainability Series,
https://doi.org/10.1007/978-3-319-70560-6_29

and participatory guidance model which responds to unfavorable factors and secures the common pool resources for the community.

Keywords Environmental and social sustainability · Communal management Governance

1 Introduction

Sustainable development can be viewed from the perspective of various schools of thought. In *The Tragedy of the Commons*, Hardin (1968) considers that the selfish behavior of a group of social actors sharing common resources leads to over-exploitation until deterioration or even exhaustion. Thus, he advises that natural common pool resources must be either privatized or nationalized (control from above) for their adequate management and use. However, this position is not shared by other authors, such as Ostrom (2005), Agrawal and Chhatre (2006) and Tetreault (2007). There are, around the world, many examples of the adequate management of natural resources by local communities. These communities have developed solid institutions that generate equitable benefits and costs from the exploitation of their common pool resources, to which no threat is generated by said exploitation. The social actors in these communities have the capacity to establish local institutions that enable them to control the mechanisms for the use of and access to natural resources, establishing rules and agreements designed by the collective. These are understood and accepted by all the actors involved, enabling the creation of cooperative strategies for the maintenance of common pool resources and, thus, achieving an efficient process of exploitation known as "institutional focus" (Ostrom 2005; Larson and Sarukhán 2003; Tetreault 2007; Poteete et al. 2010). Critical humanism, which centers on social and environmental sustainability, proposes a change in the use of natural resources with the objective of reducing poverty and establishing a new society (Pierri 2005). In this school of thought, the analysis of social sustainability is used as a bridge to the analysis of environmental sustainability by analyzing the social relations that determine interactions over the use of natural resources (Foladori and Tommasino 2005).

Ownership of the land, water and common pool resources susceptible to appropriation has a social function that must be regulated on a national level in order to guarantee fair and equitable distribution and, thus, develop and improve the life conditions of the population (Pina 1970). Due to this social function, the owner must make rational use of the goods and services provided by a territory (Sánchez 2012).

In Mexico, ownership of the land, water and common pool resources is regulated by Article 27 of the constitution. This article indicates that said ownership corresponds to the nation (DOF 2016). The state has the authority to transmit dominion to private individuals at the convenience of the public interest, recognizing four types of ownership: original; private; public; and, social (Sánchez 2012).

As a result of the Mexican Revolution (1910–1917), agrarian rights were renewed and established at a constitutional level. Thus began the agrarian reform whose central objective was to regulate the land-sharing and common pool resource procedures, redistributing large *hacienda* properties into small plots for a greater number of agricultural workers (Sánchez 2012). Rural land used for subsistence farming and destined for protection and agrarian production is considered agrarian property. Thus, the *ejidos* (traditional community-owned farmland in Mexico) and communal assets, whose territories are divided into land for human settlement for both common and parceled use, are established under constitutional protection. The community assembly is charged with resolving any type of disagreement related to the exploitation of the uses made of the resources. Common use lands are considered important for the environmental sustainability of the community, given that they are inalienable, imprescriptible and unseizable. Moreover, the rights to them are conceded in parts equal to the number of the members of the agrarian nucleus (Durand Alcántara 2005). The *ejidos* and communal assets have legal personality and proprietary rights that are regulated by agrarian law (Sánchez 2012).

The rural communities of Mexico live in a state of constant crisis. Recently, globalization and the neoliberal model, both of which concentrate and centralize capital, have led to a crisis in agrarian law, which is overseen by a legal system that does not attend to social need and which has enabled the intervention of both small producers and large multinational companies in agricultural production. To this is added the policies of the State, which are not congruent with the social reality of the country. The consequences of this are the fragmentation and disorganization of the agrarian nuclei, which has generated rural desertion and migration as former agricultural workers search for an alternative source of income (Durand-Alcántara 2005).

In recent years, the Mexican State has implemented structural reforms for the expropriation of formerly agrarian land, conceding it to foreign companies for the development of mining activity. Much of this land is characterized by biological, cultural and social wealth, as well as an abundant water supply. Water is fundamental for the extraction of minerals, a process which causes the pollution of the flora, fauna and the water itself, further to rupturing the social fabric and generating social unrest (Cárdenas 2013).

Located in the *Sierra del Tentzo*, in the central region of the State of Puebla, Mexico, *Bienes Comunales de Santo Domingo Huehuetlán* (BCSDH) is an example of a community with a high level of biodiversity and cultural wealth, in contrast with its elevated levels of poverty and marginalization. The community is currently under threat from the intention of the federal government to expropriate 700 ha of communal land to benefit a foreign mining company.

This study used a participatory methodology which aimed to systemize local knowledge on the biophysical factors of the environment, as well as the forms of community organization and the management of its resources. This information enables the empowerment of the community in response to attempts to establish unsustainable projects and enables the critical reflection on some of the strategies used for the management of its communal resources, and their associated problems, raising awareness on how to help solve these problems, and actively work towards their resolution.

The aim of this study was to analyze the community history, organization and community strategies of which have enabled *Bienes Comunales de Santo Domingo Huehuetlán* to defend and make sustainable use of its common pool resources.

2 Study Area

This research was undertaken in BCSDH, which is found in the municipality of *Huehuetlán El Grande* in the state of Puebla, Mexico, and presents a warm sub-humid climate with rains in the summer. The average temperature ranges from 17 to 23 °C, while precipitation is recorded at 700–900 mm (INEGI 2016a). The municipality forms part of the River Atoyac sub-basin in the Balsas hydrological region (Fig. 1), and has perennial and intermittent water flow, such as that found in the River Huehuetlán.

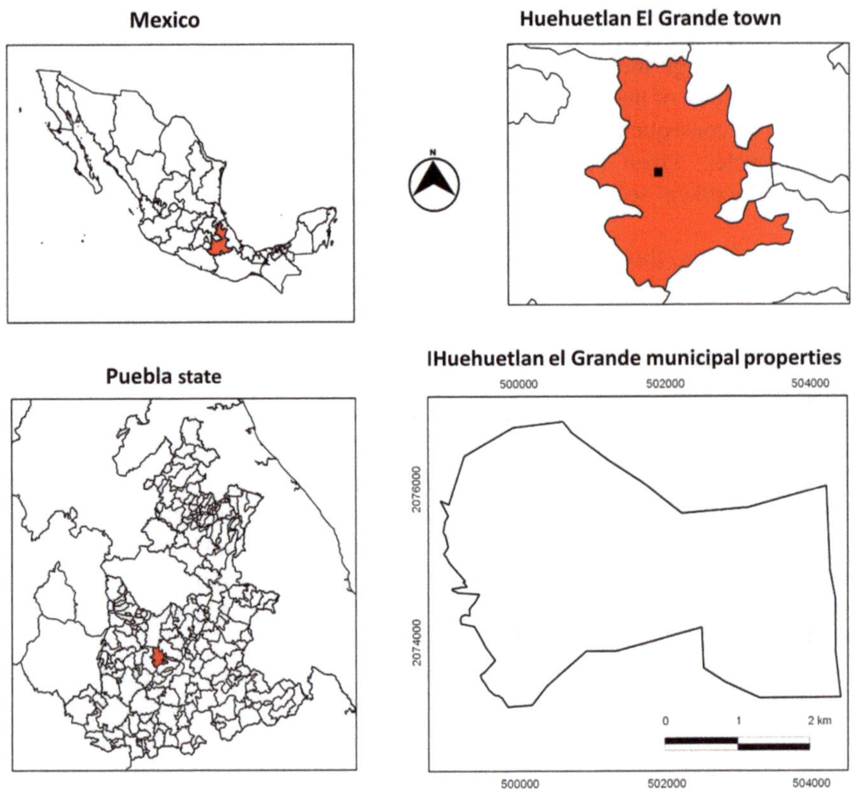

Fig. 1 Polygon of BCSDH, in the municipality of *Huehuetlán El Grande*, Puebla

The predominant vegetation is tropical dry forest, associated with secondary bushy and arboreal vegetation. Soil use comprises rainfed and irrigated agriculture (42%), rainforest (33%), forest (18%), pasture (5%) and urban areas (2%) (INEGI 2016b). There is a high level of diversity in the flora and fauna. To date, 20 reptile, 25 mammal and 55 bird species have been reported in the study area (López et al. 2015).

The population living in a state of poverty is recorded at 80.7%, of which 44.4% are found to be in moderate and 36.3% in extreme poverty, while educational lag is recorded at 44.2%. The percentage of people living in homes without the availability of basic services is 82.9% (SEDESOL 2010).

3 Methodology

This study was attributive in nature. It is based on participatory investigation, which is complemented by the application of document analysis techniques, field visits, and in-depth structured interviews. Its aim is to identify and understand, from local actors, the processes for the spatial and temporal appropriation of their territory and their methods for managing their common pool resources, considering, from the historical perspective, the community's origins, and its organization, characterization, culture, infrastructure, and productive and problematic activities. This enables solutions to be established by means of a plan for the defense, conservation and use of the community's territory.

This study was carried out from 2014 to the present day with the participation of professor-researchers and students from the Faculty of Biological Sciences at the *Benemérita Universidad Autónoma de Puebla* (Meritorious Autonomous University of Puebla).

The research was developed over four stages:

(a) **Initial Stage**—from January to April, 2014. This consisted in the compilation of bibliographical information with the objective of gaining an understanding of the social, environmental, economic and cultural context of the community. The documents were obtained from the government departments that hold official information related to the decree on the agrarian nuclei, among which are the following: the Agrarian Law Office; the Secretariat for the Environment and Natural Resources; and, the Government of the State of Puebla.

(b) **Planning**—from May to July 2014. Carried out using a participatory approach, the work plan was designed to include a total of nine workshops undertaken to document knowledge on the following: origin and history; organization; identification of the social actors; traditional festivals; description of services and education; the community's natural resources; and, the ordering into a hierarchy of the problems associated with the different aspects considered, thus identifying the type of necessities and potentialities that enable the formulation of alternative solutions. In-depth meetings and interviews were conducted with key actors in order to analyse in greater detail some aspects of the community.

(c) **Field work** was divided into two phases. The first was applied from August to November 2014. Between 110 and 130 men and women of various ages and fulfilling various roles in the community attended nine workshops. The workshops enabled the compilation, systematization and exchange of the participants' (both young and old) traditional knowledge of the area's natural resources, the origin, history, and organization of the community, and the community's traditional festivals, basic services and education. They also facilitated the identification of the community's social actors, both within and without, and its common pool resources. Using maps of the community, workshop attendees identified the rivers, landmarks, small ravines and paths, as well as identifying soil, flora and fauna use, and environmental problems. The workshops were complemented with structured interviews with those community members who had attended the workshops. The information obtained was used to calculate both the indices of cultural importance and use of the fauna (Turner 1988). The second phase (January to July 2015) comprised field visits with two or three key informants from the community at its principal landmarks in order to verify the presence of resources in the forest, fauna and agricultural systems. Aside from the field visit, profiles of the forest vegetation were prepared, in which the landmarks, soil use, associated problems and their possible solutions were characterized (Morales Mendoza 2003).

(d) **Systematization, triangulation and analysis of the information** (October to November 2015). The information obtained from the workshops, interviews and field work was organized in tables and diagrams, and was presented to the community assembly in order for it to be verified and validated. The final report containing the community diagnosis and the management of its common pool resources (soil, water, flora and fauna) was then prepared. The report forms the basis for the proposed actions to be undertaken, in the short, medium and long-term, with regard to the management of the community's territory and common pool resources.

4 Results

4.1 History

4.1.1 Legal Origin of Bienes Comunales de Santo Domingo Huehuetlán

A presidential resolution is a specific decree emanating from a unilateral decision of the President of the Mexican Republic. It is definitive in nature and cannot be modified, except at a constitutional level. Thus, BCSDH, pertaining to the municipality of Huehuetlán, Puebla, was created by means of the act of agrarian award by presidential resolution on 27th August 1986. It comprises a total of 1300.37 ha dedicated to common use, while a low percentage is dedicated to livestock, and has bodies of water classified as rivers, streams and irrigation channels, the latter being of fundamental importance for seasonal agriculture, as the lands used for this lie outside the territory of the decree. Within this territory is a 5 ha area occupied by the *Atotonilco* water park cooperative. Seven hundred hectares of the territory has been subject to dispute for four years, principally involving the multinational mining company Australian Minerals Group, the stated intention of which is to extract minerals from the common pool resource area.

4.2 Social Organization

4.2.1 Identification and Characterization of the Social Actors

BCSDH comprises 150 community members, mainly farmworkers dedicated to agriculture. They are also engaged in other activities such as trade and livestock. The majority of the informants provided a warm welcome, showed interest in visitors, and were pleased to discuss both their community and the region. In general, the social actors identified in BCSDH are people, groups or institutions who have been involved in the modifications to the territory, as well as the community and municipal environment. Through collective work and thought, the social organization of BCSDH, as described here, has contributed to its establishment as a community. A fundamental aspect in the organization of the community is the *faenas*, or *jornales*, which are a type of service or duty fulfilled by community members in the undertaking of administrative tasks and work of collective benefit (the maintenance of paths, and school, church, guard duty, etc.).The *faenas* are determined by family ties, as well as community festivals and celebrations. They are of great importance given emerging threats such as insecurity, violence, drug trafficking and the mega-projects proposed by external actors, in the face of which the community has organized itself to implement security measures, pursue legal action and establish local resistance.

4.3 Community Organization

In order ascertain how the community is organized, in-depth interviews were conducted with key informants, such as local authorities and leaders, among whom are Commissioner for BCSDH, the Security Committees and Representatives from the *Atotonilco* water park cooperative.

BCSDH community members recognize community action groups with influence at a local and regional level, whose institutional representation is based on political and administrative authority. The highest authorities at a community level are the agrarian authorities, committees and specific work groups. The Commissioner for BCSDH, along with its respective security committee, is elected every three years and is charged with the management activities related to land tenure and common pool resource projects. The committees and specific work groups are temporarily appointed to undertake various actions related to community life and production. The majority of these groups are promoted by the institutions of local government.

The most important work groups or committees are those of the administration of the *Atotonilco* water park cooperative, and the administration of the local shop and tractor.

The above described aspects reflect the complex structure of the organizational, historical and traditional life of the community, in the civil, political, administrative, religious and agrarian contexts. The community works collectively, undertaking *faenas* for the benefit of its members, the school and the church, holding meetings in which it is decided which tasks are to be undertaken. In general, those interested have a short, medium and long-term vision, reinforced with training programs for the sustainable organization and self-management of common pool resources.

It should be noted that the welfare support received from government programs has provoked the fragmentation of the habitat, principally for the opening of the agricultural frontier. These have also generated tensions and conflicts among those interested in the conservation of flora and fauna, sustainable soil use and an economic rationale supported by the demand for the development of rural tourism. These problems are reflected in significant environmental problems.

4.4 Festivals and Historical Monuments

BCSDH is located close to the municipal capital and is regularly visited by the residents of the neighboring communities, the city of Puebla, the various states of Mexico and the United States of America. The highest visitor numbers occur principally during religious festivals, the most important of which is that of the patron saint Domingo de Guzmán (4thAugust), which is celebrated with mass, prayers, processions, fireworks and live music, followed by the Holy Week festivities, the commemoration of All Souls' Day and holiday periods.

The municipality is home to the parish church dedicated to Saint Domingo de Guzmán (built by Dominican friars in the 16th Century), the temple of the Holy Child, the *Candelaria* (dating back to the 19th Century), and the temple of the Assumption (also built in the 16th Century), all of which are tourist attractions.

4.5 Atotonilco Water Park Cooperative

The establishment of the *Atotonilco* water park cooperative provided infrastructure for the refurbishment of the public swimming pools visited by people from neighboring communities and the city of Puebla. While the establishment of the cooperative has generated employment for the members of the community and other citizens, it has also generated conflict, principally around the maintenance of the swimming pools. The pools must follow the stipulations of Official Mexican Standard NOM-008-SCFI-2002 for the prevention of risks to health due to contact with pathogenic agents or chemical substances, for which reason their physico-chemical parameters must be monitored constantly.

The thermal swimming pools of *Atotonilco*, *Huehuetlán*, are built on the stream from the hypothermal spring, with telluric-type filtration and an average temperature of 33.2 °C. With a continually refreshed water supply, the pools are exempt from complying with the disinfection requirements and the other actions stipulated in the official health standards. However, community members do apply sanitary procedures in order to maintain the water quality according to the regulations, which, moreover, generates employment.

4.6 Description of the Community

4.6.1 Services and Education

Basically, the community has the following services: running water, electricity, wood and gas stoves, earth and cement floors, septic tanks, satellite telephones, medical services and education up to high school level. Public transport is available from the city of Puebla to the municipal capital.

4.6.2 Forest Resources

In order carry out an inventory of the forest species, three transects were undertaken with the community members at representative landmarks. Throughout the transects, ecologically and economically important species of the *Bursera, Polylepis and Parkinsonia* genus were registered. Like the *tepemezquite (Lysilomadivaricata)* and *pochote (Ceibaaesculifolia)*, these species can be harvested as timber. The most abundant species are *tepemezquite*, and *cubata (Acacia cochliacantha)*. Other abundant non-timber species are the agaves and cat's claw. The community members identified three types of soil: clayey, tepetate and rocky. The forest species are principally used for medicine and food, and as materials for the preparation and maintenance of fencing and forage (Fig. 2).

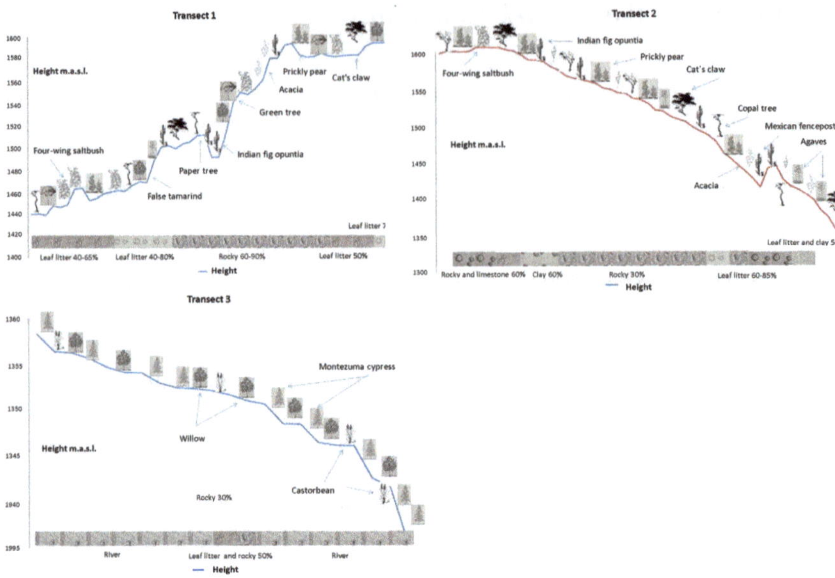

Fig. 2 Vegetation profiles in the principal landmarks of *Bienes Comunales*

4.6.3 Fauna Resources

A great wealth of fauna species was identified, principally 19 amphibians and reptiles, 35 birds, and 19 mammals, on which knowledge is held about their ecological, cultural and economic benefits. However, given the proximity of BCSDH to the municipal capital, pressure was observed on the fauna and vegetation due to hunting and illegal logging.

The Cultural Importance Index (CII) was calculated from the questionnaires applied. The most important mammal (CII = 20.13) was the coyote (*Canislatrans*), which was identified principally as a damaging species. Considered an ornamental species, the *chachalaca* (*Ortalisvetula*) was the most important bird (CII = 11.15). Of the reptiles, the rattlesnake (*Crotalusculminatus*) was given a value of 20.5. It was considered a beneficial species in that its meat has medicinal uses and its skin is used in leatherwork for the production of shoes, hats and belts etc., with economic income generated from the sale of these products.

The most frequent use of the fauna is as a foodstuff, with the white-tailed deer (*Odocoileusvirginianus*) highlighted for the consumption of its meat. Although the deer is a species key to the undertaking of any management and use proposal for the area, there are species that should not be ignored in the ecogeographical scenario which they inhabit.

4.6.4 Agrosystems

The inhabitants of BCSDH place great importance on the agrosystems they use, in that food production depends on them, as well as their economic income. The products are sold in both *Huehuetlán* and the city of Puebla. Agriculture is of great importance, given that it is considered part of the cultural heritage that is being lost as young people prefer to dedicate themselves to trade or to emigrate to either the city or abroad.

The agrosystems host a great diversity of fauna species, as well as crop diversity comprising avocado, cilantro, beans, guava, *jicama*, lime, maize, *papalo* (*Porophyllumruderale*), *pipicha* (*Porophyllumlinaria*), and sapote, among others.

For the members of the BCSDH community, it is important to establish strategies that enable the commercialization of the continued cultivation of fruit, grains and vegetables fundamental to the diet of the local population. They are also concerned about maintaining the biodiversity of the fauna associated with these crops. They consider that the use of chemical products should be reduced and substituted for natural products for pest control and improving soil fertility.

4.7 Environmental Problems

4.7.1 Environmental and Territorial Problems

Currently, the high numbers of people visiting BCSDH are causing a great number of threats to the diversity found there. The commercial activity undertaken in the town center (the sale of food, groceries, and alcoholic beverages, as well as the provision of public toilets, etc.), both by street and established vendors, is highly disorganized, with the majority of products sold in plastic bags, polystyrene containers or tin cans. This generates a large amount of solid waste, which is deposited in an illegal open air refuse dump found in the community's area of common use. The tourists visiting the water park also contribute significantly to the generation of refuse.

Rather than being recycled, the refuse is burnt, causing both atmospheric and soil pollution, and diseases in local inhabitants and fauna, both wild and domesticated. In order to attend to the problem, the community authorities have held meetings with the municipal government in order to relocate the refuse dump, a process which is currently subject to negotiation at a management level. The community members have requested a training course for the management of solid waste, which would allow them to control and reduce the quantity of refuse found in the area.

Other environmental problems are the illegal logging and hunting undertaken by both tourists and the inhabitants of neighboring towns and the municipality itself. To control these problems, the community members have organized themselves in order to carry out guard duty.

The National Forestry Commission has supported the community members in their efforts to reforest the degraded areas and detain the illegal logging.

The most significant problem that BCSDH has had to confront over the last four years is the plan of the multinational mining company, Australian Minerals Group, supported by the federal government, to extract manganese from 700 ha within the BCSDH territory. In order to defend itself, the community has sought recourse in national law, which enables it to preserve the area in dispute, and has been represented at various forums to defend the territory against megaprojects.

In general, the municipality supplies running water for use in agricultural and domestic activities. In this semi-dry region, irrigated crops are very important for the production of vegetables, fruit and basic grain. There is well-established concern over the negative effects that the proposed mine will have on the area's natural resources and water-dependent economic activities. The contamination of the water may also poison the fauna and flora of the region.

In 2011, the Government of the State of Puebla decreed the *Sierra del Tentzo* an *Área Natural Protegida* (ANP, or Protected Natural Area) under state jurisdiction.

This ANP comprises a surface area of 57,815 ha, which includes 13 municipalities, including *Huehuetlán el Grande*. This has generated discontent and resistance in the BCSDH population, as it was not consulted over the ANP proposal. The new regulations imposed by the state government, which are independent to those to which community members are held as an agrarian nucleus, have generated dispute and tensions with the government.

To the insistence of the installation of the mine and the challenge posed by the implementation of the ANP by the government, BCSDH has responded in the form of resistance based on the rules and agreements it has established as a community. This has enabled the community to ensure that the mine is not installed, as well as the establishment of certain agreements with the departments and agencies at a state level for the conservation of its resources. One of the strengths of the community is its organization, which has enabled the maintenance, despite the current disputes, of traditional knowledge and the traditional forms of organization for the management and exploitation of its resources.

4.7.2 Actions Undertaken for the Conservation of the Common Pool Resources

The inhabitants of BCSDH have a strong interest in the exploitation and conservation of their common pool resources. They also have a special interest in the development of ecotourism and the conservation of biodiversity. They believe that hiking would be an appropriate activity through which tourists visiting the water park would be able to experience the local fauna and flora and learn about the decreases in the populations of local species. They have also formed links with university students for them to carry out their obligatory social service, period of professional practice or research for their dissertation, which would enable them to undertake projects related to the conservation and management of their pool resources.

The community maintains great interest in conservation as it would enable the exploitation its natural resources, generate income and, thus, improve quality of life. It also maintains an interest in forming associations and support groups, such as the water park cooperative. This cooperative has functioned as a community organization, thus enabling the development and generation of employment and economic resources. Guarantees for the security of the common pool resources are required with regard to illegal hunting and logging, the management of solid waste, the reforestation of fragmented areas and the conservation of the territory in resistance to megaprojects such as mining.

Currently, 1000 ha have been submitted to the program of payments by the environmental authorities for the conservation of biodiversity, the administration of which has been supported by the National Forestry Commission (CONAFOR) for the next five years. Further to the reforestation program, signs have been posted which prohibit illegal hunting and logging and, most importantly, the security committee provides constant security for the area of common use.

5 Discussion

The results of this study indicate that the inhabitants of BCSDH have a wide collective and traditional knowledge of their natural resources. Moreover, they are able to identify landmarks, small ravines, the availability of water, types of soil and vegetation, as well as the presence of important fauna and flora for their use and conservation. This knowledge is reflected in the manner in which they occupy and transform their land and manage their common pool resources. They classify the territory in areas intended for conservation, forestry, restoration, agriculture, human settlements, tourism and leisure.

Based on this ancestral knowledge, the inhabitants of BCSDH have developed a series of strategies, applied through community organization, for the conservation and sustainable management of their common pool resources. While these resources are freely accessible, in that they are considered a means of sustenance, their use is regulated by standards, regulations, and the agreements put in place to avoid their over-exploitation. The tradition of community work (*faenas*) and the organization of and participation in religious festivals are fundamental to community cohesion and organization.

The creation of the *Atotonilco* water park cooperative has developed the tourism which, in turn, has converted the thermal waters into a source of income. The economic importance of the water has encouraged the community to take care of the local vegetation, as the water supply depends directly on it. Despite the economic benefits, tourism has generated new environmental problems that require the support of experts, both inside and outside the community.

Human activities generate changes to the physical environment and the biota, with notable negative effects observed in the composition, recuperation and productivity of natural ecosystems, such as the loss of species, the size and genetic variability of wild populations and the irreversible loss of habitats and ecosystems. Furthermore, alterations to the functioning of socioeconomic systems, and the health and human welfare of the community are observed. The principal threats to the conservation of the common pool resources are the proposed projects administered and managed outside the community, especially the installation of the manganese mine. Moreover, the members of the BCSDH community perceive as a threat the APN status decreed for the *Sierra del Tentzo*, and the refuse produced, principally by tourists and traders, and its mismanaged disposal. The inhabitants of *Huehuetlán* and neighboring areas also affect the flora and fauna through illegal hunting and logging.

To counteract these impacts on the environment, the community has organized itself in the attempt to reconcile the interests of the various actors involved. As, however, in many cases, this is not possible, it is necessary to seek legal recourse, leading community members to participate in forums for the defense of the land against megaprojects and commence acts of resistance. In other cases, community organization has enabled them to obtain governmental support in order to resolve specific problems, as seen in the reforestation programs. The community also knows how to seek external allies, such as university professors and students, which reflects its determination, as an agrarian organization, to conserve its common pool resources and safeguard the welfare of its members. It is, thus, of fundamental importance to incorporate outreach and environmental education processes that involve all social actors, in order that alternatives can be found to the prejudicial use of the community's common pool resources.

Recent land policies, which enable the government to expropriate communal lands in favor of foreign mining companies, constitute a threat to sustainability. Due to the idea that economic growth resolves poverty of its own accord, which environmental problems have demonstrated as erroneous, it is necessary to document and stimulate equitable development policies that close the poverty gap (Fisher and Erickson 2007).

6 Conclusions

The action-research strategy used in this study enables us to conclude that sustainable development depends on the social, cultural and political aspects that define people's actions in a historical context.

Contrary to the opinion expressed by Hardin (1968), the principal strengths of the BCSDH community are its community organization and traditional knowledge which have enabled it to develop a series of strategies, through agreements and regulation, mechanisms of self-regulation and self-management for the sustainable management of its common pool resource for their conservation and sustainable management as well as for confronting external threats. The main challenge for the participants in this project is the management of spaces for participation and inclusion in both state and national policies that affect daily life and the common pool resources of the BCSDH and their organization and territory.

Acknowledgements We thank all the people from *Bienes Comunales de Santo Domingo Huehuetlán*, for sharing their knowledge about their natural resources and social organization, and for allowing us to work in their community and support them as an academic institution. We also thank all the students and teachers who participated in the fieldwork and formalization of activities.

References

Agrawal A, Chhatre A (2006) Explaining success on the commons: community forest governance in the Indian Himalaya. World Dev 34(1):149–166. doi:10.1016/j.worlddev.2005.07.013
Cárdenas J (2013) La minería en México: despojo a la nación. Cuestiones Constitucionales 28:35–74
Diario Oficial de la Federación (DOF) (2016) Retrieved from http://www.ordenjuridico.gob.mx/Constitucion/cn16.pdf
Durand-Alcántara C (2005) El sistema de derecho agrario mexicano, su prospección científica. In: Kurczyn Villalobos P (ed) Memoria del congreso internacional de culturas y sistemas jurídicos comparados. UNAM-IIJ, México
Fisher B, Erickson J (2007) Growth and equity: dismantling the Kaldor-Kuznets Solow consensus. In: Erickson J, Gowdy J (eds) Frontiers in ecological economic theory and application. Ma, Edward Elgar Publishing, Northampton
Foladori G, Tommasino H (2005) El enfoque técnico y el enfoque social de la sustentabilidad. In: Foladori G, Pierri N (Eds) ¿Sustentabilidad? Desacuerdos sobre el desarrollo sustentable. Colección América Latina y el Nuevo Orden Mundial, Miguel Ángel Porrúa, UAZ, Cámara de Diputados LIX Legislatura, México
Hardin G (1968) The tragedy of the commons. Science 162(3859):1243–1248
Instituto Nacional de Estadística Geografía e Informática (INEGI) (2016a) Anuario estadístico del estado de puebla, México. INEGI, México
Instituto Nacional de Estadística Geografía e Informática (INEGI) (2016b) Síntesis geográfica del estado de puebla, México. Libro Electrónico, INEGI
Larson J, Sarukhán J (2003) Cuando los bienes comunes son menos trágicos: dominios eminentes y privilegios comerciales en la valoración patrimonial del México rural. Gaceta Ecológica, UNAM. 67:7–26
López Téllez MC, Yanes Gómez G, López Vivanco R, García Loza V, Fernández Crispín A (2015) Riqueza especifica de los vertebrados terrestres de la sierra del tentzo en el estado de puebla. Memorias del XXII Congreso Nacional de Zoología: Avances en el estudio, manejo y conservación de la diversidad zoológica de México, p 452
Morales Mendoza JJ (2003) Metodología de planificación ambiental participativa para formular el plan rector de producción y conservación (prpc) de la subcuenca del río jucuapa matagalpa-nicaragua. Tesis de Magister Scientiae en Manejo de Cuencas, Escuela de Posgrado del Centro Agronómico Tropical de Investigación y Enseñanza, Turrialba, Costa Rica
Ostrom E (2005) Self-governance and forest resources. In: Terracotta reader: a market approach to the environment, pp 131–154
Pierri N (2005) Historia del concepto de desarrollo sustentable. In: Foladori G, Pierri N (Eds) ¿Sustentabilidad? Desacuerdos sobre el desarrollo sustentable. Colección América Latina

y el Nuevo Orden Mundial, Miguel Ángel Porrúa, UAZ, Cámara de Diputados LIX Legislatura, México
Pina R (1970) Elementos de derecho civil mexicano, vol. 2. Porrúa, México
Poteete AR, Janssen MA, Andy Ostrom E (2010) Working together: collective action, the commons, and multiple methods in practice. Princeton University Press, Nueva Jersey
Sánchez KT (2012) Importancia de la propiedad agraria frente a las transformaciones económicas y jurídicas. El Cotidiano 27(173):81–88
Secretaria de Desarrollo Social (SEDESOL) (2010) Informe anual sobre la situación de pobreza y rezago social. Retrieved from https://www.gob.mx/cms/uploads/attachment/file/39317/Puebla_150.pdf
Tetreault DV (2007) Los proyectos de abajo para superar la pobreza y la degradación ambiental en dos comunidades de México rural: ayotitlán y la ciénaga, jalisco. Centro Universitario de La Ciénaga, Universidad de Guadalajara, México
Turner NJ (1988) The importance of a rose: evaluating the cultural significance of plants in Thompson and Lillooet Interior Salish. Am Anthropol 90(2):272–290. doi:10.1525/aa.1988.90.2.02a00020

Communal Tourism Proposal in a Common Assets Wildlife Use and Management Unit of the Mixteca Poblana as a Sustainable Development Strategy

María Concepción López-Téllez, Valentina Campos-Cabral, Gonzalo Yanes-Gómez, Antonio Fernández-Crispín, Hugo Rodolfo Molina-Arroyo and Berenice Ramírez-Vera

Abstract The aim of this work is to evaluate and detonate the sustainable tourism potential of the Wildlife Use and Management Unit, of the common assets of San Mateo Mimiapan, in the Mixteca Poblana (Puebla, Mexico, tropical lowlands) region, through a rural tourism development strategy. Participatory Workshops and field trips were done, as well as experience exchanges with other rural communities which engage in communal tourism in order to make an analysis of the potential for the development of rural sustainable tourism activities. The communal assets contain a high biological, cultural and territorial diversity, with tree and seed fossils which reflect the geologic and historical processes of the region. Therefore infrastructure was established for the creation of a community museum, flora and fauna interpretative tracks, as well as camping services for local public and students

M. C. López-Téllez (✉) · G. Yanes-Gómez · A. Fernández-Crispín
H. R. Molina-Arroyo · B. Ramírez-Vera
Facultad de Ciencias Biológicas, Benemérita Universidad Autónoma de Puebla, Blvd. Valsequillo y Av. San Claudio, Edificio 112-A, Ciudad, Universitaria, Col. Jardines de San Manuel, CP 72570 Puebla, Mexico
e-mail: biologiaclt@gmaill.com

G. Yanes-Gómez
e-mail: gonzalo_y2@hotmail.com

A. Fernández-Crispín
e-mail: anfern64@yahoo.com.mx

H. R. Molina-Arroyo
e-mail: hugo_molina_arroyo@hotmail.com

B. Ramírez-Vera
e-mail: bereramiver@hotmail.com

V. Campos-Cabral
Facultad de Derecho y Ciencias Sociales, Av San Claudio y 22 Sur, Cd Universitaria, Col. Jardines de San Manuel, CP 72570 Puebla, Mexico
e-mail: valiacc@gmail.com

W. Leal Filho et al. (eds.), *Sustainable Development Research and Practice in Mexico and Selected Latin American Countries*, World Sustainability Series,
https://doi.org/10.1007/978-3-319-70560-6_30

of different levels. Strategies are proposed for the development of tourist products in the rural environment, to be operated by local associations. It allows the creation of a business venture through promotion and commercialization plans in order to guarantee a market.

Keywords Rural tourism · Community development · Mixteca poblana

1 Introduction

Tourism, through the United Nations Conference on Environment and Development (UNCED) in Rio de Janeiro and the Agenda 21, has been established as a mechanism to achieve sustainability, because it is evidently a primordial economic activity for many countries to acquire foreign currency, create employment, and its horizontal impact on activities and various sector which directly or indirectly result in the creation of infrastructure which benefits local populations (OMT 2004; Vidal and Márquez 2007; Wearing and Neil 2009; López and Palomino 2011).

According to Pastor et al. (2011), rural community tourism is an alternative to traditional tourism because its aims are to preserve the environment and local culture, drive participatory community processes in the management of the territory, and obtain financial assets.

What matters is that rural communities share their natural resources, culture, work and production ways (farming, fishing, forestry and handcrafts), all of which, through the reassessment of the rural environment and the need to diversify economic activities, takes shape as a rural territorial development strategy from their economic, social, cultural and environmental assets (Pérez 2010).

This perspective transcends the traditional farming point of view of rural peasant communities, identifying them in their multifunctionality, integrity and complexity (Trejos 2000). In its tourist vocation, the rural environment conforms itself as a product which does not remove, but actually complements and invigorates the income obtained from primary activities through the rooting of the people, creates jobs, reassess the cultural elements and at the same time promotes the preservation of the environment (Vidal and Márquez 2007; Pérez 2010; Palomino et al. 2016).

From this perspective, rural tourism is an ally for local development from the valuation of the territory and the articulation of multiple actors. It is fundamental that the territories do not become artificial or lose their essence, and that rural tourism identifies itself as a force of territorial development, but not the only one. Likewise, negative consequences must not be ignored, such as pollution, biodiversity and local identity degradation and the desertion of other activities upon facing economic dimension over importance, among others (Burgos 2016).

The definition of rural community tourism is not homogenous, but it involves, according to Chávez and Andrade (2010: 19) "…an economically profitable tourist activity, environmentally responsible and inclusive, generator of income and dignified and qualified employment… tourist products respectful of the environment, the local people and their culture, …an alternative for local economic development". Among its main features are: "…the participation of the local people… the distribution of economic benefits through agreements…respect for the environment and valuation of biodiversity… and for the cultural identity of the people… strengthening of local societies… under the need of a common aim…". It necessarily implies that the "local community has a substantial control of the territory, participates in the development and management of its resources… the proportional distribution of the benefits in the community assumes collective responsibility… collective rights over land and resources… and the sustainable use and collective responsibility of them".

In Mexico there are many examples of communal tourism of indigenous and rural communities with social regimes of land ownership, with high biological and cultural diversity, where tourist activities are established in a collective and participatory manner, driven by the rightful social actors; they are also characterized by being based on a principle of local organization which has a bearing on the origin and history of the community, based on rules and local agreements. The communal tourism experiences at a national level are diverse, and they have established themselves mainly through the capacity to organize and strengthen their communal institutions, where a logic in the collective use and usufruct of the territory and its resources is observed (Palomino et al. 2016).

The Common Assets of San Mateo Mimiapan, from the municipality of Zacapala in the state of Puebla, Mexico, are typified for possessing a large social property territory, with a high cultural and biological diversity standing out, contrasting with margination and poverty. The implementation of alternatives for socio-economic development are required, to allow the people to have a higher standard of life, increasing the production without harming the environment. An economic—productive alternative is the sustainable and diversified use of their natural resources (flora and fauna), delivering ecological and financial benefits for the people.

This work has two purpose, one is to allow us to evaluate the possibilities of a methodology between research activities and intervention in a participatory process to link the interest, work and responsibility of the population, in an alliance with the academic sector, which engages in concrete and limited activities. The second one is to evaluate sustainable tourist potential of the BCSMM to strengthen the economic activities of the population, inserted in the territorial possibilities, in the practices and habits of the organized population.

The objective of this research is to identify the tourist potential of the Wildlife Use and Management Unit (UMA acronym in Spanish) of the Common Assets of San Mateo Mimiapan (BCSMM acronym in Spanish), located in the Mixteca Poblana region, through the development of a communal rural tourism strategy.

2 Study Area

The BCSMM are located in the southwest of the state of Puebla, Mexico, with an area of 4600 ha and an uneven relief (Fig. 1). The highest altitude is 2000 masl in the northeast, and 1250 masl in the southwest. The territory falls within the South Puebla Sierras physiographic subprovince, of the Neovolcanic axis province, and the Atoyac river basin; the climates are temperate, semi-warm and warm, all sub-humid, from east to west (INAFED 2016).

Six soil groups are present: Regosol, Feozem, Rendzina, Castañozem, Vertisol and Litosol; they are soils with less than 10 cm thickness on rock, not suitable for any kind of crop and only suitable for cattle. The main vegetation type is tropical deciduous forest, associated with secondary bush and arboreal vegetation, some induced grassland and a few patches of oak forest (INAFED 2016).

3 Methods

This work has a qualitative focus, and was done through a participatory action research method and the development of techniques such as documentary analysis, interviews, participatory observations and field trips. The team was assembled by the peasants of the BCSMM, students and professors of the Biological Sciences Faculty and the Law and Social Sciences Faculty of the Goodwill Autonomous University of Puebla (BUAP, acronym in Spanish), establishing the fundamental recognition of social actors as subjects of their daily live, with their own ways of

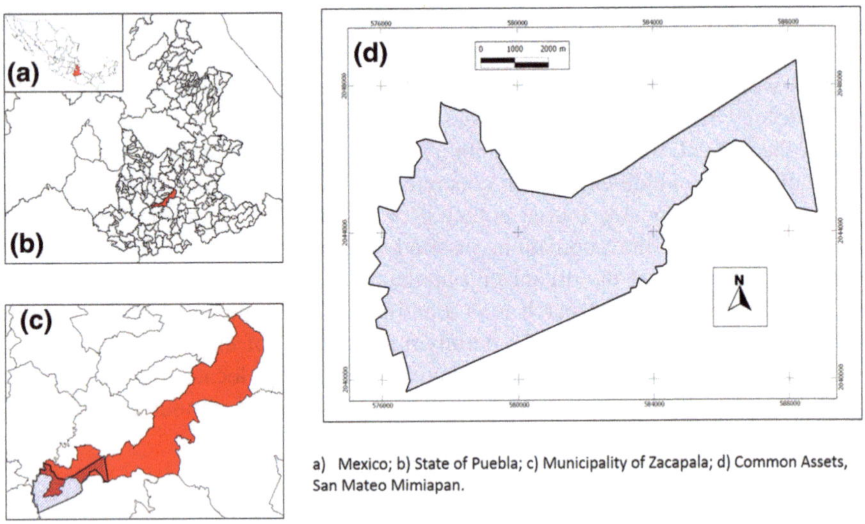

Fig. 1 BCSMM polygon, in the municipality of Zacapala, Puebla

apprehending their reality, their meanings, perceptions, decisions, intentions, motivations, ways and expectations. It was undertaken from March 2012 to December 2016 in the following stages:

3.1 Preparation Stage

From March to July 2012 the available information on the research problem and the study area was identified and organized, and the methodological and theoretical framework began to be established. The aim was to contextualize the situation of the community in the region and the development of communal tourism considering as the analytical model Ostrom's perspective (1990, 2005) on collective action and institutional systems which control the access and management of common goods. Bibliographic information was compiled in order to understand the social, environmental, economic and cultural context of the community. Official information was also obtained from government sources such as the Agrarian Public Office, Tourism Ministry (SECTUR), Environmental and Natural Resource Ministry (SEMARNAT) and the Government of the State of Puebla (Morales 2003; Giraud et al. 2005; Jalda 2008), in order to understand the organization of tourist activity, of commercialization, existing cooperation and possible supports, which would allow to understand positive and negative aspects of the territory for the development of the project.

3.2 Design Stage

From July to October 2012 methods, techniques and research instruments were established. The guideline in both stages was a constructivist view of the work and the research question on the potential for the development of rural tourism activities. An activities timeline was designed in order to recover, register and organize their knowledge and practices of the territory and their natural resources, as well as the exchange of experiences with other rural communities in the development of tourism, allowing for a joint construction of a proposal of rural community tourism strategy.

3.3 Field Work Stage

In this stage, from November 2012 to December 2013, it was fundamental to have knowledge of the features of the territory, of the social field and of power relationships between actors, as well as the working and formal and informal regulation principles of the place. Key informers were identified as those which possessed pertinent knowledge and expertise relating to the work, but also those willing to participate actively in it, for example, communal authorities, surveillance committee and older people. Thus, key informers were chosen intentionally, intensively, heterogeneously, dynamically and continuously. Through them information was registered with field diaries, photographs and tape recordings. The implementation of participatory workshops (one each month lasting two to three hours), field trips and natural resource inventories (Ayales 1991; Geilfus 1997; Brenner 2010), were accomplished with registered peasants in the agrarian census (90 people, male and female of different ages with different roles in the community). This works was interrupted once a certain information saturation level was achieved.

Systematization and analysis of information stage. Data were organized and presented to the assembly of peasants for verification and validity, obtaining in different meetings precisions and additional information which increased the value of the work on biological richness and diversity, specialization of biodiversity, practices for the management of natural resources and formal and informal institutionalization for the management of natural resources. With this information, a diagnosis of the ejido (communal property figure) and of the diversity of natural resources was obtained. Experience exchanges (January–July 2014) and training activities (August 2014–March 2015) were made, in order to allow the peasants to know the pros and cons of other experiences of rural tourism development; cases of rural communal tourism were visited. In order to generate and fortify capacities of self-government in the alternative sustainable tourism proposal, training of the peasants from SECTUR and BUAP professors was engaged. This stage was supported by the fulfillment of three workshops, by establishing contact with the main actors involved in the commercialization of communal rural tourist services, to establish business meetings and agreements for the acquisition communal rural tourist services.

A tourist activities proposal was prepared (April–August 2015) and finally the information was coordinated and presented to the peasants, and in workshops and meetings, different activities were defined, emphasizing mainly on natural, cultural, social and economic resources, presenting the final proposal in January 2016.

4 Results

Potential analysis for the development of rural community tourism.

The BCSMM have eight important landscape sites: El Huizache, El Timbre (a population center), El Algodón, El Pochote, El Pelado, the fossil area and the Cantera (quarry). Several gullies, ravines, streams and rivers are also present (Agua de León, La Presa, Bordo Nextepetl, Agua Nanche and Atoyac river) (Fig. 2).

The community uses wildlife flora and fauna widely, 69 species of plants are used, 30 of which are medicinal (for kidney, stomach and pancreas ailments), 20 are edible, 12 are used for cattle fodder, nine for construction materials, and the rest as firewood and others. Fauna is used as food (eight species) and medicine (seven species).

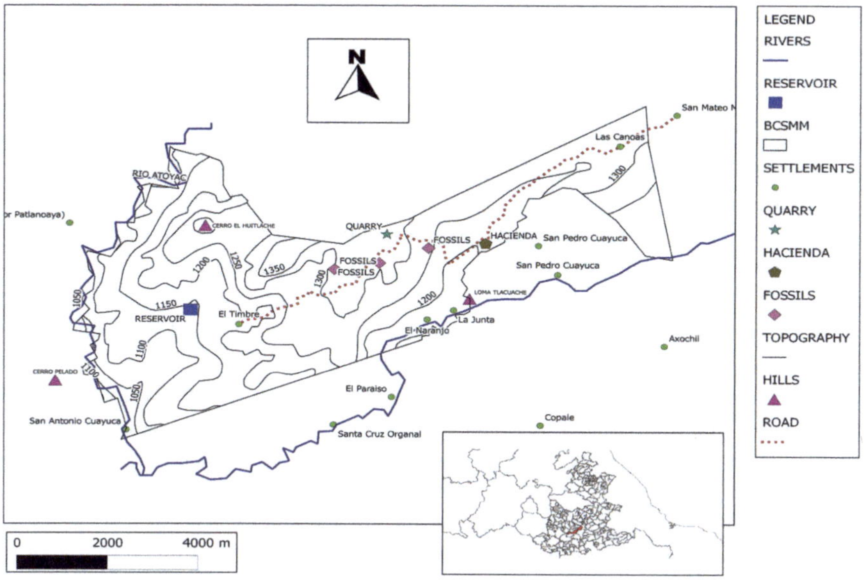

Fig. 2 Territorial representation map of the BCSMM, Zacapala, Puebla

The most representative species of flora are: tepehuaje (*Leucaena pueblana*), poppy (*Pseudobombax palmeri*), colorín (*Eritrina spp.*), pochote (*Ceiba parvifolia*), palo blanco (*Conzattia multiflora*), cuajilote (*Acacia acatlensisi*), copal (*Bursera bipinnata*, *B. excelsa*), cueramo (*Cordia spp.*), cuachalalate (*Amphiterygium adstringens*), guaje (*Leucaena glauca*), huizache (*Acacia farenesiana*), cubata (*A. cymbispina*). Other secondary species are: maguey (*Agave spp*), cazahuate (*Ipomoea spp*), biznaga (*Ferocactus spp* y *Mammilarria spp.*), mezquite (*Prosopis juliflora*) and prickley pear (*Opuntia spp*).

The most representative species of fauna are white-tailed deer (*Odocoileus virginianus*), rabbit (*Sylvilagus floridanus y S. cunicularis*), jackrabbit (*Lepus spp*), raccoon (*Procyon lotor*), gray fox (*Urocyon cineroargenteus*), skunk (*Spilogale putorius*), coati (*Nassua nassua) and* coyote (*Canis latrans*). Some edible species of bird are quail (*Colinus virginianus*), dove (*Zenaida macroura*), chachalaca (*Ortalis poliocephala*). Among the reptiles are rattlesnake (*Crotalus spp.*), Gila monster (*Heloderma horridum*) and black iguana (*Ctenosauria pectinata*).

Several fossil deposits are also present, mainly large mineralized tree trunks from the Jurassic and Oligocene, several meters apart, as well as fern impressions from the Jurassic. These findings contribute to the enrichment of paleontological knowledge in Mexico, and increases the number of fossiliferous localities.

The BCSMM owners also have a quarry where marble extraction is licensed to a mine company.

The community took the agreement to register as a UMA having interest in preserving the common use resources and in reforesting. Financial resources would be obtained through controlled deer hunting, as part of a tourist project which also involves scientific research and environmental education.

As any other community, there are problems with land and common use resources use. The origin seems to be the population and demographic structure of both sections of the community, since the El Timbre population is younger, being the sons of BCSMM owners, which therefore have no agrarian rights.

They cannot participate in decision making, whereas in Mimiapan the population is older and they will not or cannot get involved in activities which are no longer their responsibility. This has hindered social development and the opportunity of integration and production between both sections. In spite of this, the community raises proposals through collective solutions and agreements to maximize the available resources and strengthen the agrarian unit.

In the workshops and field work, community strength and potential were evident, to develop tourist activities through natural and human resources, infrastructure to develop security, habitat management, camping services, hiking and plant production.

Additionally, as a product of the relationship with the University, white-tailed deer population monitoring is being done to allow for hunting authorized by SEMARNAT, as well as obtaining funding from the UMA promotion program of the same ministry in 2013 for infrastructure development and training.

4.1 Diagnosis of Communal Tourist Potential

The BCSMM have a Popoloca ethnic origin, from the records of some relics. However there is a loss of traditional ceremonies and of the dialect, because of the interruption of intergenerational transmission mechanisms, as a result of homogenizing national projects which led to sanction and embarrassment of adults and youngsters when trying to speak and teach it to new generations.

The origin of the community dates to the seventeenth century; evidence comes from the church and remains of a country estate. The present-day inhabitants do not know their exact origin, however one of the most traditional festivities is the "Toro Calabacero", which remits to 1800, when ranchers arrived with animals from Santa Inés Ahuatempan to brand the cattle.

As a farming unit, the origin remits to July 2, 1965, when a surface of 4682 ha was assigned to them. This land is classified in terms of farming, environmental and socio-economic development. They possess common-use land to obtain everyday economic sustain; the human settlements are established in two sections, Mimiapan and El Timbre, where daily activities take place. This land is not transferable or reduceable, unless the people or the law establish it. It is included the national system of Wildlife Use and Management Units (UMAs) since 2005; this has allowed the preservation of the tropical deciduous forest in the almost 4600 ha which they possess.

The prevailing guild organizations are community oriented: the Vigilance Council, the Communal Land Council ("ejido") and the Auxiliary City Hall Major. Other organized sectors are migrants which participate in the development of constructions and cultural activities, the church post system and the saint's celebration, the parent-teacher association, the livestock producers association (registered in the government), a group responsible for the extraction of marble in the local quarry, and finally the UMA committee.

Most of the people of the BCSMM are native of the region, therefore, as a statutory agreement, women are not allowed to be land owners so that land would be lost through marriage.

Both sections of the village, Mimiapan and el Timbre, belong to the same farm unit, however productive and daily productive dynamics are maintained. The total number of inhabitants in is 370 people, 267 of which live in Mimiapan, where 128 are male and 139 female; 86 are under aged and 181 are adults; most are over 60 years of age. In el Timbre 103 people live, of which 57 are male, 46 female, 45 underaged and 58 adults, of which 13 are over 60.

The main economic activities are rain-dependent and mechanized subsistence farming. Corn, pumpkin, bean and sorghum are mostly grown. They rotate their crops and use organic fertilizer; the livestock is mainly goat and cows, but also donkeys, horses, pigs and chicken. Livestock is free to roam in the wilderness, foraging a great variety of plants. The people use natural flora and fauna for their subsistence.

As in many other towns in the country and the state, financial remittances are sent from the family living in the United States, which are an important source of domestic income; there is also migration to other parts of the country. In both cases migrants return to the town's saint festivities and for Christmas.

Federal government programs which exist in the region are from SEDESOL (social needs), CONAFOR (forestry), and SEMARNAT (environment). Of the latter, reforestation, conservation, watchtowers, wildlife monitoring and temporary employment programs have been applied.

4.2 Experiences Exchange

Experiences were exchanged with rural communities which have developed community tourism:

1. San Juan Raya. This community belongs to the Tehuacán Cuicatlán Biosphere Reserve (RBTC). They engage in communal tourist activities in the "Centro Ecoturístico de San Juan Raya", offering hiking, cycling, horse riding, marine fossil and dinosaur footprints sightings, as well as the observation of a high diversity of cacti and local fauna. They offer accommodation in rustic cabins and regional food.

2. Tlayúa Regional Museum and Pie de Vaca in Tepexi de Rodríguez, the former, in charge of the N.A. University of Mexico (UNAM). The latter is administered by local people, who offer hikes to explain the origin of different fossils in the quarries which are exploited in the municipality of Tepexi.
3. Santa Cruz Achichipilco UMA in Teotlalco. They offer legal hunting services, accommodations and hikes in interpretative tracks.

This allowed the owners of BCSMM to visualize the experiences of other communities and to propose tourist activities matching the characteristics of their territory.

4.3 Training

To choose the training themes, the BCSMM owners reflected upon the workshops and exchanges, giving priority to themes such as client attention, hiking, administration and environmental legislation. During training, the following questions were asked: What do you want and expect from your land? How do we do it? How to represent the territory and the common use resources? How to identify tourist service supply potential and problems? How to solve these problems?

Issues related to the measurement of environmental values from an economical perspective were considered. This implied the analysis of concepts such as economic value related to direct use value, indirect use, option, existence, territorial legacy and common use resource value, to do a cost-benefit analysis of the possible services to offer. SECTUR offered training in entrepreneurial organization, business administration, visitor attention and legal aspects. The University offered themes such as responsible hunting, identification and conservation of fossils, tree nursery maintenance and cactus propagation, solid waste management, hiking, furniture construction and wildlife monitoring for hunting.

4.4 Marketing of Communal Rural Tourist Services

Three experience exchange workshops with the social actors involved in the marketing of communal rural tourist services were done, where business meetings and agreements for the acquisition of these services with alternative tourist operators were established, as well as with hunting clubs and educational institutions of the state, such as public and private universities (BUAP, U. Iberoamericana, UDLA, U. Mesoamericana, etc.).

One of the main agreements was with Produce Foundation (Fundación Produce Puebla A.C.), which made a commitment to give support for the development or lináloe tree (*Bursera linaloe*) use, as well as giving publicity. The participating academic institutions commited to establish agreements to visit the community in order to do field work, as well as support in training and scientific disclosure. Hunting clubs El Sable and Los Halcones commited to give disclosure to the tourist and hunting services, as well as to the yearly hunting permits. SECTUR commited to arrange financing to continue with the training program to strengthen the communal technicians, tourist service guides, marketing of tourist products and administration. Finally SEMARNAT established a commitment to support the disclosure of the program and to continue with the actions to preserve the common use resources.

4.5 Proposals

The owners of BCSMM established several opportunities for the development of tourist activities. One of them was to develop a rural community tourist program for eco-paleontological and wildlife use, to contribute to the integral development and improvement of the quality of life of the people, through the rational use of common use resources.

The proposal was delivered to and approved by the UMA promotion program of the SEMARNAT. Resulting from this, the necessary infrastructure for the creation of a community museum was established, as well as two cabins, paleontological interpretative pathways, biocultural flora and fauna pathways, as well as camping service for local public and students of different levels.

5 Discussion

The rural community tourism Project which has begun in the BCSMM represent an option of use and value of the territory, stemming from the organization system presented. The limitations which were identified relate to unlawful hunting and forestry, however the peasants mention that they can control it through their organization. This highlights the existence of an organization and an institutionality that, in spite of its contradictions, manage their territory collectively, and hereby search and produce projects where tourist activities represent an income opportunity and the conservation of the environment through the use of common resources and the establishment of alliances with external actors, in order to improve the standard of life of the population. The external factors have promoted the diversification and increase of income, as well as the creation of employment, the ingrainment of the population, the display of the role of women and youth (although much remains to be done in this dimension), the revaluation of cultural, architectural, paleo-archaeological and environmental patrimony, notwithstanding that in this process, an acting capacity as subjects in projects which feed confidence and community cohesion has been developed. The community proposal of the BCSMM includes developing medium and long range strategies to develop tourist products with identity and originality, operated by local organization; these products focus mainly on natural, cultural and landscape richness. The diagnosis has allowed to know the reality of the community, identifying their needs and problems, and proposing strategies to allow the improvement of their life standard through these proposals. Rural communal tourism is considered as an alternative to mass tourism. In Mexico these are recent and highly heterogeneous experiences with different levels of consolidation which respond to external and internal factors. In spite of this, it is an important alternative option, in the sense that they are projects where their construction stems from local assets and from the participation of the owners of communal goods and as owners of the territory and its resources, who as a function of this and their knowledge, take decisions about the project and distribute the gains among themselves. It is clear that the future of these kind of projects faces challenges mainly because of the complex organization system, where power relationships prevail even on a community level. These exercises are important to allow for the structuring of local governance, as well as to cover the needs to improve the administration of tourist activity, and to have market projection in order to establish an internal and external position.

6 Conclusions

This work pertains to three types of conclusions. The first is on the possibilities of the method. This work is an example of a positive result of the link between research activities and intervention in a participatory process, as well as of the operational advantages that bring a better prediction of the project when it is a result of interest, work and responsibility of the population, in an alliance with the academic sector, which engages in concrete and limited activities. The second one is on tourist potential of the BCSMM, established initially by local resources, be they natural or human, as well as the kind of activities-products-services proposed, which will be inserted in the territorial possibilities, in the practices and habits of the organized population, and that, overall, allow to strengthen the economic activities of the population. And last but not least, the basic conditions for success are the maintenance of a horizontal, respectful and inclusive participation, and the limitations of the exercise must reside in the respect of reproductive cycles and maintenance of the ecosystem through authorized permits for hunting activities to look after the territory.

References

Ayales I (1991) Haciendo camino al andar: Guía metodológica para la acción comunitaria. OEF Internacional

Brenner L (2010) Gobernanza ambiental, actores sociales y conflictos en las áreas naturales protegidas mexicanas. Rev Mex de Sociología 72(2):283–310. Retrieved from http://www.unwto.org/español/frameset/frame_sustainable.html

Burgos DR (2016) El turismo comunitario como iniciativa de desarrollo local. Caso localidades de Ciudad Bolívar y Usme zona rural de Bogotá. Hallazgos 13(26):193–214 doi:http://dx.doi.org/10.15332/s1794-3841.2016.0026.08

Chávez DRM, Andrade RE (2010) Introducción. In: Chávez DRM, Andrade RE, Espinoza SR, Navarro GM (eds) Turismo comunitario en México. Distintas visiones ante problemas comunes, México: Universidad de Guadalajara, pp 19–23

Geilfus F (1997) 80 herramientas para el desarrollo rural participativo: diagnóstico, planificación, monitoreo, evaluación. IICA.SAGAR, México

Giraud C, Gouley C, Hernandez G, Laats E (2005) Manejo de conflictos y recursos naturales en un área protegida: el ejemplo del SHM Centro Bartolomé de Las Casas. Colegio Andino, Universidad para La Paz, Cusco, Perú

Instituto para el Federalismo y el Desarrollo Municipal (INAFED) (2016) Enciclopedia de Los Municipios y Delegaciones de México Estado de Oaxaca, México: Secretaría de Gobernación (SEGOB)

Jaime Pastor V, Casas Jurado C, Soler Domingo A (2011) Desarrollo rural a través del turismo comunitario. Análisis del valle y cañón de Colca. Gestión turística, p 15

Jalda (2008) Manual de Técnicas Participativas, Sucre, Bolivia. Guías y Manuales, p 10

López G, Palomino B (2011) El turismo sustentable como estrategia de desarrollo. In: Delgadillo J (ed) Los terrenos de la política ambiental en México. México, Miguel Ángel Porrúa., pp 263–297

Morales MJJ (2003) Metodología de Planificación Ambiental Participativa para formular el Plan Rector de Producción y Conservación (PRPC) de la Subcuenca del Río Jucuapa Matagalpa-Nicaragua. Magister Scientiae thesis in Manejo de Cuencas. Escuela de Posgrado del Centro Agronómico Tropical de Investigación y Enseñanza. Turrialba, Costa Rica

OMT (2004) Desarrollo sostenible del turismo. Definición conceptual. Retrieved from http://unrn.edu.ar/blogs/maydt/files/2012/09/OMT-DesSostTuris.pdf

Ostrom E (1990) Governing the Commons: The Evolution of Institutions for Collective Action. Cambridge: UK: Cambridge University Press

Ostrom E (2005) Self-governance and forest resources. Terracotta reader: A market approach to the environment. New Deli: Academic Foundation, pp 131–154

Palomino VB, Gasca ZJ, López PG (2016) El turismo comunitario en la Sierra Norte de Oaxaca: perspectiva desde las instituciones y la gobernanza en territorios indígenas. El Periplo Sustentable (30):06–37. http://rperiplo.uaemex.mx/index.php/elperiplo/article/view/2465/2618

Pérez SF (2010) El valor estratégico del turismo rural como alternativa sostenible de desarrollo territorial rural. Agronomía Colomb 28(3):507–513

Trejos RA (2000) Nueva ruralidad. Temas emergentes, nuevos condicionantes y viejos problemas. Perspectivas Rurales Nueva Época (8)

Vidal ÁL, Márquez D (2007) Turismo y territorio: el desarrollo local sostenible al centro del debate. Revista@ local. glob. Centro Internacional de Formación (CIF) de la Organización Internacional del Trabajo (OIT) (4):1–84

Wearing S, Neil J (2009) Ecotourism: impacts, potentials and possibilities?. Butterworth-Heinemann, Amsterdam

Transhumance: Sustainable Strategy for Human and Ecological Conservation

María Isabel Mora Ledesma

Abstract Transhumance is the seasonal migration of livestock in search of pastures and water. This migration allows for intensive, rotational, diversified and optimal use of the land. The functional value of this mobility is not exempt from certain confusion arising from a lack of knowledge of this thousand-year old system. Transhumance, as a livestock management system, has therefore generated debate due to the interest and controversy it provokes, particularly with respect to the environment, geography and cultural aspects of arid and semi-arid areas. The purpose of this presentation is to demonstrate how the goat herders of the San Luis Potosí highlands plan their grazing practices based on a cyclical migration between two points with ecological complementarity: the mountains and the valley. The difference between these two points is caused by the cyclical combination of each point, which involves the climate, season, type of vegetation and topography. This system, called transhumance, has allowed the area's goat herding families to survive, persist and resist. Due to the relationship between ecological complementarity, planned practices, and the tie between the animals and the land, there is a society and culture of sustainable transhumance in the region.

Keywords Transhumance · Territory · Grazing · Sustainability
Ecology

The dictionary defines the term transhumance (Latin trans- and humus earth) as a form of pastoralism consisting of the seasonal movement of livestock to make better use of grazing grounds and water based on the physical and climatic characteristics of the land. This concept is discussed throughout this article.

M. I. M. Ledesma (✉)
Anthropological Studies Program, El Colegio de San Luis,
San Luis Potosí, Mexico
e-mail: imora@colsan.edu.mx

W. Leal Filho et al. (eds.), *Sustainable Development Research and Practice in Mexico and Selected Latin American Countries*, World Sustainability Series,
https://doi.org/10.1007/978-3-319-70560-6_31

1 Introduction

This article is part of the results of the project: *La ganadería trashumante en el desierto Chihuahuense. El sistema de reproducción de cabreros en el Altiplano potosino: Territorio, diversificación y ecología. (Livestock Transhumance in the Chihuahua Desert. The Goat Reproduction System in the San Luis Potosi Highlands: Territory, Diversification and Ecology)*, sponsored by the Basic Sciences section of the National Board of Science and Technology (CONACYT). The general objective of this project was to study the historical, social and symbolic configuration of the land as a reproduction system by the transhumant farmers of the arid north of San Luis Potosí. Based on the results of the first stage of the project,[1] the purpose of this article is to demonstrate how the transhumant system implemented by the goat herders of the region is a sustainable system given its conditions of conservation and reproduction based on the orderly planning production implemented by the transhumant livestock farmers. This sustainability allows the inhabitants of the San Luis Potosí desert to create a territorial identity. First, we will address the concepts of territory and transhumance with respect to resources, work, social organization and livestock. We will then describe the productive grazing system of the region and analyze the transhumant system based on the seasonal cycles that define the activities of the herders. Lastly, we will address certain problems faced by the herders in carrying out their activities that place the sustainability of this system at risk.

2 The Environment—Society Relationship

The notion of space and territory has been addressed in anthropological literature to explain, from different perspectives, the relationship between nature and society. The classic Evans-Pritchard study on *The Nuer* (1977) is a pioneering work that explains how the Nuer society organizes life, and movement of livestock and residency based on the ecological timeline, which is cyclical. Mauss (1979), who studied the Eskimos, also stressed how ecological and environmental conditions affect the social conditions related to hunting, forms of residence and annual seasonal cycles. Steward (1972) of the American school founded the theory of cultural ecology, indicating that population density, sedentary lifestyles and division of

[1]*Los caminos de la trashumancia: Territorio, persistencia y representaciones de la ganadería pastoril en el altiplano potosino. (The Paths of Transhumance: Land Persistence and Representations of Pastoral Livestock Grazing in the San Luis Potosi Highlands)* (Coord. Mora, María Isabel, Colsan 2013). The National Board of Science and Technology (CONACYT) supported the continuity of the project under the title: *Sistema pastoril, conocimientos locales e identidad territorial. Alternativa alimentaria para el desarrollo regional en el desierto. (Grazing System, Local Knowledge and Territorial Identity: Feeding Alternative for Regional Development in the Desert).*

labor follow the influence of this aspect. Harris (1994), the founder of cultural materialism, also established the relationship between material and spiritual well-being and the cost/benefit of various systems for increasing production and controlling the population growth. Although Harris takes ritual factors into consideration, this does not arise from ecological determinism, since, as indicated by Tomé (1996), his reasoning of the cause and effect is based on generalizations, considering a conjectural ecology without explanatory value.

From another perspective, Descola (1996), in his studies on the Achuar of Ecuador, indicates that the relationships that a society has with the environment are not unambiguous and cannot be exclusively reduced to adaptive responses, meaning that the contribution of an ecological approach consists of demonstrating the creativity of each culture in the way they socialize their nature (22). From this same perspective, Viveiros de Castro (1996), with his study on the Amazonian Arawete, together with the work of Descola, opens up another perception of nature, where animals and plants possess attributes of social life, thus avoiding the disassociation of the natural sphere with human action. For example, for Viveiros, hunting among the Arawete has a significant value in their cosmology, establishing a relationship with nature, giving favor to social and symbolic interactions between the animal world and shamanism, the central institution of this group. We have taken these last perspectives as the starting point for our research.

3 Transhumant Territory: Appropriation and Representation

Turning first to the characterization of Godelier (1989) of territory as:

> A portion of nature and therefore space over which a given society lays claim and ensures, for all or some of its members, stable rights of access, control and use regarding all or some of the resources found there, and which it is desirous and capable of exploiting (107).

For this author, space is considered to be an extension of land, water, exploitable resources, and resources for the production and reproduction of a society, based on two conditions:

> The first, that it can directly or indirectly satisfy a human need, and the second, having the technical means to separate it from the rest of nature and use it for its purposes. This implies a society with a certain interpretation of nature that combines its intellectual representations with the material means for acting with nature for physical and social reproduction" (ibid.)

Godelier indicates in this regard that all societies require a social organization that determines behaviors that are socially visible and accepted by all, both in the material aspects and as symbolic aspects for intervening with the intangible. To appropriate a territory, in terms of access and control, it is necessary to establish rules that make it possible to have a code of social order that guarantees the reproduction conditions of man and the resources they depend on. From this perspective, for grazing societies, livestock is not just a production material (meat, milk, leather), but also the content of what the environment and resources represent for the production and reproduction of the society itself. Giménez (2005: 430) considering the proposals of new geography conceives territory as: "the space appropriated by a social group to ensure its reproduction and satisfaction of its vital needs, whether material or symbolic."

With reference to these authors, we analyze the transhumant territory in the case in question as the space where the herders engage in their transhumant practice, the appropriation of which occurs in two ways: (1) in the functional sense, regarding exploitable resources, land, water, pastures, as indispensable flora for maintenance of the livestock and themselves, and (2) in the social historical sense as the location and method by which the herders have carried out these practices for generations. A space rich with history, landscapes and symbolism, tattooed by the livestock paths, with local technologies, affection and loyalty ("without the land, we are nothing"), solidarity and reciprocity (with humans and non-humans) and natural heritage. These appropriations function as elements that allow them to identify themselves as being "goat herders": "we who live from this", namely, as a territorial and regional identity. In this way, we can indicate that it is symbolic-cultural appropriation. These two forms of appropriation define the limits of the transhumanist territory, which we can define as the portion of space appropriated by the herders who claim it as the place where they have found the ecological and functional resources to live that are in turn significant due to the symbols and typical features required to maintain their conditions of existence and reproduction over several centuries.

Goat herding is fundamental to the economies of arid and semi-arid societies that practice it due to its use of xerophytes and shrubs. Gutiérrez (2009) mentions the importance of this production, which attributes value to certain types of land not apt for typical agricultural use. The environmental conditions of this arid environment (scarcity of water) make it difficult for the herders to have a sedentary lifestyle. In this context, the production of life is subject to the climatic conditions determined by the annual cycles of rainy and dry seasons that appear in these torrid climates, with rain precipitation of less than 500 mm per year. The mobility required to endure in the desert marks the rhythm and social representations related to the transhumant life of the San Luis Potosí herders.

4 Production and the Grazing System in the San Luis Potosí Highlands

The zone under study is located in the northern portion of the San Luis Potosí highlands, in the great Chihuahua desert, and it covers 11 municipalities[2] and 225 areas of commonly held land (*ejidos*), consisting of communities of less than 500

[2]The municipalities are: Vanegas, Cedral, Catorce, Matehuala, Villa de Guadalupe, Charcas, Venado, Villa de Arista, Guadalcázar and Villa Hidalgo.

inhabitants. These settlements have a mixed economy based on seasonal activities such as grazing, agricultural scarcity (corn and beans), hunting and gathering. The size of the herds is determined by the number of members, particularly the young, available to care for them. Production consists of the seasonal production of goat meat and milk to prepare cheeses for sale.

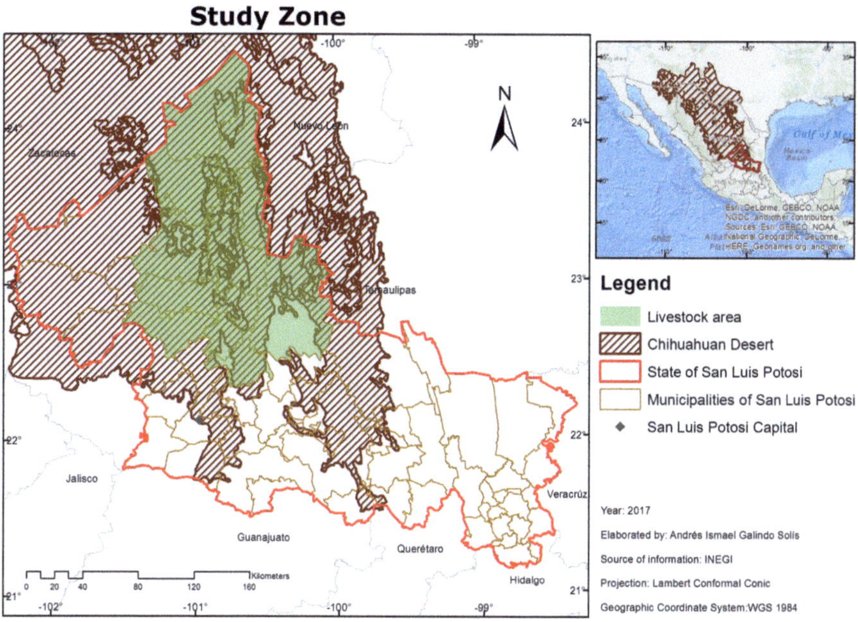

Historically, since the Spanish brought goats to Mexico in the 16th century, San Luis Potosí has been one of the country's main producers of goats.[3] Mexico was also recognized as one of the main producers of goats at the global level, with an average production of 10,439,000 in the nineties, ranking as the fifth location with the highest production at the global level. This ranking has been pushed downward due to the effects of neoliberal policies: in 2013 production dropped to 4,000,000 with this area ranking 20th on the global scale (FAOSTAT 2009–2013).

Most of the production of goats in the state of San Luis Potosí is located in the north, in the San Luis Potosí highlands. It is in this area that we find the highest number of exploitations, characterized by family productions with herds of between 20 and 500 animals, with a small amount of infrastructure.

[3]Goat herds were brought into the country in 1521 and were herded to the northern lands in 1492 (see Mora 2013).

In parallel to the national trend at the end of the nineties, the state of San Luis Potosí ranked first in production in Mexico, with 1,074,276 heads (11.2% of the national total). In 2009, the number of heads produced in San Luis Potosí dropped to less than half (FAOSTAT 2009–2013). See graph.

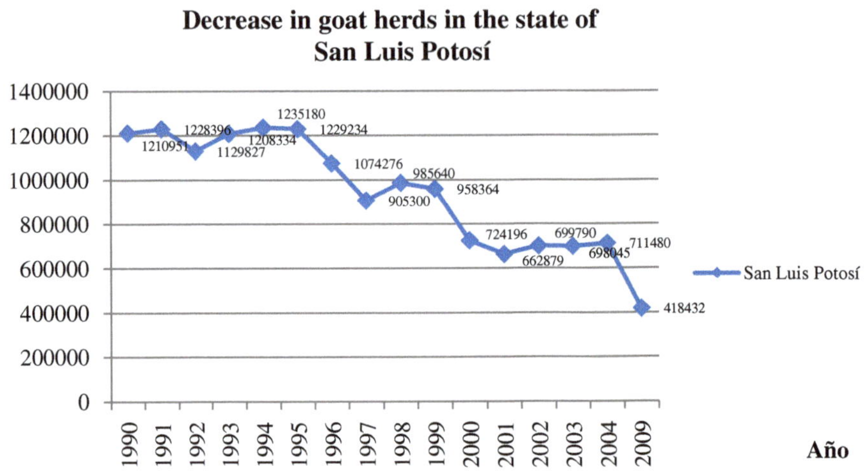

Source: Prepared by the author, based on information from FAOSTAT 2009-2013

This is a significant decrease and it owes to the change in ownership structures established in the Agricultural Law Reform in 1992, under the Program for Certification of Common Land (*Ejido*) Rights and Titling of House Plots (PROCEDE), a legal instrument aimed at the regulation and privatization of land. This new standard reduced the territory of common use available for grazing and circumscribed it to the limits of the common land (*ejidos*), affecting the practice of annual transhumance, which is of vital importance to the herders and their herds, particularly in seasons of drought and low water levels. The inhabitants, unable to carry out the transhumant work needed for herds larger than 100 heads, have had to sell and gradually reduce the numbers of their herds. This has affected the conditions for their reproduction, both the vital conditions, and the social and symbolic conditions as well. It has limited the conditions and possibilities of living within the common land (*ejidos*) and domestic units and it has driven the new generations to leave these areas. This dynamic has caused the predominant lifestyle option to be the emigration of the young, and the result has been a depopulation of the region. This agricultural policy has affected not only the economy of the region, but has also resulted in the loss of a society with a system based on thousands of years of knowledge in the handling of livestock and the territory that had allowed this activity to be sustainable.

5 Transhumance as a Sustainable System

Transhumance has been observed in different times and spaces throughout history.[4] Various authors in different contexts have studied this grazing system. Sánchez Moreno (2007) defines these movements as the alternative and periodic migration of herds of livestock between two environmental regions in order to make better use of the vegetable complementarity throughout the seasonal cycle to which the economies, cultures and biological cycles of these territories are tied. Bendini et al. (2004) have contributed to the interest in knowing and understanding these systems in Latin America, which have been relatively unstudied. Their studies on the Crianceros of northern Patagonia were recognized by the Ministry of Agriculture of Argentina, resulting in a new Goat Law regulating the use and control of the territory for the corresponding transhumant work. Based on their observations, the authors defined transhumance as a recurring, pendulous and functional movement tied to the climate, altitude and receptiveness of the fields, separated into two climatic periods: "summer" and "winter", contributing to the use of different ecological levels (24). We can confirm that the transhumant work extends beyond merely the exploitation of the land, as documented by Cassigoli (in Salas 2004: 163) with respect to the cases of the Colla transhumant families from northern Chile, where he explains how the life of the group is based on their seasonal transhumance. These are people who follow transhumant movements between gullies and waters and organize their world view based on the contrast of mountain-water-origin on the one hand, and valley-land-harvest on the other.

The functional, social and cultural value of the transhumant work has been subject to very little study and recognition, but it has generated controversy related to the deterioration of the environment generated by grazing. The players of an ecosystem can consciously act within the limits imposed by the medium. This is the framework in which transhumance make sense (Sánchez 2007).

6 The Grazing Societies of Northern San Luis Potosí

As indicated above, the grazing of goatherds is fundamental for societies situated in arid and semi-arid regions since it allows for the use of xerophytes and shrubs (Gutiérrez 2009).

[4]There is evidence of transhumance in the Bronze Age, in the Mediterranean era in the IV and II century A.C. In the XII and XIII centuries in Spain it was of great importance for controlling the pastures formed by the first groups of livestock farmers. Livestock paths have generated cultural heritage in Europe; however, in recent decades, between 20% and 40% of these livestock paths have been lost, which has contributed to the progressive abandonment of these paths by livestock farmers (Ambientum 38/2001).

For the case under study, the livestock territory is comprised of 1,607,968 hectares that belong to the vast region of the Chihuahua Desert, replete with grasslands (bushes) and a wide variety of cacti and fibers (lechuguilla and catkin), with reliefs that range from flatlands to mountains with altitudes from 2000 to 3000 meters above sea level, wells, berms and watering troughs, minerals, caves and pens with great significance and value to the transhumant herder, which to those on the outside signify nothing. It is common to hear people say things like "How do people live there, if there is nothing there?"

The transhumant work of the herders of San Luis Potosí is based on what Giménez (2005) defines as *targeted* and *internalized* forms. The former are tied to the mountains, the valleys, the pastures, sources of water, flora, paths and elements that are used and appropriated based on certain functional rules of the arid nature that the herders know and understand. These factors are internalized by abstract rules (Godelier 1989) based on knowledge, its transfer and reproduction, which are seen in the society's rituals, seasonal cycles, food, myths, language, whistles, and systems of reciprocity. All of this forms a territorial livestock herding identity through a shared symbolic system based on a historical grazing activity with standards, values and knowledge that are both acquired and inherited.

Based on these factors, we can define the transhumant work of the herders of the San Luis Potosí highlands as a pendulous (leaving and returning) and seasonal migration between two points with ecological complementarity: the mountains and the valley. The migration between these two points is due to the climatic cycles based on the rainy and dry seasons. By paying attention to these elements, the herders plan their practices and sync them to the reproductive cycles of the animals.

The drives up the mountain begin in October, and the descents to the valley take place in February, when the low water levels can be surmounted, with this period of uncertainty in wait for rain (May and June) representing a period for the herders that is just as important and vital as the journey up the mountain. Remaining at the valley location out of season is not viable. It is important to reach the range lands of the mountain in time before the pastures and water in the valleys become scarce. A sedentary lifestyle is not viable in these societies, so transhumance becomes a vital demand.

7 The Mountains and the Valleys as Complementary Spaces

The sierra is the mountainous zone where, due to its altitude, the pens are located, and there are rangelands at altitudes above 2000 meters above sea level. It is in this area that the type of substrate, water sources and pastures are found that can best nourish the goats. The pens are seasonal settlements built as refuges for the herders and as locations for gathering the livestock during the fall-winter season that runs from October to February. Some permanent pens have been built where family groups congregate and create settlements far from urban centers and without access to electricity, water or schools. This settlement pattern marks the landscape of the highlands as geo-symbols related to toponymies such as: The pen of Chon, the hairy pen, the dappled pen, the little ranch pen, the goat berm, the crowned herder (Hernández in Mora 2013:101).

On the other hand, the valley is a flatland area where the ranches of origin are located and where the social, cultural, economic, religious, educational, political, family and out-of-family lives of the herders are led. The families often stay at these locations, particularly those with school-age children.[5] The centers of the common land area (*ejidos*) and parcels are located in the valley, and the work of returning the transhumant movement to the mountains begins. It is important to note that some herders also make shorter movements (10–15 km), which are carried out on a daily basis from the flatlands to the mountains, either during the rainy season, or throughout the entire year. These herders usually have herds of less than 100 animals. With larger herds it is necessary to carry out long transhumant movements to keep the herds healthy, particularly in the dry months. That is why those who make short transhumant movements have to limit the size of their herds.

Alternating between spaces allows the pastures in one territory and the other to recover from their use during the season. There is clearly environmental and social complementarity between the two spaces, since the ecological characteristics of each contribute different resources. However, the balance depends on being able to access the mountain and valley areas in the corresponding periods in order to carry out the transhumance.

Based on a planned activity, the herders establish the seasonal periods based on the reproductive cycles of the animals, and plan most of the births for the summer-winter months with the most abundant pastures to ensure the health of the does and their kids, which helps ensure that the kids are not lost before or after their birth. This relationship is also affected by the other rhythm the goat herders are exposed to, which is that of the market, since they also have to plan the births to ensure availability of product during the droughts, so as to have kids and cheese in the months with the highest demand due to their scarcity. More than just the interaction of the livestock, sometimes transhumance also defines the rhythm of life of the herders, particularly with regard to the birth of children during the rainy season, so they can be born on the flatlands.

The breeding of goats is also complemented by other activities, such as the seasonal planting of crops on steep land, where the pens are placed above the fields of corn, using a terraced system that makes use of the "sirri" or droppings of the goats to fertilize the crops below. These types of technologies allow the herders to make use of steep land that would be very difficult to make agricultural use of in any other way.

[5]Previously, the entire family followed the migration, but currently, because of access to schools and health and other services, only the father and the older children follow the fold to care for the livestock.

In addition, the goats act as reserves for other types of needs or activities: education, illness, procedures, contributions, parties and ceremonial events, or as a platform to enable migratory travel. When economic resources are needed, the families can sell some goats to get money.

8 Use of Geographic Space

Based on the geographic study we performed as part of our research (see Mora 2013), we found a close relationship between goat farming, vegetation and the type of soil. Our starting point was that this type of livestock farming is generally carried out in spaces where sheep and cattle farming are unsustainable. Goats eat plants located on the scrublands at the base of mountains, such as thickets, and they also must be moved with extreme care, which involves daily work that affects the daily life of the herders, which is why the herding families base their lives on "goat time" (Hernández in Mora 2013: 92).

As a result of the reform to Article 27 of the Mexican Constitution, these activities are carried out in the agricultural centers, even though not all of the common lands (*ejidos*) have the conditions required to handle transhumance. There is a close relationship between transhumance, its ecological handling, the patterns of settlement and animal density (ibid.). This relationship shows a pattern of specialization in the different ecological niches. When the topography has inclines and mountains, it is more likely to find the type of vegetation located at the foot of the mountain, such as thickets or forests, which are made use of in transhumance, so this movement between the different ecological niches can be explained, since the fact that there is no particular ecosystem but instead a transhumant area means that there is interaction between various ecological niches, unlike in the valleys where the composition is more homogenous in terms of semi-desert vegetation, pastures or agriculture, which is why these areas have the least dense occupation of goat livestock, although as indicated above, transhumance also allows for recovery of the high areas and avoids overgrazing. In contrast, the higher areas are more apt for grazing due to the type of vegetation, and reciprocally speaking, this occurs at the times when the foliage and pastures are depleted in the valleys (Mora 2013: 96). The study considered elevation to determine the different activities:

At the first level, which is the valleys or the plains, dry land or watered agriculture is carried out to make use of the runoff from the mountains. The second level is the area at the base of the mountain or range for goat herding, which, due to the vegetation or topographical conditions, does not make it difficult to bring in cattle or sheep. This area is predominated by bushes and thickets, both of which are used by the goats. The third level is the Meseta de Catorce (plateau) with an altitude of 3000 meters above sea level, which allows for a proliferation of high-mountain grasslands for breeding cattle and sheep (Mora 2013: 97).

The figure below shows the different ecological levels and their use in different activities.

Ecological zones and productive activities

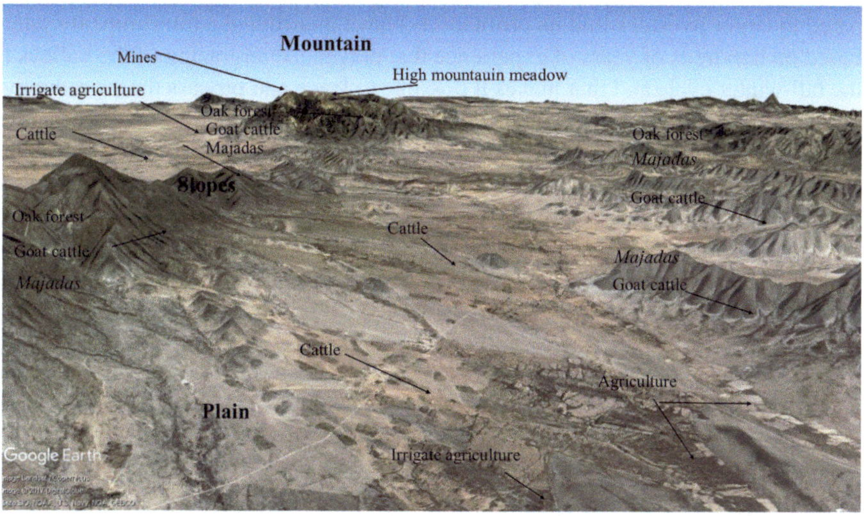

Plain activities: Cattle, agricultura, irrigate agriculure

Slope activities: Goat cattle

Mountain activities: mines, goat cattle.

Digital Elevation Model with Google Eart.
By Gerardo Hernandez

9 Conclusions

What we want to communicate in this document is a description of a thousand-year-old system that, because of unawareness, has not captured the interest of the policies of regional development for Mexico's rural areas. As has been demonstrated throughout this article, the perspective of maintaining the transhumant system as a productive space for the reproduction of families in the desert is uncertain as a result of neoliberal reports regarding ownership. For the past two decades, there has been an exponential abandonment of the transhumant system, caused by factors such as: private ownership of the land based on the amendment to Article 27 of the Agricultural Law, which limits the rangeland available to the common land boundaries and also allows for the sale of common land (*ejidos*). The effect of this has been that, through limiting the territory available for extensive livestock farming and transhumance, many goat herders have had to sell their livestock because they have insufficient space for transhumant grazing, with visible consequences in the form of emigration, particularly of the young population, with the domestic units remaining conformed solely of older people and children, which, given the work required for grazing, restricts the possibility of having flocks of more than 20 goats.

The lack of knowledge, appreciation and prejudice against this activity (compared to larger livestock and its profitability) have interrupted a transhumant system based on hundreds of years of experience that prevents overgrazing of the valleys in the fall-winter cycle and that makes it possible to leave and process the reproductive germination (in the sirri) with an elevated reserve of seeds, contributing to the rapid and abundant recovery of the plants when the rains come. We consider that extensive and seasonal grazing contributes significantly and sustainably to the diversification and conservation of habitats with a very high environmental value in these desert areas. We therefore consider that it is necessary to recover the ecological and cultural productive basis of the transhumant work and conserve these societies and this system.

The transhumant practice is also tied to making use of common land (*ejidos*), which, in many cases has now become private property or is restricted to the common land holders, and there are now fences that impede and make it impossible to maintain the routes and paths that the goats have followed for hundreds of years for the sustainable reproduction of the livestock and the region. Government involvement via the Ministry of Agriculture, Livestock, Rural Development, Fishing and Food (SAGARPA) has established a limit on the number of heads each herder can maintain. Although this is necessary, in some cases the experience the herders have acquired over generations on how to handle the territory and the common land is not taken into consideration. Some common land areas have their own standards agreed on at common land meetings where they determine the number of animals the land can maintain, particularly in times of drought. One example is the case of the Cerro Gordo common land (*ejido*) in the municipality of Charcas, where the goat herders agreed on a strategic plan to face the severe drought experienced in 2012. They decided that they could only maintain a maximum of 100 heads per producer due to the conditions of the common use rangeland. As a reference, they considered that each unit of goats, corresponding to six animals (equal to on cow), required an average minimum of between 15 and 20 hectares. The results of this strategy were very favorable for this *ejido*, which suffered very few losses. The *ejidos* that did not plan their animal burden and instead followed the indications of the government agencies experienced significant losses.

The sustainability of goat herding culture is that it can be carried out under even the most adverse conditions, provided the herders have the knowledge and social culture required to handle it, based on local knowledge that is tied to technological innovations. We consider it to be a mistake to assume that the herders are individuals incapable of having an environmentalist culture, and to consider transhumance to be a predatory activity with low economic value and significant uncertainty, as certain government agencies and environmental groups have claimed. Even in the face of these perceptions, herding continues in spite of the attacks of these policies that, far from fostering a sustainable activity (based on history, knowledge and experience), instead encourage abandonment of the goats and promote a restructuring of production. Neoliberal policies and profitability have driven this change from herding activity to livestock production of larger livestock, which, as demonstrated above, is ecologically incorrect for the region.

This scenario has become much more complex and adverse with the granting of mining concessions, mega solar plant projects, agro industry and toxic waste dumps in areas with extensive livestock and transhumance.

Lastly, we consider that the transhumant livestock processes of the region under study is a technique that has had strong historical roots of sustainability since the arrival of the goats and that, in addition to protecting the biodiversity of the desert, maintains ties to cultural values that would be lost if transhumance were to disappear in the region.

References

Bendini M, Tsakoumagkos P, Nogues C (2004) "Los crianceros trashumantes del Neuquén" ("The Transhumant Crianceros of Neuquén"). In: Bendini M, Alemany C (coordinadores) Crianceros y chacareros en la Patagonia. (Crianceros and Small Farmers of Patagonia) Editorial La Colmena (Cuadernos de GESA 5), Buenos Aires

Cassigoli R (2004) "Familias collas trashumantes en el norte de Chile" ("Transhumant Colla Families in Northern Chile"). In: Salas H, Pérez-Taylor R (eds) Desierto y frontera. El norte de México y otros contextos culturales. V Coloquio Paul Kirchhoff. (Desert and Border: Northern Mexico and Other Cultural Contexts. V Discussion Paul Kirchhoff). UNAM, Instituto de Investigaciones Antropológicas/ Plaza y Valdés, Mexico

Descola P (1996) La selva culta: simbolismo y praxis en la ecología de los Achuar. (The Cultured Jungle: Symbolism and Praxis in the Ecology of the Achuar). Ecuador, Editorial Abya-Yala

Evans-Pritchar EE (1977) Los Nuer (The Nuer). Anagrama, Barcelona

Food and Agriculture Organization of the United Nations (FAOSTAT, 2009–2013)

Giménez G (2005) Teoría y análisis de la cultura. (Theory and Analysis of Culture). CONACULTA/ICOCULT, Mexico

Godelier M (1989) Lo ideal y lo material (The Ideal and the Material) Anagrama, Madrid

Gutiérrez H (2009) "La economía de pastoreo en La Montaña". ("The Economy of Grazing in the Mountain"). Mexico. Estado de desarrollo económico y social de los pueblos indígenas de Guerrero, publicado por el Programa Universitario México nación multicultural. ("State of Economic and Social Development of the Indigenous Peoples of Guerrero, Published by the Multicultural Nation Mexico University Program"). UNAM and Secretaría de Asuntos Indígenas del Gobierno del Estado de Guerrero

Harris M (1994) El materialismo cultural. (Cultural Materialism). Alianza Editorial, Madrid

Hernández G (2013) "Tras el sendero de las cabras. El contexto geográfico y la ganadería trashumante en el altiplano potosino". ("Behind the Path of the Goats: The Geographical Context and Transhumant Livestock in the San Luis Potosi Highlands"). In: Mora MI (coordinadora) Los caminos de la trashumancia. Territorio, persistencia y representaciones de la ganadería pastoril en el altiplano potosino. (The Paths of Transhumance: Territory, Persistence and Representations of Livestock Grazing in the San Luis Potosi Highlands). El Colegio de San Luis, Mexico

Mauss M (1979) Sociología y antropología. Ensayo sobre las variaciones estacionales en las ciudades esquimales. Un estudio de la morfología. (Sociology and Anthropology: Essay on Seasonal Variances in Eskimo Cities. A Study of Morphology). Editorial Tecnos 1979, Madrid

Mora MI (coord) (2013). Los caminos de la trashumancia. Territorio, persistencia y representaciones de la ganadería pastoril en el altiplano potosino. (The Paths of Transhumance: Territory, Persistence and Representations of Livestock Grazing in the San Luis Potosi Highlands). El Colegio de San Luis, Mexico

Sánchez E (2007) "De ganados, movimientos y contactos. Una nueva aproximación al debate sobre la trashumancia en la Hispania Antigua". ("Livestock, movements and contacts: a new approach to the debate on transhumance in ancient Spain"). Universidad Autónoma de Madrid, Departamento de Historia Antigua, Madrid. Available at: http://ffil.uam.es/antigua/piberica/ganado/ganado1.html

Steward JH (1972) Theory of culture change: the methodology of multilinear evolution. University of Illinois Press, Urbana and Chicago

Tomé P (1996) Antropología ecológica. Aportaciones, influencias e insuficiencias. (Ecological Anthropology: Contributions, Influences and Deficiencies). Ávila, IGDA, Madrid

Urban Sprawl, Environmental Justice and Equity in the Access to Green Spaces in the Metropolitan Area of San Luis Potosí, Mexico

Adrián Moreno-Mata

Abstract The objectives of this research are: to analyze the effects of the urban sprawl on spatial distribution and social accessibility to green spaces in the metropolitan area of San Luis Potosí, México and to assess the level of environmental justice in the Metropolitan Area of San Luis Potosí, Mexico. The study evaluates the equity in the distribution of public green spaces and their accessibility. The research questions are: how accessible are public parks or green spaces to different social groups? and, which socio spatial areas need more public green spaces or parks? The methodology relies on the environmental justice approach, through a system of indicators which allows to assess the distribution, disposability and accessibility of the population to the green spaces. The analysis of this information allows to determine the extent to which the population of the metropolitan area of San Luis Potosí, benefits from equal access to the parks of the city, through its distribution, surface, disposability and accessibility. The paper concludes that the current distribution of green spaces in the city studied presents an unequal spatial distribution pattern, which benefits population of high-income brackets and excludes the population of lower income brackets. In Mexico there are few urban studies that analyze the problem of green areas and urban parks from the perspective of environmental justice and the connections between their spatial distribution and accessibility in time and distance, and the quality of life of the population. The contribution of this work is to advance in this line of research and to propose a methodology that could be applied in other Mexican cities. One of the limitations of the study is that it was not possible to analyze the interaction between the indicators and other concepts, like biophilic cities. This could be a line of future research that would allow to delve into the complex relationship between the human being and the environment.

Keywords Environmental justice · Equity in access to green spaces
Urban sprawl

A. Moreno-Mata (✉)
Habitat Faculty, Autonomous University of San Luis Potosi,
San Luis Potosí, Mexico
e-mail: adrian.moreno@uaslp.mx

W. Leal Filho et al. (eds.), *Sustainable Development Research and Practice in Mexico and Selected Latin American Countries*, World Sustainability Series,
https://doi.org/10.1007/978-3-319-70560-6_32

1 Introduction

Loss of natural landscape and green spaces due to rapid urbanization is occurring which might be detrimental to human health (Andrews 2014). The increase of the physical urban expansion is not produced in a balanced proportion between free areas, public spaces and green areas, concerning the areas urbanized and the population needs. This can oppose the same urban development, since the presence of green areas represents one of the major actives of the social and individual welfare.

In addition, most public urban green space is not distributed equally and fairly. Access to public green space is often stratified based on class, socioeconomic status or income level. Many authors hypothesized that socioeconomically deprived neighborhoods or those areas with concentration of population with low income levels were underexposed to parks and green spaces as "environmental goods", testing the suitability of the environmental justice framework for positive environmental outcomes. Empirically, environmental inequality research has focused heavily on exploring disproportionate burdens on low-income and minority groups by residential proximity and exposure to environmental hazards, and less research has been conducted to investigate spatial inequalities of underexposure to "environmental goods," such as green spaces and urban parks (Miyake et al. 2010).

In Mexico there are few urban studies that analyze the problem of green areas and urban parks from the perspective of environmental justice and the connections between their spatial distribution and accessibility in time and distance, and the quality of life of the population. The contribution of this work is to advance in this line of research and to propose a methodology that could be applied in other Mexican cities. One of the limitations of the study is that it was not possible to analyze the interaction between the indicators and other concepts, like biophilic cities. This could be a line of future research that would allow to delve into the complex relationship between the human being and the environment.

With this aim, this paper, proposes an integrative theoretical framework to better illustrate the relationship between green spaces accessibility, and use, within the context of urban open space such as parks. The objectives of the research are: (i) to analyze the effects of urban sprawl on spatial distribution and social accessibility to green spaces in the metropolitan area of San Luis Potosí, México and (ii) to assess the level of environmental justice in the Metropolitan Area of San Luis Potosí, Mexico.

1.1 Theorical Approach and Analytical Framework

The environmental justice framework offers a useful conceptual paradigm for studying the spatial distribution of green spaces and urban parks. In short, the environmental justice framework embraces the principle that all people and communities, regardless of their sociodemographic background, are entitled to equal distributions of environmental amenities, and no group should be disproportionately affected by environmental hazards (Zhang et al. 2009). As in other social sciences, inquiries related to place, class, socioeconomic status and income level occupy a central position in theoretical and empirical investigations of uneven distribution of environmental "goods" or "bads" across social groups (Boone et al. 2009).[1]

The approach to environmental justice constitutes a valuable analytic category to assess the conditions of equity in the access to green areas in cities. This line of thought considers that the distribution of the natural resources and therefore its benefits must be equitable among all of the groups in the society (Hervé Espejo 2010). Following this idea, the concept of environmental justice has been incorporated into the study of the urban green areas, with the same importance as the classic indicator of quality of life.

1.2 Data and Measures

In green spaces literature two classical measures are used: disposability and accessibility. The first is related with green area and vegetation covers (Shlomo et al. 2012). From a regulatory perspective, green space disposability refers to the international standards stablished by the World Health Organization (WHO), which recommends between 9 and 12 m^2 of green area by inhabitant (WHO 2012). But the urban green space coverage index is too general and not entirely consistent with the environmental justice framework. Other factors or variables commonly considered in assessing the equitable distribution of green areas are the size, the quality and the distance between the population and these spaces (Omer 2006; Sotoudehnia and Comber 2010).

For evaluating the green space disposability, a useful instrumental element can be the degree of accessibility that the population has to parks, understood as a public urban good (Omer 2006, 2005; Dai 2011). Commonly, green spaces accessibility refers to physical distance or travelling time from a residential house to green spaces (Sotoudehnia and Comber 2010). Some complementary measures to the classical concept, widely used in the researches of accessibility to green space, are the distance and travel time from the origin of the user to the parks (Boone et al. 2009).[2] The United Kingdom benchmark standard recommends that to make the best of

---

[1]An emphasis on class or socioeconomic status as possible antecedents of environmental inequalities is consistent with Weber's notion of "life chances" addressing race and class dynamics leading to individuals' life circumstances and social mobility trajectories (Weber 1946). In addition, this emphasis on class or socioeconomic status accords with a fundamental cause theory, which contends that race and class are both fundamental causes of health disparities because they are closely bound up with a wide range of resources that promote health and hazards that harm health (Phelan et al. 2010).

[2]In the case of Mexico the standard of green spaces accessibility was established in 1976 by the extinct Secretariat of Human Settlements and Public Works (1976), and updated by Social Development Secretary (SeSoDe 1999), regarding the location, equipment, coverage and size of urban green areas, which must not be found to a maximum distance of 500 m as regards the location of the areas of residence of the population.

green spaces, no-one should live more than 300 m from nearest green area of at least 2 ha in size.

Other authors (Sotoudehnia and Comber 2010, 2011) show the relationship between the distribution of green spaces and the spatial networks (road systems, travel routes, accessibility and economic importance linking them with residential neighborhoods) as well as how the ease of access to a particular site can be measured by the cost of the trip or its duration. De Vries et al. (2003) has focused on the ease of access to green spaces, researching residents' needs and their selection of a neighborhood which is close to services for easy access.

On the base of that criterion, several studies had applied different qualitative and/or quantitative methods to measure accessibility to green spaces (Barbosa 2007; Omer 2006). Their findings show inequality in the provision and access to green spaces for different social groups. This means, that another complementary indicator of accessibility to green spaces is the social distance, established by the spatial location of the population and its level of population income (Byrne et al. 2011). This indicator also considers the issue of differential density of the population at Metropolitan level, neighborhoods, districts and sectors.

Following the theoretical approach and analytical framework described, in this work evaluates the environmental justice depending on the accessibility differential of the residents in the different sectors or areas of metropolitan area of San Luis Potosí (MASLP) to the green spaces of the city depending on their spatial distribution, disposability and accessibility. In order to quantify physical access to green spaces, a network analysis was applied to measure -bias road distance-, in the hinterland or cover zone of the main green spaces. This was done within 3 distance constraints: (i) distance up to 300 m ("good access"); (ii) distance between 300 to 1000 m ("average access"), and (iii) distance more than 1000 m ("poor access"). Using the census tract information about urban marginalization levels (NACOPO 2010), also evaluates differentiated accessibility to urban parks by intra-urban distribution of the population according to socio-economic strata.

2 Results

The metropolitan area of San Luis Potosí (MASLP) is located in the Centre—West region of Mexico. It covers parts of the area of the municipalities of San Luis Potosí and Soledad de Graciano Sánchez (see Fig. 1). Recent studies locate the MASLP in the 14th place, within the classification of the more competitive cities at the national level, and the 207th at the global level, thanks to its capacity of attraction of investments in the industrial sectors—in special, because of its automotive industry—and services sector (Sobrino 2012; Moreno Mata and Parra Rodríguez 2017).

The city of San Luis Potosí was founded in 1592, in the colonial era. During the first three hundred years, the city grew under the classic urban model of the New Spain cities, practically unchanged in all that time, and in which the old Center established the limits of it city space. From the 20th century, this model began to

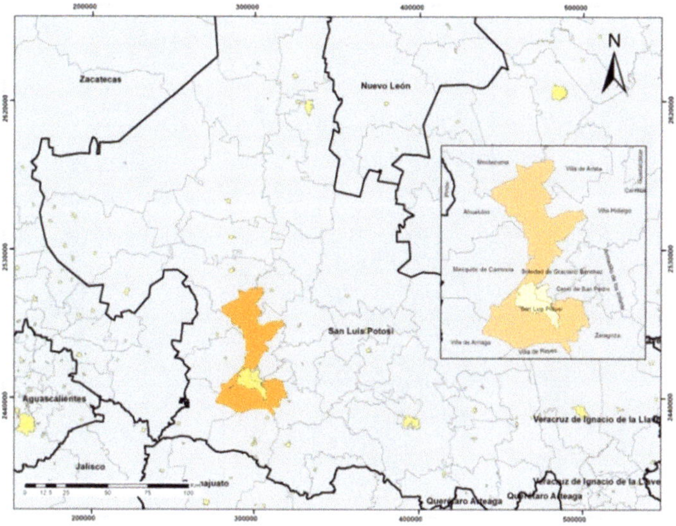

Fig. 1 State of San Luis Potosí: principal cities, 2010. *Source* Own elaboration based in Moreno Mata and Villasis Keever (2015)

turn slowly and in the middle of that century it adopts a growth pattern of the concentric type, combined with other sectoral radial patterns produced by the growing of the road system and the use of the automobile as the main way of transport (see Fig. 2). Since then, and particularly in the last five decades (1960–2010), this growth has become expansive, messy and unsustainable, either for lack of a planning system, or by the lack of respect to the reservation or mandated protected areas, and the ignorance of the effects that the prevailing urban models can produce. With an expansion of the space and the growing population, by 2010 the MASLP reached 1,040,443 inhabitants (NIEGI 2010).

Between 1960 and 2000 the MASLP multiplied four times its population and almost eight times its surface. The acceleration of population growth together with urban sprawl and fragmentation, occurred predominantly during the second half of the 20th century (see Table 1).

2.1 Urban Green Space Accessibility and Environmental Justice

Environmental justice and accessibility to green spaces in the MASLP the socio-spatial inequality described in the preceding paragraph is also expressed in a historical pattern of inequity in access to green spaces. Towards 1959, the unique urban parks, gardens or green areas of the city were concentrated in traditional neighborhoods and in the Central Alameda (1653), located in the historical Center,

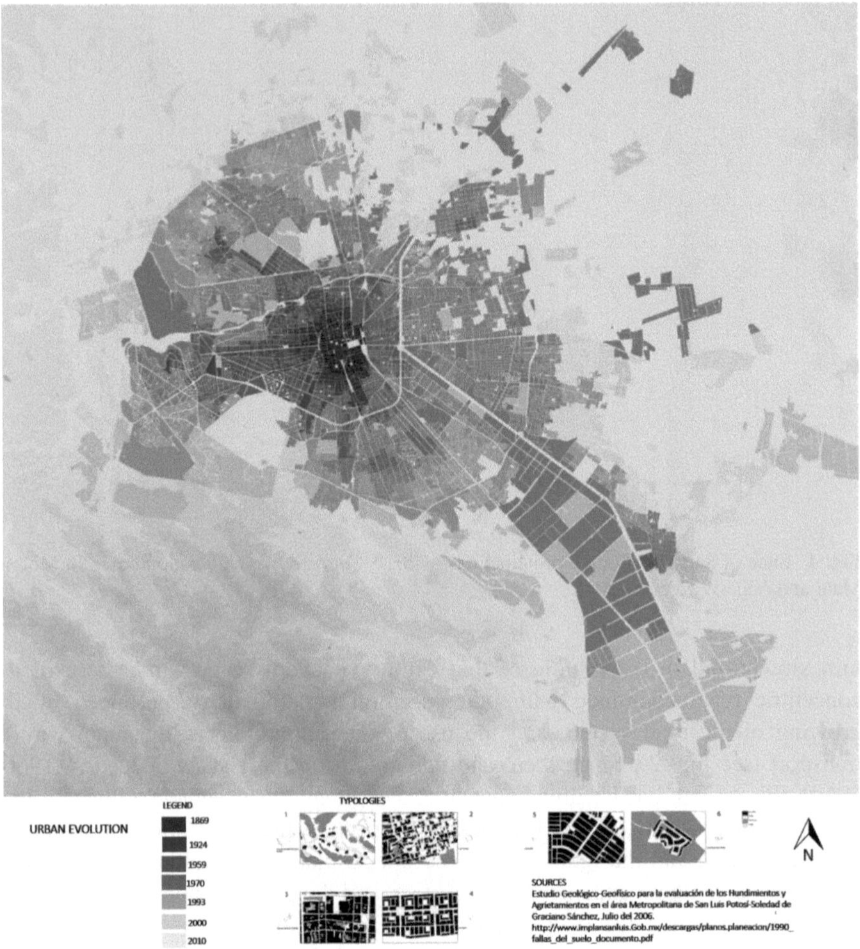

Fig. 2 MASLP. Historical evolution & urban structure, 1959–2005. *Source* Own elaboration adapted from GCC/AHO/UASLP (2016), after modification by Alva (2017)

and the Juan H. Sanchez Park (1924) located to the West of the urban spot. Until 1972 the surface of green areas of the city was 33.4 ha (Leyva and Segura 1995; cited by Rodríguez Rangel 2010), see Fig. 3.

It was until 1982, during the State administration of Carlos Jonguitud Barrios (1979–1985), when on the expropriated land of the Garita de Jalisco, the most important urban parks of the city were built: the parks Tangamanga I and II with 450.0 and 215.3 ha, respectively (Rodríguez Rangel 2010). The first is located in the southern—west sector of the city and the second in the northern sector. Since the creation of these areas, so far there hasn't been built another city park of this dimensions (see Fig. 4).

Table 1 MASLP. Population and urban surface growth, 1990–2010; 2000–2010

Municipality	Population				Growth rate		Urban surface (ha)		
	1990	1995	2000	2010	1990–2000	2000–2010	2000	2010	2015
Central	429,238	586,585	638,533	772,604	1.70	1.43	–	–	–
Peripheral	123,943	147,188	212,294	267,839	4.16	4.04	–	–	–
Total	613,181	733,773	850,828	1,040,443	2.04	2.03	12,859	14,893	19,642[a]

Note [a]This data refers to the surface of the compact city. If the fragmented urbanized area is also considered, the total area of the MASLP reaches a total of 23,445 ha (Moreno Mata and Villasis Keever 2015)

Source SeSoDE (1999), NIEGI (1990), (1995), (2000), (2010), MI of Planning San Luis Potosí (2015), Moreno Mata and Villasís Keever (2015), Moreno Mata y Cárdenas Nielsen (2015)

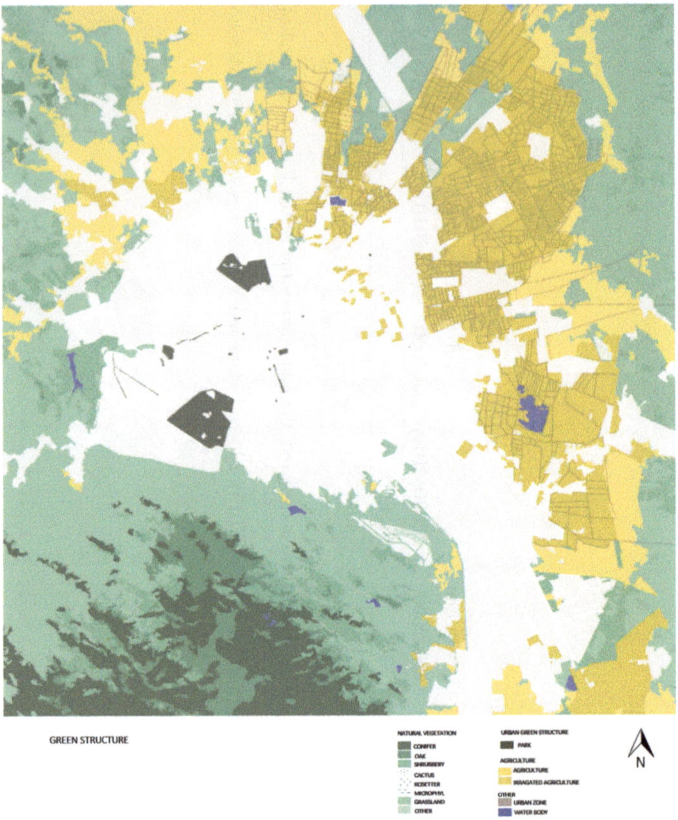

Fig. 3 MASLP. Green areas and historical evolution to the city, 1959–2005. *Source* Own elaboration adapted from GCC/AHO/UASLP (2016), after modification by Alva (2017)

Accordingly, at the beginning of the 1980s the total number of green areas in the MASLP reached a surface area of approximately 746.56 ha.

It is worth mentioning, that by 2012, the Mining Company Mexico began to rehabilitate contaminated areas by it's over one hundred years of operation, through the project of the urban Bicentennial Park, which is a small linear urban park with a surface of over 5.5 ha located in the vicinity of the company. However, this green space has little accessibility to the open population because its location and the private access control (see Fig. 4 again).

Also, there is a partially enabled linear urban park known as Paseo de la Presa located in the Western periphery, with a planned total area of 344.03 ha—(Sandoval Mendoza 2007) and several failed projects promoted during the states administrations of Fernando Silva Nieto (1997–2003) and Marcelo de los Santos Fraga (2003–2009), as the purported Parque Tangamanga III, started in 2000 that would had been be located on the eastern part of the city, and the urban park named Ejido of San Juan de Guadalupe referred in the Plan of Strategic Center of

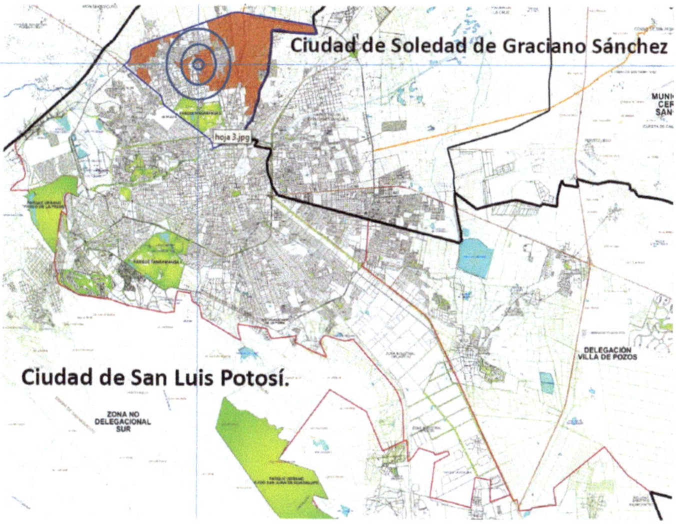

Fig. 4 MASLP. Green areas: metropolitan location, 2015. *Source* Adapted from Municipal Planning Institute of San Luis Potosí (2012)

Population San Luis Potosi—Soledad de Graciano Sánchez (PSCPSLP 2003), with a total area decreed from 1200.00 ha, and whose implementation has been slowed down several years by the interests of different groups -social, community and private-, who oppose his execution (see Fig. 4 again).

Thus, by the year 2010 the total area of urban parks and green areas has reached 1062.8 ha, and the projections to the year 2025, assuming the consolidation of the projects in process, would reach a total of 1623.7 ha (Rodríguez Rangel 2010).

2.1.1 Disposability to Green Spaces and Urban Parks

Finally, several studies on the inventory and coverage of green areas in the MASLP (Leyva and Segura 1995; Rodríguez Rangel 2010), reveal that, according to the international standard (9–12 m per inhabitant), at a general level, the rate of green areas per capita in the period 2000–2025 will range from 10.08, 11.78 to 12.83 m^2 by inhabitant, for 2000, 2010 and 2025, respectively (see Table 2).

In a retrospective of the historical evolution of the surface of urban parks and green areas in the MASLP, there can be seen a point of break in the trend of physical growth of green spaces built starting in 1982, with the construction of the urban parks Tangamanga I and Tangamanga II. Thereafter, the total surface has stabilized, and stays in a range that goes from 10.0 to 12.8 m^2 of green area per inhabitant, during the period 2000–2015. According to these data, in a first approximation, arguably the surface of effective green area per capita required to meet the international standard and the quality of green spaces and urban parks in the MASLP is satisfactory. However, if the area of areas is related to the total

Table 2 MASLP. Main green areas: spatial distribution, surface and green areas index, 2000–2025

Sector			Population			Surface green areas			Green areas index m^2/ inhabitants)		
Num.	Name	Surface (ha)	Year			Year					
		2010	2000	2010	2025*	2000	2010	2025*	2000	2010	2015*
1	Centro	1625.1	138,065	156,329	239,396	33.4	41.8	41.8	2.4	2.6	1.74
2	Lomas—Tangamanga	1783.0	55,080	64,215	84,555	467.8	503.3	1047.0	84.9	78.3	123.8
3	Morales—Ind Aviación	1806.2	111,318	122,604	170,888	56.2	79.6	79.6	5.0	6.4	4.6
4	Saucito—Terceras	2234.7	59,939	78,554	92,015	215.3	221.4	221.4	35.9	28.1	24.0
5	Satélite—Progreso	2313.9	207,648	213,168	290,745	14.4	46.7	46.7	0.6	2.1	1.6
6	Zona industrial	2635.6	447	687	1033	0.0	86.4	86.4	0.0	1258.2	836.7
7	Delegación Pozos	3623.5	56,711	68,295	87,058	9.2	38.0	38.0	1.6	5.5	4.3
8	Soledad Norte	1508.1	50,268	57,183	88,845	1.6	16.2	33.4	0.3	2.8	5.8
9	Soledad Sur	1561.0	119,309	158,785	210,866	7.5	29.1	29.1	0.6	1.8	1.3
Total		19,091.5	798,785	901,902	1,265,055	805.6	1062.8	1623.7[a, b]	10.0	11.7	12.8
AGACI (Green area surface/Urbanized urban area) %	**2000**	12,859.0							**6.2**		
	2010	19,091.50								**5.6**	
	2015	27,319.1									**5.9**

Source Own elaboration

Note The supply parameter of green areas green it sets by the Social Development Secretary (Sodese 2000), in the Guide for the formulation of plans of development urban.[a]Considered the projections made for the horizon of long term. [b]In the projections to 2025 horizon (Rodríguez Rangel 2010) is possible to observe the consolidation of the linear urban park Paseo de la Presa urbanized urban area, the actual green area coverage index (AGACI) is obtained, whose behavior is different from the traditional index established by the international standard. The AGACI goes from 6.2% in 2000 (the highest level) to 5.56% in 2010 and finally to 5.94% in 2015. If these values (in bold in Table 2) are compared with those of the traditional index for the same years (10.0, 11.7 and 12.8, respectively), the proportions obtained show marked differences (see Graphic 1). If this operation were carried out at the level of sectors, neighborhoods or Basic Geostatistical Areas (BEAs), the effective coverage of the green areas obtained would also be much lower than that established by the international standard.

2.1.2 Spatial Inequalities in Access to Green Spaces

On the basis of the information in Table 1, but analyzed at the level of each urban sector in a recent period, there are marked disparities in the rate of green areas per capita (see Table 1 again): in 2015, sector 2 named Lomas—Tangamanga, located at southwest of the metropolitan area, had the best rate of green areas per capita (123.8 m^2 by inhabitant), while the sector 5, named Satélite—Progreso, had the lowest rate with 1.6 m^2 by inhabitant. Thus, 85.9% of the total metropolitan population did not have great access to public green space, in respect that they have access to only the 16.5% of total green areas. Meanwhile, 14.1% of total population had access to 83.5% of total green spaces and parks of the MASLP.

The retrospective and prospective approaches of cover of green areas during the period of 2000–2025 confirms the observed tendencies: Sector 1 has a descending and well below the standard behavior (2.41, 2.67 and 1.74 m^2 per inhabitant in 2000, 2010 and 2025, respectively). The sectors 3 and 5, have a similar behavior

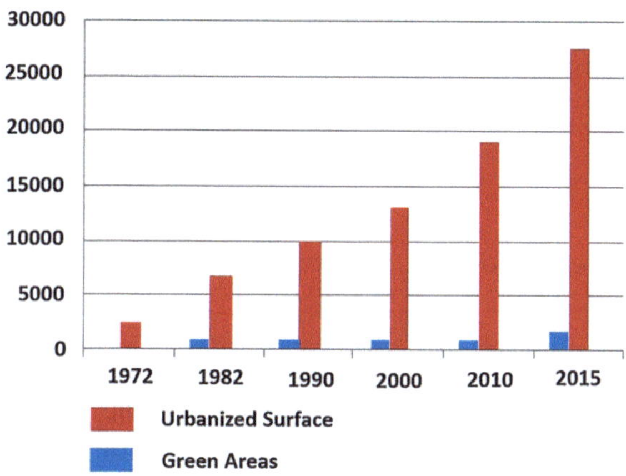

Graphic 1 MASLP. Greeen Areas Surface versus Metropolitan Surface, 1972–2025. *Source* Own elaboration adapted from Lárraga Lara et al. (2015), Leyva and Segura (1995), Rodríguez Rangel (2010)

(top-down and well below the norm); sectors 7, 8 and 9 have a slight increase, but in all cases and years fall below 6 m^2/inhabitant, far away from the national and international standards. Only sectors 2 (Lomas—Tangamanga) and 4 (Saucito—Terceras), presented a favorable behavior of the index, passing in the first case of 84.9, 78.3 and 123.8 m^2 per inhabitant in 2000, 2010 and 2025, respectively. In the second case, the index went from 35.91, 28.1 and 24.0 m^2 per inhabitant in those years. It should be noted that, in both cases, their closeness with the parks Tangamanga I and II, increase its index of coverage and accessibility to green spaces. Together, these parks concentrated 718.6 ha, representing 67.6% of the total number of green areas of the MASLP. It is also important to note that the sector 2 has concentrated the most affluent suburban population.

Now, if it is considered in the analysis the variable density of areas—that is the relationship between population over the urban surface- to level of BGAs), the existing differentials are even more notorious, regardless of the general level, in that the MASLP expresses a very low coefficient of the density of green spaces. We can appreciate how the bigger urban parks of the city—and almost the only ones-, are concentrated in the areas or sectors with the lower population density. So, while in Soledad Sur sector the green area per capita is 1.83 m^2 per inhabitant in 2010, in the Satélite—Progreso sector it is of 2.19 m^2 per inhabitant, even though both are urban settlements with the highest rate of urban marginalization in the MASLP. The same indicator of demographic density, in relation with the index of green areas by inhabitant, reveals that the areas with better coverage, sectors 2 (Lomas—Tangamanga) and 4 (Saucito—Terceras) correspond with two of the areas with lower population density: 36.8 and 35.1 inhabitants by hectare, respectively (see again the Fig. 5).

These variations and differences in the spatial distribution and disposability of green spaces and parks at the intra urban level of MASLP allows, in a first analysis, to note a great number of urban sectors that does not have the access to public green space than other sectors do. On this matter, 5 urban sectors are below of the international standard (12 m^2/inhabitant) while 2 urban sectors are above the standard. The measure of the physical disposability of public urban green space for the nine urban sectors reveals a certain level of spatial inequality on this metropolitan area.

2.2 *Social Inequalities in Access to Green Spaces*

To analyze the green areas and urban parks under the perspective of equity and environmental justice, there are some factors that must be known such as the distribution in the MASLP, its quality according to the vegetation cover and equipment presented and measure that caters to the population of these services; at the same time, it is necessary to know the relationship between the distribution of green spaces and: (i) residential location of socio-economic groups; (ii) physical distance or (iii) travelling time from a residential house to green spaces, and (iv) spatial networks (road systems, travel routes and spatial connectivity).

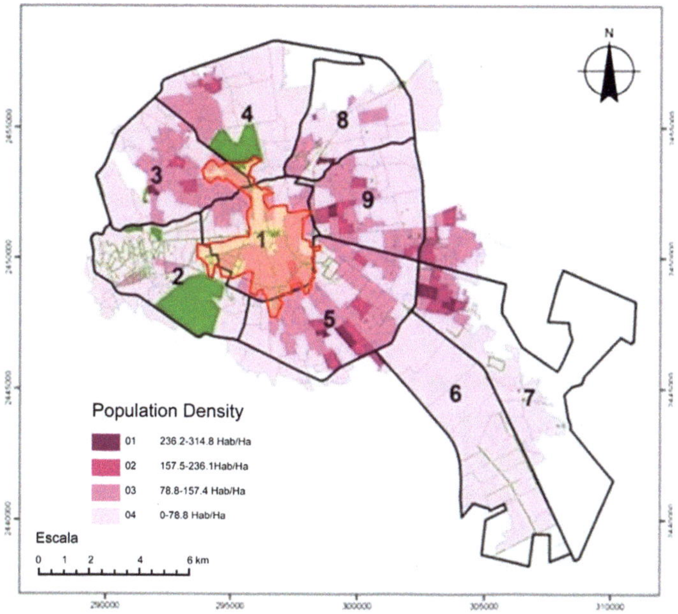

Fig. 5 MASLP. Spatial distribution of urban parks and demographic density by Basic Geostatistical Areas, 2010. *Source* Adapted from Lárraga Lara et al. (2015)

Using Basic Statistical Areas (BEAs), the analysis of income levels by socio-economic strata and average level of education, in relation with conditions of habitability (grade of overcrowding and deterioration of the housing), the endowment of basic services (water drinking and drainage) and educational levels (access to knowledge society and technologies) shows that practically none of sectors of the MASLP observed a complete homogeneity (see Fig. 6).

In some cases, this phenomenon produces a differential of residential consolidation, that reveals, to this time, at least five levels: very high, high, medium, low, very low (NaCOPO 2012).

Nevertheless, if analyzed, the pattern of spatial accessibility to green structure depending on the socioeconomic strata of the population and its physical distance over three distance constrains (good access, average access and poor access) in respect to the location of existing urban parks it is possible to observe a clear difference in the urban sectors of the MASLP. For example, for the Southern West sector (sector 2), the accessibility has a maximum of 3 km for the majority of population. The same situation is observed in the Northeastern, Northern and Central sectors (sectors 3, 4 and 1). In general, those urban sectors have a level of accessibility to green spaces between good and regular. However, the most notorious in this analysis is that the rest of the residents in the metropolitan area (about 421,702 inhabitants), that is, 53.24% of the population, have poor or no accessibility to green spaces, in terms of coverage radius or physical distance (see Fig. 6 again).

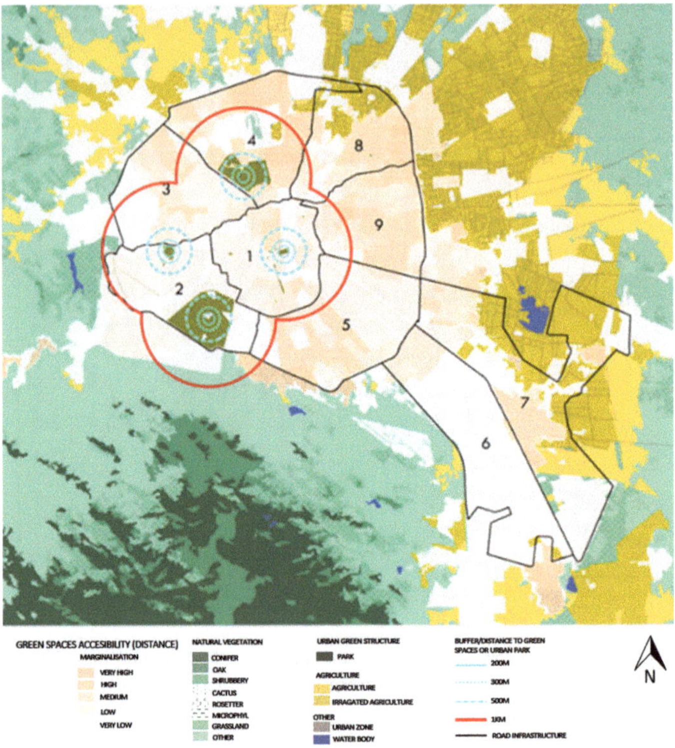

Fig. 6 MASLP. Residential stratification by income level and habitability and accessibility to Green Areas, 2010. *Source* Own elaboration adapted from GCC/AHO/UASLP (2016), after modification by Alva (2017)

Similarly, if you insert the time factor for the accessibility indicator, using the spatial networks and level of connectivity of road system (high, medium, low) it is possible to evaluate the accessibility of green spaces in the city, in terms of time and distances of traveled with respect to the location of the existing urban parks and residential location of social groups (see Fig. 7 again). For example, for the Southern West sector, the relative accessibility has a maximum of 15 min for the medium and high, socio-economic groups, in addition to availability to different means of transport. In contrast, much of the population of the city—especially the residents of the peripheral areas (located at Northwest, North, Northeast and Southeast) are located in a ratio that exceeds far from the norm of 500 m, with 1–5 km to distance and 20–45 min of travel time to be able to move to the few green spaces available (see Fig. 7).

It should be noted that accessibility based on ratio distance and travel times might also have to do with the differences observed in relation to the availability of public transport or the privilege of owning a private automobile. This situation reinforces the spatial disparities in the distribution of green spaces and urban parks in the MASLP. The urban sectors best served in terms of coverage and space

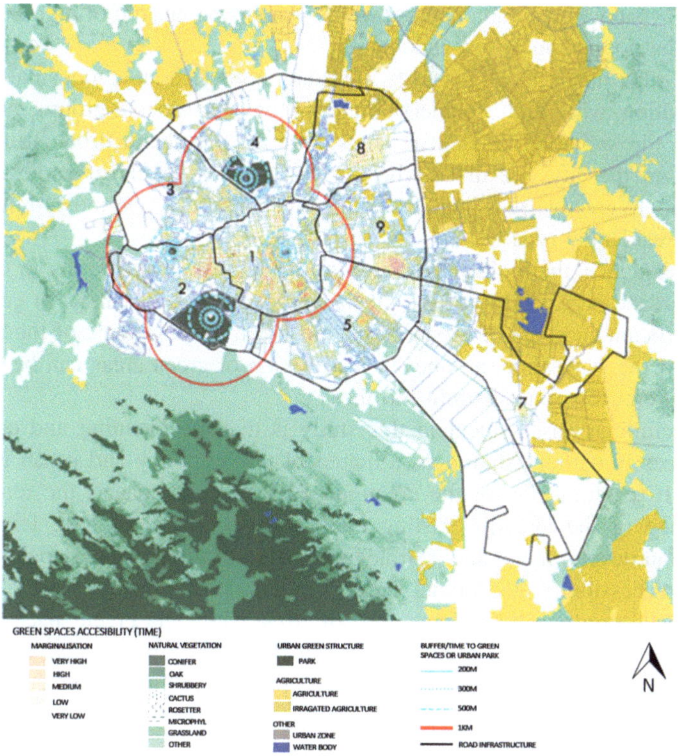

Fig. 7 MASLP. Residential stratification by income level, accessibility to Green Areas and connectivity of road system, 2010. *Source* Own elaboration adapted from GCC/AHO/UASLP (2016), after modification by Alva (2017)

accessibility are those that also have the best levels of connectivity of the mobility and transportation system, and greater availability of different modes of transportation. In contrast, much of the population of the city—especially the residents of the peripheral areas (located at Northwest, North, Northeast and Southeast)—are located in a ratio that far exceed widely the norm and these areas are also characterized by low connectivity levels of the mobility and transport networks and, therefore, the time and cost to move to green spaces and urban parks is greater.

3 Conclusions, Main Lessons and Future Prospects of Research

The analysis of this information allows to determine the extent to which the population of the metropolitan area of San Luis Potosí, benefits from equal access to the parks of the city, through its distribution, surface, disposability and accessibility. It′s concluded that the current distribution of green spaces in the city studied presents an unequal spatial distribution pattern, which benefits population of high-income and excludes the population of low income.

On the other hand, the larger green areas are more accessible to a small sector of the population with high socioeconomic status. The social groups, which are located in the suburban residential areas over the peripheral urban fringe in the southwest of the city, are those who possess the best levels of endowment and accessibility to green spaces in their own residential spaces, and most of them are of the gated residential type.

In consequence, the indicators of environmental justice and equity in the spatial distribution and accessibility of the green areas are not satisfactory. In general, we see a significant shortage of urban parks and green areas in most of the sectors of the city. At the same time, an uneven distribution of green areas can be seen with marked differences in accessibility to them.

The preliminary assessment in this study is that the planning and creation of green spaces in the city should be a matter of public policy and urban land management. It recommends for the creation of more green spaces, the renewal and maintenance of the already existing ones under a vision of long term for the metropolitan development, and a strategic location of urban parks depending on its universal accessibility and social equity at the neighborhood scale. The more affluent social groups already have enough urban parks and green spaces; now it's up to the population with greater shortages far away—physically, socially and economically—of those spaces. A study in depth, with a finer scale, could provide additional elements to be exposed, such as strategic axes and criteria suitable for the planning of green areas in the case study.

One of the main lessons of the study refers to the need to use mixed indicators, incorporating the quantitative and qualitative dimensions in the measurement of equity in the spatial distribution of green spaces. A second aspect has to do with the complex character of environmental justice, so it is recommended the combined use traditional indicators such as surface, distance, time of transfer, with nature, social and demographic indicators, this would apply a more integrated methodological approach, and most attached to the urban reality analysis. A third aspect is related to the need to conduct deeper studies in urban space, with a more fine scale (as the bringing the basic geostatistical Areas), which generate information in greater detail, in order to support plans and strategies to improve the social and spatial equity in the allocation and distribution of green spaces in the city. These three elements constitute both the future prospects of research, which are considered indispensable support in future work on the topic.

References

Alva B (2017) A motorway named rio. Universidad Autónoma de San Luis Potosí, San Luis Potosí, México

Andrews A (2014) Quantifying the well-being benefits of urban green space. Doctoral Thesis of Doctor of Philosophy, School of Environmental Sciences at the University of East Anglia

Barbosa O (2007) Who benefits from access to green space? A case study from Sheffield, UK. Landscape Urban Plann 2007:187–195

Boone ChG, Buckley GL, Morgan GJ, Chona S (2009) Parks and people: an environmental justice inquiry in Baltimore, Maryland. Annals of the Association of American Geographers, Taylor & Francis, LLC 99(4):767–787

Byrne J, Wolch J, Zhang J (2011) Planning for environmental justice in an urban national park. J Environ Plann Manage 52(3):365–392

Dai (2011) Racial/ethnic and socioeconomic disparities in urban green space accessibility: where to intervene? Landscape and Urban Plann 102(4):234–244

De Vries S, Groenewen PP, Spreeuwenberg P (2003) Natural environment sö healthy environments? An exploratory analysis of the relationship between greenspace and health. Environ Plann A 35:1717–1731

GCC/UASLP/AHO (2016) Workshop "A motorway named rio". German Cultural Center—University Autonomous of San Luis Potosí—The Oslo School of Architecture and Design, Oslo, Work Paper

Hervé Espejo D (2010) Notion and elements of the justice environmental: guidelines for its application in the planning territorial and in the evaluation environmental strategic. J Law 23(1):9–36

Human Settlements and Public Works (1976) General Law of Human Settlements. Secretariat of Human Settlements and Public Works, Mexico

Lárraga Lara R, Moreno Mata A, González Hernández R (2015) Human security, socio-spatial segregation and inequality in access to green areas: the case of the metropolitan area of San Luis Potosí. Magazine Contributions to the Social Sciences, 27. April 2015. Online: http://www.eumed.net/rev/cccss/2015/01/security-human.htm

Leyva ME, Segura D (1995) Management of green areas of gardens and ridges in the cities of San Luis Potosí and Cuautla, Morelos (working paper)

MI of Planning San Luis Potosí (2015) Municipality of San Luis Potosí, Prospecting 2015. Municipal Planning Institute of San Luis Potosí, San Luis Potosí, Mexico

Miyake KK, Maroko AR, Grady KL, Maantay JA, Arno PS (2010) Not just a walk in the park. Methodological improvements for determining environmental justice implications of park access in New York City for the promotion physical activity. American J Prev Med 34:269

Moreno Mata A, Cárdenas Nielsen A (2015). Chapter IV. Metropolitan fragmentation in peri-urban fringe of San Luis Potosí. Residential segregation and differentiation of the space social: 1990–2015. In: Moreno Mata A (ed) Urban Environment, Sustainability and Territory. Volume II: Metropolitan Systems in Crisis, San Luis Potosí, Mexico: University Autonomous of San Luis Potosí/ ISBN: 978-607-9453-49-7

Moreno Mata A, Parra Rodríguez O (2017) Globalization and local development? The automotive cluster in Villa de Reyes, San Luis Potosí. In: Acosta Collazo A, Moreno Mata A (eds) (Coords) Anthropic space called city: infrastructure + heritage + interdisciplinary creativity. Autonomous University of Aguascalientes, University of Alcalá de Henares, Autonomous University of San Luis Potosí (in press)

Moreno Mata A, Villasis Keever R (2015) Fragmentation urban and socio-residential segregation in the cities of the region West Central of Mexico. Cases Aguascalientes, León, San Luis Potosí and Zacatecas. In: Layuno Rosas A (ed) (Coord) Heritage, Territory and Landscape, International Research Journeys. School of Architecture, Alcalá de Henares, University of Alcalá de Henares, Spain

Municipal Planning Institute of San Luis Potosí (2012) Metropolitan Area of San Luis Potosí. Green areas: metropolitan location, Municipal Planning Institute of San Luis Potosí, San Luis Potosí, Mexico

NACOPO (2010) Delimitation of the Metropolitan areas in Mexico. National Council of Population, Mexico

NACOPO (2012) Delimitation of the Metropolitan areas of Mexico. National Council of Population, Mexico

NIEGI (1990, 2000, 2010) Censuses general of population and housing. National Institute of Statistics, Geography and Informatics, Mexico

Omer I (2005) Distributive environmental justice in the city: differential access in two mixed Israeli cities. Royal Dutch Geog Soc 2005:1–11

Omer I (2006) Evaluating accessibility using house-level data: a spatial equity perspective. Comput Environ Urban Sys Israel 30(3):254–274

Phelan JC, Link BG, Theranifar P (2010) Social conditions as fundamental causes of health inequalities: theory, evidence, and policy implications. J Health Soc Behav 51:S28–S40

PSCPSLP (2003) Plan of the strategic center of population of San Luis Potosi—Soledad de Graciano Sánchez. Municipal Government of San Luis Potosí, San Luis Potosí, Mexico

Rodríguez Rangel R (2010) Inventory of parks and gardens in the suburbs of the city of San Luis Potosí, S.L.P. Mexico, Undergraduate Thesis in Geography, Coordination in Social Sciences and Humanities. UASLP, México

Sandoval Mendoza C (2007) Current and potential vegetation and its experimental restoration in the area 'urban park Paseo de La Presa' San Luis Potosí, S.L.P. Master's Thesis in Environmental Sciences. UASLP, PMPCA-Mexico

Secretariat of Social Development (SeSoDe) (1999) Regulatory system of urban equipment, tomo V, subsystem, recreation and sports recreation in structure of the regulatory system. http://www.sedesol2009.sedesol.gob.mx/archivos/802402/file/recreacion_y_deporte.pdf. Accessed 22 Oct 2011

Shlomo A, Parent J, Civco DL, Blei AM (2012) Atlas of urban expansion. Lincoln Institute of Land Policy, Cambridge, Mass

Sobrino J (2012) Urban competitiveness. A global and for Mexico perspective. El Colegio de México, Mexico

Sodese (2000) Methodological guide to develop territorial planning plans. Social Development Secretariat, Mexico

Sotoudehnia F, Comber A (2010) Poverty and environmental justice: a GIS analysis of urban greenspace accessibility for different economic groups. In: 13th AGILE International Conference on Geographic Information Science, Portugal. http://agile2010.dsi.uminho.pt/pen/ShortPapers_PDF%5C128_DOC.pdf, consultado el 3 de junio de 2011

Sotoudehnia F, Comber L (2011) Measuring perceived accessibility to urban green space: an integration of GIS and participatory map. In: AGILE 2011. 18–22 April

Weber M (1946) From max weber: essays in sociology. Oxford University Press, New York

World Health Organization (WHO) (2012) Addressing the social determinants of health: the urban dimension and the role of local government, 47 pp. http://www.euro.who.int/__data/assets/pdf_file/0005/166136/UrbanDimensions.pdf. Accessed 16 Aug 2012

Zhang J, Wolch J, Bryne JA (2009) Planning for environmental justice in an urban national park. J Environ Plann Manage 52(3):365–392

Sustainability Skills for High School Graduates; Case Study in the University of Guadalajara

Ruth Padilla Muñoz and Teresita Serna Enciso

Abstract In the context of the Comprehensive Reform for Higher Secondary Education, and the creation of a High School National System in Mexico, the establishment of a Common Curriculum Framework was decided, within which basic characteristics that secondary education graduates should achieve, outlined by eleven generic skills, are gathered together. One of these skills, aims for the graduate to be able to combine knowledge, abilities and attitudes in order to contribute to sustainable development. With the purpose to measure the achievement level of this skill for high school graduates, and if such skills are trasfered to the surrounding where students are growing, a group of new students to the University Center of Tonala of the University of Guadalajara, located in Jalisco, Mexico was taken as a case study and a 62-question survey took place. In this article, we show the results of the survey. Results show that these young scholars have enough knowledge for a sustainable development, however, shifting knowledge to action is differentiated according to each student intrinsic characteristics and their surroundings.

Keywords Environmental sustainability · Higher secondary education
Skills and abilities

R. P. Muñoz (✉)
Department of the CUCBA, University of Guadalajara, Camino Ramón Padilla Sánchez 2100, ZP: 45200 Nextipac, Zapopan, Jalisco, Mexico
e-mail: ruthpadillam@hotmail.com

T. S. Enciso (✉)
Department of the CUCSH, University of Guadalajara,
1060 Jalisco Street, ZP: 44260 Guadalajara, Jalisco, Mexico
e-mail: teresitaserna@hotmail.com

W. Leal Filho et al. (eds.), *Sustainable Development Research and Practice in Mexico and Selected Latin American Countries*, World Sustainability Series,
https://doi.org/10.1007/978-3-319-70560-6_33

1 Introduction

For the past few years Higher Secondary Education (HSE) in Mexico has had great enforcement, all in a constant search to make this education level more effective and a high quality one. For that matter, the Comprehensive Reform for Higher Secondary Education (CRHSE), implemented in 2008, created the bases for the establishment of the High School National System (HSNS) and the development of a skills based Common Curriculum0 Framework (CCF). The February 9th, 2012 reform would result in a legislative recognition of enforceability and universalization for all Mexicans, which demands the Mexican government to look for alternatives in order to comply with the responsibility of the State to educate the population between 15 and 17 years of age, which pressures the support capacity of educational institutions.

The 2013 reform, aims to guarantee an efficient administrative design with the appropriate educational personnel. The last proposal for changes points out as an objective, to "teach students to think on their own and facilitate what they need to learn in order to move with a sense of security and trust in a changing and diverse environment" (SEP 2016, p. 8). In this frame, the educational institution is recognized as only guarantor of educational training, so the school is expected to interact with the environment, as mentioned in Roegiers (2016) "with the local and regional environment, but with the global environment as well" (p. 7); at the same time it encourages for a permanent participation with international movements and a better understanding of the world's political, economic, social and cultural functioning, with a local vision, but never losing the perspective of sustainable development.

Perspective that has its records in the school curriculum greening ever since the end of the seventies of the XX century, this approach was described in a synthesized way in González (2012), as education for life, with a strong ethical component and an active participation of problem prevention and solution that starts with "a vision of the environment as a whole (natural, human, ecological, political, economic, social, legal, cultural and esthetical); that will affect all areas of knowledge (transversality); with an interdisciplinary perspective and of the complex systems as well" (p. 17).

All along the path travelled by the consecutive education reforms, the concern has always been present to add an environmental component in the school curriculum. In the framework of CRHSE (2008), it is given a specific emphasis when it was announced as one of the eleven generic skills that all graduates of HSE should have. In the lights of the evaluation as a methodological tendency and mandatory action, in order to measure the scope of the project and to present results, this article aims to represent the level of achievement in this skill, within a group of students that have finished high school and are continuing with their school career. The case study was on a group of new students of the University Center of Tonala of the University of Guadalajara, because this is the only public education multidisciplinary center located in Guadalajara's urban area in the state of Jalisco, Mexico.

2 High Secondary Level in Mexico

The transformation process of the national education system, initiated in 2007–2012 of the federal administration, had as some of its basic goals for the HSE: an increase of quality, improvement of relevance, and to provide organization, articulation and consistency.

The above lead to a series of actions, like the creation of the High School National System (HSNS) among others, that continue to have repercussions in all public and private HSE subsystems, along with the permanence of the basic concepts of the CRHSE in the National Development Plan 2012–2018 and in the Education Sectoral Program for the same period.

With the creation of the HSNS, the heads of the reform are determined; the first one being the establishment of a Common Curriculum Framework CCF), that pretends to gather together what it calls "shared final performance", as well as a series of common components to all subsystems and HSE institutions, starting from a skilled based approach, expressed in the graduate profile, with the understanding that: the skills approach considers that knowledge by itself is not the most important thing, but it is the use that is given to it, in real life, personal, social and professional situations. In this way "skills need a solid base of knowledge and certain abilities, which are integrated for a same purpose in a particular context" (DOF 2008, p. 32).

With the proposal of the CCF, concepts of generic and professional skills are resumed, and the concept of disciplinary skill is added to describe the profile of the high school graduate, they are reviewed with the 2013 reform and the aim is to be able to complete them with the revision and updating actions done in 2016; in order to "ensure their alignment to achieve a graduate profile, and particularly to establish clear parameters of graduality during their development" (SEP 2016, p. 281).

3 Skills Assessment

Since 2008, year were the CCF was approved, there has been different approaches for skills assessment, nevertheless, results obtained until know do not enable knowledge regarding if the graduates actually have the whole capacities involving the skills: knowledge, abilities, attitudes and values, and if they could transfer them effectively into different contexts, in particular those students that are being accepted or want to go through higher education; that depends largely, on the success or failure during their time taking the chosen plan of studies. Regarding the matter, Le Boterf (in Perrenoud 2007, p. 11) determines that even if the skills described are not the skills by themselves "… they mobilize, integrate and inter-relate such resources, for a determined situation" by the use of complex mental operations that are expressed in concrete actions facing a certain situation (located action).

We can find background information regarding the assessment of environmental skills in studies such as Kollmuss and Agyeman (2002), which describes historical records of the theoretical models that have been built to explain the variables and the relationship between pro-environmental knowledge-conduct. These authors point out the oldest and simplest model of the 1970's, which is based on a linear progression of environmental knowledge. This rationalistic model assumed that educating people regarding environmental issues (providing them with knowledge) will automatically drive them towards an environment attitude, and in their case, a pro-environmental behavior (Kollmuss and Agyeman 2002, p. 241).

Kollmuss and Agyemman point out that during the 1980s, models were built based on the theory of rational action; the most influential in social psychology Ajzen and Fishbein, will set the bases for the model suggested by Hines, Hungerford and Tomera, called "responsible environmental behavior model" where situational factors are another trigger of pro-environmental behavior. They also mentioned that later on other models will be provided, based on altruism, empathy and pro-social behavior, such as Eisenberg and Miller, and the one proposed by Stern. Another group was created by models that recovered sociological characteristics and a few psychological ones to explain pro-environmental behavior (or the lack of it), the one standing out was Fietkau and Kessel, and the one for Blake, "who starts to differentiate single behaviors, responsibility and practicality" (Kollmuss and Agyeman 2002, p. 247).

After this analysis, Kollmus and Agyeman (2002), propose their own model where internal factors stand out (knowledge, attitudes, feelings, fears, motivations), demographic and exterior ones (institutional, economical, social and cultural) that determine environmental behavior. It is acknowledged that there is only one model able to explain the gap between knowledge and pro-environmental behavior.

Pro-environmental behavior will be a consequence of skills achievements related with sustainability, quoted in the graduate profile (CCF): "Contribute to sustainable development in a critical way with responsible actions". Attributes for such skill are the following: (a) Assumes an attitude that favors the solutions of environmental problems in a local, national and international scope; (b) Acknowledges and understands biological, economic, political and social implications of environmental damage in an interdependent global context and (c) Contributes to reach a balance between short and long-term interests in relationship to environment.

In regards to Basic Disciplinary Skills, we explicitly found the ones related to sustainability in the field of Experimental Science, and they are:

- Establishes the relationship between science, technology, society and the environment in specific social and historical contexts.
- Justifies opinions on the impact of science and technology in an everyday life, assuming ethical considerations.
- Analyses general laws that rule the functioning of physical environment and values human actions of environmental impact.
- Associates the levels of chemical, physical and ecological organization of living systems.

- Implements security regulations in the handling of substances, devices and equipment, to carry out everyday activities.

Nevertheless, thanks to the transversality of generical and disciplinary skills, they can be located indirectly in other skills of different fields of knowledge. Follow up, scope and assessment of skills acquired, is the main objective of this research, which brings us an overview of which are the knowledge, attitudes, motivations and actions of the student as soon as the graduate finishes school, so this research contributes to delimit the real transference from theory to practice.

Now, the federal education authority considers the implementation of the *New Education Model*, starting the school cycle 2018–2019. In order to describe the graduation profile of each school level, it determines eleven areas in which the mainstay of socioemotional abilities and life plan is added, included implicitly or explicitly in the transversal skills of precedent education models and in the current one. For the area where environmental care is mentioned, the HSE hopes that at the end of the studies, the student will understand the importance of sustainability and will assume a proactive attitude to find solutions; to think globally and to act locally; valuing the social and environmental impact of scientific advances and innovations (SEP 2017).

With such proposal, the aim is to articulate in the Mexican curriculum, the "comprehensive vision curriculum" described by the UNESCO (Roegiers 2016), which perceived three phases: inclusive education, holistic education and the no fragmentation education (willingness to articulate the different dimensions of the curriculum); based on generic skills (that provide the potential of the "knowing how to be"), disciplinary and professional ones (that provide the "knowing how to do"), in addition to situational skills, that the IBE determines as the "knowing how to act" in particular situations of life; together with a set of *life skills* that determine the graduate student's actions.

4 Research Methodology

The research starts from a quantitative approach with an explainatory reach (Hernández, Fernández and Baptista 2010, p. 85) and it has the objective to learn if newly registered students at the University Center of Tonala have the skills related to sustainaibility that are pointed out in the CCF and if such skills are transferred to their own surroundings. To analize the level of skills achievement and environmental behaviour of a student that has finished High School and continues his education, a measuring instrument was designed.

The survey has 62 questions, divided in two sections. The first one aims to obtain general data of the student, and the second section is subdivided in three: one represents 40 statements to be answered using a Likert scale, according to a level of agreement and disagreement, whith proenvironmental attitudes; another one has 8 questions with concrete answers, positive or negative activities that implicate actions for environmental care, and a third one has 14 questions that present the Likert scale as well, with the aim to measure how often certain actions, that favors or damages the environment, are taken. In order to analyze the comprehension of instructions and questions, is worth mentioning that a pilot test with 45 students was performed, as well as an assessment from experts on the educational and environmental field, affiliated to the University of Guadalajara; at the end a 62-question survey, from the initial one of 84, was achieved.

For the layout of each question and to facilitate their analysis and assessment, questions were initially grouped into three large skill components: Knowledge, abilities and attitudes (values). Even though this classification allowed us an initial outline, the Holton Model was used to assess transference of knowledge: *Learning Transfer System Inventory* (LTSI). This enables to learn about the factors that contribute for transference success after education and to identify possible weaknesses in a developing surrounding. Holton (Padilla 2013, p. 133), suggests that the LTSI is added because of the 16 factors that influence transference of knowledge. For this research, the 62 statements are grouped in only ten constructs (Willingness to learn, Achievement of implementation, Motivation for transference, Expectations of transference, Expectations regarding result achievement, Resilience for transference, Positive results, Chances to implement, Ability for transference and Acquired knowledge (see Fig. 1).

In addition to the ten constructs of LTSI, four phases are proposed to specify level of achievement: (A) Knowledge and understanding: determines the basic general level of acquired knowledge, determines learning willingness, as well as the

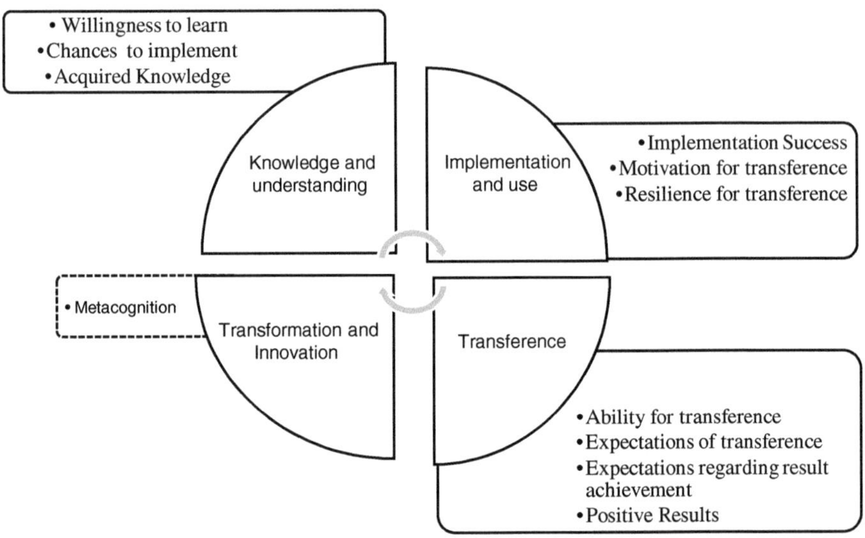

Fig. 1 Phases for adopting skills for sustainability. Own figures based on Holton (2005) Model dimentions: *Learning Transfer System Inventory* (LTSI)) and levels suggested by Padilla, Becerra and Serna (2014)

vision of common use of basic knowledge; (B) Application and use: refers to a conscious use of knowledge acquired regarding sustainability, starting some spontaneous experiences but being discarded by individual if found out complex or problems appear that cannot be dealt with. (C) Transference: phase in which there is the capacity and disposition to implement knowledge in different contexts through time; knowledge is applied efficiently in a timely manner, in order to achieve sustainability, and (D) Transformation and innovation: phase that requires a credible recovery of what was learned and development of new knowledge to create new metacognitive processes and attain innovative and proactive practices within the frame of sustainability.

5 Analysis Results

5.1 Students Characterization

The most common method used by researchers to prove reliability is the Alfa Cronbach index, which allows to estimate the average of correlations between items of the instrument, so it was established that in order to consider it credible, a minimum or higher of 0.70 should be obtained. For this study, an Alfa Cronbach of 0.913 was collected, which allows us to confirm that the instrument is reliable.

The survey was done to 556 students out of 810 (68.6%) that got in the University Center of Tonala (Cutonala) of the University of Guadalajara during the school cycle 2015B (Chart 1). The average age is 19.9 years old. According to statistic tests of self-reliance, it was found that differences in students' answers regarding age were not significant, but they were regarding gender or study programs being taken (with values Chi-Square, minor or same as 0.05). So, we see the existence of answer dependency on nine items, according to their choice of studies, and four by gender of the surveyed students (see Chart 2). It can be observed that these items do not make reference to specific knowledge but it makes reference to certain conducts or concrete actions.

An example of the differences founded according to the study program being taken, can be noted in the statement of *I actively participate in decision making, within associations and organizations, in regards to environmental care*, where the average answer of 75% of Public Health, Handicraft Design, Medicine, Nutrition and Energy students, answered that they agree or partially agree with it, while Business Administration, Computer Science and Accounting students, expressed in the same magnitude, to be partially or totally in disagreement with such statement. In the case of gender, the answer to item *I am a consumer of technological trends* stands out, 68% of women answered to fully agree, differing from an 80% from surveyed males that declared the same thing.

Chart 1 Surveyed students' layout, according to studying degree

Degree	Gender		Total	Representation (%)
	Female	Male		
Public health	22	10	32	5.8
Law	40	27	67	12.1
Business administration	44	23	67	12.1
Computer science	1	28	29	5.2
Public accounting	26	32	58	10.4
Handicraft design	15	7	22	4.0
Energy	9	38	47	8.5
Liberal studies	4	5	9	1.6
Gerontology	22	3	25	4.5
Art History	9	6	15	2.7
Medicine	44	23	67	12.1
Nanotecnology	14	53	67	12.1
Nutrition	36	15	51	9.2
Total	286	270	556	100.0

In answers to concrete questions of acquiescence or negation, (see Chart 3), it was found that students consider that the school where they are studying (Cutonala) has a healthy environment, a contrasting perception with the complex reality of this campus, since it is socio-environmentally located in an area of low income and with a "biodiversity formed by the wetlands of Las Rucias and El Cajon dams, but on the other hand, an area fully contaminated because of the presence of brick companies and water pollution" (Guerrero, Arellano and Padilla 2016). Projects such as the Environment Sustainability Comprehensive Program (ESCP) implemented since 2014 by school authorities, contribute to have this optimistic vision, by developing skills for sustainability through environmental education strategies in the school community.

To learn about prioritization of students in regards to high importance issues related to sustainability, the following question was asked: *Indicate from 1 to 5, from bottom to top, which you believe are priority issues for sustainability in the world.* Out of the 5 possible options, answers were given in priority order: (a) Reduce unemployment, followed by (b) Poverty, then (c) Human Rights Protection, the fourth relevant problem is (d) Environmental care, and at last (e) Access to education. Even though only few of the many problems affecting society at this moment were mentioned, the environment issue ranked fourth and far placed, since the students think economic issues are more relevant; this is explained by the country current conditions, with unemployment rates of 4.5% and with 46.1% of the population in poverty, while in Jalisco is of 35.4% (INEGI 2016).

Structural conditions are urgent for students, but when being questioned *Which areas of knowledge do you relate to sustainability?* the given options were: (a) ecology (b) economy (c) education (d) politics and (e) all the above; 70% of

Chart 2 Self-reliance test results, according to studying degree and gender

Question	Pearson Chi-squared Test (sig)
Studying degree	
If I see garbage on the street or at the school, I pick it up and put it in a garbage container	0.047
I reuse paper	0.004
I separate recyclable materials to protect the environment	0.040
I have planted trees	0.014
Climate change is a serious matter because it is an environmental, social and economical probrem	0.001
I actively participate in decision making, within associations and organizations, in regard to environmental care	0.004
I believe that the career I am studying is related to the responsable use and care of the environment	0.004
I am interested in learning about the environmental impact of products I use, from their manufacture, packaging, use, and recycling or reuse	0.038
I find fulfilling to implement what I have learned at the school to improve environmental conditions	0.048
Gender	
I am willing to take care of the trees I planted	.002
I am a consumer of technological trends	.002
The use of vehicles is one of the factors with more impact on environmental pollution	.005
I do home composting	0.00

surveyed students marked the last one. Although it is recognized that sustainability is a term with multiple meanings, that was institutionalized and represented with concrete actions, ever since the Brundtland Report and the Agenda 21 proposed in 1992 (Rauch 2002), this study shows that students know four key components of sustainable development.

5.2 Skills Adoption Level for Sustainability

To determine the skills adoption level for sustainability, results were presented according to the four described phases, taken into consideration a Likert scale from 1 to 5 (that goes from total disagreement to total agreement, respectively), in 54 questions. The followed criteria for result interpretation was to set answers by intervals. Highest answers or the same 4, are considered acceptable, and from 3.9 or less, are considered non-acceptable for the partial achievement of the analyzed characteristic.

Chart 3 Answers to items, according to first phase: knowledge and understanding

		Average	S.d.
1	The right to an adequate quality of life, health, food availability and safe water, are human rights	4.72	0.896
4	I believe that the career I am studying is related to the responsible use and care of the environment	4.67	0.732
7	I know the importance of waste separation and what is it about	4.84	0.650
8	I am aware that not all solid waste is garbage	4.51	0.712
11	Batteries should not be thrown to the garbage because they contain acids	3.86	1.199
14	I accept that the mass consumption of wood, has deteriorated and deforested hundreds of hectares	4.65	0.717
16	Throwing oil direct into the drainage system is harmful for rivers	3.94	1.010
17	The method of extraction unconventional gas or oil found stuck in source rock is harmful for the environment	4.68	0.861
18	Renewable energy contributes to improve environment in my community and in a global scale	4.20	0.900
20	I know how to save energy in my cellular phone	3.76	0.954
24	Paper recycling is good for the environment	4.62	0.783
34	Before entering high school, I understood the importance of environmental care to improve my life and that of my community	4.45	0.722
35	I am able to explain repercussions in the environment due to advancements of science	4.38	0.835
38	Climate change is a serious matter because is an economic, environmental and social problem	3.96	0.942
40	Polluting emissions to the atmosphere of fumes and combustion gases are environmental crimes	4.40	0.840

In the initial phase that refers to adoption of knowledge and understanding, questions were posed to analyze the willingness to learn and the opportunity of use to implement knowledge. As seen in Chart 3, out of the 15 items, the average answer was found to be higher than 4, except from four questions: 11, 16, 20 and 38; which represent the greatest standard deviation (S.d.), which shows that some of the everyday knowledge is unfamiliar to the surveyed.

To analyze the adopting and usage phase, students were questioned on how often they perform some actions in favor of environmental care (Chart 4), it can be seen that there is great scattering of answers, considering that the simplest everyday activities are the ones they really do; such as switching of lights when they are not needed, respect established rules in public areas, or avoiding throwing garbage. This can be outlined by the social commitments they could acquire to undertake more relevant actions which are not been carried out by the surveyed students.

Regarding environmental education, 73% of students answered not to have attended extracurricular classes, but when questioned about some of everyday activities such as recycling, 86% answered that they do it, as well as planting trees.

Chart 4 Frequency of actions taken in favor of the environment

Question		1	2	3	4	5
49	I sort solid recyclable materials in order not to affect the environment	9.2	41	25.2	19.8	4.9
50	I switch off lights in my home when I don't need them	0.7	3.1	9.5	22.7	64.0
51	I place batteries in special containers for their recycling	14.7	19.6	17.3	17.8	30.6
52	I throw batteries to the garbage when they are of no use	57.9	18.7	10.1	5.9	7.4
53	I place a bucket in the shower to reuse cold water after I bathe	39.7	20.9	11.0	7.4	21.0
54	I respect rules for entering a historic building or a natural reserve in order to take care of our cultural and natural heritage	0.5	3.4	9.9	11.7	74.5
55	I recharge phones and mobile devices during night time and unplug them in the morning	23.6	26.6	15.5	15.6	18.7
56	I unplug all electronic devices when they are not in use	17.1	26.8	17.1	15.5	23.6
57	I reuse paper	15.6	33.3	18.7	17.1	15.3
58	I do not throw garbage on the street	0.9	1.3	5.2	17.1	75.5
59	If I see garbage on the street and in the school, I pick it up and put it in a trash can	12.8	34.7	20.0	21.6	11.0
61	I take advantage of rain water by reusing it for other activities in my home	33.5	30.6	12.8	13.1	10.1
62	I feel encouraged to take actions to improve the relationship between society and nature	4.7	21.2	22.7	26.8	24.6
63	I take electronic devices I use no more, to recycling deposits	29.0	23.7	28.2	10.8	8.3

1: never, 2: hardly ever, 3: occasionally, 4: almost always and 5: always

Out of 66% of students that answered this activity, 42.5% do it because they like tree-lined surroundings and only 13.2% of students show a strong concern for the environment; these students could be considered in a transference phase because they have the knowledge and the take it into practice in their surroundings. It is distressing to see the low percentage of students that take this practices into actions, after having gone through basic education (9 years) and middle education (3 more), where such skills for sustainaibility, and particulary those for environmental care, are present in all education levels in Mexico.

To measure transference capabilities, as well as expectations on performance and results that could be achieved to reach a sustainable development, 17 questions were asked (Chart 5). Even though students show positive attitudes and broad expectations that the concrete actions taken will be reflected in a sustainable development of the school and their surroundings, they never come to practice, as mentioned before.

Finally, 8 questions were asked to analyze if some students have gotten to the transference and innovation phase, meaning, they have come up with some personal or collective practices focused on innovation and generation of new knowledge, as

Chart 5 Answers to items, according to third phase: transference

Question		Average	S.d.
6	Natural resources are everyone's responsibility	3.43	0.996
9	I consider important saving electric power	4.41	0.854
12	I am willing to plant trees in my community and in my school	4.77	0.704
13	I am willing to take care of the trees I plant	4.53	0.779
19	I am a consumer of technological trends	4.64	0.724
22	I believe that the use of non-motorized transportation improves air quality in the city and preserves our health	4.52	0.756
25	I use garbage containers located on the street and at the school	4.77	0.639
26	I urge my family and friends the reuse of crystal and plastic bottles	4.79	0.617
29	I implement what I learn at school in regards to environmental care, to upgrade the surrounding where I am growing	4.2	0.871
30	There are programs and activities in my municipality where I can implement my knowledge	4.35	0.759
31	I do home composting	3.60	0.865
32	I find fulfilling to implement what I learn at the school to improve environmental conditions	3.14	1.206
33	I believe I have the abilities to create/transform/act in behalf of the environment	4.34	0.779
36	I see the diversity present in my group as an asset for changing our surroundings	4.01	0.853
37	There are extracurricular activities in my school where I can implement my theoretical and practical knowledge	4.33	0.802
39	I believe the issue of sustainability is relevant	4.66	0.744
41	Disappearance of endangered species is inevitable	4.56	0.623

Chart 6 Answers to items, according to third phase: transformation and innovation

Question		Average	S.d.
2	I have done an innovative action in favor of the environment of my community	4.72	0.896
3	I am willing to make the necessary changes in my habits, on behalf of the environment	3.78	0.958
5	I actively participate in decision making, within associations and organizations, in regards to environmental care	4.22	0.927
15	I have planted a tree because the forest mass retains oxygen and eliminates contaminants	4.73	0.695
27	I am interested in learning about the environmental impact of products I use, from their manufacture, packaging, use and recycling or reuse	4.04	0.91
28	I share with family and friends the information I have about environmental care	4.15	0.855
60	A better surrounding where I live, means a higher quality of life for me and my family	4.65	0.740
62	I feel motivated to take actions that will improve the relationship between society and nature	3.46	1.204

well as using and acting in ways that will decisively and determinedly benefit sustainable development, (Chart 6). Favorable answers to participate in the transformation were found, even though there is a large scattering regarding motivation and willingness (questions 3 and 62 with a lower average of answers and a larger standard deviation).

6 Conclusion

Assessment of the level of skills achievement of higher secondary education graduates is still an emerging and unclear field. This project presents the assessment of generic and situational skills that refer to the adoption of knowledge, capacities, abilities, values and attitudes in order to reach sustainable development, always aiming for enforcement of relevance and viability, as well as validity and reliability of the instrument consisting on 62 questions.

It was observed that students have specific knowledge regarding climate change, human actions affecting the environment and destroying nature, as well as activities they should take to take care of their surroundings. Transference to concrete actions happens in a differentiated manner, because there is a large gap between what they say they know and how they use their knowledge, this gap increases according to the professional studies taken (career being studied) and in some cases, by gender role, which indicates some active involvement conditions in society. Therefore, while the reach of the skills described in the CCF creates knowledge and skills needed to think, act and work, it needs to be blended with basic and ordinary actions in order to get to metacognitive processes and an active and innovative participation, in the endeavor to reach sustainable development.

It is challenging to determine which characteristics lead to an appropriate environmental behavior. Among the different proposed models by the theoreticians, there is not one that is able to explain the gap between knowledge and action, especially when the goal is to associate learning in school and behavior of individuals, once they have finished their studying cycle. In the case of Higher Secondary Education in Mexico, to determine the level of achievement, internal and external factors that influence conduct need to be analyzed, such as motivation, expectation to implement knowledge, legal and attitudinal considerations as well as subjective norms that lead to a conduct intention that will finally translate into a permanent conduct. Giving the existing complexity to determine if a skill was reached or not, it is necessary to continue the creation of research material that will allow a better assessment of the field.

References

Federal Official Gazette (DOF 2008). Agreement 442 within which is established the High School National System in a diversity frame

González EJ (2012) Greening of school curricula: brief count of an eventful history. Professors. Curriculum and teacher training magazine. vol 16, un, 2. University of Granada, Spain

Guerrero AA, Arellano MA, Padilla R (2016) Comprehensive program of environmental sustainability, University Center of Tonala (PISACUT). Presented lecture at the First National Congress of Environmental Education for sustainability

Hernández R, Fernández C, Baptista P (2010) Research methodology. McGraw-Hill, Mexico

Holton EFIII (2005) Holton´s evaluation models: new evidence and construct elborations. Adv Developing Human Resour 7(1):37–54

INEGI (2016) National institite pf statistics and geography. http://www.inegi.org.mx/

Kollmuss A, Agyeman J (2002) Mind the gap: why do people act environmentally and which are the barriers to pro-environmental behavior? Environ Educ Res 8(3):2002

Padilla R (2013) Teacher training for higher secondary education. Design and assessment of a model. University of Guadalajara, Mexico

Padilla R, Becerra M, Serna T (2014) Technological capabilities in high school. University of Guadalajara, Mexico

Perrenoud P (2007) Ten new skills to teach. Graó/Colofón.Editorial, Mexico

Rauch T (2002) The potential of education for sustainable development for reform in schools. Environ Educ Res 8:43–51

Roegiers X (2016) Conceptual framework for skill assessment, N° 4 of the serie de la serie Fundamental and actual issues of curricula and learning. Looked up on February 1st 2017 in http://www.ibe.unesco.org/sites/default/files/resources/ipr4-roegiers-competenciesassessment_spa.pdf

SEP (2016) Educational model 2016. Teaching approach of the educational reform. Secretary of Public Education, Mexico

SEP (2017) Purposes of education in Mexico. Available in: http://www.gob.mx/cms/uploads/attachment/file/114503/Los_Fines_de_la_Educacio_n_en_el_Siglo_XXI.PDF

The Perception of Urban Insecurity and Its Implications for Sustainable Development

R. Villasis Keever and L. Arista Castillo

Abstract The phenomenon of the increasing violence has intensified the perception of urban insecurity, generating diverse manifestations ranging from social, economic, and institutional categories and those related to the urban environment. The concept of urban safety and its relationship with the quality of life has awakened interest and vision about the sustainability of cities, as seen in the sustainable development objectives of the United Nations Agenda 2030 which seeks to promote peaceful and inclusive societies, significantly reduce the forms of violence, and guarantee equal access to justice, among others. The study of the perception of insecurity is very important for the sustainable development of the cities as the perception of insecurity affects the coexistence and implications on the quality of life of the inhabitants of vulnerable areas because it constitutes an obstacle to the sustainable development of the city. According to the above, this research deals with the perception of insecurity to intra-urban scale in the city of San Luis Potosi, in order to know the current situation and trends of urban insecurity, and identify vulnerable areas as a basis for the design of preventive policies and urban design. The method applied is the model: Generation, Demonstration and Attention (GMA); It includes quantitative and qualitative indicators that make up the proposed information system and register according to the census information with reference to scale AGEB polygons, used as a unit of space. Different patterns of perception were obtained that exhibit a different reality in each sector of the city as data obtained do not correspond with official statistics on urban insecurity. Therefore, it may be concluded that there are gaps in the levels of inequality and attention to safety that reflect conflicts of social coexistence and generating areas of increased incidence of violence threatening the urban sustainability.

Keywords Urban insecurity · Vulnerability · Citizen perception

---

R. Villasis Keever (✉) · L. Arista Castillo
Autonomous University of San Luis Potosí, San Luis Potosí, Mexico
e-mail: villasis@fh.uaslp.mx

L. Arista Castillo
e-mail: leticia.arista@uaslp.mx

W. Leal Filho et al. (eds.), *Sustainable Development Research and Practice in Mexico and Selected Latin American Countries*, World Sustainability Series,
https://doi.org/10.1007/978-3-319-70560-6_34

1 Introduction

Understand the fear of crime as a factor that affects the quality of life of inhabitants in urban areas has been the subject of studies based on several perspectives, ranging from studies of criminology, social work, psychology, government and more recently environmental and spatial, generating important debates on situations of high levels of crime in the cities, with a diversity of theoretical positions. However, the official data of insecurity do not coincide with the citizen perception, in this sense the perspective of the citizen deserves review by researchers and civilian groups of the problems of insecurity that are realized by surveys as part of the knowledge and analysis of a non-governmental vision.

The cities of Latin America and the Caribbean have a high concentration of inhabitants, as the percentage of the urban population that lives in them (2010) is 80%, simultaneously economic activities have been concentrated in these areas since 60 and 70% of the gross domestic product is produced in the cities (IBD 2011). Despite these factors, the escalation of urban violence has spread in the cities showing a greater advance in rates of violence (Carrión 2005:30). As a response of this, the society shows a perception of urban insecurity much greater to those reported by official institutions. Analysis on the citizen perception enables us to see that there are urban areas of increased vulnerability, and others seeking to provide security through private residential urbanizations of high costs that separate and privilege certain areas of the city.

Since mid-sixties of the twentieth century, cities—at the international level—began to experience an accelerated growth of delinquency. Progressively, the choices about where to live, where to work, where to register children and where to buy, were becoming decisions that took into consideration the city's perception of delinquency (Polese and Stren 2000). Urban insecurity in Latin America has increased in the last decades, to the point of conditioning the daily life of the inhabitants in the cities, modifying their habits of behavior, mobility, and consumption, that impacts their way of living, and appropriating the city.

Par example, studies realized in Montevideo, argues that the neighborhoods with the highest rates of population growth are those with the highest probability of high crime rates. This demographic growth is linked to processes that determine the social composition of neighborhoods and crime rates (Retamoso and Corbo 2003). It should be noted that there is an important debate about situations that favor high levels of crime in cities, with a diversity of theoretical positions. Basically, citizen security is now conceived as a key element in the quality of life precept in large urban areas (Sillano et al. 2006).

The studies of the perception of insecurity have great importance for the sustainable development of the cities, because the perception of insecurity affects the coexistence and implications on the quality of life of the inhabitants of vulnerable areas because it constitutes an obstacle to the sustainable development of the city, according to the Inter-American Bank of Development (2011), insecurity is a factor which constantly threatens the well-being of the citizens, raises the economic costs of individuals and enterprises, the peaceful co-existence of a democratic society, destroy the virtuous share capital and sometimes encourage parallel economies typical of organized crime, with all of which erodes the trust of institutions, especially those responsible for ensuring the rights and duties of individuals. (IBD 2011).

So, the objective lack of security and perception of insecurity impacts negatively on social life and deteriorates the everyday environment, being the public safety one of the factors that most influence the quality of life and sustainable development in cities.

2 Conceptual Framework: The Paradigm of the Insecure City

In the production of the unequal, fragmented and exclusionary city, it has been considered as a necessity to understand the fear of crime, not only from a purely criminological point of view, but as part of a larger theme, that is, the perception of insecurity that includes a series of "environmental, labor and citizen participation, among others" (Ospina 2006). In the last two decades, there has been a significant increase in insecurity in Latin American cities, the general impoverishment of the broad strata of the population of the cities, the generalized increase of the middle and lower classes and the growth of insecurity areas are associated with crime and violence (Auyero et al. 2014; González 2011; Velázquez and Giraldo 2009; Petrella and Vanderschueren 2003). In addition, with the increase of urban violence, the citizen develops a perception of individual and collective insecurity.

Recent studies on the subject identify the role of urban inequality and exclusion as a key factor in explaining socio-demographic and cultural processes linked to the development of unsafe cities, stimulating profound changes in scheduling patterns and the use of public space as strategies for families to develop their daily lives, as they stimulate the emergence of situations of vulnerability and social risk, which particularly affects certain groups, such as children, young people, women and older adults (Mingione 1994; Veiga 2004). On the other hand, Arriagada and Godoy (2000) identify the age, as a central factor to explain the local perceptions of poor people related to insecurity at the micro scale level.

It is important to emphasize that there is an significant debate about the situations that favor high levels of insecurity or delinquency in the cities (Navarrete et al. 2015; Moreno 2011), there are four basic aspects that can be highlighted: (i) decomposition of the social capital; (ii) social inequality; (iii) socio-spatial exclusion and (iv) characteristics of the urban environment. Some indicators, in which these conditions are manifested, are related to demographic and socioeconomic variables of the population [structure of age groups, economic dependency, employment and underemployment, etc.], living conditions of the dwelling and its environment [overcrowding, quality and accessibility of urban services, etc.] or levels related to the conditions of income, education and health. A key aspect to be considered is gender equity indicators [misogyny, machismo, women's participation in household headship and labor markets, among others].

At the present time, cities generally observe severe intra-urban segmentation between prosperous areas and lagging areas. This differentiation has generated a controversial urban structure, with direct consequences on the living conditions of the population, bringing with it difficulties in access to services, loss of identity of neighborhoods and colonies, and daily manifestations of urban insecurity and violence. This phenomenon has become a serious problem located, both in the big cities and in the medium cities of the world.

At the World Urban Forum in Vancouver (2006), it was recognized that urban security is an increasingly important global issue. In some studies, conducted in urban centers of northern countries with more than 100,000 inhabitants, it was observed that crime grew between 3 and 5% annually between 1970 and 1990. In Latin America and the Caribbean, the youth crime urban and violence has increased.

Since the mid-1970s, cities experienced accelerated growth in delinquency. The evidence gathered in several countries in the region also points to a significant increase in crime, victimization, and citizen insecurity (UN-Habitat 2011). However, in the 1990s, due to prevention and reinforcement policies, the urban crime rate has begun to stabilize, except for youth crime (12–25 years), and in particular that of minors (12–18). The criminality has become more violent and the age of onset has decreased from 15 to 12 years (UN-Habitat 2012).

The high inequality observed in Latin America, in a context of increasing wealth and modern consumption that characterizes the high socioeconomic strata generates the structural conditions that underlie the serious wave of urban crime and violence. And if we add the increment in the isolation of the poor and the segmentation of social opportunities, this reinforces the problems of the marginal population, such as family violence, school dropout, vagrancy, and drug addiction.

In this way, unlike the previous stages, in which the city and its urban sectors were defined from the income of the inhabitants or other forms, they are described as divisions brought by the conscious acts of their residents and those who hold power (Marcuse 2006). Paradoxically, it is the poor neighborhoods where urban vulnerability is most acutely to processes of social, economic, cultural, physical and urban exclusion. This set of dimensions constitutes a series of processes that deteriorate the living conditions of the excluded groups, and maintain or increase the advantages and inequalities.

Beginning with the planning of urban development, it is necessary to review the paradigm of safe cities; this implies looking at the vulnerable and critical neighborhoods since the exclusion of the poor neighborhoods of the city has created serious difficulties so that its inhabitants can live in the city in a safe way. Certain neighborhoods or colonies require specific attention because their inhabitants feel violated in their rights to appropriate the city and to enjoy security in it.

There are vulnerable spaces to violence that are expressed through the emergence of violent groups, especially youth gangs, and less demonstration of a positive and proactive community development model. Perpetuating a paternalistic model with The State and without link with the rest of the city (Saraví 2005).

Some of the indicators that allow evaluating certain urban contexts with risks of insecurity, violence, and criminality, refer to the demographic and socioeconomic characteristics. The concentration of some of these characteristics permits identifying urban sectors or areas with a greater degree of vulnerability in relation to these problems. Some population groups are also particularly vulnerable to crime, either as potential victims or as potential offenders.

3 Model of Generation-Manifestation-Attention. Methodological Approach

The need for safer cities is a constant worldwide. Starting in 2015, the Millennium Development Goals (MDG) were transformed into the Sustainable Development Objectives (SDG) within the framework of the United Nations Agenda 2030. One of the challenges of the SDG is the participation of the different actors, especially civil society. This work is focused on the 16 Sustainable Development Goal: *Promote peaceful and inclusive societies for sustainable development, provide access to justice for all and build effective, accountable, and inclusive institutions at all levels.*

Following the proposal of the Organization for Economic Co-operation and Development (OECD) on a system of pressure, state and response, the GMA Model of Generation, Manifestation and Attention was developed in 2010 by the University of Caribe based on the United Nations Human Settlements Programme; indicators such as those set out in the Un-Habitat Safer Cities Program were incorporated into its 2008–2013 strategic plan, and the Un-Habitat urban sustainability indicators and of course all the above classified according to the categories of understanding of violence of the World Health Organization (WHO), all of them were the base elements of the GMA model (González et al. 2011).

It is important to mention that with the increase of urban violence and delinquency, an individual and collective perception of insecurity is developed in the city, not always objective; this sense of insecurity has caused the abandonment of entire neighborhoods, the development of closed housing clusters and the stigmatization of areas or population groups. Certain international researches establish various categories and types manifestations of urban violence as a social, institutional and those related to the physical environment.

The approach that underlies this research highlights one of the least explored fields in the empirical studies developed in Mexico regarding the relation of delinquency with urban development. This article talks about the perception of urban insecurity from the citizen perspective, through applying 1200 surveys to an open population of adults between 18 and 65 years of age, and data from official sources at local, and national level. The use of georeferenced systems as a tool for the spatial location of the affected urban areas, and explores the socio-spatial and socioeconomic conditions which operates as an explanatory factor in the phenomenon of the perception of urban insecurity.

The objective of the citizen security observatories in Mexico—created since 2010—is to know the current situation and trends of security in the cities of Mexico, and in this case, the city of San Luis Potosí. The census information at AGEB scale is used as a space unit (Basic geo-statistical urban-area), made up of the San Luis Potosí Citizen Security Observatory (OCS-SLP) and the perception of insecurity derived from the survey conducted in the period 2011–2014. The method used for observatories was based on the model GMA which identifies several key factors that generate urban insecurity composed of three subsystems:

(a) The Generation subsystem: composed of the set of variables accounting the human activities, of the society and the individual at different levels, identifying the possible causes of the phenomenon [G] to be studied [28 indicators and 34 variables].
(b) The Manifestation subsystem: is composed of variables about the accountability of the events in which the lack of security materializes, understood as the manifestation [M] of the phenomenon in reference [21 indicators].
(c) The Attention subsystem is the response of the State [population, territory and government] to the society, related to the events recorded in the previous subsystem. This subsystem [A] records and reports general or sectorial policies, programs, projects or institutional actions about prevention or correction [8 indicators].

In the GMA model, socioeconomic data are analyzed to characterize social categories, age groups of the population and vulnerable areas in socio-spatial terms, and their relationship with the indicators of the GMA Model.

4 Context and Results of the Citizen Perception of Insecurity Survey

Through the implications and analysis of the perception and victimization surveys, it is possible to evaluate subjective factors such as the sense of security that the individual constructs in scenarios in which he/she develops his daily life [public space, transportation, and neighborhood] and deepen the characterization of criminal patterns and the level of involvement in the population. For this reason, these tools are used to complement the official information produced by the citizen's complaint and the follow-up of cases in the judicial system. Thus, they become an additional input for the design and implementation of public policies, specialized in reducing crime and in intervening factors that affect the perception of citizen security.

An important element derived from the studies of citizen perception about insecurity is related to new factors or factors unpublished in recent research on the subject, from which the behavior and trends of insecurity and violence are analyzed, in particular with a "safe habitat" (Un-Habitat 2012). The relationship between the perception of insecurity and its sequels with morphological, structural and scale characteristics of the city constitute a fundamental vein for understanding the complex interaction between the phenomenon of the urban environment and the dynamics of risk factors in the city.

In the mid-1960s, two experimental surveys were conducted in households in the United States and England, in order to know if any household member had been a victim of an offense in the previous year and, if so, whether it had been reported. In this way, these surveys represented an alternative mechanism through which information could be accessed that the official sources did not report through the complaints.

In Latin America, countries such as Chile, Venezuela and Argentina have developed surveys of this type. The International Crime Victimization Survey (Icvs) is the one with the greatest coverage. It is applied in more than 35 countries through the United Nations Interregional Crime and Justice Research Institute (Unicri).

In Colombia, the history of victimization surveys refers to the modules dedicated to this topic in the National Survey of Households of 1985, 1991 and 1995. In 2003, The National Administrative Department of Statistics (DANE) responsible for conducting research, and statistics on population and quality of life, developed a study to characterize and measure aspects related to crime reported or not in the cities (Ciafardini 2010).

In the case of Mexico, the governmental Agency Inegi developed the National Survey of Victimization and Perception on Public Safety, however the studies of perception insecurity on urban scale are incipient, highlighting among others, those carried out by Navarrete et al. (2015) where they analyze the case of the city of Leon, and the one realized by Moreno (2012) in the metropolitan area of San Luis Potosí. In both cases, there is a coincidence that the increase in urban insecurity and the sense of fear in the cities have led to urban inequality as a predominant model in the development of the city.

4.1 Area of Study and Application of the GMA Model

The area of study corresponds to the city of San Luis Potosí, which is situated in the center of the Mexican country, in the State of S.L.P., the population of this city in 2010 was 760,746 inhabitants, composed of 366,086 men and 394,660 women (Inegi Census 2010), see Fig. 1.

In this case, the data obtained through the insecurity perception surveys were performed with open population between 2011 and 2014 by the OCS-SLP, they revealed 4 types of crime reported: sexual offenses, homicides, injuries and

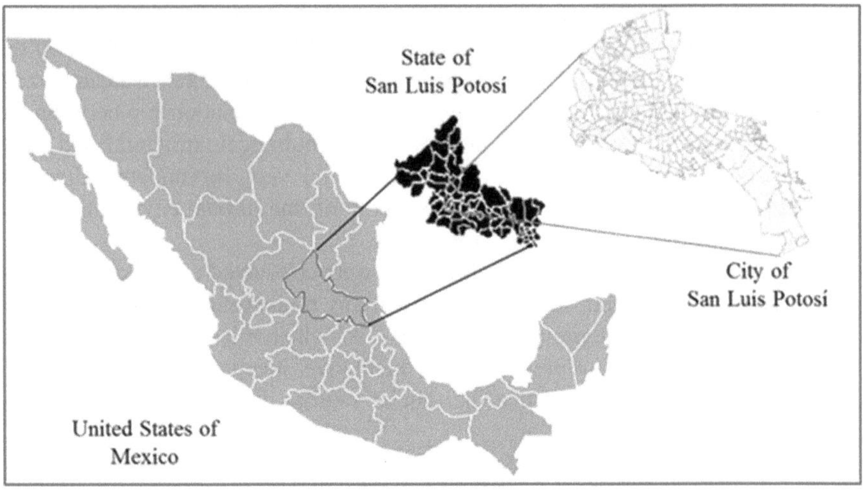

Fig. 1 Location of the study area. *Source* Own elaboration based on INEGI, 2014

suicides, 2 of them [sex offenses and injuries] had a high incidence [above 40 per 100,000 inhabitants], while the other two [homicides and suicides] observed a relatively low incidence [below 30 per 100,000 inhabitants]. In the case of the information derived from the survey, a classification pattern of situations of greater risk was observed in the following subjects: theft, accidents and contact with alcoholics and drug addicts, corresponding to the highest exposure data indicated by the respondents.

The information obtained through the application of surveys also allows estimating the characteristics of the crime, and let us know those environments that are propitious to victimization, and the frequency of these phenomena [quantitative data]. In addition, the crossing of this type of information with the Geographic Information Systems (GIS) allows the geo-referencing of the data and thematic maps, in order to explore the relationship between crime, insecurity and urban space conditions (Villasís-Keever et al. 2015).

4.2 Geographic Information Systems as a Tool to Analyze the Perception of Citizen Insecurity

The systems (GIS) are an ideal bridge for the comprehension of the study represented in maps, and also the levels of the variables of insecurity in the space, allowing its classification. It is a tool that enables the active participation of the community of researchers, academics, authorities, and citizens, since it allows defining the baseline in which strategies can be planned to identify or reduce the rate of insecurity. It facilitates the evaluation of local intra-urban perception using maps with crossing information that shows the spatial units resulting from indicators obtained through surveys on problems of insecurity. The maps constitute a base document on which any desired theme can be mapped, this allows representing a system of orientation according to the variables of insecurity that characterize each place in the space and is easily understandable. These maps guide and recognize the area of the affected community and its territory, see Fig. 2.

However, it is important to note that the data resulting from studies of perceived insecurity are not always objective, since they have a high emotional component. This emotive dimension is easily instrumented and can lead to biased interpretations of reality. According with the above, it is proposed that the perception indicators can serve as a complement to the documentary information, to obtain both: objective and subjective values. For this case, the map (Fig. 2) shows the results of the survey from the citizen perspective to the perception of urban insecurity. The spatial distribution of the zones of vulnerability to insecurity was: In the center of the city, medium and high levels; In the western periphery the level is lower; In the north and south the level is from high to very high.

As a summary of the interpretation of the most important results of the Citizen Perception Survey applied to citizens, the occurrence or manifestation of the main problems of insecurity, violence, and crime in the city of San Luis Potosí was

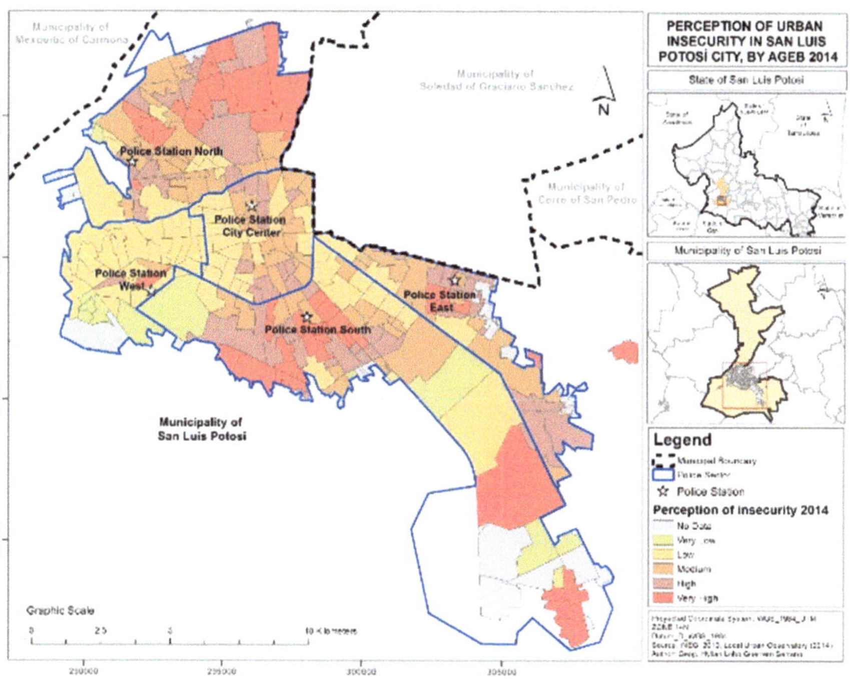

Fig. 2 Spatial distribution of the vulnerability zones to insecurity in the city of San Luis Potosí. *Source* Own elaboration from OCS-SLP, México, 2014

characterized by 10 elements as shown in Table 1. With the results of the survey, the spatial behaviour of these factors is wide-ranging and in some cases very pronounced in certain neighborhoods. Based on the spatial behaviour of each of these factors, a multiple correlation analysis should be performed to identify and confirm the association between factors, such as the relationship between the first four factors, as well as the fifth and sixth factors, and those related to the seventh to the eighth factors, and between the last two factors.

There could also be several associations between factors, such as alcoholism, street racing and speeding, or gang and street assaults and other options. In addition, it is possible to analyze the spatial trends of this perception, depending on the spatial distribution of economic activities and land uses and the residential typology —among others—that would surely allow a wider interpretation of the phenomenon.

4.3 Discussion and Considerations

Considering the perception of insecurity implications during the night, very interesting trends are observed, citizen survey reflects that temples and schools are safe during the day, but insecure through the night, the same perception is placed in parks, gardens and plazas, as well as shops and the city center; In general, the streets and the neighborhood are unsafe. Likewise, the typology of the perception of insecurity, according to the spatial dimension could be established according to different typologies: residential spaces, public spaces, private spaces, institutional spaces, etc.

Another way of reviewing the results is a comparison of the citizen's perception of insecurity at different territorial scales. For this purpose, the national survey

Table 1 Occurrence of urban insecurity

Survey of citizen perception of insecurity 2011–2014	Frequency of response with the most significant results
Manifestation of problems of insecurity	1. Assaults on the street
	2. Theft with violence
	3. Drug addiction
	4. Alcoholism
	5. Traffic speeding
	6. The races on the streets
	7. Graffiti
	8. Pandillerismo (gangs)
	9. Vehicle theft
	10. Express hijacking

Source Own elaboration based on citizen survey 2014

Table 2 Comparison of the citizen perception of insecurity (CPI) at different scales

Entity or territorial scale	Perception of insecurity (RPI) 2014 (%)	Victims 2013 (rate per 100,000 inhabitants)
National or country	73.30	28.22
State of S.L.P.	73.10	25.71
City of San Luis Potosí	84.00	30.00

Source Own elaboration based on: ENVIPE of INEGI, 2014; Citizen Survey 2014

carried out in Mexico by the INEGI was utilized; this Institute developed the National Survey of Victimization and Perception on Public Safety (ENVIPE 2014). One of its objectives is to measure the perception of the inhabitants of the country about the security of the place where they live and where they carry out their daily activities. The population surveyed in the ENVIPE is 18 years and older, on a wider national scale [95,516 households] and with the spatial distribution of Mexican states, is the most important reference of this type, but does not yield results at the city level, for this reason this study is focused in urban areas.

In order to validate the results, Table 2 was elaborated to compare the Rate of the Citizen Perception of Insecurity (CPI 2014) among national [country], regional [state] and local [city] perception; it was unquestionable that the national rate (73.3) is very similar to that of the state of San Luis Potosí (73.1), but at the city level, the citizen perception of insecurity is higher (84.0); it is evident that the spatial distribution of the results is important and is differentiated according to its scale. Concerning to the concept of the Victims Rate per 100,000 inhabitants (2013), the results are similar to the perception of the national (28.22) and state scale (25.71) but smaller than the local perception (30.00).

According to the current results, for this study it is conclusive that detailed analysis is required at the intra-urban and neighborhood scale. In this sense, some of the data derived from this perception study, allow evaluating a certain degree of relationship of the inhabitants within the neighborhood or housing complex levels. The results are worrying, since they reveal that a significant percentage of the population does not express a greater intention or need of contact with their neighbors or their community, showing that—behind their fears and perception of insecurity—there is a trend of individualization and community anonymity as defensive factors, that undermines social cohesion and the development of social networks towards the advancement of urban insecurity and crime.

5 Conclusions

Some of the indicators obtained demonstrate that the unit studied faces risks inherent to its demographic, socioeconomic and environmental characteristics. The spatial concentration of some of these factors also makes it possible to identify some urban sectors with a greater degree of vulnerability to insecurity and crime. Certain groups of the population are particularly vulnerable to crime, either as potential victims or as potential offenders. It is therefore important to identify such groups, to understand the specific problems and the measures that can be taken in the area of prevention.

The spatial categories that are particularly vulnerable to certain types of crime [as the source or recipient of the crime] are concentrated in the young population, the low-income population and the elderly population, which is more exposed than other age groups. This would allow the identification of vulnerable urban areas, which would serve as a basis for the design of preventive policies and urban design, with a view to addressing these problems.

The results of the survey of perception and victimization in San Luis Potosí City reflect the need to advance in the design and execution of alternatives focused on strengthening citizens' imaginary about security. It is important to focus efforts on the streets and avenues as the main scenario associated with the perception of insecurity and the occurrence of crimes that most affect citizens.

Strategies to reduce victimization should consider the development of comprehensive self-protection campaigns against theft, aimed at adopting a culture of prevention, as an example may be targeted to young people regarding with the theft of cell phones and the elderly 60 years in relation to the handling and transport of cash.

In the case study, the changes of authorities did not allow us to continue with the model of the citizen survey, consequently there is no updated data, due to a change in the municipal public safety policy model. Therefore, we must continue to promote plans and programs with the participation of civil society in the different scales of the presence of the phenomenon. In the spatial dimension-based on various disciplines—we can work in the physical dimension or morphology of cities: like urban planning, landscape design, architecture, industrial design, engineering and infrastructure design, among other professions.

All this will turn out in a better sustainability policy for cities, understanding governance as an active citizen participation and attention to crime, which will lead a more successful approach in cities in order to achieve the 16 Goal (SDG). Finally, one question remains: *Is this goal achievable?* For this, it is necessary to constantly monitor the performance of cities in this area, for the evaluation of public policies from the perspective of citizens.

Acknowledgements For this article, we thank the support of the Subsemun Program of San Luis Potosí Municipality (2011–2014); the Habitat Faculty of the UASLP, and the research team of the Local Urban Observatory of San Luis Potosí, México (OUL_SLP).

References

Arriagada I, Godoy L (2000) Prevent or suppress: false dilemma of citizen security. J CEPAL, Numb. 70 (LC/G.2095-P), April

Auyero J, De Lara A, Berti M (2014) Uses and forms of violence among the urban poor. J Lat Am Stud 46(3):443–469. doi:10.1017/S0022216X14000698

Carrión M, Fernando (2005) The insecurity in Latin America, J Quorum, Núm. 12, pp. 29–52, University of Alcala, Madrid, Spain

Ciafardini M (2010) Inseguridad y sensación de Inseguridad. In: Perspectivas y enfoques sobre percepción de seguridad Ciudadana. Artículos especializados, Alcaldía de Bogotá, Bogotá

González F (2011) Ciudad fragmentada: reflexiones en torno a la espacialización de la violencia en la ciudad de México. In: Revista Geográfica de América Central, vol 2. Universidad Nacional Heredia, Costa Rica, pp 1–15, jul–dec. 2011

González F, Valtierra A, Izquierdo J, Franco M (2011) Manual para la Implementación y Fortalecimiento de los Observatorios. Cancún, Quintana Roo, México: Observatorio de Violencia Social y de Género de Benito Juárez Q. Roo, Observatorio Urbano Local de la Ciudad de Cancún, Universidad del Caribe y Bufete de Estudios Interdisciplinarios A.C

Interamerican Bank of Development (IBD) (2011) Urban sustainability in Latin America and the Caribbean, Office of External Relations

Marcuse P (2006) Enclaves yes, ghettoes, no: segregation and the state. In: Varady D (Coord.) Desegregating the city. Ghettos, enclaves and inequality. State University of New York Press, Albany, pp 15–31

Mingione E (1994) Polarización, fragmentación y marginalidad en las ciudades industriales. In: Alabart J, García S, Giner S (eds.) Clase, poder y ciudadanía. Siglo XXI, Madrid

Moreno MA (2011) Reporte del Observatorio Ciudadano de Seguridad de San Luis Potosí. Eje: Generación, OCS-SLP/H. Ayuntamiento de San Luis Potosí, México

Moreno MA (2012). Segregación residencial, exclusión socio espacial y el modelo de ciudad cerrada e insegura en México: el caso de la ZM de San Luis Potosí. In: VII Cátedra Nacional de Arquitectura, CUMEX. Mayo 2012, Monterrey

Navarrete D, De la Torre M, Pérez-Vega A (2015) Patrones espaciales del delito en centralidades patrimoniales. In: Topofilia, Revista de Arquitectura, Urbanismo y Territorios. I.C.S. y H., BUAP. Segunda Época. vol V. Numb. 1. Aug–Dec. 2015 Puebla

Ospina JM (2006) La percepción de inseguridad en Bogotá. Primer Seminario Internacional de Percepción de Seguridad Ciudadana. 18 y 19 de abril, Bogotá, Colombia

Petrella L, Vanderschueren F (2003) Ciudad y violencia. Seguridad y ciudad. In: M. Balbo et al.: La ciudad inclusiva, Cuadernos de la CEPAL, Numb. 88, CEPAL/Cooperazione Italiana, Santiago de Chile, pp 29–58

Polese M, Stren M (2000) The social sustainability of cities diversity and the management of change. University of Toronto Press

Retamoso A, Corbo G (2003) La evolución del sistema urbano uruguayo: una aproximación al fenómeno de la delincuencia y criminalidad en Montevideo. Working Paper, The Center for Migration and Development, Princeton University

Saraví G (2005) Urban segregation and public space: young people in enclaves of structural poverty, in ECLAC Review. English version Nº 83, UN, CEPAL, Santiago de Chile, pp 31–46

Sillano M, Greene M, Ortúzar J (2006) Cuantificando la Percepción de Inseguridad Ciudadana en Barrios de Escasos Recursos. EURE (Santiago) 32(97):17–35. https://dx.doi.org/10.4067/S0250-71612006000300002

UN-Habitat (2011) Planificación urbana en proceso de cambio. Planificación, diseño y gestión de ciudades más seguras. Herramientas e intervenciones, Un-Habitat, Nairobi

UN-Habitat (2012) Presentación Programa Ciudades más Seguras (PCS): http://es.unhabitat.org/iniciativas-urbanas/iniciativas-programas/ciudades-mas-seguras/

Veiga D (2004) Desigualdades sociales y fragmentación urbana: obstáculos para una ciudad democrática. In: El rostro urbano de América Latina. CLACSO, Consejo Latinoamericano de Ciencias Sociales, Buenos Aires, pp 193–209

Velázquez E, Giraldo F (2009) Hábitat y seguridad urbana. Tendencias, prevención y gobernanza de la seguridad. UN-Habitat-Pnud-Alcaldía Mayor de Bogotá, Bogotá

Villasís-Keever R, Vázquez G, Marroquín F (2015) Análisis y caracterización de polígonos de inseguridad urbana mediante encuestas de percepción y sistemas de información urbana. In: Realidad, Datos y Espacio. Revista Internacional de Estadística y Geografía. Inegi. vol 6, Numb 2, May–Aug. 2015, pp 24–39

Author Biographies

Professor Ricardo Villasís Keever Ph.D. in Environmental Sciences and Architect by the Autonomous University of San Luis Potosí, Master in Gestion Urbaine et Logement by l'Université d'Aix-Marseille, France. Researcher of the U.A.S.L.P. Guest lecturer at several institutions of Higher Education. Responsible of the Local Urban Observatory of San Luis Potosí (Conacyt, Sedatu, UN-Habitat).

Professor Leticia Arista Castillo Ph.D. in Architecture, having also completed a post-doctorate program on urban development of historic sites. She is currently Researcher at the Autonomous University of San Luis Potosí, working in the Master Program in Science of Habitat. Her current research interests include sustainability in the cities and governance and management of cities cultural heritage. Member of ICOMOS México.

MIX
Papier aus verantwortungsvollen Quellen
Paper from responsible sources
FSC® C105338

FSC
www.fsc.org

If you have any concerns about our products,
you can contact us on
ProductSafety@springernature.com

In case Publisher is established outside the EU,
the EU authorized representative is:
Springer Nature Customer Service Center GmbH
Europaplatz 3, 69115 Heidelberg, Germany

Printed by Libri Plureos GmbH
in Hamburg, Germany